T0331113

Thermal Radiation

Thermal Radiation: An Introduction is a complete textbook for a one-semester introductory graduate course on radiative energy transfer. It bridges the gap between a basic introduction and comprehensive coverage of thermal radiation, focusing on insight into radiative transfer as practiced by engineers. Covering radiative transfer among surfaces, with an introduction to the effects of participating media between surfaces, the book includes surface and medium property characteristics and solutions to the radiative transfer equation in simple geometries.

- Tailored and organized specifically to suit a one-semester graduate course in radiative heat transfer.
- Emphasis is placed on insight into radiative transfer as practiced by engineers.
- Discusses how radiation is incorporated into finite element analysis (FEA) codes.

The textbook is intended for instructors and graduate students in a first-year course on radiative heat transfer or advanced heat transfer. Supplementary resources for students and instructors are available online.

Thermal Radiation

An Introduction

John R. Howell, M. Pinar Mengüç, and Kyle J. Daun

CRC Press is an imprint of the
Taylor & Francis Group, an **informa** business

First edition published 2023
by CRC Press
6000 Broken Sound Parkway NW, Suite 300, Boca Raton, FL 33487-2742

and by CRC Press
4 Park Square, Milton Park, Abingdon, Oxon, OX14 4RN

CRC Press is an imprint of Taylor & Francis Group, LLC

© 2023 John R. Howell, M. Pinar Mengüç, and Kyle J. Daun

Library of Congress Cataloging-in-Publication Data

Names: Howell, John R., author. | Mengüç, M. Pinar, author. | Daun, Kyle J., author.
Title: Thermal radiation : an introduction / John R. Howell, M. Pinar Mengüç, and Kyle Daun.
Description: First edition. | Boca Raton, FL : CRC Press, 2023. | Includes bibliographical references and index.
Identifiers: LCCN 2022037368 | ISBN 9781032015316 (hbk) | ISBN 9781032015347 (pbk) | ISBN 9781003178996 (ebk) | ISBN 9781032422541 (eBook+)
Subjects: LCSH: Heat--Radiation and absorption. | Materials--Thermal properties. | Energy conservation.
Classification: LCC QC331 .H69 2023 | DDC 536/.3--dc23/eng20230106
LC record available at https://lccn.loc.gov/2022037368

ISBN: 978-1-032-01531-6 (hbk)
ISBN: 978-1-032-01534-7 (pbk)
ISBN: 978-1-003-17899-6 (ebk)
ISBN: 978-1-032-42254-1 (eBook+)

DOI: 10.1201/9781003178996

Typeset in Times
by Deanta Global Publishing Services, Chennai, India

Access the Support Material: www.routledge.com/9781032015316

Contents

The Cover

The book cover is a NASA Goddard Space Flight Center Solar Dynamics Observatory photo of solar flares that occurred on September 28, 2021.

To view a movie of the flare event, scan the QR code or go to:

https://svs.gsfc.nasa.gov/vis/a010000/a013900/a013982/13982_ActiveOctober_1080_Best.mp4

Preface

This book is structured to give a complete resource for a one-semester introductory graduate-level course on radiative energy transfer. It is based on the classic reference text *Thermal Radiation Heat Transfer* (7th ed.), Taylor & Francis, 2021. However, much material in that book that relates to advanced research areas (nanomaterials, inverse problems, advanced methods for treating participating media, extensive references) has been removed. Material generally taught in a one-semester graduate course is retained. Emphasis is on insight into radiative transfer as practiced by engineers. This includes radiative transfer among surfaces, with an introduction to the effects of participating media in enclosures, and the background necessary for treating these cases.

Teachers may obtain a full solution manual from the publisher. In addition, the website (www.ThermalRadiation.net/webappendix/webappendixindex.htm) has supplementary homework problems, a history of the early development of the theory and practice of thermal radiation energy transfer, a list of important reviews and seminal references, and continually updated Errata. A full catalog of configuration factors is also available.

John R. Howell

M. Pinar Mengüç

Kyle J. Daun

PROPOSED ONE-SEMESTER SYLLABUS

The authors of this text have each taught courses in radiative energy transfer many times. Each has different areas of emphasis that they espouse, and it is expected that adopters of this text will have their own special areas of interest and emphasis. A proposed syllabus can be found in Appendix J of the online appendix.

LINKS TO ONLINE RESOURCES

Appendix to this book (errata, supplementary homework, history of radiation, bios, and timeline, historical and seminal references, and syllabus)

www.ThermalRadiation.net/appendix/html

Videos on this book, the lunar excursion model story, radiation history; and PowerPoint slides on radiation history:

www.ThermalRadiation.net/videos.html

Catalog of configuration factors: QR code or

www.ThermalRadiation.net/indexCat.html

MATLAB® is a registered trademark of The MathWorks, Inc. For product information, please contact:

The MathWorks, Inc.
3 Apple Hill Drive
Natick, MA 01760-2098 USA
Tel: 508 647 7000
Fax: 508-647-7001
E-mail: info@mathworks.com
Web: www.mathworks.com

List of Symbols

This is a consolidated list of symbols for the text. Some symbols only used in local development are defined where they are used and not included here. The symbols listed here are typical of those used for engineering energy transfer and follow, where possible, those adopted formally by the major energy transfer journals (Howell 1999). Some typical units have been indicated. Some quantities may have multiple units, such as a spectral bandwidth that can be in terms of wavelength, wave number, or frequency; some of these are designated by (mu), meaning multiple units. A length could, for example, be in nm, μm, cm, m, or other units, so that only a typical unit is shown. Some quantities are nondimensional; these are designated by (nd).

a_{kj}	Matrix elements (mu or nd)
$\mathbf{a}$	Matrix of elements a_{kj}
$\mathbf{a}^{-1}$	Inverse matrix (mu or nd)
A	Surface area, m²; fraction of incident radiation absorbed by a group of multiple layers
A_l^m	Coefficients in spherical harmonics expansion
b	Width of a base, m; a dimension, m; damping coefficient
B	Length dimension, m
c	Speed of electromagnetic radiation propagation in medium other than a vacuum, m/s
c_0	Speed of electromagnetic radiation propagation in vacuum, m/s
c, c_p	Specific heat, J/(kg·K)
C	Cross-section; correction to mean beam length
C_1	Constant in Planck's spectral energy distribution (Table A.2), W · μm⁴/(m²·sr)
C_2	Constant in Planck's spectral energy distribution (Table A.2), μm · K
C_3	Constant in Wien's displacement law (Table A.2), μm · K
C_4	Constant for maximum intensity at peak wavelength in blackbody distribution, Table A.2.
C_i	Concentrations of component i in a mixture (nd)
C_{CO_2} C_{H_2O}	Pressure-correction coefficients (nd)
CDF	Cumulative distribution function
D	A dimension, m; diameter of cylinder or hole, m; diameter of an atom, molecule, or particle, m
e	Energy level of a quantized state or photon, J; energy carried by a Monte Carlo bundle; electron charge
$\hat{e}$	Orthogonal unit vector

E	Emissive power (usually with a subscript), W/m^2; the amplitude of electric intensity wave, N/C; the energy of a given quantum state.
E_n	Exponential integral (online Appendix M) (nd) and Equation (9.71)
E	Electric field intensity
f_v	Volume fraction of particles
F	Configuration factor (nd)
$F_{0 \to \lambda}$	Fraction of total blackbody intensity or emissive power in spectral region 0 to λ (nd)
$\mathscr{F}_{i-j}$	Exchange factor between surfaces I and j
g	Asymmetry factor
$g(k_\eta)$	Cumulative distribution function in k-distribution method, Equation 8.15
G	Incident radiative flux onto a surface, W/m^2
h	Planck's constant, J·s (Table A.1); height dimension, m; convective energy transfer coefficient, W/(m^2·K)
H	Dimensionless height
H	Magnetic field intensity
i	Radiation intensity, W/(m^2·sr)
$\hat{I}$	Source function, W/m^2, Equation (9.10)
j	Emission coefficient, Equation (2.57)
J	Radiosity, outgoing radiative flux from a surface
k	Thermal conductivity, W/(m·K); absorption index for electromagnetic radiation, m^{-1}
k_B	Boltzmann constant (Table A.1), J/K
K	Kernel of the integral equation (mu or nd); spring constant
l	A length, m
l_m	Mean penetration distance, m
L	Length dimension, m
L_e	Mean beam length of gas volume, m
$L_{e,0}$	Mean beam length for the limit of very small absorption, m
M	Mass of a molecule or atom
m	Rest mass of an electron
n	Index of refraction of a lossless material c_0/c (nd); ordinate directions in S_n approximation (nd)
$\bar{n}$	Complex refractive index $n - ik$ (nd)
N	Number of surfaces in an enclosure (nd); the number of shields; the number of Monte Carlo sample bundles per unit time, s^{-1}; the number of particles per unit volume, m^{-3}
N_{CR}	Conduction-radiation parameter, $k/\kappa\sigma T_{ref}^3$
Nu	Nusselt number hD/k (nd)
p	Partial pressure of gas in the mixture, atm; probability density function
P	Perimeter, m; total pressure of gas, atm; probability; radiation pressure
R	Dimensionless radius
Pr	Prandtl number $c_p\mu_f/k$ (nd)
q	Energy flux, energy per unit area and per unit time, W/m^2

$\dot{q}$	Internal energy generation per unit volume, W/m^3
$\mathbf{q}_r$	Radiative flux vector, W/m^2
Q	Energy per unit time, W
Q	Efficiency factor (nd)
r	Radial coordinate, m; radius, m
r_e	Electrical resistivity, N·m^2·s/C^2 = Ω · m
R	Radius, m; overall reflectance of a translucent plane layer or a group of multiple layers (nd); random number in range 0 to 1 (nd); dimensionless radius
Re	Reynolds number $Du_m\rho_f/\mu_f$ (nd); real part
S	Coordinate along the path of radiation, m; distance between two locations or areas, m
S_{ij}	Spectral line intensity (mu)
$\mathbf{S}$	Poynting vector, $\mathbf{E} \times \mathbf{H}$
St	Stanton number, Nu/(Re·Pr) (nd)
t	Time, s
$\tilde{t}$	Dimensionless time (nd)
$t(S)$	Transmittance of a medium (nd)
T	Absolute temperature, K; overall transmittance of a plane layer or a group of multiple layers (nd)
u, v, w	Velocity, m/s
U	Total number of unknowns for an enclosure (nd); radiant energy density, J/m^3
V	Volume, m^3
w	Weighting factors (nd)
$\mathbf{x}$	Displacement vector
x, y, z	Coordinates in Cartesian system, m

GREEK SYMBOLS

α	Absorptivity of a surface (nd); the angle between two sides of a groove
α, β, γ	Constants in dielectric reflectivity Equations (4.43) and (4.44)
$\alpha(S)$	Absorptance of a medium (nd)
β	Extinction or attenuation coefficient $\kappa + \sigma_s$, m^{-1}; angle in x-y plane, rad; coefficient of volume expansion, K^{-1}
γ	Half-width of a spectral line (mu); Electrical permittivity
ε	Emissivity of a surface (nd)
$\varepsilon(S)$	Emittance of a medium (nd)
ϵ	Dielectric constant, γ/γ_0
$\bar{\bar{\epsilon}}$	Complex dielectric constant, Equation (4.16)
ζ	The quantity $C_2/\lambda T$ (nd); damping constant
η	Fin efficiency (nd); wave number, $1/\lambda$, m^{-1}
θ	Polar (cone) angle measured from the surface normal, rad
Θ	Scattering angle, rad

ϑ	Dimensionless temperature T/T_{ref} (nd)
κ	Absorption coefficient, m^{-1}
λ	Wavelength, m
μ	Magnetic permeability, N/A^2; the quantity $\cos\theta$ (nd)
ν	Frequency, c_0/λ_0; c_0/λ; c/λ_m, $s^{-1} = Hz$
ξ	Length coordinate, m; parameter $\pi D/\lambda$ for scattering (nd)
ξ_C	Clearance parameter for particle separation criteria, $\pi C/\lambda$
ρ	Reflectivity (nd); density, kg/m^3
σ	Stefan–Boltzmann constant, Equation (2.33) and Table A.2, $W/(m^2 \cdot K^4)$
σ_e	Electrical conductivity
σ_s	Scattering coefficient, m^{-1}
σ_0	RMS surface roughness, m
τ	Transmissivity (nd); optical thickness (nd); transmittance (Chapter 11) (nd); Relaxation time
τ_D	Optical thickness for path length D (nd)
ϕ	Circumferential (azimuthal) angle, rad; dimensionless temperature (nd)
Φ	Scattering phase function (nd)
Φ_d	Viscous dissipation production, $J/(kg \cdot m^2)$
χ	Angle of refraction, rad
ψ	Dimensionless energy flux (nd); stream function
ω	Albedo for scattering (nd); angular frequency, rad/s
ω_0	Resonant frequency
ω_p	Plasma frequency
Ω	Solid angle, sr

SUBSCRIPTS

A	Area
b	Base surface; at the base of a fin
bo	At a location on the boundary
b,	Blackbody condition
bi-d	Bidirectional
c	Evaluated at cutoff wavelength; collision broadening; at a collector (absorber) plate; cylinder
D	Doppler
d–h	Directional hemispherical
e	Emitted or emitting; environment; electrical
eq	at thermal equilibrium
fc	Free convection
g	Gas; interval index for k-distribution method
i	Incident; inner
j, k	Gas index in SLSWGG method
m	Mean value; in a medium
m, n	Number of identical semitransparent plates in a system
max	Corresponding to maximum energy; the maximum value

n	Normal direction
N	Number of quadrature points; the number of discrete ordinates
o	Outer; outgoing
p	Projected
P	Planck mean value; pressure
r	Reflected; radiative
R	Rosseland mean value
s	Scattering; shield; specular
solar	Solar
t	Transmitted
V	Volume
w	Wall
η	Wave number dependent
λ	Wavelength (spectrally) dependent
$\Delta\lambda$	for wavelength band $\Delta\lambda$
$\lambda_1 \longrightarrow \lambda_2$	in wavelength region from λ_1 to λ_2
λT	Evaluated at λT
υ	Frequency dependent
0	In vacuum
$\perp$	Perpendicular component
$\parallel$	Parallel component
$\cap$	Hemisphere of solid angles

SUPERSCRIPTS

T	Transpose of matrix
+	Along directions having positive $\cos\theta$
$-$	Along directions having negative $\cos\theta$
$\overline{}$	(Overbar) averaged over all incident or outgoing directions; mean value

Authors

John R. (Jack) Howell received his academic degrees from Case Western Reserve University (then Case Institute of Technology), Cleveland, Ohio. He began his engineering career as a researcher at NASA Lewis (now Glenn) Research Center (1961–1968) and then took academic positions at the University of Houston (1978–1988) and the University of Texas at Austin, where he remained until retirement in 2012. He is presently Ernest Cockrell, Jr., Memorial Chair Emeritus.

He pioneered the use of the Monte Carlo method for the analysis of radiative energy transfer in complex systems that contain absorbing, emitting, and scattering media. Jack concentrated his research on computational techniques for radiative transfer and combined-mode problems for over 60 years. He has adapted inverse solution techniques to combined-mode problems and to radiation at the nanoscale. His awards in heat transfer include the NASA Special Service Award (1965), AIAA Thermophysics Award (1990), ASME Heat Transfer Memorial Award (1991), ASME/AIChE Max Jakob Medal (1998), the Poynting Award from JQSRT (2013) and the International Center for Heat and Mass Transfer Luikov Medal (2018). He is an elected member of the US National Academy of Engineering (2005) and the Russian Academy of Sciences (1999) and is an Honorary Life Fellow of the American Society of Mechanical Engineering (ASME).

M. Pinar Mengüç completed his BSc and MS in mechanical engineering from the Middle East Technical University (METU) in Ankara, Turkey. He earned his PhD in Mechanical Engineering from Purdue University in 1985. He joined the faculty at the University of Kentucky the same year, and was promoted to the ranks of associate and full professor in 1988 and 1993, respectively. In 2008, he was named an Engineering Alumni Association Chair Professor. In 2009, he joined Özyegin University in Istanbul as the founding head of the Mechanical Engineering Department and also established the Center for Energy, Environment, and Economy (CEEE), which he still directs.

He is currently one of the three editors-in-chief of the *Journal of Quantitative Spectroscopy and Radiative Transfer* and an editor of *Physics Open*. He has organized several conferences, including five International Symposia on Radiative Transfer, as the chair or co-chair. His research expertise includes radiative transfer, light scattering-based particle characterization and diagnostic systems, nanoscale transport phenomena, near-field radiative transfer, applied optics, and sustainable energy and its applications to buildings and cities. He was the co-owner of a start-up company on particle diagnostics, co-author of 5 books, and has 7 patents. He is a fellow of both the ASME and ICHMT and received the ASME Heat Transfer Memorial Award in 2018 and Purdue University Outstanding Mechanical Engineering Alumni Award in 2020. He was elected to the Science Academy, Turkey, in 2017, and since 2018 he serves on its executive committee. In 2022, he was named Fiba Holding Renewable Energy (FYE) Chair Professor at Ozyegin University.

Kyle J. Daun received his BSc from the University of Manitoba (1997), his MASc from the University of Waterloo (1999), and his PhD from the University of Texas at Austin (2003). From 2004 to 2007, he worked at the National Research Council Canada as a Natural Sciences and Engineering Research Council of Canada Postdoctoral Fellow and later as a Research Officer. While at NRC he investigated radiation energy transfer in solid oxide fuel cells and helped develop and improve laser-based combustion diagnostics. In 2007 he returned to the University of Waterloo, where he is now a Professor in the Department of Mechanical and Mechatronics Engineering. Professor Daun's research interests include inverse problems in radiative transfer, laser-based combustion diagnostics, nanoscale transport phenomena, and energy transfer in manufacturing. He has published over 250 contributions, including 90 refereed journal papers. In 2010 he received the JQSRT Ray Viskanta Young Scientist Award. Professor Daun is a fellow of the Humboldt Foundation, a DFG Mercator Fellow, and a Fellow of the American Society of Mechanical Engineering.

1 Why Study Radiation Energy Transfer?

Ludwig Eduard Boltzmann (1844–1906) *made seminal contributions to the kinetic theory of gases and radiation energy transfer but is probably best known for his invention, independently of J. Willard Gibbs, of statistical mechanics, and the formulation of entropy on a microscopic basis. He derived Stefan's fourth power law for radiation emission by considering a heat engine with light as the working fluid. Boltzmann's epitaph in the Central Cemetery in Vienna reads*
Ludwig Boltzmann
1844–1906
S = k log W

William Thomson, Lord Kelvin (1824–1907) *became Professor of Natural Philosophy in 1846 at the University of Glasgow at age 22. He suggested his eponymous absolute temperature scale in 1848 based on Sadi Carnot's ideal thermodynamic cycle. While working on the laying of the Atlantic cable, he found time to publish work in 1849 (at age 25) that included the first use of the terms* thermodynamic *and* mechanical energy.

1.1 INTRODUCTION

In undergraduate and graduate courses on heat transfer or unit operations, radiation is often relegated to the last few weeks of the course. The importance of conduction and convection to many engineering applications is obvious; moreover, these modes are described by differential equations and data correlations in terms of familiar dimensionless groups. Radiation depends on integral relations, and undergraduate students are often less well-versed in this area of engineering mathematics. Radiation seems mysterious to many, and some shy away from studying it. But it has many key attributes distinct from conduction or convection. It is the only one of the energy transfer mechanisms that operates without needing a propagating medium (e.g., from the Sun to the Earth), and allows energy transfer over large distances (just look outdoors on a moonlit or starry night).

In this chapter, we will show that radiation: 1) has important engineering applications as well as in the natural world; and 2) is an emerging and fertile field for research, particularly as it affects climate change and other key technological challenges of the 21st century.

Let's start with some basic ideas and definitions.

DOI: 10.1201/9781003178996-1

Energy always radiates from all types of matter under all conditions, and this *emission* of thermal radiation arises from random fluctuations in the quantized internal energy states of the emitting matter. Temperature is a measure of the internal energy level of matter, and the nature of the fluctuations can be related to an object's temperature. Once matter radiates energy, the energy propagates as an electromagnetic (EM) wave. If the waves encounter matter, they may partially lose their energy and increase the internal energy of the receiving matter. This process is called *absorption*. The amount of emitted and absorbed radiation are functions of the physical and chemical properties of the participating material as well as its internal energy level as quantified by its temperature.

An EM wave can also undergo *scattering* as it propagates through a heterogeneous medium, e.g., a porous ceramic, a layer of freshly fallen snow, or even the molecules of a gas. As they encounter these scattering centers, the waves may be reflected, refracted, or diffracted, or any combination of these phenomena. Each phenomenon redirects the wave, usually without increasing the internal energy of the scattering medium. The Maxwell equations govern the propagation and scattering of EM waves, including all effects of reflection, refraction, transmission, polarization, and coherence.

Since all matter emits and absorbs radiation under all conditions, there is always a radiative transfer of energy, even within an isothermal system. If two objects are at different temperatures, there will be *net* radiation energy transfer between them, even if there is no matter between the objects. An obvious example is radiation emitted by the Sun, which travels through the vacuum of space and is partially absorbed and scattered upon encountering the Earth's atmosphere. A significant part of this radiation reaches the Earth's surface, where again some is absorbed and some is reflected (Figure 1.1). These three distinct physical mechanisms: emission, absorption, and scattering, are all *spectral* in nature; that is, their magnitudes all depend on the wavelength or frequency of the EM wave. Global warming, for example, is driven by spectral variations in the radiation emitted by the Sun, the radiation emitted by

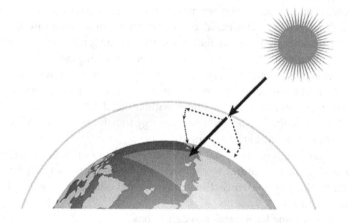

FIGURE 1.1   Atmospheric scattering effects on radiation from the Sun.

the Earth, and the fact that the atmosphere absorbs and scatters radiation differently at different wavelengths. (Section 1.4 has further discussion of this.)

The strength and spectral distribution of emitted energy is correlated with the internal energy state of the emitter. The temperature of the emitting matter connects the spectral distribution of emitted radiation through the distribution derived by Max Planck in 1901 (Section 2.6). For objects containing many atoms and molecules, the object's internal energy can be described using the principles of statistical thermodynamics. Statistical averaging allows for the definition of average thermodynamic and thermophysical properties, including internal energy and temperature. We can use simple laws such as Planck's distribution (Section 2.7) and Kirchhoff's law (Section 3.3) to explain these phenomena. Each emitting object exchanges radiant energy with the others that it can "see." Some of the incident energy that an object receives is absorbed and converted into its internal energy.

The quantitative calculation of the exchange of EM energy among matter at different energy levels constitutes radiative transfer—the focus of this book. We start our discussion with this chapter on the importance of radiative transfer and its contemporary applications. Then in Chapter 2, we discuss the ties between the traditional radiation exchange formulation in enclosures with the radiative transfer equation (RTE) in participating media. In this context, the word "participating" refers to a medium, such as the Earth's atmosphere, that alters the propagating radiative energy because of absorption, emission, or scattering mechanisms (Figure 1.1). We introduce the fundamentals of blackbody radiation, define radiation intensity, and outline the phenomena of emission, absorption, and propagation, including reflection, refraction, and scattering. The fundamentals and applications of radiative energy transfer are emphasized throughout the book.

EM radiation can be emitted through many mechanisms. Spontaneous fluctuations of atoms and molecules between quantized energy levels cause *thermal* radiation emission of radiation. When the atom or molecule transitions to a lower energy level, the First Law of Thermodynamics requires some energy to be emitted to "balance the books." The size and rapidity of these "jumps" (and thus the wavelength and intensity of emission) depend on the average internal energy of the matter, as evidenced by its temperature. This phenomenon is called *incandescence*.

Not all radiation is thermal in nature. For example, chemical reactions can stimulate non-thermal radiation emission. This phenomenon is called *chemiluminescence*, e.g., the reaction stimulated by luminol when it meets the hemoglobin in blood, triggering a blue emission (as seen in many TV crime dramas) and *bioluminescence* (fireflies and foxfire). Chemiluminescence associated with the recombination of chemical species (free radicals) causes the blue color of natural gas flames during combustion. Chemical and biochemical reactions do not constitute thermal radiation because the transitions aren't induced by random thermal transitions or the temperature of the emitter. Neither does *electroluminescence*, which is induced in some materials by electrical inputs (e.g., light emitting diodes), nor the light emitted by a fluorescent light bulb, which occurs when electrons from an electrically stimulated gas within the bulb collide with the phosphorescent coating on the bulb's inner surface. These phenomena are not *thermal* radiation and will only be mentioned in a few cases of special interest.

1.2   THERMAL RADIATION AND THE NATURAL WORLD

It is difficult to overstate the importance of thermal radiation in the natural phenomena that shape our daily lives, starting with the Sun. The Sun converts hydrogen fuel into helium; energy is emitted from this highly exothermic reaction as thermal radiation. The immense gravitational body forces acting toward the center of the Sun are balanced by the radiation pressure (Section 2.8.5) acting outwards; when these forces become out-of-balance, stars can collapse under their own gravitational force.

Radiation is the only mechanism for transporting thermal energy through a vacuum. This is how energy is conveyed from the Sun to the planets in the solar system. Solar energy governs the climate of Earth. Solar radiation evaporates water from the oceans, which irrigates the land masses and induces currents in the ocean. The differences in solar angle of incidence at different latitudes because of the Earth's tilt on its axis and the annual variations during its orbit cause climate zones and seasonal variations. Solar irradiation is converted to chemical energy via photosynthesis, which is the basis for all plant and animal life on Earth. The temperature of the Earth, which allows the chemical reactions that support life, is determined by a balance between absorbed solar radiation and emitted radiation. This balance is changing due to the increased absorption of solar irradiation because of anthropogenic pollutants such as CO_2, methane, and particulate matter.

On a clear day, the blue sky is due to the scattering of solar irradiation by nitrogen and oxygen molecules as well as extremely small dust and pollen particles in the air. Most animals on Earth see by reflected solar radiation, which, not coincidentally, has a peak emission intensity aligned with the center of the visible spectrum (approximately 550 nm, green light). Figure 1.2 shows that the visible spectrum also corresponds to the wavelengths at which water is least absorbing, and therefore most transparent; this fact is critical for aquatic species that possess eyes. The short-wavelength (blue) part of the visible spectrum is weakly absorbed compared with the longer-wavelength (red) part, which accounts for the blue hue common to underwater scenes.

1.3   THERMODYNAMIC ASPECTS OF THERMAL RADIATION

From a thermodynamic viewpoint, heat is the mechanism through which systems reach a state of equilibrium with their surroundings. For conduction and convection, energy transfer between two objects depends approximately on the difference in the temperature of the objects as well as the presence of an intervening physical medium. For natural (free) convection, or when variable property effects are included, the energy transferred can be a function of the difference between T_1^n and T_2^n, where the power n may be slightly larger than one and can reach two.

In contrast, thermal radiation transfer between two bodies depends on the difference between the fourth power of their absolute temperatures ($n = 4$).

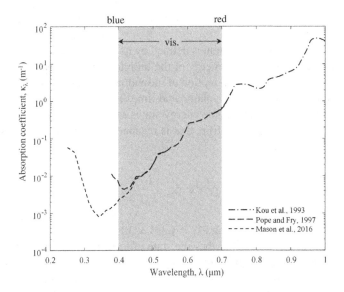

FIGURE 1.2 Representative data for the spectral absorption coefficient, κ_λ, of pure water. Data from: Kuo et al., 1993; Pope and Fry, 1997; Mason et al.,2016.

If temperature-dependent material properties are accounted for in the calculations, the radiative flux can be proportional to an even higher power n of absolute temperatures, and consequently the importance of radiation transfer is significantly enhanced at high temperatures.

The laws that govern energy transfer, including thermal radiation, descend from the First Law of Thermodynamics (conservation of energy) and the Second Law of Thermodynamics. The latter broadly concerns the principle of the increase in entropy or alternatively the direction of time.

Application of the First Law is usually quite straightforward. Radiation energy must be conserved across interfaces, i.e., applying the First Law over a control surface, which envelops no mass, shows that the radiation incident upon a surface must exactly balance the absorbed, reflected, and, in the case of semitransparent surfaces, transmitted radiation. By extending the control volume to include mass, it is possible to relate radiative transfer to other modes of energy transfer. The net rate of radiation absorption by a surface is linked to the rate at which the temperature of a surface rises, or the rate at which energy is conducted through the surface. In the case of a participating medium, the difference between absorbed and emitted radiation in a particular volume element becomes a volumetric radiative energy source term in a coupled radiation–advection–diffusion equation.

Concepts based on the Second Law are just as important, if more nuanced. Energy transfer is the way by which systems reach thermal equilibrium with their surroundings. When a macroscopic system is at thermal equilibrium with its surroundings, it is in a state of uniform thermal potential, and there can be no net, spontaneous energy transfer across the system boundaries without work input. This *Clausius*

Statement of the Second Law of Thermodynamics is used for proofs and discussions throughout this text.

The thermal potential of a system is defined by its temperature, which is a macroscopic averaging of the internal energy of the individual microscopic states (microstates) of matter, e.g., the random motion of individual atoms and molecules induced by the temperature, over a given volume and time. The connection between macroscopic temperature and microscopic energy states requires the assumption of some probability density function (PDF) p_i that is reached by their energy states. This is often a Boltzmann distribution,

$$p_i \propto \exp\left(-\frac{E_i}{k_B T}\right) \tag{1.1}$$

where E_i is the energy of state i and $k_B = 1.381 \times 10^{-23}$ J/K is Boltzmann's constant. The Boltzmann distribution describes the state of local thermal equilibrium (LTE). This occurs when the constituent atoms and molecules have undergone enough random interactions so that their energy states have reached a maximum state of disorder subject to the constraints of an average value (i.e., the temperature), and the probability distribution no longer changes. For most practical engineering systems, the LTE assumption is valid for thermal radiation, since the times needed for a system to reach local radiative equilibrium are much faster than those typical of the macroscopic processes, e.g., temperature change.

Another Second Law concept that impacts radiative transfer is *microscopic reversibility*. Loosely stated, microscopic reversibility requires a process reduced to its simplest and deterministic form must be reversible. This is because irreversibility arises from random processes, and, in the absence of randomness, the process must be reversible. For example, suppose a gas molecule undergoes adiabatic scattering (reflection) at a perfectly smooth surface; its incident direction and energy can be exactly determined from its scattered direction and energy by applying the conservation of linear momentum. Because this process is deterministic, it must also be reversible, so, were the scattered molecule to "reverse course," it would recreate the incident trajectory. Were the scattering influenced by the random thermal motion of the surface, it would only be possible to calculate a distribution of probable incident trajectories and the interaction would be irreversible. This idea is explored in later chapters as applied to thermal radiation incident on a surface.

1.4   ENGINEERING APPLICATIONS OF THERMAL RADIATION

We have seen that thermal radiation plays a central role in the natural world, in ways ranging from the atomic and subatomic scale to virtually every natural process on Earth, and even in the structure of stars. We now turn our attention to some examples of how thermal radiation may be used to solve engineering problems.

In Section 1.3, we said that thermal radiation emission is proportional to absolute temperature to the fourth power, and consequently radiation is the dominant heat transfer mode in high-temperature applications that include furnaces, combustion chambers, fires, rocket plumes, spacecraft atmospheric entry, high-temperature heat exchangers, hydrocarbon distilling and cracking, and during nuclear and chemical explosions. Infrared processing furnaces use radiation to anneal metals, dry paint, and cure composites. Thermal radiation can monitor and control the temperature of these processes via sensing by infrared cameras and pyrometers.

Thermal radiation plays an important role in power generation. At the time of writing, approximately 80% of the world's generated energy is via the combustion of fossil fuels and radiation energy transfer within boilers and combustion chambers. Understanding the principles of radiative transfer is key to the design and operation of efficient fossil-fuel-burning combustion systems, which means less energy use and minimum effects on the environment. Increasingly, environmentally sustainable solar energy systems are replacing combustion-based devices. Understanding the spectral and directional nature of solar radiation, and the conversion of incident solar energy to other energy forms by an absorbing surface, is also crucial in optimizing the design of these systems.

It is a commonly held misconception that as radiation dominates energy transfer at high temperatures it must be unimportant at lower temperatures. This is not so. For example, when designing heating, ventilation, and air conditioning (HVAC) systems for buildings, engineers must consider both convection and radiation between people and their surroundings, which are often similar in magnitude. Furthermore, HVAC systems function to counteract undesirable heating of buildings in the summer and cooling in the winter, which is becoming more pronounced with climate change. Because of their large energy consumption (usually obtained from the power grid), HVAC systems themselves are indirectly one of the growing sources of greenhouse gas emissions. While architecturally appealing, windows consume between 1% and 2% of the total world-generated energy because HVAC systems must counteract both transmitted solar radiation in the summer and emission losses in the winter. Accordingly, thermal radiation is a crucial element in developing more efficient building energy systems and making more energy-efficient architectural choices.

Radiation is a key element of spacecraft thermal control and cryogenic insulation systems. These systems are often under vacuum, so radiation is the only mode of energy transfer. Orbiting spacecraft undergo extreme variations in radiative heating from the Sun and the Earth, and cooling by emission to deep space. The Skylab space station orbited the Earth between 1973 and 1979, although it was in operation for only 24 weeks after its launch in May 1973. During its launch, Skylab lost a crucial radiation shield intended to protect the space station from solar irradiation. Ground control was forced into emergency maneuvers to orient the station to prevent overheating. A reflective Mylar film replacement shield was boosted up to the space station 11 days after the incident.

Waste energy from onboard spacecraft electronics must be radiated to space through specially designed radiant emitter panels. Deep-space probes (e.g., Voyager 1 and 2, Pioneer 10 and 11) incorporate radiant emitters into the radioisotope-based thermoelectric generator systems that power the onboard electronics because solar irradiation is too weak at the outer reaches of the solar system to provide sufficient power from solar cells.

These general examples represent only a small subset of the many engineering systems that feature thermal radiation energy transfer. (A more detailed account of the early history of the development of the theory and practice of thermal radiation energy transfer is available in online Appendices G and H at www.ThermalRadiation. net/appendix.html/webappendix/webappendixindex.)

We now consider some key examples of thermal radiation engineering in more detail.

1.4.1 SOLAR ENERGY

Solar energy is applied over a range of scales from individual homes to large central power installations that extend over hectares. Solar energy may be used to heat or cool buildings, generate steam or electricity, or promote other types of thermochemical or photochemical reactions.

1.4.1.1 Central Solar Power Systems

Solar energy may be converted to electricity in various ways, including photovoltaic and thermal pathways. In thermal generation, solar irradiation is an input to some form of heat engine. According to the Carnot principle, the harvested solar energy must be supplied at a high enough temperature to ensure high thermal efficiency in terms of the fraction of energy converted to work. Thus, solar thermal power systems typically incorporate concentrating solar collectors of some type, such as parabolic concentrators or large fields of steerable mirrors (Central Solar Power or CSP systems, such as in Figure 1.3).

The high temperatures reached in the tower absorber of CSP systems have spurred research into various collector designs, including spectrally selective surfaces that can withstand high temperatures; using direct absorption into a porous medium with energy removal by a gas or liquid flow; endothermic chemical conversion of the

FIGURE 1.3 The *Themis* CSP research tower, near Targassonne, France (left). The mirror array as viewed from the tower (right).

particles or fluid passing through the absorbers; and absorption of radiation into working fluids that can then be stored at high temperatures to overcome the intermittency of the incident solar energy.

1.4.1.2 Photovoltaic (PV) Conversion

In contrast to thermal systems, photovoltaic cells convert solar radiation directly into an electrical current via the photovoltaic effect. PV cells consist of two types of semiconductors: an n type and a p type, separated by a p–n junction. An electric circuit is completed by connecting the two semiconductors externally. Incident radiation causes an electron within the valence band of a semiconductor to overcome its bandgap and transition into the conduction band. Some of the excited electrons diffuse to the p–n junction, where they are accelerated into the p-type material by the Galvanic potential difference between the p and n semiconductors. This produces an electromotive force and current.

An efficient PV system relies on the semiconductor material to absorb, as opposed to reflect, incident sunlight. Thermal considerations are also important, since the probability that radiation can promote a valence electron into the conduction band depends strongly on the temperature of the cell via the temperature sensitivity of the bandgap. Nano-tailored structures can enhance the surface radiative properties of PV materials, based on near-field radiation techniques (see Chapter 12).

Figure 1.4 shows a typical multilayer thin-film silicon solar cell. This configuration consists of a thin film of amorphous silicon (a-Si) cells, a rear reflecting film of aluminum, an intermediate aluminum oxide layer, a silica (SiO_2) surface layer, and an outer protective layer of indium-tin oxide (ITO). The thicknesses of these layers are optimized to maximize absorptance (and minimize reflectance) over the wavelengths corresponding to the silicon cell bandgap to produce electrons by radiative energy absorption, while also maximizing reflectance over the IR spectrum.

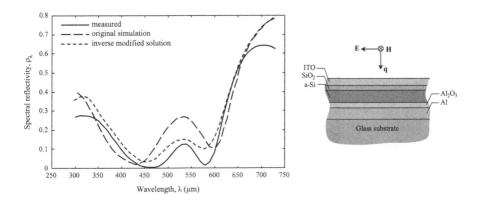

FIGURE 1.4 (a) Multilayer solar cell assembly. (b) Spectral normal reflectivity of the optimized multilayer solar cell. (Hajimirza, S. and Howell, J.R.: Computational and experimental study of a multilayer absorptivity enhanced thin-film silicon solar cell, *JQSRT*, 143, 56–62, 2014.)

Maximizing IR reflectance minimizes radiative heating of the cells, which reduces the photovoltaic conversion efficiency.

Nanoscale layering, patterning, and particle inclusions and morphology in surface layers can improve PV cell efficiency significantly, but the cost of these treatments per unit area currently precludes commercial deployment. Their use is presently limited to special applications such as spacecraft, where improved efficiency translates into smaller arrays of PV cells and therefore reduced launch mass. Recent advancements in nanofabrication techniques may fundamentally change this situation, leading to a new generation of high-performance PV designs that may further revolutionize decentralized energy production.

1.4.1.3   Solar Enhanced HVAC Systems

A wide range of low-cost solar HVAC systems have been developed for residential and commercial spaces; many of these systems are economically competitive with non-solar systems, particularly for providing domestic hot water, heating swimming pools, and space heating. Some thermally driven cooling cycles, such as absorption cooling, can operate using input energy from low-temperature sources (~100°C), making solar air conditioning cycles feasible, but so far they are economically marginal because the low coefficient of performance of these cycles require large solar arrays per unit of cooling.

Thermal balances on large structures are affected by the choice of coatings that augment or decrease solar absorption by structure surfaces. Recent work has focused on developing pigments and coatings that reflect solar radiation, thereby reducing the energy required to cool the building and reducing harmful greenhouse gas emissions. Moreover, by carefully tailoring the spectral properties of a surface to match high emissivity IR spectral regions with the transmitting spectral windows in the atmosphere (that is, avoiding the CO_2 and H_2O atmospheric bands by keeping spectral emissivity/absorptivity low in those regions, Figure 1.5), it is possible to link a surface with the cold absolute temperature of space that acts as a radiative energy

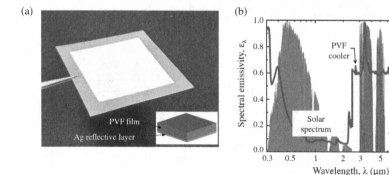

FIGURE 1.5   (a) Photo of a radiative cooler consisting of a PVF film and Ag reflective layer. (b) measured spectral emissivity of the PVF cooler, solar spectrum, and atmospheric transmittance. (From S. Meng, et al., 2020.)

sink. This allows for the cooling of a surface to below local ambient temperatures on Earth (Meng et al. 2020).

We often think of engineered nanostructures only in technological applications, but they are also found in nature. For example, Tsai et al. (2020) describe how butterflies modify the nanoscale surface structure of their wings to prevent local overheating.

1.4.2 ARCHITECTURE AND VISUAL COMFORT

Many architecture texts discuss daylight design principles, but they do not always consider energy efficiency. Light shelves and anidolic ceilings direct sunlight into interior spaces without glare, allowing the natural light to penetrate deeply into architectural spaces to reduce artificial light use. Similarly, solar tubes can direct sunlight to lower floors which would otherwise receive little or no daylight. Figure 1.6 shows an example of how architectural designs can apply fundamental radiative transfer principles for effective solar lighting.

1.4.3 GLOBAL WARMING AND CLIMATE CHANGE

The greenhouse effect and its implications for climate change have been widely discussed (WMO 2020, 2022; IPCC 2022). In the greenhouse effect, the atmosphere acts as a selective transmitter for the solar radiation flux incident on the Earth. The spectral properties of the atmosphere allow the visible part of solar energy to pass through unobstructed on a clear day (neglecting cloudiness, pollution, etc.) In the infrared (IR) portion of the spectrum, absorption of sunlight occurs by water vapor,

FIGURE 1.6 Using architectural design to use solar lighting to reduce energy costs.

CO_2, and other polyatomic gases such as methane. Using the solar temperature of 5,780K and the breakpoint between visible and IR spectra as 0.7 µm, about half of the incident solar radiation is in the visible plus UV solar spectrum. The remaining half is in the IR and some of this is absorbed by greenhouse gases. However, very nearly 100% of the radiation *emitted* by the Earth's surface is in the IR portion of the spectrum; thus, if greenhouse gas concentrations are increased, much more of the Earth's emitted energy is absorbed into the atmosphere rather than being transmitted to outer space, thus contributing to global warming.

Although the greenhouse effect clearly is a major factor in global warming, complete modeling of the Earth's energy budget is complex. The budget is affected by increased absorption of CO_2 into the oceans when global CO_2 concentrations increase, partially balanced by the release of CO_2 from the oceans when their mean temperature increases; changes in the vegetative and ice cover with global tempera-ture increase causing changes in the Earth's albedo; particle emissions from volcanic eruptions; and many others. Particulate matter has a strong influence on climate change through a variety of mechanisms. Most of these factors combine to drive pre-dictions toward long-term warming as greenhouse gas concentrations and other air pollutants increase. Changes in sea ice coverage reflects this trend (Maslowski et al. 2012). The World Meteorological Statement (2020) gives measured global average yearly temperature changes and the direct correlation with increasing atmospheric CO_2 levels. The picture is further complicated by the important roles played by meth-ane and particulate matter, which are still not fully understood. Methane has a global warming potential approximately 86 times that of CO_2 on a per-kilogram basis over a 20-year span, so anthropogenic (i.e., due to human activity) emissions must be reduced urgently to avoid the worst outcome. Adding further to the complexity is the fact that approximately half of methane emissions come from natural sources, e.g., the decay of organic matter in wetlands. These natural pathways are changing as the planet warms, leading to the frightening prospect of runaway global warming.

Controversial proposals have been put forth for geoengineering the properties of the atmosphere to reduce the greenhouse effect. Based on observations of significant local and global cooling following major volcanic eruptions, one idea is to exploit the scattering of various particle types and concentrations (e.g., titanium dioxide, sulfates) by injecting them into the stratosphere to act as a solar radiation shield/reflector, thereby reducing the greenhouse warming effect (Ocean Studies Board of the National Research Council 2015, Robock 2020, Maruyama 2020). The NRC report outlines various options and needed research on the effects of changing the Earth's albedo.

1.4.4 COMBUSTION AND FLAMES

Predicting radiative transfer during combustion is challenging on many levels. Combustion chemistry, fluid mechanics, and energy transfer are closely coupled, making the governing relations highly non-linear and "stiff" because of the strong temperature dependence of the chemical kinetics. High-fidelity modeling may involve tens to hundreds of intermediate serial and parallel chemical reactions, even for simple fuels like methane. When combustion occurs in a turbulent flow, the coupling of radiation with turbulent temperature and species concentration fluctuations increases the modeling difficulty. A further complication is the accurate accounting for the spectral radiative properties of the gases and particulates produced by combustion.

Flames can be categorized as non-sooting (non-luminous), or sooting (luminous), as shown in Figure 1.7. Non-sooting examples include hydrogen flames and hydrocarbon flames operated under near-stoichiometric or fuel-lean conditions. Combustion under these conditions does not tend to produce particulate matter, so radiative transfer is predominantly by emission from the gaseous products of combustion, mainly CO_2 and water vapor, although other species may sometimes be important. Non-luminous flames radiate a comparatively small fraction of the enthalpy of combustion to the surroundings, but it is nonetheless often important to account for radiation in combustion modeling of non-luminous flames. For example, the formation of the respiratory damaging oxides of nitrogen (NO_x) are highly sensitive to temperature, so seemingly insignificant changes in the predicted temperature field caused by neglecting radiant losses can lead to very large errors in NO_x predictions.

Emission from particulates dominates radiation from luminous flames, particularly soot. Soot arises from the pyrolysis of fuel, which forms small hydrocarbon fragments. These fragments can assemble into more complex polycyclic aromatic

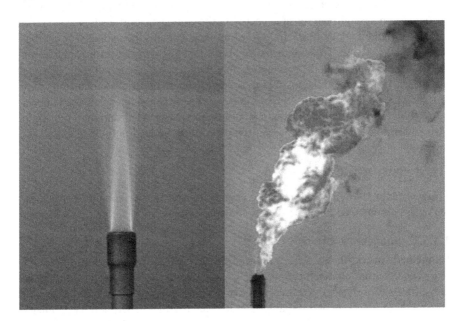

FIGURE 1.7 Examples of (left) a non-luminous flame; and (right) a luminous flame. Note that the visible (blue) emission in the non-luminous flame, a propane torch, is from chemiluminescence. The gaseous combustion products also emit thermal radiation, mostly in the infrared. Radiation from the luminous flame, an oilfield flare, is due to thermal emission from soot particles as well as gas species.

hydrocarbons, which continue to grow and polymerize. Soot production in fuel-rich flames can increase radiant emission vastly, which may be desirable for energy transfer applications, or undesirable if the soot fouls energy transfer surfaces or because of augmented radiation from unplanned fires and explosions.

Coal-fired large utility boilers are particularly difficult to design and model. The complex coal combustion process, spotty knowledge of exact fuel composition, the presence of coal ash causing scattering, the prediction of the varying composition of combustion products and their effect on radiative transfer, the complex geometry, and the resulting complicated flow environment, all make exact modeling a challenge.

Porous media and packed beds are used in combustion, as well as high-temperature heat exchangers, regenerators and recuperators, insulation systems, and packed and circulating bed combustors and reactors. At elevated temperatures, porous media absorb and emit radiant energy while interacting by convection with any fluid passing through the structure. The flowing medium is usually a gas, which can often be assumed transparent to radiation because the dimensions between the elements of the porous medium are usually much less than the radiative mean free path for absorption or scattering by the fluid. Porous ceramic burners offer several advantages in combustion applications, in terms of flame stability and combustion efficiency. In conventional flames, some of the enthalpy of combustion is lost by radiation to the surroundings. In the best-case scenario, the entire enthalpy of combustion is converted to the sensible energy of the combustion products, the so-called "adiabatic flame temperature." In porous burners, some of the enthalpy of combustion is conducted and radiated *upstream* of the combustion front through the porous matrix, where it preheats the reactants before they reach the reaction zone. Through this effect, in principle it is possible to achieve "superadiabatic" combustion conditions in which the local flame temperature exceeds the free flame adiabatic temperature. Observed benefits of porous media burners are extended flammability limits, high power density, and pollution control, particularly reduction in NO_x emissions.

1.4.5 Space Applications

Thermal radiation is the only mode of energy transfer in the vacuum of space, and thus plays an outsized role in space-related applications.

1.4.5.1 Spacecraft Thermal Control

Unmanned spacecraft often have internal energy sources for powering instrumentation and/or communication systems. These systems have stringent temperature range limits for reliable operation. The thermal designer must therefore balance the absorption of incident solar radiation and internal thermal generation with radiative energy emission. Control of solar absorption is done by careful choice of spectrally selective surfaces or the use of radiation shields, while the magnitude of required emission drives the size of radiating surfaces used for energy rejection.

Multilayer radiation shields can protect temperature-sensitive satellites from incident solar energy. Two contemporary examples of long-term protection of orbiting satellites are those of the Webb telescope and the Parker solar fly-by craft.

The Webb telescope has a large aperture meant to sense radiation in the near-to-mid IR spectral range (0.6 to 25 μm) and must be kept at a very low temperature to avoid emission from the telescope structure itself from interfering with the sensed signals. The five-layer sunshield of thin Kapton film coated with aluminum is shown in Figure 1.8. The two outer layers also have a doped silicon layer to protect them from degradation by solar UV radiation. The shield is designed to keep the mirror and structure at about 36 K. After delays, the telescope was launched on December 25, 2021.

On August 12, 2018, the Parker solar mission was launched to study the solar atmosphere (corona) and to explore the phenomena that accelerate the solar wind. The Parker probe will use seven Venus fly-bys over nearly seven years to gradually shrink its orbit around the Sun, coming as close as 6.16 million kilometers, well within the orbit of Mercury. It will be the first satellite to fly into the corona. The 8-foot-diameter heat shield protects everything within its umbra. At the closest approach to the Sun, temperatures on the outer surface of the heat shield are predicted to reach 1,370°C, but the spacecraft and its instruments must be kept at about 30°C. The heat shield, shown in Figure 1.9, is made of two panels of carbon–carbon

FIGURE 1.8 Five-layer metalized Kapton radiation shield for the Webb infrared telescope. The shield has the area of a regulation tennis court. Observe the technicians at the far lower left and right for scale. (NASA/Chris Gunn).

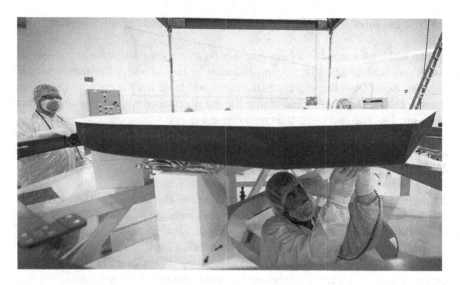

FIGURE 1.9 Testing the heat shield for the Parker Solar Probe. (NASA/Johns Hopkins APL/Ed Whitman).

composite sandwiching a lightweight 4.5-inch-thick core of low thermal conductivity carbon foam. A sprayed white coating is on the Sun-facing side of the heat shield to reflect as much of the Sun's energy from the spacecraft as possible. The thermal protection system connects to the truss on the spacecraft at only six points to minimize energy conduction.

The NASA/European Space Agency Solar Orbiter launched in January 2020. It will remain in a near-solar orbit for some years, allowing long-term and close-up observation and measurement of solar characteristics. Rather than the white reflective coating of the Parker probe, the Solar Orbiter heat shield uses a proprietary coating composed of bone char and alumina that is electrically conducting to prevent electrical arcing among instruments.

Spacecraft re-entry into planetary atmospheres causes extreme frictional heating. Re-entry heat shields, like the one shown in Figure 1.10, protect the spacecraft payload through a combination of ablation of the heat shield surface and very low thermal conductivity of the shield material. Radiation and convection to/from the ablative surface must also be considered, and the radiation from the shock layer preceding the ablating surface is complicated by the presence of many ablated species, thermal non-equilibrium in the gas, and the coupled chemistry and fluid mechanics.

1.4.6 Advanced Manufacturing and Materials Processing

Radiative heating is a key element of many industrial processes, particularly those that take place at high temperatures. Advantages of radiation over conduction and convection include the fact that radiative heating does not require physical contact with the heated product, thereby avoiding surface damage or contamination, and it is also possible to heat a surface uniformly by carefully controlling the spatial

FIGURE 1.10   Heat shield for the Orion re-entry capsule. The shield is composed of an ablative material covered with a layer of white epoxy and aluminum tape to prevent degradation from exposure to solar radiation prior to re-entry. The ablative shield is a highly cross-linked amorphous epoxy thermoplastic resin with special additives in a fiberglass honeycomb matrix. (Courtesy of NASA: Kim Shiflett).

irradiation. A wide range of applications use industrial infrared furnaces, including industrial baking, rapid thermal processing (RTP) of integrated circuits, paint drying, curing thermoset plastics and polymers, and metallurgical annealing (Figure 1.11). Some 3D printers use laser radiation (i.e., additive manufacturing) and welding processes.

Many of these processes require precise temperature control to obtain the desired properties at the end of the process. In the case of RTP chambers, for example, nonuniform heating of silicon wafers during chip lithography may lead to thermally induced stresses within the wafer, and defective integrated circuits. Likewise, recent advanced high-strength steel alloys require precise temperature control during annealing to obtain the microstructures responsible for their superior mechanical properties. In these and many other processes, it is paramount to accurately model the radiative transfer to the product being processed. Other manufacturing processes, such as photolithography, also involve irradiating a surface, although with the objective of inducing a photochemical, as opposed to thermal, transformation.

1.4.7 BIOMEDICAL APPLICATIONS

Thermal radiation plays an increasingly important role in medicine (Koba and Matsumoto 2017), including diagnostic and therapeutic applications. Infrared thermography is used as a diagnostic to find biomedical anomalies in blood flow, detect

FIGURE 1.11   Roller hearth furnace used for manufacturing high-strength automotive parts. The hot steel sheets exiting the roller hearth furnace are robotically transferred to a press, where they are simultaneously quenched and formed into shape. (Alex Schmidt, Cosma Int'l (left). Jonasson J., Billur E., and Ormaetxea, A. (right): *A Hot Stamping Line*. In: Billur, 2019.)

tumors, particularly in breast cancer, monitor body thermoregulation, and detect feverish and thus possibly infectious people at airports.

Thermal radiation can treat various diseases and disorders. Lasers are commonplace for a variety of surgeries, including those involving soft tissues, eye surgeries, dermatology, denture forming, and even tattoo removal. Hyperthermia cancer treatment uses targeted EM radiation to locally increase the temperature of a tumor to denature the proteins and destroy the blood vessels that supply flow to the tumor.

1.5   CONCLUSIONS

In this chapter, we briefly reviewed the key aspects of thermal radiation. Far from being just "the third mode of heat transfer," radiation is responsible for producing the unique conditions that support life on Earth, including wind, precipitation, and climate. Thermal radiation is also the mechanism responsible for climate change, which is likely to be the most important challenge ever faced by humankind. On the other hand, thermal radiation presents potential solutions in the form of more efficient combustion processes, more efficient heating, cooling, and illumination systems, and sustainable energy sources like solar heating. Thermal radiation is also key to the manufacturing processes used to make the objects we use in our daily lives, including the steel in our cars and the microchips in our phones.

The examples provided in this chapter are by no means exclusive; other examples include remote sensing of phenomena on Earth; maximizing signal propagation in optical fibers; analysis of the provenance of oil paintings; modeling forest and building fires; remote IR sensing for military target identification; identifying building energy leaks and monitoring pollution sources; weather prediction; insect adaptations to sunlight; and many others.

While thermal radiation has been studied for roughly two centuries, the underlying theory is by no means "a closed book." Fundamental radiation theory continues to evolve, and each new development presents a parallel engineering application.

Recent advancements include the observed radiative attraction between black-body (perfect radiation absorbing) objects at the atomic scale (Sister et al. 2017; Haslinger et al. 2018), far-field super-blackbody exchange between nanoscale particles (Thompson et al. 2018), and photon wave packet phase transitions with potential applications in encryption and multi-participant messaging (Kurtscheid et al. 2019, Öztürk et al. 2021). Future engineering applications of these and other ongoing theoretical studies, predictions, and experiments promise to produce still more surprising and useful innovations for applying thermal radiation energy transfer.

2 The Basics of Thermal Radiation

Wilhelm Carl Werner Otto Fritz Franz (Willy) Wien (1864–1928) *studied mathematics and physics at the Universities of Göttingen and Berlin. In 1893, he announced the Law of Displacement, which states that the product of wavelength and absolute temperature for a blackbody is constant. In 1896, he proposed a formula which described the spectral composition of radiation from an ideal body, which he called a black-body. This earned him the 1911 Nobel Prize in physics, and later impelled Max Planck to propose quantum effects to bring Wien's distribution into agreement with experimental measurements.*

Max Planck (1858–1947) *arguably laid the basis for quantum mechanics and was one of the forerunners of modern physics. He originally developed his blackbody spectral distribution based on the observation that the denominator in Wien's classically derived distribution needed to be slightly smaller to fit the experimental data. His attempts to explain the theoretical basis of his proposed spectral energy equation led him to hypothesize the existence of quantized energy levels, at odds with all of classical physics and thermodynamics.*

2.1 NATURE OF THE GOVERNING EQUATIONS

While the equations governing conductive, convective, and radiative transfer are all derived from the First and Second Laws of Thermodynamics, the equations that describe radiation are distinct and reflect fundamental differences in the underlying physics.

In the case of a material having a thermal conductivity k, the conduction energy flux in direction S is given by *Fourier's Law*,

$$q_{\text{cond}}(S) = -k\frac{dT}{dS} \tag{2.1}$$

where q has units of W/m^2. Within a purely thermally conductive medium, this equation can be applied to each surface of the infinitesimal cube shown in Figure 2.1a to derive an energy balance that quantifies the time-dependent energy change in the infinitesimal element dV,

$$\rho c_{\text{P}} \frac{\partial T}{\partial t} dV + \nabla(k\nabla T)dV = \dot{q}dV \tag{2.2}$$

DOI: 10.1201/9781003178996-2

Equation (2.2) includes a volumetric energy source term, $\dot{q}$, which can result from internal electrical heating, nuclear or chemical reactions, or through net absorption and emission of thermal radiation in scenarios where conduction and radiation occur simultaneously. At steady state, Equation (2.2) simplifies to:

$$\nabla(k\nabla T) = \dot{q} \qquad (2.3)$$

With appropriate boundary conditions, Equations (2.2) or (2.3) can be solved for the temperature distribution in a conducting medium. Key observations are that: (i) conduction requires an intervening medium; and (ii) the local energy flux depends only on the local temperature distribution, and therefore the equations governing the temperature distribution are differential (and, at steady state, elliptical). When these equations are discretized into matrix form, they have a sparse "banded" structure that is convenient to solve. A similar analysis can be made for convection, again demonstrating that the energy balance depends only on the conditions in the immediate vicinity of the location being considered.

In contrast, radiative energy is transmitted among distant area and volume elements without requiring an intervening medium. As an example, consider a heated enclosure with surface A and volume V, filled with a semitransparent radiating material such as a hot gas, a cloud of hot particles, or molten glass, as shown in Figure 2.1b. If $q_A dA$ is the rate of radiant energy arriving at dA_1 from a surface element dA and the energy rate $q_V dV$ arrives at dA_1 from a volume element dV, then the rate of radiative energy from the entire enclosure arriving at dA_1 is:

$$Q_{rad}(dA_1) = \int_A q_A dA + \int_V q_V dV \qquad (2.4)$$

Equation (2.4) highlights that, in contrast to the equations that govern conduction and convection, which are differential and elliptical in space, those governing radiations are *integral* in nature. Equation (2.4) does not explicitly account for the radiation that is absorbed between the emitting volume or surface element and dA_1; when this is done, the governing equations become *integro-differential*. Discretized integral-type equations produce a full matrix (i.e., every element in the matrix may be nonzero), reflecting the fact that the incident radiation on any element on the surface is influenced by the radiation leaving every other element on the surface, and, in the case of a participating medium, every element in the volume. The integral nature of radiation makes numerical solution procedures computationally expensive compared with similarly sized conduction and convection problems.

In addition to the complexity caused by the directional nature of radiation, we must also consider its spectral aspect. Radiation transfer calculations must be performed at each wavelength or frequency, and the *total* radiative energy is then determined by integrating over the entire EM spectrum. Radiative properties of surfaces, particles, and gases usually vary substantially over the wavelength (or frequency) spectrum, necessitating a finely resolved spectral quadrature. Boundary conditions may be specified in terms of total energies, but detailed spectral energy flows at the

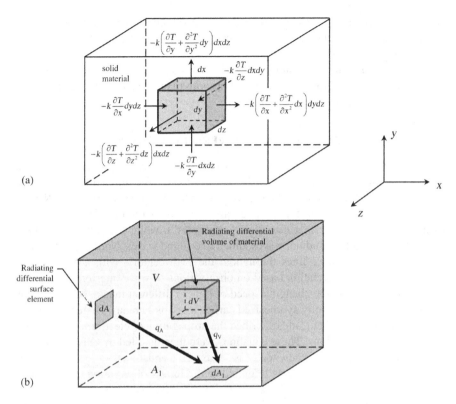

FIGURE 2.1 Terms for conduction and radiation energy balances: (a) energy conduction terms for volume element in solid and (b) radiation terms for enclosure filled with radiating material.

boundaries are usually unknown. This means that solutions must be obtained iteratively to satisfy the total energy boundary conditions.

In addition to these mathematical complexities, radiative transfer calculations are often further complicated by poorly defined physical properties. Radiative properties for solids and surfaces depend on many factors, such as surface roughness and degree of polish, purity of the material, and thickness of any thin film (e.g., oxide) or paint coating on the surface. The coating thickness, its material properties, and temperature may affect the radiation absorbed by the material or radiation leaving the surface, and, moreover, the surface state may change with time. Available experimental data for surface radiative properties may not include the necessary information on the coatings, making finding their values guesswork at best.

Radiating gas properties are especially challenging due to their spectrally complicated characteristics (Chapter 8). These properties are functions not only of the constituent molecules but also of gas pressure and temperature. In addition, atmospheric or combustion gases may also be laden with various particles, including soot, dust, pollen, fly ash, or char particles. The particles absorb and scatter radiation in a way that depends on their relative size with respect to incoming radiation, their shape, and their composition.

To quantify radiative transfer between volume and surface elements, and how this is complicated by how radiation interacts with matter, we must first examine the fundamental nature of radiative transfer.

2.2 ELECTROMAGNETIC WAVES VERSUS PHOTONS

Radiative energy propagation is often considered from one of two viewpoints: the flow of discrete quanta of energy, or the propagation of electromagnetic waves. The conflict between these viewpoints originates with the dispute between Isaac Newton and Christiaan Huygens in the 17th century. Newton favored the interpretation of light as discrete quanta (that he called "corpuscles"), while Huygens saw light as the propagation of longitudinal waves, much like sound waves. Natural philosophers were divided along these lines throughout the 19th century, with a gradual shift toward the wave interpretation based on observations of wave interference, diffraction, polarization, and the changing speed of light in different media. In 1865 James Clerk Maxwell successfully synthesized earlier work by Faraday, Ørsted, Coulomb, Ampere, Gauss, and others, and described the propagation of electromagnetic waves using a set of 20 equations. These equations were then distilled by Oliver Heaviside into the four equations we know today as "Maxwell's equations."

The wave–photon debate was reignited with Hertz's observation of the photoelectric effect in 1885, which appeared to be incompatible with the strictly wave interpretation of light. Max Planck struggled to explain his heuristically derived distribution of emitted energy from an ideal surface. He found that assuming discrete energy states rather than a continuum of states was a necessary ingredient and introduced the controversial idea of quantized energy states in 1901. Albert Einstein (1905) suggested the existence of "light quanta" (*Lichtquanten*), which explained the photoelectric effect and allowed Planck to interpret the basis of the spectral distribution of blackbody radiation, which is often regarded as the birth of quantum theory. The equivalence of these two competing perspectives was postulated by de Broglie in 1924, who showed that all matter could be interpreted as *either* particles or waves.

Envisioning radiation as the propagation of photons, much like a gas, is an appealing conceptual simplification. Nevertheless, it fails to explain some of the fundamental physics predicted from full-fledged quantum mechanics that include polarization, coherence, and tunneling; these phenomena are important in a range of key applications, including radiation at the nanoscale. There is a growing awareness that the conventional interpretation of "photons" in the Einstein sense is an oversimplified and inaccurate model, and a more sophisticated approach (based on quantum electrodynamics, QED) is needed to describe the true nature of radiation. Moreover, concepts like "photons" and "photon bundles," which are often used to solve the governing radiative transfer equation (and are thus sometimes called "RTE photons"), are often conflated with QED photons that are more like "quantized disturbances in the local electromagnetic field" as opposed to massless particles propagating at the speed of light. To avoid confusion between these concepts, there is a growing trend to refer to RTE photons as something else, e.g., energy "packets" or "bundles."

The classical EM wave view of radiation, in most cases, provides conservation equations akin to those obtained from quantum mechanics. Within the framework of wave theory, EM waves are comprised of coupled oscillations in the electric and magnetic fields perpendicular to the direction of wave propagation. In other words, EM waves are transverse waves, as opposed to the longitudinal waves envisioned by Huygens. Maxwell's equations describe this wave propagation in a simple, beautifully symmetric way.

EM waves propagate with the speed of light in a vacuum; indeed, light itself is an EM wave within the narrow visible range of the spectrum. The speed of light in a vacuum, c_0 (subscript "0" denotes a vacuum), is 2.9979×10^8 m/s. This speed is related to the speed of light c in matter through the refractive index n, where $n = c_0/c$. The refractive index, which is the real part of the complex index of refraction, $\bar{n} = n - ik$, is greater than unity for almost all natural materials within the visible and infrared spectra. For glass, it is about 1.5 in the visible spectrum; for water, it is about 1.33, and for most gases, it is close to unity. The imaginary component k is the *absorption index* and is a measure of how the energy of an EM wave is absorbed as it propagates through a conducting medium.

Any property used for radiative transfer calculations is spectral in nature, as it is defined at a given wavelength, λ, or frequency, ν. These quantities are related to the wave speed c by:

$$c = \lambda \nu \qquad (2.5)$$

while the energy of the wave is given by:

$$E = h\nu = hc / \lambda \qquad (2.6)$$

where $h = 6.626 \times 10^{-34}$ J·s is Planck's Constant (Table A.1 in the Appendix).

As an EM wave propagates from one medium into another, its wavelength changes as a function of the index of refraction, as $n = \lambda_0/\lambda = (c_0/\nu)/(c/\nu) = c_0/c$. The energy of a wave, and thus the frequency of the oscillations, ν, remains the same in each medium; for that reason, frequency (in units of s^{-1} or Hz) is often used to describe the spectral behavior of matter. When wavelength is used, it is customary to specify the wavelength in a vacuum, $\lambda_0 = c_0/\nu$. The *wavenumber* is often used to represent spectral effects in quantitative spectroscopy applications and is defined as $\eta_0 = 1/\lambda_0$. Finally, in some applications it is convenient to work in units of *angular frequency*, $\omega = 2\pi\nu$, having units of rad/s. In this case, Equation (2.6) is rewritten as $E = \hbar\omega$, where $\hbar = h / 2\pi$ is the reduced Planck's Constant.

Wavelength is usually given in units of micrometers (μm) or nanometers (nm), where 1 μm is 10^{-6} m or 10^{-4} cm and 1 nm is 10^{-9} m. Sometimes, the units are expressed in Angstroms (Å), where 1 μm = 10^4 Å or 1 nm = 10 Å. Wavenumber is customarily given in units of cm^{-1}; the wavelength spectrum can be converted to wave number by using the multiplier $10,000/\lambda$, where λ in μm yields η in cm^{-1}. Spectral units of electron volts (eV) are also sometimes used, where eV = $h\nu$ = hc/λ.

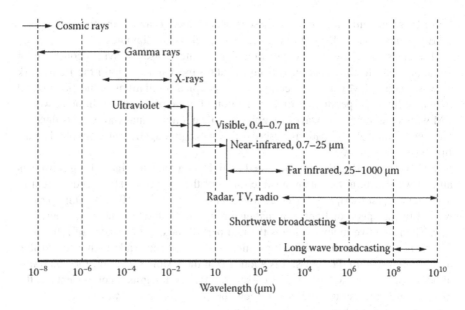

FIGURE 2.2   Spectrum of EM radiation (wavelength in a vacuum).

The extended EM spectrum is shown in Figure 2.2. Radio and television wavelengths may extend to thousands of meters. At the other end of the spectrum, the wavelengths of cosmic rays can be as small as 10^{-14} m. Within this vast range spanning more than 18 orders of magnitude, visible light is confined to 400–740 nm (i.e., 0.4–0.7 μm). However, even this "visible" span can differ among individuals, as some people (and other species) may "see" beyond this range.[1] For most practical purposes, thermal radiation is concerned with a wavelength range of 0.1 to 100 μm.

2.3 RADIATIVE ENERGY EXCHANGE AND RADIATIVE INTENSITY

Before further discussing the fundamentals of radiative energy transfer, we consider a few physical problems where radiative exchange is important. A common objective in radiative transfer calculations is to determine the amount of energy leaving one surface and reaching another after traveling through an intervening space. If the intervening space is a vacuum, the problem is chiefly one of geometry. On the other hand, if the space is filled with a participating medium, such as the mixture of water vapor and carbon dioxide common in combustion applications, the medium may selectively absorb radiation at certain wavelengths, which attenuates the radiation as it travels along its path. These same gases also emit at infrared wavelengths, which augments the radiation along the path. The extent of augmentation depends on the local temperature and gas concentration, which can vary along the path. Particles in the medium (e.g., soot, char) may also absorb and emit radiation. In addition, they may scatter the radiation in a different direction. A complete radiative transfer analysis must consider all these interactions, which are highlighted in Figure 2.3.

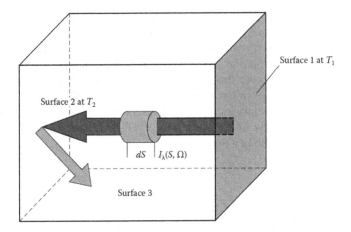

FIGURE 2.3   General schematic for radiative transfer discussions.

Emitted radiation leaving a surface is always spectral and directional in nature, meaning that it may change both as a function of wavelength and direction (as in the case of surface 1, in Figure 2.3). If the medium is participating, the radiative energy will change along the path from surface 1 to surface 2 due to absorption, emission, and scattering. The small cylindrical volume element shown in Figure 2.3 will act as a control volume for the conservation of radiative energy, which is expressed only in a specific direction Ω as indicated. Once the radiative energy is incident on the second surface, a portion of the energy may be absorbed (and converted to internal energy in the body), and the remainder reflected in another direction, for example, toward surface 3.

The arrow in Figure 2.3 could represent the propagation of a laser beam through a semitransparent medium, or it may represent emission from surface 1. Alternatively, we can consider the emitting body as the sun; then, the problem is the same as the propagation of solar rays through the Earth's atmosphere, which includes absorbing and scattering gases and particles. In that case, the temperature of the air is relatively low so that the radiative energy emitted by the atmosphere is small compared to the incident solar radiation or that reflected by the Earth. As a more practical engineering problem, we can visualize a three-dimensional (3D) configuration with several surfaces. This could be a kitchen oven, which contains air that neither absorbs, scatters, nor emits (provided you don't burn the pizza). The problem, however, is still complicated as the exchange of energy from one surface element to another needs to be carefully accounted for to determine the energy balance everywhere. For a combustion chamber, the flame or hot gases or particles inside will absorb, emit, and scatter radiation, and therefore must be accounted for when calculating radiative transfer. In this case, energy may be emanating from any point in the medium or on any surface; therefore, radiation propagating in any direction carries the radiative fingerprints of all particles, gas molecules, and surfaces in the enclosure. Radiative transfer needs to be determined after accounting for all directional effects, making the governing equation a complicated one.

It is obvious that, to quantitatively analyze any of these problems, we need to properly define and quantify the radiative properties, including those that govern the emission and absorption of radiation by surfaces, as well as absorption, emission, and scattering in the media. For this, we start with the concepts of solid angle and radiation intensity, which are fundamental to all radiation analyses. After that, we consider emission and absorption by an idealized surface, called a "blackbody." We then discuss absorption, emission, and scattering in the medium. These effects are combined into the *radiative transfer equation*, or RTE, which describes the conservation of radiative energy along a line of sight.

2.4   SOLID ANGLE

To develop the understanding of radiative energy propagating from one surface to another, consider the spectral radiative energy propagating along a direction S and incident on a small control volume dV at $S(x,y,z)$. This energy is confined to a small conical region, which is called the *solid angle*, Ω, and has units of *steradians* (sr).

The solid angle is defined in 3D space in analogy with a 2D planar angle, as shown in Figure 2.4. An infinitesimal *planar* angle is the ratio of the infinitesimal arc length to the distance from the apex to the base, while the infinitesimal *solid* angle is the ratio of the *projected infinitesimal area*, dA_n, to the square of the chord length r from the apex to the area:

$$d\Omega = \frac{dA_n}{r^2} \tag{2.7}$$

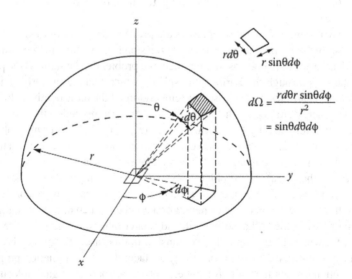

FIGURE 2.4   Hemisphere showing solid angle relations. Schematic for the definition of solid angle.

It is often convenient to work in spherical coordinates with an origin centered on elemental surface area dA, as shown in Figure 2.4. Direction is measured by the zenith and azimuthal angles, θ and ϕ, respectively, where θ is measured from the normal to dA. The angular position for $\phi = 0$ is arbitrary but is usually set at the x-axis. The dA_n is a differential area that is normal to a vector in the θ,ϕ direction. An infinitesimal solid angle $d\Omega$ can be related to infinitesimal increments $d\theta$ and $d\phi$. As shown in Figure 2.4, an area element on the hemisphere is given by $dA_n = r^2\sin\theta d\phi d\theta$, so that:

$$d\Omega = \frac{dA_n}{r^2} = \sin\theta d\theta d\phi \tag{2.8}$$

and

$$\int_{\cap} d\Omega = \int_{\theta=0}^{\pi/2}\int_{\phi=0}^{2\pi} \sin\theta d\phi d\theta = 2\pi \tag{2.9}$$

where the symbol "$\cap$" denotes a hemispherical domain. Thus, a hemisphere subtends a solid angle of 2π steradians when viewed from its base.

While solid angles may first seem like an abstract mathematical concept, they define how we perceive the world around us through vision (remembering that visible light is EM radiation). For example, our brains interpret the size of a viewed object based largely on the solid angle subtended by the object relative to our field-of-view, which is roughly a hemisphere. For this reason, objects rapidly appear larger as they become nearer, due to the $1/r^2$ term in Equation (2.7). We have all experienced this phenomenon when rowing a boat or paddling a canoe across a lake; when we set out, it feels as if we are making little progress for our exertion, but our speed appears to quicken as we approach the opposite shoreline. For a similar reason, as we drive down the highway, we seem to approach objects in the distance more slowly than objects close to the road.

2.5 SPECTRAL INTENSITY

We can use the definition of solid angle, along with the coordinate system shown in Figure 2.4, to quantify the spectral and directional nature of the radiation emitted by dA, and radiation incident upon dA. The spectral intensity leaving a surface in the $\Omega_o = (\theta_o,\phi_o)$ direction and at a wavelength λ is defined by:

$$I_{\lambda,o}(\Omega_o) \equiv \frac{dQ_{\lambda,o}}{dA\cos\theta_o d\lambda d\Omega_o} \tag{2.10}$$

where $d\Omega_o$ is an infinitesimal solid angle centered on Ω_o, $dA\cos\theta_o$ is the projected area of dA in the Ω_o direction, $d\lambda$ is an infinitesimal wavelength interval centered on λ, and $dQ_{\lambda,o}$ is the rate at which radiation leaves dA into $d\Omega_o$ over $d\lambda$. These quantities give spectral intensity units of $W/(m^2\cdot\mu m\cdot sr)$.

The intensity leaving a surface consists of emitted intensity (i.e., due to thermal emission as the constituent atoms and molecules spontaneously drop to lower energy states), as well as the reflected portion of the spectral intensity incident upon the surface. The incident intensity is given by:

$$I_{\lambda,i}\left(\Omega_i\right) = \frac{dQ_{\lambda,i}}{dA\cos\theta_i d\lambda d\Omega_i} \tag{2.11}$$

where $dQ_{\lambda,i}$ is the rate that radiation is incident upon dA within an infinitesimal solid angle $d\Omega_i$ centered on the $\Omega_i = (\theta_i,\phi_i)$ direction, and for wavelengths contained within an infinitesimal interval $d\lambda$ centered on λ.

At first glance, intensity appears to be a cumbersome way to define radiative transfer. For example, why is intensity defined relative to the surface area projected in the direction of radiation propagation, as opposed to simply the surface area? To understand why this is, consider a thought experiment in which your professor is holding a hot black rectangular plate at the front of the classroom, and you are holding a detector aimed at the plate. The amount of signal you observe is proportional to the incident intensity on the detector. This, in turn, depends on the plate temperature and the solid angle subtended by the plate as viewed by the detector. This solid angle depends on the *projected* area of the plate, not the true area, and diminishes as the professor tilts the plate away from you. Put another way, were the professor to be holding a tilted rectangle or a trapezoid at the same temperature, the detector will see the same incident energy. We shall see other examples throughout the book that justify intensity as a useful way to quantify radiative transfer.

It is also important to appreciate the meaning of the infinitesimal quantities: in a probabilistic sense, the actual probability of a photon bundle being emitted at any given wavelength λ is zero, just as the probability of encountering a person who is exactly 1.800000... m tall is zero, while the probability of finding someone who has a height H between 1.8 m +/– a small amount dH is small, but finite. Accordingly, the infinitesimal quantity $I_{\lambda,o}\cdot dA\cdot\cos\theta\cdot d\lambda\cdot d\Omega\cdot dt$ is proportional to the infinitesimal but nonzero probability that a photon bundle will be emitted by the surface element dA having energy contained between $d\lambda$, within the solid angle $d\Omega_o$, and over a time interval dt. We will revisit such probabilistic interpretations.

Equations (2.10) and (2.11) define the spectral intensity leaving or incident upon a surface at a given wavelength, λ. The *total* intensity is found by integrating the spectral intensity over all wavelengths. For example, the total intensity incident on the surface from direction Ω_i is given by:

$$I_i\left(\Omega_i\right) = \int_0^\infty I_{\lambda,i}\left(\Omega_i\right)d\lambda \tag{2.12}$$

2.6   CHARACTERISTICS OF BLACKBODY EMISSION

The key aspect that distinguishes thermal radiation from other types of radiant emission is that it is caused by matter undergoing random, spontaneous transitions in quantized energy levels due to its finite temperature. What, precisely, is the relationship between emitted intensity, wavelength, and temperature? Finding the answer to this question had a profound impact on the history of modern physics.

In principle, emission from a given body is a function of its material properties, its temperature, and direction:

$$I_{\lambda e} = f(T, \lambda, \theta, \phi) \tag{2.13}$$

But it is not clear up until this point exactly how these parameters influence radiant emission. The quest to clarify this relationship mirrors the concept of the idealized thermodynamic cycle. Motivated by the quest to improve the efficiency of steam engines, Sadi Carnot laid the foundations for the Second Law of Thermodynamics in 1824 by asking the question, "What is the maximum amount of work that can be extracted from a source of heat?" The answer is a cycle comprised of reversible processes. The cycle is independent of the working fluid. An analogous question in thermal radiation is, "What is the greatest rate at which a surface can emit radiation for a given wavelength and temperature?" The concept of such an ideal radiating body, called a *blackbody*, is a thermodynamic limiting case, as is the Carnot cycle. In contrast to real surfaces, we will show that the blackbody absorbs all incident radiation, and that its properties are independent of direction. As an ideal emitter and absorber, it serves as a hypothetical standard with which the properties of real surfaces can be compared. (The discussion of the radiative properties of real surfaces, and how they depart from the idealized blackbody, is in Chapters 3 and 4.)

The blackbody, conceived by Gustav Kirchhoff in 1860, derives its name from observing that good absorbers of incident visible light appear black to the eye (i.e., they don't reflect). However, except for the visible region, the eye is not a good indicator of absorption capability over the entire wavelength range corresponding to thermal radiation. For example, a surface coated with white oil-based paint is a very good absorber of infrared radiation, although it is a poor absorber of the shorter wavelengths of visible light, as evidenced by its white appearance. Only a few materials, such as carbon black, some specially formulated black paints on absorbing substrates, and some nanostructured surfaces, approach the blackbody in their ability to absorb radiant energy (Sapritsky and Prokhorov 2020). Vantablack (derived from "Vertically Aligned Nanotube Array black"), which consists of a "forest" of carbon nanotubes grown through chemical vapor deposition, absorbs 99.96% of incident visible radiation, very closely approaching blackbody behavior in the visible spectrum.

2.6.1 PERFECT EMITTER

Consider a blackbody at a uniform temperature contained within a perfectly insulated, evacuated, enclosure of arbitrary shape, whose walls are also blackbodies at a uniform temperature but different from that of the object (Figure 2.5). After some time, the blackbody object and enclosure will reach thermal equilibrium by means of radiative exchange. In equilibrium, the blackbody must emit as much energy as it absorbs; otherwise, there would be net energy transfer between two objects

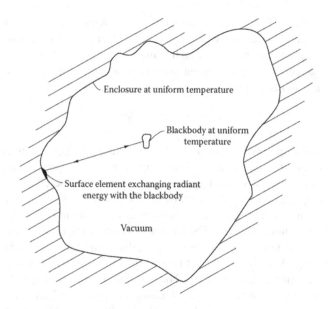

FIGURE 2.5 Enclosure geometry for derivation of blackbody properties.

at the same temperature, violating the Clausius statement of the Second Law of Thermodynamics. Because the blackbody, by definition, absorbs the maximum possible radiation incident from the enclosure at each wavelength and from each direction, it must also be emitting the maximum total amount of radiation. This deduction becomes clear by considering any less-than-perfect absorber, which must consequently emit less energy than the blackbody to remain in equilibrium.

This example highlights a misconception about thermal radiation. It is often assumed that the Second Law prohibits radiative transfer between objects at thermal equilibrium. This is not the case; the Second Law requires that there be no *net* energy transfer between objects at thermal equilibrium. Since all objects emit and absorb thermal radiation, there must always be radiative transfer between objects, even those in a state of thermal equilibrium. The fact that a body emits radiation even when it is at thermal equilibrium with its surroundings is called *Prevost's Law*.

2.6.2 RADIATION ISOTROPY OF A BLACK ENCLOSURE

Consider the isothermal enclosure with black walls and arbitrary shape in Figure 2.5; now, we move the blackbody to another position and rotate it to another orientation. The blackbody object must still be at the same temperature because the whole enclosure remains isothermal. Consequently, the blackbody must be emitting the same amount of radiation as before. To be in equilibrium, the body must still receive the same amount of radiation from the enclosure walls. Thus, the total radiation received by the blackbody is independent of its orientation or position within the enclosure. Hence, the radiation traveling through any point within the enclosure is independent of the position or direction. This means that the radiation filling the black enclosure is *isotropic*.

2.6.3 PERFECT EMITTER FOR EACH DIRECTION AND WAVELENGTH

Consider an area element on the surface of the black isothermal enclosure and an elemental blackbody within the enclosure. Some of the radiation emitted by the surface element strikes the elemental body at a given direction. All this radiation, by definition, is absorbed by the blackbody. To maintain thermal equilibrium and isotropic radiation throughout the enclosure, the radiation emitted back into the incident direction must be equal to that received. Since the body is absorbing the maximum amount of radiation from any direction, it must also be emitting the maximum amount in any direction. Because the radiation filling the black enclosure is isotropic, the radiation received and hence emitted in *any* direction by the enclosed black surface, per unit projected area normal to that direction, must be the same as that in any other direction.

By considering the detailed balance of absorbed and emitted energy in each small wavelength interval, we also reach the conclusion that the blackbody is a perfect emitter and absorber at every wavelength.

2.6.4 TOTAL RADIATION INTO A VACUUM

If the enclosure temperature is changed, the enclosed blackbody temperature also changes and becomes equal to the new temperature of the enclosure. The absorbed and emitted energy by the blackbody is again equal, although the new equilibrium temperature differs from the original temperature of the enclosure. Because a blackbody absorbs, and hence emits, the maximum amounts of radiation corresponding to its temperature, the characteristics of the surroundings do not affect its emissive behavior. Hence, the *total radiant energy emitted by a blackbody into a vacuum is a function only of its absolute temperature.*

As mentioned earlier, the Second Law of Thermodynamics forbids spontaneous net energy transfer from a cooler surface to a hotter surface. If the radiant energy emitted by a blackbody increases with a decreasing temperature, we could build a device to violate this law. Without considering the proof in detail, we observe that radiant energy emitted by a blackbody must *increase* with temperature. The total radiant energy emitted by a blackbody is therefore proportional to some monotonically increasing function of temperature.

2.6.5 BLACKBODY INTENSITY AND ITS DIRECTIONAL INDEPENDENCE

As we found previously, intensity is the fundamental quantity to describe radiation propagating in any direction. The spectral intensity of a blackbody is $I_{\lambda b}$, where the subscripts denote, respectively, that it is a spectral quantity and that the properties pertain to a blackbody. The emitted spectral intensity from a surface is the radiative energy emitted per unit time per unit infinitesimal wavelength interval around the wavelength λ, per unit elemental projected surface area normal to (θ,ϕ) direction and into the elemental solid angle centered on the direction (θ,ϕ). The total blackbody intensity, I_b, may be found by integrating the spectral intensity over all wavelengths per Equation (2.12).

The directional independence of $I_{\lambda b}$ and I_b can be shown by considering a spherical isothermal blackbody enclosure of radius R with a blackbody element dA at its center (Figure 2.6a). The enclosure and the central element are in thermal equilibrium, and therefore all radiation throughout the enclosure is isotropic. Consider radiation in a wavelength interval $d\lambda$ about λ emitted by an element dA_s on the enclosure surface that travels toward the central element dA (Figure 2.6b). The emitted energy in this direction per unit solid angle and time is $I_{\lambda b,n}\, dA_s\, d\lambda$.

The normal spectral intensity of a blackbody is used because the energy is emitted normal to the black wall element dA_s of the spherical enclosure. The amount of energy per unit of time that strikes dA depends on the solid angle that dA occupies when viewed from dA_s. This solid angle is the projected area of dA normal to the (θ,ϕ) direction, $dA_p = dA\cos\theta$, divided by R^2. Then, the energy absorbed by dA is $I_{\lambda b,n}dA_s(dA\cos\theta/R^2)d\lambda$. The energy emitted by dA in the (θ,ϕ) direction and incident on dA_s on solid angle dA_s/R^2 (Figure 2.6c) must equal that absorbed from dA_s, or the equilibrium would be disturbed. Hence,

$$I_{\lambda b}(\theta,\phi)dA_p\left(\frac{dA_s}{R^2}\right)d\lambda = I_{\lambda b}(\theta,\phi)dA\cos\theta\left(\frac{dA_s}{R^2}\right)d\lambda = I_{\lambda b,n}dA_s\left(\frac{dA\cos\theta}{R^2}\right)d\lambda \quad (2.14)$$

which shows that:

$$I_{\lambda b}(\theta,\phi) = I_{\lambda b,n} = I_{\lambda b} \quad (2.15)$$

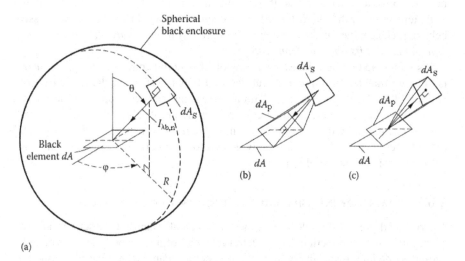

(a)

FIGURE 2.6 Energy exchange between the element of enclosure surface and element within enclosure. (a) Black element dA within the black spherical enclosure, (b) energy transfer from dA to dA_p, and (c) energy transfer from dA_p to dA_s.

The intensity of radiation from a blackbody, $I_{\lambda b}$, as defined based on its projected area, is *independent* of the direction of emission. Neither the subscript n *nor the* (θ, ϕ) *notation is therefore needed for the intensity emanating from a blackbody.* Since the blackbody is always a perfect absorber and emitter, its properties are independent of its surroundings. Hence, these results are independent of both assumptions used in the derivation of a spherical enclosure and thermodynamic equilibrium with the surroundings.

The isotropic nature of blackbody intensity means that the amount of radiation emitted by a blackbody that is intercepted by another surface depends only on the solid angle subtended by the blackbody as viewed by the intercepting surface, and thus its projected area, and not the true area of the surface that matters in the radiative exchange between surfaces. Think again of the professor holding a black trapezoid at the front of the lecture hall.

Some exceptions exist for most of the blackbody "laws." These are rarely encountered in practical engineering scenarios but need to be considered when the assumption of local thermal equilibrium is not satisfied, e.g., extremely rapid transients or extreme temperature gradients (e.g., near shock layers). If the transient period is of the order of the timescale of the process that governs the emission of radiation from the body in question, then the emission properties of the body may lag the absorption properties. In such a case, the concepts of temperature used in the derivation of the blackbody laws no longer hold rigorously. The treatment of such non-local thermodynamic equilibrium (non-LTE) problems is generally outside our scope.

2.6.6   BLACKBODY EMISSIVE POWER: COSINE-LAW DEPENDENCE

A common misconception is that the rate of energy emitted by a blackbody is independent of the angle due to its isotropy. This is *not true*. Recall that intensity is defined relative to the surface area projected in the direction of propagation. The spectral emission from dA per unit time and unit surface area passing through the element on the hemisphere shown in Figure 2.4 is given by Equations (2.10) and (2.8) as:

$$dQ_{\lambda e}(\theta, \phi) = I_{\lambda b}\cos\theta dA d\Omega d\lambda = I_{\lambda b}\cos\theta dA\sin\theta d\theta d\phi d\lambda \qquad (2.16)$$

The directional spectral emissive power of a surface, $E_{\lambda b}$, is defined as the rate of spectral emission in each direction, per solid angle, $d\Omega$, per wavelength interval, $d\lambda$, and per unit area, dA:

$$E_{\lambda b}(\theta, \phi) = \frac{dQ_{\lambda e}(\theta, \phi)}{dA d\Omega d\lambda} = I_{\lambda b}\cos\theta = E_{\lambda b}(\theta) \qquad (2.17)$$

Since blackbody intensity is isotropic, this means that the blackbody emissive power is a maximum if the surface is normal to a receiving surface or a detector.

Equation (2.17) is known as Lambert's Cosine Law, proposed by Johann Heinrich Lambert (Lambert, 1760). Surfaces having a directional emissive power that follow this relation are called Lambertian, Diffuse, or Cosine-Law surfaces.

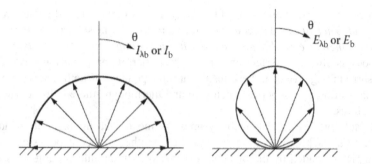

FIGURE 2.7 Cosine Law: The angular distribution of blackbody intensity $I_{\lambda b}$ (independent of θ) and blackbody directional emissive power $E_{\lambda b}$ (cosθ function).

A blackbody, because it is always diffuse, serves as a standard for comparison with the directional properties of real surfaces that may not follow the Cosine Law. Figure 2.7 shows how both the blackbody intensity and blackbody energy depend on zenith angle θ, although $I_{\lambda b}$ is independent of θ. Emission from nonblack surfaces may depend on both the zenith and azimuthal angles and can be expressed more properly as $E_\lambda(\theta,\phi)$.

Real surfaces emit less energy than the ideal blackbody, discussed further in Chapter 3. For many practical engineering problems, the intensity leaving a surface can be considered isotropic, and does not have (θ,ϕ) dependence. As an example, ceiling-mounted fluorescent light boxes are often equipped with glazing (e.g., opal glass) to produce an approximately Lambertian distribution of light; there is a bright spot immediately under the light box, and the visible light will diminish in proportion to cos θ as the viewing angle increases. On the other hand, with the advances in micro- and nanomanufacturing techniques, novel structures are currently being built to obtain very directional emissions that may be exploited (Chapter 12).

2.6.7 HEMISPHERICAL SPECTRAL EMISSIVE POWER

Equation (2.17) provides the radiant energy flux emitted by a blackbody in each direction θ, per unit surface area, per unit solid angle, and per unit wavelength interval centered on λ. The *hemispherical spectral emissive power*, $E_{\lambda b}$, is the energy emitted by a black surface per unit of time, per unit area, and per unit wavelength interval around λ, into *all* directions. This quantity is found by integrating Equation (2.17) over all solid angles of the hemisphere to give:

$$E_{\lambda b} = \frac{dQ_{\lambda e \cap}}{dA d\lambda} = I_{\lambda b} \int_{\phi=0}^{2\pi} \int_{\theta=0}^{\pi/2} \cos\theta \underbrace{\sin\theta \, d\theta d\phi}_{d\Omega} = \pi I_{\lambda b}. \tag{2.18}$$

So, *the blackbody hemispherical emissive power is π times the blackbody intensity.*

2.7 PLANCK'S LAW: SPECTRAL DISTRIBUTION OF EMISSIVE POWER

So far, we have inferred blackbody characteristics from basic thermodynamic reasoning, but this is insufficient to obtain the spectral nature of blackbody emission. Indeed, this was Planck's impetus for exploring a more exact theoretical expression, which eventually became the foundation of quantum theory. The blackbody formulation by Planck (1901) is built upon Boltzmann's (1884) derivation of Equation (1.1) using statistical thermodynamics and the concept of thermal equilibrium.[2] We now adopt Planck's blackbody formulation to express the spectral distribution of hemispherical emissive power and radiant intensity in a vacuum. This expression is based on wavelength λ and the absolute temperature T of the blackbody:

$$E_{\lambda b}(T) = \frac{2\pi h c_0^2}{n_\lambda^2 \lambda^5 \left[\exp\left(\dfrac{h c_0}{n_\lambda k_B \lambda T} \right) - 1 \right]} = \frac{2\pi C_1}{n_\lambda^2 \lambda^5 \left[\exp\left(\dfrac{C_2}{n_\lambda \lambda T} \right) - 1 \right]} \qquad (2.19)$$

Here, λ is the wavelength in the medium of refractive index n_λ. Equation (2.19) is the Planck spectral distribution of emissive power. The blackbody spectral intensity is given by

$$I_{\lambda b}(T) = \frac{E_{\lambda b}(T)}{\pi} = \frac{2 C_1}{n_\lambda^2 \lambda^5 \left[\exp\left(\dfrac{C_2}{n_\lambda \lambda T} \right) - 1 \right]} \qquad (2.20)$$

For most engineering applications, radiative emission is into air or other gases, for which the index of refraction n_λ is near unity. Note that Planck's blackbody function is expressed in terms of two universal constants: Planck's Constant, h, which defines the smallest energy quantum in terms of wavelength, and Boltzmann's constant, k_B, which relates temperature to the average energy of the microstates. Two auxiliary radiation constants are $C_1 \equiv h c_0^2$ and $C_2 \equiv h c_0 / k_B$. The values of these constants are given in Table A.2. Planck initially derived Equations (2.19) and (2.20) for the restricted case of blackbody emission into a vacuum, using the speed of light in a vacuum c_0 which appears in both constants C_1 and C_2 (Planck 1913, pp. 43–44). He assumed correctly that air and other gaseous media had a simple refractive index near unity, so little error was introduced in most cases by using this assumption. Correcting for emission into substances with non-unity refractive index (which is important in some cases, such as radiative transfer in glass) requires that the speed of light in the medium $c = c_0/n_\lambda$ be substituted into the relations, resulting in the forms given.

Example 2.1

A black surface at a temperature of 1000°C is radiating into a vacuum. At a wavelength of 6 μm, what are the emitted intensity, the directional spectral emissive power of the blackbody at an angle of 60° from the surface normal, and the hemispherical spectral emissive power?

From Equation (2.20),

$$I_{\lambda b}\left(6\ \mu m\right)=\frac{2\times0.59552\times10^{8}\ W\cdot\mu m^{4}/m^{2}}{6^{5}\ \mu m^{5}\left(e^{14.388/[6\times1273]}-1\right)sr}=2746\ W/(m^{2}\cdot\mu m\cdot sr)$$

From Equation (2.17), the directional spectral emissive power is:

$$E_{\lambda b}\left(6\ \mu m,\ 60°\right)=2746\cos60°=1373\ W\ //(m^{2}\cdot\mu m\cdot sr)$$

The hemispherical spectral emissive power is $E_{\lambda b}(6\ \mu m) = \pi I_{\lambda b}(6\ \mu m)$ = 8627 W/ (m². μm).

Example 2.2

To a good approximation, the Sun can be considered to emit as a blackbody at 5780 K. Determine the Sun's intensity at the center of the visible spectrum.
From Figure 2.2, the wavelength of interest is 0.55 μm. Then, from Equation (2.20),

$$I_{\lambda b}\left(0.55\ \mu m\right)=\frac{2\times0.59552\times10^{8}\left(W\cdot\mu m^{4}/m^{2}\right)}{0.55^{5}\left(\mu m^{5}\right)\left[e^{(14.388/0.55\times5780)}-1\right](sr)}=0.256\times10^{8}\left[W/(m^{2}\cdot\mu m\cdot sr)\right]$$

Alternative forms of Equation (2.19) result if frequency or wave number is used rather than wavelength. Working in terms of frequency is advantageous when spectral radiation travels between media with different refractive indices because frequency remains constant while wavelength changes with propagation speed. We can rewrite Equation (2.19) in terms of frequency, noting that the wavelength in the medium is related to the wavelength in a vacuum and the frequency through $n\lambda = \lambda_0 = c_0/v$, and hence $d\lambda_0 = -(c_0/v^2)\ dv$. Then, the emissive power in the wavelength interval $d\lambda_0$ becomes:

$$E_{\lambda b}\left(T\right)d\lambda_{0}=\frac{2\pi n_{\lambda}^{2}C_{1}}{\lambda_{0}^{5}\left[\exp\left(\dfrac{C_{2}}{\lambda_{0}T}\right)-1\right]}d\lambda_{0}=\frac{2\pi n_{v}^{2}C_{1}}{\left(\dfrac{c_{0}}{v}\right)^{5}\left[\exp\left(\dfrac{C_{2}v}{c_{0}T}\right)-1\right]}\left|\frac{d\lambda_{0}}{dv}\right|dv$$

$$=\frac{2\pi n_{v}^{2}C_{1}v^{3}}{c_{0}^{4}\left[\exp\left(\dfrac{C_{2}v}{c_{0}T}\right)-1\right]}dv=E_{vb}\left(T\right)dv$$

(2.21)

The quantity $E_{vb}\left(T\right)$ is the emissive power per unit frequency interval at v. Likewise, in terms of wavenumber,

$$E_{\eta b}\left(T\right)d\eta_{0}=\frac{2\pi n_{\eta_{0}}^{2}C_{1}}{\left(\dfrac{1}{\eta_{0}}\right)^{5}\left[\exp\left(\dfrac{C_{2}\eta_{0}}{T}\right)-1\right]}\left|\frac{d\lambda_{0}}{d\eta_{0}}\right|d\eta_{0}=\frac{2\pi n_{\eta_{0}}^{2}C_{1}\eta_{0}^{3}}{\left[\exp\left(\dfrac{C_{2}\eta_{0}}{T}\right)-1\right]}d\eta_{0}$$

(2.22)

As before, the blackbody spectral intensity is given by $I_{\nu b} = E_{\nu b}/\pi$ and $I_{\eta b} = E_{\eta b}/\pi$.

The spectral nature of Planck's Law is shown in Figure 2.8, where the hemispherical spectral emissive power, Equation (2.19), is plotted as a function of wavelength for several temperatures. We have shown that total radiated energy, which is the spectral energy integrated over all wavelengths, increases with temperature. Figure 2.8 shows that the spectral emissive power also increases monotonically with the temperature at each wavelength. Also, the maximum spectral emissive power shifts toward smaller wavelengths as temperature increases.

Conceptually, this aligns with our understanding of the Second Law: Thermal radiation is emitted as atoms and molecules fluctuate from higher to lower energy states, and the fluctuation rate increases with temperature. Moreover, according to Boltzmann's distribution, the population of atoms and molecules shifts toward higher energy states with increasing temperature, corresponding to bigger drops in energy, and thus more energetic/higher frequency EM waves emitted with each transition. These are the reasons why the curves in Figure 2.8 increase in magnitude and shift to shorter wavelengths with increasing temperature.

Planck's distribution also connects the visible appearance of a blackbody to its temperature. For a blackbody at 300 K, only a small amount of energy is emitted within the visible region, which is undetectable by the eye. Red light first becomes visible to the unaided eye from a heated object in darkened surroundings at about the Draper Point of 798 K (Draper 1847). At higher temperatures, additional wavelengths in the visible light spectral range appear, and the object's apparent color will gradually shift from red, to orange, and then to yellow. At a sufficiently high temperature,

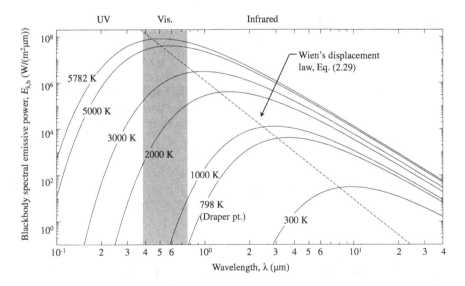

FIGURE 2.8 Hemispherical spectral emissive power, $E_{\lambda b}$, of a blackbody at several temperatures.

the visible emitted light appears white as it is composed of a mixture of radiation at all the visible wavelengths. The brightness of an object increases substantially with its increased temperature due to the overall increase in intensity.

It is not a coincidence that our eyes are most sensitive to the peak spectral emissive power of the Sun, which resembles that of a blackbody at 5780 K. Although human vision is well-tuned to the solar spectrum, this is not true of all life on Earth. Turtles have eyes sensitive to infrared but not to blue; bees have eyes sensitive to ultraviolet but not red. If we find life in other solar systems with a sun with an effective temperature different from ours, it will be interesting to discover what wavelength range encompasses the "visible spectrum" if beings there possess sight. While we perceive our Sun as emitting near-white light, and stars that are hotter or colder than our Sun as blue or red, respectively, we will never encounter a green star because the wavelength corresponding to green light lies in the middle of our visible spectrum.

Incandescent light bulbs emit radiation near to a blackbody spectrum. Most tungsten filament lamps are limited to a temperature below 3000 K due to vaporization of the filament at higher temperatures, and they therefore emit a large fraction of their energy in the infrared (centered at about 1 µm). In contrast, compact fluorescent bulbs operate by electrically exciting fluorescence in a chemical coating, and light-emitting diodes (LEDs) work by electroluminescence rather than thermal emission. Neither LEDs nor compact fluorescent bulbs emit significant amounts of *thermal* radiation. Because of their low efficiency, incandescent light bulbs are now routinely replaced by compact fluorescent bulbs and LEDs.

Note the similarity of the curves shown in Figure 2.8. It is possible to collapse these into a single curve by dividing Equation (2.19) by T^5 to obtain:

$$\frac{E_{\lambda b}(T)}{T^5} = \frac{\pi I_{\lambda b}(T)}{T^5} = \frac{2\pi C_1}{n_\lambda^2 (\lambda T)^5 \left[\exp\left(\dfrac{C_2}{n_\lambda \lambda T}\right) - 1\right]} \tag{2.23}$$

which is a univariate function of the product λT. The single plot in Figure 2.9 then replaces the multiple curves in Figure 2.8. Illustrative numerical values are shown in Table A.3 for $n_\lambda = 1$. Equation (2.22) can also be placed in a more universal form in terms of the variable η/T (= $1/\lambda T$):

$$\frac{I_{\eta b}}{T^3} = \frac{E_{\eta b}(T)}{\pi T^3} = \frac{2C_1 \left(\dfrac{\eta}{T}\right)^3}{n_\eta^2 \left[\exp\left(\dfrac{C_2 \eta}{n_\eta T}\right) - 1\right]} \tag{2.24}$$

2.7.1 Approximations to the Blackbody Spectral Distribution

Some approximate forms of Planck's distribution can be useful because of their relative simplicity. Care must be taken to use them only in the range of acceptable accuracy.

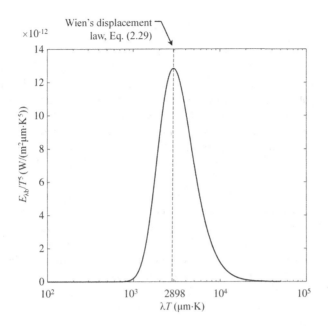

FIGURE 2.9 Spectral distribution of blackbody hemispherical emissive power as a function of λT.

If the term $\exp(C_2/\lambda T)$ is much larger than unity, Equation (2.23) reduces to

$$\frac{E_{\lambda b}(T)}{T^5} = \frac{2\pi C_1}{n_\lambda^2 (\lambda T)^5 \exp\left(\dfrac{C_2}{n_\lambda \lambda T}\right)} \tag{2.25}$$

which is known as *Wien's approximation*. In 1896 Wien conjectured the $\exp(a/\lambda T)$ relationship, where a is a constant, based on the Maxwell-Boltzmann distribution for the velocities of gas molecules, which in turn descends from Equation (1.1). This is the earliest attempt to provide a theoretical basis for the spectral distribution of blackbody emission. Wien's approximation is accurate to within 1% for λT less than 3000 μm·K. In two-wavelength pyrometry, for example, the emitted radiation from an object measured at two wavelengths is taken to be proportional to $I_{\lambda b}(T)$ at each wavelength. The ratio of Equation (2.25) at the two measurement wavelengths can be rearranged into an explicit function of temperature, which is not possible using Equation (2.24) due to the "−1" in the denominator.

Another approximation is found by expanding the exponential in the denominator of Equation (2.23) in a series (for $n = 1$) to obtain:

$$e^{C_2/\lambda T} - 1 = \left[1 + \frac{C_2}{\lambda T} + \frac{1}{2!}\left(\frac{C_2}{\lambda T}\right)^2 + \frac{1}{3!}\left(\frac{C_2}{\lambda T}\right)^3 + \cdots \right] - 1 \tag{2.26}$$

For λT much larger than C_2, the term $e^{C_2/\lambda T} - 1$ is approximated by the single term $C_2/\lambda T$, and Equation (2.23) becomes:

$$\frac{E_{\lambda b}(T)}{T^5} = \frac{2\pi C_1}{C_2 \cdot (\lambda T)^4} \tag{2.27}$$

This is the Rayleigh–Jeans formula. It was first postulated by Lord Rayleigh (1900), and later improved upon by Rayleigh (1905) and James Jeans (1905). Like Wien's distribution, the Rayleigh–Jeans formula is derived from statistical thermodynamics but follows a route based on the concept of equipartition of energy. Equipartition is the principle that at thermodynamic equilibrium the internal energy of a system is evenly distributed across the different types of microstates. In a molecular gas, the microstates consist of the energy levels of the individual gas molecules, while in the case of radiation Rayleigh and Jeans envisioned the microstates as individual frequencies. The distribution is only accurate to within 1% for λT greater than 7.78×10^5 µm·K. At lower wavelengths (higher energies) the error increases, resulting in the so-called *ultraviolet catastrophe*, named because the Rayleigh–Jeans formula fails to predict the blackbody spectral emissive power at wavelengths shorter than the visible spectrum. (Observe what happens in Equation (2.27) as $\lambda \to 0$.)

2.7.2 WIEN'S DISPLACEMENT LAW

A useful quantity is the wavelength λ_{max} at which the blackbody intensity $I_{\lambda b}(T)$ is a maximum for a given temperature. This maximum shifts toward shorter wavelengths as the temperature increases (Figure 2.8). The value of $\lambda_{max}T$ is at the peak of the distribution curve in Figure 2.9. It is obtained by differentiating Equation (2.23) with respect to λT and setting the left side equal to zero. This gives the transcendental equation (assuming $n = 1$):

$$\lambda_{max}T = \frac{C_2}{5} \frac{1}{1 - e^{-C_2/(\lambda_{max}T)}} \tag{2.28}$$

The solution to this equation for $\lambda_{max}T$ is a constant,

$$\lambda_{max}T = C_3 = 2897.7729 \text{ µm} \cdot \text{K} \tag{2.29}$$

which is one form of Wien's Displacement Law. Values of the constant C_3 in other units are listed in Table A.2. Equation (2.29) indicates that with increased temperature, the maximum emissive power is achieved, and the intensity shifts to a shorter wavelength in inverse proportion to T. Wien proposed a form of Equation (2.29) in 1893, which predates Planck's distribution (1901), and even Wien's own approximation to $E_{\lambda b}$ (1896). Wien derived the Displacement Law from a conceptual argument centered on how radiation contained within a volume would change as it slowly expanded or contracted.

Substituting Wien's Displacement Law Equation (2.29) into Equation (2.20) gives the maximum intensity at T as

$$I_{\lambda\text{-max},b} = \frac{E_{\lambda\text{-max},b}}{\pi} = \frac{2C_1}{C_3^5 \left(e^{C_2/C_3} - 1\right)} T^5 = 4.09570 \times 10^{-12} \left(\frac{W}{m^2 \cdot \mu m \cdot K^5 \cdot sr}\right) T^5 = C_4 T^5$$

(2.30)

where $C_4 = 4.09570 \times 10^{-12}$ W/($m^2 \cdot \mu m \cdot K^5 \cdot sr$) and $n = 1$. Equation (2.30) shows that the maximum spectral blackbody intensity increases as absolute temperature to the *fifth power.*

Example 2.3

What temperature is needed for a blackbody to radiate its maximum intensity at the center of the visible spectrum?

Figure 2.2 shows that the visible spectrum spans the range $\lambda = 0.4–0.7$ μm with its center at 0.55 μm. From Equation (2.29),

$$T = \frac{C_3}{\lambda_{max}} = \frac{2897.8\,\mu m \cdot K}{0.55\,\mu m} = 5269\ K$$

This compares with the observed effective radiating surface temperature of the Sun of 5780 K.

Example 2.4

At what wavelength do you observe the maximum emission from a blackbody at room temperature?

Using a room temperature of 21°C (294 K), Wien's Displacement Law gives

$$\lambda_{max} = \frac{C_3}{T} = \frac{2897.8\,\mu m \cdot K}{294\,K} = 9.86\ \mu m$$

which is in the middle of the near-infrared region. About 25% of radiative emission at room temperature is at wavelengths smaller than this value.

2.7.3 TOTAL BLACKBODY INTENSITY AND EMISSIVE POWER

The previous discussion provided the energy per unit wavelength interval that a blackbody radiates into a vacuum at each wavelength. Now the *total intensity* is determined, where a "total" quantity includes radiation for all wavelengths.

The intensity emitted in the wavelength interval $d\lambda$ is $I_{\lambda b}(T)d\lambda$. Integrating over all wavelengths gives the *total blackbody intensity*:

$$I_b(T) = \int_0^\infty I_{\lambda b}(T)d\lambda$$

(2.31)

Substituting the Planck distribution from Equation (2.20) with $n = 1$ and transforming variables using $\zeta = C_2/\lambda T$, Equation (2.31) becomes:

$$I_b(T) = \int_0^\infty \frac{2C_1}{\lambda^5\left(e^{\frac{C_2}{\lambda T}} - 1\right)}d\lambda = \frac{2C_1 T^4}{C_2^4}\int_0^\infty \frac{\zeta^3}{e^\zeta - 1}d\zeta = \frac{2C_1 T^4}{C_2^4}\frac{\pi^4}{15} = \frac{\sigma}{\pi}T^4 \quad (2.32)$$

where the constant σ is:

$$\sigma \equiv \frac{2C_1\pi^5}{15C_2^4} = 5.670367\times10^{-8}\ \text{W/(m}^2\cdot\text{K}^4) \quad (2.33)$$

The *hemispherical total emissive power* of a blackbody radiating into a vacuum is then:

$$E_b = \int_0^\infty E_{\lambda b}d\lambda = \int_0^\infty \pi I_{\lambda b}d\lambda = \pi I_b = \sigma T^4 \quad (2.34)$$

which is the emitted blackbody energy flux (W/m²). Equation (2.34) is the Stefan–Boltzmann Law, and σ is the Stefan–Boltzmann Constant.

This Stefan–Boltzmann Law was postulated by Josef Stefan in 1879 based on empirical observations and confirmed theoretically by Ludwig Boltzmann in 1884. The astute observer will notice that Boltzmann theoretically verified Equation (2.34) well before Planck derived his distribution in 1901. Boltzmann did this by using the concept of radiation pressure within an isentropically expanding volume along with the theory of Carnot, much as Wien used similar principles (including the Stefan–Boltzmann Law) to derive his Displacement Law without detailed knowledge of how $E_{\lambda b}$ varies with wavelength. For a more complete history with short biographies of important contributors and a timeline of events, see Appendices G and H at the online site www.ThermalRadiation.net/appendix.html.

Example 2.5

Energy emitted in the normal direction from a blackbody surface is found to have a total radiation per unit solid angle and per unit surface area of 10,000 W/(m²·sr). What is the surface temperature?

The hemispherical total emissive power is related to the total intensity in the normal direction by $E_b = \pi I_b$. Hence, from Equation (2.34), $T = (\pi I_b/\sigma)^{1/4} = (10,000\pi/5.67040\times10^{-8})^{1/4} = 862.7$ K

Example 2.6

A black surface is radiating with a hemispherical total emissive power of 20 kW/m². What is its surface temperature? At what wavelength is its maximum spectral intensity?

From Stefan–Boltzmann Law, the temperature of the blackbody is $T = (E_b/\sigma)^{1/4} = (20,000/5.67040) \times 10^{-8})^{1/4} = 770.6$ K. Then, from Wien's Displacement Law, $\lambda_{max} = C_3/T = 2898/770.6 = 3.76$ μm.

2.7.4 BLACKBODY RADIATION WITHIN A SPECTRAL BAND

The hemispherical total emissive power of a blackbody radiating into a vacuum is given by the Stefan–Boltzmann Law, Equation (2.34). In calculating radiative exchange between two objects, it is often necessary to find the fraction of the total emissive power emitted in a wavelength band, as illustrated in Figure 2.10. This fraction is designated by $F_{\lambda_1 T \to \lambda_2 T}$ and is given by the ratio

$$F_{\lambda_1 T \to \lambda_2 T} = \frac{\int_{\lambda=\lambda_1}^{\lambda_2} E_{\lambda b}(T)d\lambda}{\int_{\lambda=0}^{\infty} E_{\lambda b}(T)d\lambda} = \frac{\int_{\lambda=\lambda_1}^{\lambda_2} E_{\lambda b}(T)d\lambda}{\sigma T^4}$$

(2.35)

$$= \frac{1}{\sigma} \int_{\lambda T=(\lambda T)_1}^{(\lambda T)_2} \frac{2\pi C_1}{(\lambda T)^5 \left[\exp(C_2/\lambda T)-1\right]} d(\lambda T)$$

The integrals in Equation (2.35) can be expressed as the difference between two integrals, each with a lower limit $\lambda = 0$, so that

$$F_{\lambda_1 T \to \lambda_2 T} = \frac{1}{\sigma T^4}\left[\int_0^{\lambda_2} E_{\lambda b}d\lambda - \int_0^{\lambda_1} E_{\lambda b}d\lambda\right] = F_{0 \to \lambda_2 T} - F_{0 \to \lambda_1 T}$$

(2.36)

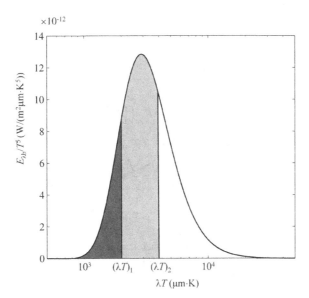

FIGURE 2.10 Emitted energy in a wavelength band.

The fraction of the emissive power for any wavelength band can therefore be found by having $F_{0\to\lambda_1 T}$ as a function of λT. The $F_{0\to\lambda_1 T}$ function is illustrated by Figure 2.11a, where it equals the light gray area divided by the total area (shaded) under the curve.

For a blackbody, the hemispherical emissive power is related to intensity, Equation (2.20), and the $F_{\lambda_1 T\to\lambda_2 T}$ function gives the fraction of the total intensity within the wavelength interval λ_1 to λ_2. Expression of the F function in terms of a single variable

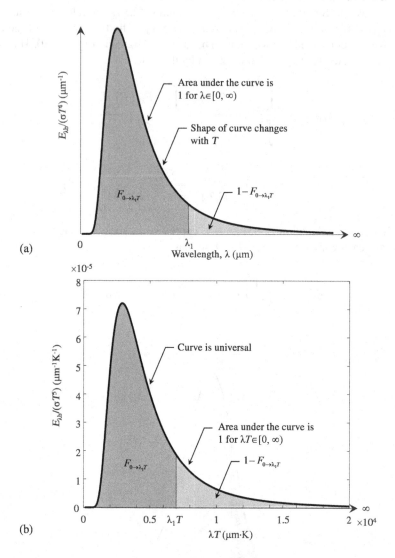

FIGURE 2.11 Physical representation of F factor, where $F_{0\to\lambda 1}$ or $F_{0\to\lambda 1 T}$ is the ratio of light gray to total shaded area, (a) in terms of curve for specific temperature.

λT simplifies the calculations significantly (Figure 2.11b). This is done by rewriting Equation (2.36) as:

$$F_{\lambda_1 T \to \lambda_2 T} = \frac{1}{\sigma} \left[\int_0^{\lambda_2 T} \frac{E_{\lambda b}}{T^5} d(\lambda T) - \int_0^{\lambda_1 T} \frac{E_{\lambda b}}{T^5} d(\lambda T) \right] = F_{0 \to \lambda_2 T} - F_{0 \to \lambda_1 T} \qquad (2.37)$$

As shown by Equation (2.23), $E_{\lambda b}/T^5$ is a function only of λT, so the integrands in Equations (2.36) and (2.37) depend only on the λT variable. A convenient series form for $F_{0 \to \lambda T}$ is found by using the substitution $\zeta = C_2/\lambda T$ to obtain, from Equation (2.37),

$$F_{0 \to \lambda T} = \frac{2\pi C_1}{\sigma T^4} \int_0^{\lambda} \frac{d\lambda}{\lambda^5 \left(e^{\frac{C_2}{\lambda T}} - 1 \right)} = \frac{2\pi C_1}{\sigma C_2^4} \int_{\zeta}^{\infty} \frac{\zeta^3}{e^{\zeta} - 1} d\zeta \qquad (2.38)$$

Using the definition of σ and the value of the definite integrals in Equation (2.38), this reduces to:

$$F_{0 \to \lambda T} = 1 - \frac{15}{\pi^4} \int_0^{\zeta} \frac{\zeta^3}{e^{\zeta} - 1} d\zeta \qquad (2.39)$$

By using the series expansion $(e^{\zeta} - 1)^{-1} = e^{-\zeta} + e^{-2\zeta} + e^{-3\zeta} + \cdots$ and then integrating by parts, the $F_{0 \to \lambda T}$ becomes (Chang and Rhee 1984):

$$F_{0 \to \lambda T} = \frac{15}{\pi^4} \sum_{m=1}^{\infty} \left[\frac{e^{-m\zeta}}{m} \left(\zeta^3 + \frac{3\zeta^2}{m} + \frac{6\zeta}{m^2} + \frac{6}{m^3} \right) \right] \qquad (2.40)$$

where $\zeta = C_2/\lambda T$. The series in Equation (2.40) converges very rapidly, and using the first three terms of the summation gives good results over most of the range of $F_{0 \to \lambda T}$. As λT becomes large (small ζ), a larger number of terms is required. The series is useful in computer solutions. Values of $F_{0 \to \lambda T}$ from Equation (2.40) are in Table A.3, and a program for Equation (2.40) is available at www.ThermalRadiation.net/blackbody.html.

A plot of $F_{0 \to \lambda T}$ as a function of λT is in Figure 2.12.

When working with wave numbers or frequency, the fraction $F_{0 \to \eta/T} = F_{0 \to \nu/c_0 T} = F_{0 \to 1/\lambda T}$ is needed. The inverse relation between wave number and wavelength gives $F_{0 \to \eta/T} = 1 - F_{0 \to \lambda T}$. The use of the $F_{0 \to \lambda T}$ function is illustrated in these examples:

Example 2.7

A blackbody is radiating at a temperature of 3000 K. An experimenter wants to measure total radiant emission by using a radiation detector. The detector absorbs all radiation in the range of $\lambda = 0.8 - 5$ μm but detects no energy outside that

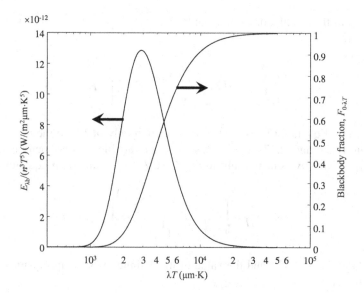

FIGURE 2.12 Fractional blackbody emissive power in the range 0 to λT.

range. What percentage correction does the experimenter need to apply to the energy measurement? If the sensitivity of the detector could be extended in the range of λ by 0.5 μm at only one end of its sensitivity range, which end should be extended?

Using $\lambda_1 T = 0.8 \times 3000 = 2400$ μm·K and $\lambda_2 T = 5 \times 3000 = 15{,}000$ μm·K in Equation (2.40) gives the fraction of energy outside the sensitivity range as $F_{0 \to \lambda_1 T} + F_{\lambda_2 T \to \infty} = F_{0 \to \lambda_1 T} + \left(1 - F_{0 \to \lambda_2 T}\right) = \left(0.14026 + 1 - 0.96893\right) = 0.17133$, or a correction of 17.1% of the total incident energy. Extending the sensitive range to the longer-wavelength side of the measurement interval adds little accuracy because of the small slope of the curve of F against λT in that region, so extending to shorter wavelengths provides the greatest increase in the detected energy. The quantitative results to demonstrate this are found from the modified factor at the short wavelength of 0.3 μm of $F_{0 \to 0.3 \times 3000} = 0.00009$ and at the long-wavelength end to 5.5 μm of $F_{0 \to 5.5 \times 3000} = 0.97581$. The correction factors after extending the short- and long-wavelength boundaries of the interval are then $0.00009 + (1 - 0.96893) = 0.03116$ and $0.14026 + (1 - 0.97581) = 0.16445$.

Example 2.8

The experimenter of the previous example has designed a radiant energy detector that can be made sensitive over only a 1 μm range of wavelength. She wants to measure the total emissive power of two blackbodies, one at 2500 K and the other at 5000 K and plans to adjust the 1 μm interval to give a 0.5 μm sensitive band on each side of the peak blackbody emissive power. For which blackbody would you expect to detect the greatest percentage of the total emissive power? What will the percentage be in each case?

TABLE 2.1
Fraction of Blackbody Emission Contained in the
Region 0 to λT or 0 to η/T

λT				
$\mu m \cdot R$	$\mu m \cdot K$	$F_{0 \to \lambda T}$	$\eta/T, (\mu m \cdot K)^{-1}$	$F_{0 \to \eta/T}$
2,606	1,448	0.01	0.4369×10^{4}	0.01
$5,216 = \lambda_{max} T$	$2,898 = \lambda_{max} T$	0.25	1.627×10^{-4}	0.25
7,393	4,107	0.50	2.435×10^{-4}	0.50
11,067	6,148	0.75	3.451×10^{-4}	0.75
41,200	22,890	0.99	6.906×10^{-4}	0.99

Wien's Displacement Law tells us that the peak emissive power exists at $\lambda_{max} = 2897.8/T$ μm in each case. For the higher temperature, a wavelength interval of 1 μm will give a wider spread of λT values around the peak of $\lambda_{max} T$ on the normalized blackbody curve (Figure 2.9), so the measurement should be more accurate for the 5000 K case. For the 5000 K blackbody, $\lambda_{max} = 0.5796$ μm, and $\lambda_1 T = (0.5796 - 0.5000) \times 5000 = 398$ μm·K. Similarly, $\lambda_2 T = 1.0796 \times 5000 = 5398$ μm·K. The percentage of detected emissive power is then $100(F_{0 \to 5398} - F_{0 \to 398}) = 100 \times (0.6801 - 0) = 68.0\%$

A similar calculation for the 2500 K blackbody shows that 48.6% of the emissive power is detected.

Some commonly used values of the $F_{0 \to \lambda T}$ and $F_{0 \to \eta/T}$ are listed in Table 2.1. Very nearly one-fourth of the total emissive power lies in the wavelength range below the peak of the Planck spectral distribution.

Example 2.9

A light-bulb filament is at 3000 K. If the filament radiates as a blackbody, what fraction of its emission is in the visible region?

The visible region is within $\lambda = 0.4 \to 0.7$ μm. The desired fraction is
$F_{0 \to 2100 \mu m \cdot K} - F_{0 \to 1200 \mu m \cdot K} = 0.1220 - 0.0049 = 0.1171$

This very low efficiency (12%) of converting energy into visible light is a compelling reason for replacing incandescent bulbs with more efficient lighting.

2.7.5 SUMMARY OF BLACKBODY PROPERTIES

The blackbody possesses fundamental properties that make it a standard for comparing real radiating bodies. These properties are:

1. The blackbody is the maximum absorber and emitter of radiant energy at all wavelengths and in all directions.

2. The total (including all wavelengths) radiant intensity and hemispherical total emissive power of a blackbody into a medium with a constant index of refraction n are given by the Stefan–Boltzmann Law:

$$\pi I_b = E_b = n^2 \sigma T^4$$

3. The blackbody spectral and total intensities are independent of the direction so that the emission of energy into a direction θ away from the surface normal direction is proportional to the projected area of the emitting element, $dA \cos\theta$. This is Lambert's Cosine Law.

4. The blackbody spectral distribution of intensity for emission into a medium with refractive index n is given by Planck's distribution:

$$\frac{E_{\lambda b}(T)}{\pi} = I_{\lambda b}(T) = \frac{2C_1}{n^2 \lambda^5 \left[\exp\left(\dfrac{C_2}{n\lambda T} \right) - 1 \right]}$$

5. The wavelength at which the maximum spectral intensity of radiation into a medium with refractive index n for a blackbody occurs is given by Wien's Displacement Law:

$$\lambda_{max,n} = \frac{\lambda_{max,0}}{n} = \frac{C_3}{nT}$$

For vacuum or air, n equals 1, so $\lambda_{max,vacuum} = \lambda_{max,0}$.

A calculator for these properties can be accessed at *www.thermalradiation.net/blackbody.html*.

The definitions for blackbody properties introduced in this chapter are summarized in Table 2.2. The formulas for the quantities are given in terms of either the spectral intensity $I_{\lambda b}$, which is computed from Planck's Law, or the surface temperature T.

2.8   RADIATIVE ENERGY ALONG A LINE OF SIGHT

Now, we return to the discussion of radiative transfer as in Figure 2.3, and we discuss the fate of radiative transfer in a participating (absorbing, emitting, and/or scattering) medium as it propagates after emission from a surface.

We begin by extending our definition of radiant intensity, defined for a surface element in Equations (2.10) and (2.11), to any point in a volume. If we consider radiative energy propagating at a rate $dQ_\lambda(S,\Omega,t)$ in direction Ω along path S, we can define the radiant intensity in a volume element as:

$$I_\lambda(S,\Omega,t) = \frac{dQ_\lambda(S,\Omega,t)}{dA\,d\lambda\,d\Omega} \tag{2.41}$$

where dA is a surface element with its unit normal vector aligned with S.

TABLE 2.2
Blackbody Radiation Quantities ($n = 1$)

Symbol	Name	Definition	Geometry	Formula
$I_{\lambda b}(T)$	Spectral intensity	Emission in any direction per unit of projected area normal to that direction, and per unit time, wavelength interval about λ, and solid angle		$2C_1\left[\lambda^5\left(e^{C_2/\lambda T}-1\right)\right]$
$I_b(T)$	Total intensity	Emission, including all wavelengths, in any direction per unit of projected area normal to that direction, and per unit time and solid angle		$\sigma T^4 / \pi$
$E_{\lambda b}(\theta, T)$	Directional spectral emissive power	Emission per unit solid angle in direction θ per unit surface area, wavelength interval, and time		$I_{\lambda b}(T)\cos\theta$
$E_b(\theta, T)$	Directional total emissive power	Emission, including all wavelengths, in direction θ per unit surface area, solid angle, and time		$\left(\sigma T^4 / \pi\right)\cos\theta$

(Continued)

TABLE 2.2 (CONTINUED)
Blackbody Radiation Quantities ($n = 1$)

Symbol	Name	Definition	Geometry	Formula
$E_{\lambda b}\left(\theta_1 \to \theta_2, \phi_1 \to \phi_2, T\right)$	Finite solid-angle spectral emissive power	Emission in solid angle $\theta_1 \le \theta \le \theta_2$, $\phi_1 \le \phi \le \phi_2$ per unit surface area, wavelength interval, and time	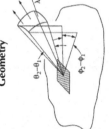	$I_{\lambda b}(T)\left[\left(\sin^2\theta_2 - \sin^2\theta_1\right)/2\right] \times (\phi_2 - \phi_1)$
$E_b\left(\theta_1 \to \theta_2, \phi_1 \to \phi_2, T\right)$	Finite solid-angle total emissive power	Emission, including all wave lengths, in solid angle $\theta_1 \le \theta \le \theta_2$, $\phi_1 \le \phi \le \phi_2$ per unit surface area and time.		$\dfrac{\sigma T^4}{\pi} \dfrac{\sin^2\theta_2 - \sin^2\theta_1}{2}\left(\phi_2 - \phi_1\right)$
$E_{\lambda b}\left(\lambda_1 \to \lambda_2, \theta_1 \to \theta_2, \phi_1 \to \phi_2, T\right)$	Finite solid-angle band emissive power	Emission in solid angle $\theta_1 \le \theta \le \theta_2$, $\phi_1 \le \phi \le \phi_2$ and wavelength band $\lambda_1 \to \lambda_2$ per unit surface area and time		$\dfrac{\sigma T^4}{\pi} \dfrac{\sin^2\theta_2 - \sin^2\theta_1}{2}\left(\phi_2 - \phi_1\right) \times \left(F_{0 \to \lambda_2} - F_{0 \to \lambda_1}\right)$
$E_{\lambda b}(T)$	Hemispherical spectral emissive power	Emission into hemispherical solid angle per unit surface area, wavelength interval, and time	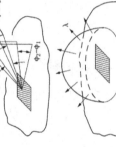	$\pi I_{\lambda b}(T)$

(Continued)

TABLE 2.2 (CONTINUED)
Blackbody Radiation Quantities ($n = 1$)

Symbol	Name	Definition	Geometry	Formula
$E_{\lambda b}(\lambda_1 \rightarrow \lambda_2, T)$	Hemispherical band emissive power	Emission in wavelength band $\lambda_1 \rightarrow \lambda_2$ into hemispherical solid angle per unit surface area and time		$(F_{0 \rightarrow \lambda_2} - F_{0 \rightarrow \lambda_1})\sigma T^4$
$E_b(T)$	Hemispherical total emissive power	Emission, including all wavelengths, into hemispherical solid angle per unit surface area and time		σT^4

As it propagates along direction Ω over path interval dS, the intensity loses some of its strength due to scattering and absorption. On the other hand, it gains additional strength because of emission and in-scattering of the radiation incident from other directions into direction Ω. Along this direction and path, we can tally the changes in radiative energy, which provides a conservation equation of radiative energy within a small directional cone along a given direction, for a small wavelength interval around a fixed wavelength, and within an infinitesimal time duration at a given time (see Figure 2.13). This formulation leads us to the radiative transfer equation, expressed in terms of I_λ.

Before discussing the RTE further, we must quantify absorption, scattering, and emission. Once we obtain the conservation of radiative energy along a line of sight, it can be written over a finite length of the medium, from one surface to another. After these fundamental expressions are outlined, we are poised to model and solve the expressions for engineering calculations.

2.8.1 RADIATIVE ENERGY LOSS DUE TO ABSORPTION AND SCATTERING

Consider spectral radiation of intensity I_λ incident on a volume element of length dS along the direction of propagation S. As radiation passes through dS, its intensity is reduced by absorption and scattering. This reduction is also proportional to the magnitude of the local intensity, since the matter contained along dS absorbs and scatters a fixed fraction of I_λ incident on the volume. The decrease in radiation intensity is:

$$dI_\lambda(S,\Omega) = -\beta_\lambda(S)I_\lambda(S,\Omega)dS \tag{2.42}$$

where β_λ is the spectral *extinction coefficient* of the medium and includes losses due to both absorption and scattering. The extinction coefficient is a physical property, has units of reciprocal length, and depends on the local properties of the medium, including temperature T, pressure P, composition of the material (e.g., the concentrations C_i of the i components), and wavelength of the incident radiation. Therefore, it depends on local parameters, as indicated by S dependency, and in explicit form, it is

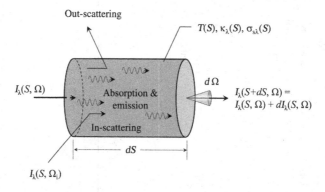

FIGURE 2.13   Radiation intensity propagating through a volume element.

expressed as $\beta_\lambda(S) = \beta_\lambda(T,P,C_i)$. If β_λ in the medium is constant, it can be interpreted as the inverse of the mean penetration distance (or *mean free path*) of radiation, as will be shown later.

The extinction coefficient consists of two parts: *absorption* coefficient κ_λ and *scattering* coefficient $\sigma_{s,\lambda}$,

$$\beta_\lambda = \kappa_\lambda + \sigma_{s,\lambda} \tag{2.43}$$

The subscript s on the scattering coefficient is included to avoid confusion with the Stefan–Boltzmann Constant. All these parameters have units of reciprocal length and are called *linear*, or *volumetric*, *coefficients*.

Absorption describes how the radiative energy is converted to the internal energy of the absorbing matter. It is one of two mechanisms coupling the radiative energy propagation with the thermodynamic state of the matter. The other is radiative emission, which is proportional to the internal energy of matter as indicated by its temperature.

Scattering causes a change in the direction of radiation propagating along S. During an elastic scattering event, no radiative energy is converted to internal thermal energy (scattering in radiation is assumed to be perfectly elastic except in some very special cases). However, scattering changes the balance of the energy propagating in each direction.

In a nonparticipating (transparent) medium, β_λ is zero, and there is no change in intensity along a path. In a translucent, or participating, medium, however, the initial radiative intensity changes as it propagates along a line of sight. To determine how the energy decreases, integrate Equation (2.42) over a finite path length S to obtain:

$$\int_{I_\lambda = I_\lambda(0)}^{I_\lambda(S)} \frac{dI_\lambda}{I_\lambda} = -\int_{S^*=0}^{S} \beta_\lambda(S^*)dS^* \tag{2.44}$$

where $I_\lambda(0)$ is the intensity incident on the control volume at $S = 0$, and S^* is a dummy variable of integration. Carrying out the integration gives the intensity at location S for a given direction Ω:

$$I_\lambda(S) = I_\lambda(0)\exp\left[-\int_0^S \beta_\lambda(S^*)dS^*\right] \tag{2.45}$$

Equation (2.45) is known as Bouguer's Law, Beer's Law, The Bouguer–Lambert Law, The Beer–Lambert Law, and various permutations and combinations of these names. This law shows that the intensity of spectral radiation is attenuated exponentially as it passes through an absorbing–scattering medium. It has extensive use in many fields, from astrophysics to spectroscopy. Historically, it was first proposed by Pierre Bouguer (bōo'gâr) in 1729, and later cited by Johann Heinrich Lambert in 1760. August Beer recognized the relationship between the attenuation of light and the concentration of the medium (i.e., the connection between β_λ and C_i) in 1852.

We call Equation (2.45) Bouguer's Law in recognition of his priority, and to avoid confusion with Lambert's Cosine Law, Equation (2.17).

For a homogeneous medium, Equation (2.45) simplifies to:

$$I_\lambda(S) = I_\lambda(0)\exp(-\beta_\lambda S) \tag{2.46}$$

If the medium is only absorbing, then the intensity is converted to internal energy of the matter along the line of sight of the beam without being redirected to other directions due to scattering. If $\sigma_{s\lambda} = 0$, then $\beta_\lambda = \kappa_\lambda$ and Equation (2.42) becomes:

$$\frac{dI_\lambda}{dS} = -\kappa_\lambda(S)I_\lambda \tag{2.47}$$

which again is in direction Ω. This expression can be integrated to determine the intensity leaving the boundary after it travels over a distance S:

$$I_\lambda(S) = I_\lambda(0)\exp\left[-\int_0^S \kappa_\lambda(S^*)dS^*\right] \tag{2.48}$$

and, if κ_λ is uniform within the medium,

$$I_\lambda(S) = I_\lambda(0)\exp(-\kappa_\lambda S) \tag{2.49}$$

We can obtain similar expressions for a purely scattering medium by substituting $\beta_\lambda = \sigma_{s\lambda}$ into Equations (2.47–2.49). The decrease in intensity due to the scattering of radiative energy out of the direction in propagation is often called "out-scattering" to distinguish from the augmentation of intensity caused by "in-scattering" from other directions, which is discussed later.

Bouguer's Law only accounts for the attenuation in intensity along S due to absorption and scattering. This implicitly assumes: (i) emission from the medium is negligible, which is reasonable only if the incident intensity $I_\lambda(0)$ is much larger than $I_{\lambda b}$ within the medium; and (ii) we neglect the in-scattering of radiation from other directions in the S direction. Removing these assumptions requires a more thorough analysis of the conservation of radiative energy.

2.8.2 Mean Penetration Distance

We can use the previous analysis to find out how much radiation would penetrate to a location S in the medium. Starting from Equation (2.46), the fraction of the radiation attenuated in the small volume element from S to $S + dS$ is:

$$-\frac{dI_\lambda(S)}{I_\lambda(0)} = \beta_\lambda(S)\frac{I_\lambda(S)}{I_\lambda(0)}dS = \beta_\lambda(S)\exp\left[-\int_0^S \beta_\lambda(S^*)dS^*\right]dS \tag{2.50}$$

A mean penetration distance of the radiation, l_m, is obtained by multiplying the fraction absorbed at S by the distance S and integrating over all path lengths from $S = 0$ to ∞:

$$l_m = \int_{S=0}^{\infty} S\beta_\lambda(S)\exp\left[-\int_0^S \beta_\lambda(S^*)dS^*\right]dS \tag{2.51}$$

If β_λ is uniform along the path, carrying out the integration yields:

$$l_m = \beta_\lambda \int_{S=0}^{\infty} S\exp(-\beta_\lambda S)dS = \frac{1}{\beta_\lambda} \tag{2.52}$$

demonstrating that the average penetration distance is the reciprocal of β_λ. This is a very useful expression for the interpretation of experimental measurements. Penetration depths can be obtained for absorbing-only or scattering-only media by replacing the extinction coefficient β_λ with either the absorption coefficient κ_λ or the scattering coefficient $\sigma_{s,\lambda}$.

2.8.3 OPTICAL THICKNESS

The exponential factor in Equation (2.45) is often written as a dimensionless quantity, $\tau_\lambda(S)$, called the *optical thickness* or *opacity* along the path length S:

$$\tau_\lambda(S) = \int_0^S \beta_\lambda(S^*)dS^* \tag{2.53}$$

Equation (2.48) becomes:

$$I_\lambda(S) = I_\lambda(0)\exp\left[-\tau_\lambda(S)\right] \tag{2.54}$$

which is an integral function accounting for all the values of β_λ between 0 and S. Because β_λ is a function of the local parameters P, T, and C_i, the optical thickness is also a function of these variables along the path between 0 and S. For a medium with uniform properties, Equation (2.53) reduces to $\tau_\lambda(S) = \beta_\lambda \cdot S$

The optical thickness indicates how strongly a medium attenuates radiation at a given wavelength. If $\tau_\lambda(S) \ll 1$, the medium is called *optically thin*, whereas if $\tau_\lambda \gg 1$, the medium is *optically thick*. With decreasing optical thickness, the participating medium approaches a nonparticipating one; on the other hand, with increasing τ_λ, less and less energy is transferred and the medium eventually becomes *opaque*.

The optical thickness can also be interpreted as the number of mean penetration distances within the path length, l_m:

$$\tau_\lambda(S) = \frac{S}{l_m} \tag{2.55}$$

(The notation for optical thickness τ_λ should not be confused with the surface transmissivity defined in Chapter 3, which uses the same symbol τ.)

2.8.4 RADIATIVE ENERGY GAIN DUE TO EMISSION

As radiative energy propagates in a medium along a given direction, its strength increases due to volumetric emission from every point along its path. If the strength of the emission at a given location along S is defined as $j_\lambda(S,\theta,\phi)$, then the incremental increase in radiative energy due to emission is:

$$dI_\lambda(\theta,\phi) = j_\lambda(S,\theta,\phi)dS \qquad (2.56)$$

Here $j_\lambda(S,\theta,\phi)$ is any function representing the physics of the emission mechanism (e.g., incandescence, chemiluminescence, electroluminescence) and has units of W/m³·μm·sr. Under LTE, the thermal radiation strength is directly related to how much energy is absorbed by the medium and is correlated with the Planck blackbody function:

$$j_\lambda\left(S,\theta,\phi\right) = \kappa_\lambda I_{\lambda b}(S) \qquad (2.57)$$

This relationship can be intuited by remembering that κ_λ defines the coupling between the electromagnetic wave and the internal energy states of the matter.

For a more rigorous analysis, consider an elemental volume dV with a spectral absorption coefficient κ_λ, which can be a function of local temperature, pressure, and species concentration (T,P,C_i). Let dV be at the center of a large black hollow sphere of radius R at uniform temperature T (Figure 2.14). The space between dV and the sphere boundary is filled with nonparticipating (transparent) material. Then, the spectral intensity incident at the location of surface element dA_s on dV, from element dA on the enclosure surface (at $S = 0$), is:

$$I_\lambda\left(0\right) = I_\lambda\left(S=0\right) = I_{\lambda b}\left(S=0\right) = I_{\lambda b}(0) \qquad (2.58)$$

The change of this intensity across dV because of absorption is:

$$-I_\lambda(0)\kappa_\lambda dS = -I_{\lambda b}(0)\kappa_\lambda dS \qquad (2.59)$$

The energy absorbed by the differential subvolume $dV = dSdA_s$ is $I_{\lambda b}(S=0)\,\kappa_\lambda dSdA_s d\lambda d\Omega$, where $d\Omega = dA/R^2$ is the solid angle subtended by dA when viewed from dV, and dA_s is the projected area normal to $I_\lambda(0)$. The energy emitted by dA and absorbed by all dV is found by integrating over dV (over all dA_s and dA elements) to obtain $\kappa_\lambda I_{\lambda b}dVd\lambda d\Omega$. To account for all energy incident upon dV from the entire spherical enclosure, integration over all solid angles gives $4\pi\kappa_\lambda I_{\lambda b}dVd\lambda$.

To maintain equilibrium in the enclosure, dV at location S must emit energy equal to that absorbed. Hence, the *spectral emission* by an isothermal volume element is:

$$4\pi\kappa_\lambda I_{\lambda b}\left(S\right)dVd\lambda = 4\kappa_\lambda E_{\lambda b}\left(S\right)dVd\lambda \qquad (2.60)$$

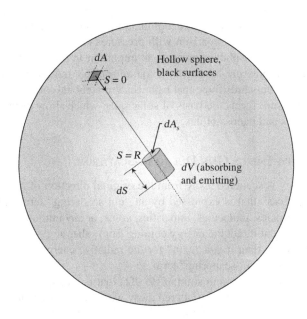

FIGURE 2.14 Emission from a volume element.

The shape of dV is arbitrary, but it must be small enough that energy emitted within dV escapes before reabsorption occurs within dV. The emitting material must be in thermodynamic equilibrium with respect to its internal energy.

For all conditions discussed here, the thermal emission is uniform over all directions (isotropic), so the spectral intensity emitted by a volume element into any direction, $dI_{\lambda,e}$, is obtained by dividing by $4\pi d\lambda$ and the cross-sectional area to give:

$$dI_{\lambda,e}(S) = \kappa_\lambda(S)I_{\lambda b}(S)dS = j_\lambda(S,\theta,\varphi)dS \tag{2.61}$$

If the refractive index of the medium is $n \neq 1$, the spectral volumetric emission is modified by a factor of n_λ^2:

$$dI_{\lambda,e}(S) = n_\lambda^2(S)\kappa_\lambda(S)I_{\lambda b}(S)dS \tag{2.62}$$

2.8.5 RADIATION PRESSURE

When radiation energy is emitted from a surface, an opposite reaction force provides pressure on a surface due to its emission. The pressure on a body from a surrounding isotropic blackbody radiation field is (Lebedev 1901; Feynman 1963):

$$P = \frac{4\pi}{3c}I_b = \frac{4}{3c}E_b \tag{2.63}$$

Because of the speed of light c in the denominator of Equation (2.63), radiative pressures are usually small in comparison with pressures caused by molecular interactions. Nevertheless, RTE photon pressure is important in some applications: most notably, the photon pressure arising from the anisotropic radiation field within the Sun results in a net outwards force that balances gravitational forces acting inwards. Also, photon pressure forms the basis of solar sails, which show promise to propel deep space probes and nanosatellites.

2.8.6 RADIATIVE ENERGY GAIN DUE TO IN-SCATTERING

If radiative energy is scattered away from the original direction of radiative energy, it constitutes a "loss" that is expressed by an "out-scattering" term in the RTE. If the scatterers (particles, molecules, impurities, voids, or any inhomogeneities) in the small volume element direct the energy coming from other angles into the direction of the original beam, then it is a "gain" for the radiative energy propagating in the direction S. This is the "in-scattering" term.

Consider the radiation within solid angle $d\Omega_i$, that is incident on an element dA. The portion of incident intensity scattered away within the increment dS is $dI_{\lambda,s}(S) = -\sigma_{s,\lambda}I_\lambda dS$. More specifically, the term $dI_{\lambda,s}(S)$ is the spectral energy scattered within path length dS per unit incident solid angle and area normal to the incident direction. The drop in intensity along dS due to scattering manifests as an increase in intensity in a direction defined by azimuthal angles $\Omega_s = (\theta_s,\phi_s)$ measured relative to the direction $\Omega_i = (\theta_i,\phi_i)$ of the incident intensity. The scattered intensity in the Ω_s direction is defined as the energy scattered in that direction per unit solid angle of the scattered direction and per unit normal area and solid angle of the incident radiation:

$$dI'_{\lambda,s}\left(S\right) = dI'_{\lambda,s}\left(\Omega_s\right) = \frac{\text{Spectral energy scattered in direction } \Omega_s = \left(\theta_s,\phi_s\right)}{d\Omega_s dA d\Omega_i d\lambda} \qquad (2.64)$$

The "prime" notation is used to distinguish the increase of intensity into direction Ω_s, $dI'_{\lambda s}$ due to scattering, caused by the scattering of intensity out of Ω_i, dI_s.

Now we introduce *the scattering phase function*, Φ, which is a measure of the amount of radiative energy propagating in incident direction $\Omega_i(\theta_i,\phi_i)$ that is redirected into $\Omega_s(\theta_s,\phi_s)$:

$$\Phi\left[\Omega_i\left(\theta_i,\phi_i\right),\Omega_s\left(\theta_s,\phi_s\right)\right] = \Phi\left(\Omega_i,\Omega_s\right) = \Phi\left[\left(\theta_i,\phi_i\right),\left(\theta_s,\phi_s\right)\right] \qquad (2.65)$$

in units of sr^{-1}. The angular profile of $\Phi(\Omega_i,\Omega_s)$ is directly related to the size (relative to the wavelength λ), shape, and material properties of the scatterer and varies with the wavelength of the incident radiation.

Using the definition of the phase function, the directional magnitude of $dI'_{\lambda s}(\Omega_s)$ is related to the entire out-scattered intensity $dI_{\lambda,s}(\Omega_i)$ scattered away from the incident radiation as:

$$dI'_{\lambda,s}(\Omega_s) = \frac{1}{4\pi}\Phi_\lambda(\Omega_i,\Omega_s)dI_{\lambda s} = \frac{\sigma_{s,\lambda}}{4\pi}\Phi_\lambda(\Omega_i,\Omega_s)I_\lambda(S)dS \qquad (2.66)$$

The spectral energy scattered into all $d\Omega$, per unit $d\lambda$, $d\Omega_i$, and dA is:

$$dI_{\lambda,s} = \int_{\Omega_{s=0}}^{4\pi} dI'_{\lambda,s}(\Omega_s)d\Omega_s \qquad (2.67)$$

Using Equation (2.67) to eliminate $dI_{\lambda,s}$ in Equation (2.66) gives the phase function as:

$$\Phi(\Omega_s) = \frac{4\pi dI'_{\lambda,s}(\Omega_s)}{dI_{\lambda,s}} = \frac{dI'_{\lambda,s}(\Omega_s)}{(1/4\pi)\int_{\Omega_s^*=0}^{4\pi} dI'_{\lambda,s}(\Omega_s^*)d\Omega_s^*} \qquad (2.68)$$

Thus $\Phi(\Omega_s)$ is the scattered intensity in a direction Ω_s, divided by the mean intensity scattered into all directions. Integrating Equation (2.68) over all possible directions allows the normalization of Φ such that:

$$\frac{1}{4\pi}\int_{\Omega_s=0}^{4\pi}\Phi_\lambda(\Omega_s)d\Omega_s = \frac{1}{4\pi}\int_0^{2\pi}\int_0^\pi \Phi_\lambda(\Omega_s)\sin\theta_s d\theta_s d\phi_s = 1 \qquad (2.69)$$

Finally, the increase of intensity in each direction Ω due to in-scattering from all other directions is given by:

$$dI_{\lambda,\text{in-scat}}(\Omega) = \frac{\sigma_{s,\lambda}}{4\pi}\int_{\Omega_i=0}^{4\pi}\Phi_\lambda(\Omega_i,\Omega)I_\lambda(S,\Omega_i)dS \qquad (2.70)$$

If the scattered intensity is *isotropic*, that is, the same for all scattered directions, Φ is equal to unity. Isotropic scattering rarely occurs in real-world problems, but it is often assumed to be the case to significantly simplify radiative transfer calculations.

2.9   RADIATIVE TRANSFER EQUATION

So far, we have discussed the change in radiative energy along a path in absorbing and scattering media. We have introduced the radiation intensity, absorption, scattering and extinction coefficients, and the scattering phase function. We have also provided a formulation for the attenuation of the radiative energy of a collimated beam (like a laser beam) along the direction of its propagation. This theoretical background is enough to describe many practical problems and the working principles of several diagnostic tools where a single beam is incident on a medium. Equipped with this formulation, we can now determine the amount of radiative energy reaching the location of a surface element or a detector.

We write the conservation of radiative energy along a line of sight for a given wavelength. This conservation equation constitutes the radiative transfer equation (RTE), which is a fundamental relation of radiation energy transfer. In deriving the RTE, we consider a small volume element dV with macroscopic properties such as absorption, scattering, and extinction coefficients, κ_λ, $\sigma_{s,\lambda}$, and β_λ, and the scattering phase function $\Phi(\Omega_i,\Omega_s)$. Also, we consider emission by the matter within each small volume element. We can write conservation of radiative energy within a small volume element dV for a small path increment dS along S within a small solid angle $d\Omega$ around the direction (θ,ϕ) for a small time interval dt at time t and within a small wavelength interval $d\lambda$ around a given wavelength λ as:

Change in radiative energy in dS = gain due to emission − loss due to absorption
  − loss due to outscattering + gain due to inscattering

This conservation relation over infinitesimal path dS is expressed mathematically as:

$$
\begin{aligned}
I_\lambda\left(S+dS,\Omega,t+dt\right)-I_\lambda\left(S,\Omega,t\right) \\
= \kappa_\lambda I_{\lambda b}\left(S,t\right)dS - \kappa_\lambda I_\lambda\left(S,\Omega,t\right)dS \\
-\sigma_{s,\lambda}I_\lambda\left(S,\Omega,t\right)dS + \frac{\sigma_{s,\lambda}}{4\pi}\int_{\Omega_i=4\pi} I_\lambda\left(S,\Omega_i,t\right)\Phi_\lambda\left(\Omega_i,\Omega\right)d\Omega_i dS
\end{aligned} \tag{2.71}
$$

We can recast this expression into a differential form:

$$
\begin{aligned}
\frac{\partial I_\lambda\left(S,\Omega,t\right)}{c\partial t}+\frac{\partial I_\lambda\left(S,\Omega,t\right)}{\partial S} = \kappa_\lambda I_{\lambda b}\left(S,t\right)-\kappa_\lambda I_\lambda\left(S,\Omega,t\right)-\sigma_{s,\lambda}I_\lambda\left(S,\Omega,t\right) \\
+\frac{\sigma_{s,\lambda}}{4\pi}\int_{\Omega_i=4\pi} I_\lambda\left(S,\Omega_i,t\right)\Phi_\lambda\left(\Omega_i,\Omega\right)d\Omega_i
\end{aligned} \tag{2.72}
$$

where ∂ represents the partial differential with respect to time or space. This energy balance is the radiative transfer equation (RTE). Note that in Equation (2.72), the first term is important only for ultrafast phenomena due to the large value of the speed of light c appearing in the denominator and is negligible for most engineering applications.

2.10 RADIATIVE ENERGY TRANSFER IN ENCLOSURES WITH NONPARTICIPATING MEDIA

The RTE expresses the general conservation of radiative energy along a line of sight in an absorbing, emitting, and scattering medium. The solution of this integrodifferential equation is by no means trivial, especially in 3D geometries. It is the stepping stone to determining the contribution of radiation to an overall energy conservation relation. Among all complications, it is the in-scattering integral that couples the intensity in one direction to the intensity in all other directions that is most difficult to resolve. If scattering can be neglected (e.g., if the medium can be modeled as homogeneous), the RTE is reduced to a partial differential equation. Applications abound for nonscattering media, for example, radiative propagation through pure crystals like glass. Depending on the wavelength, even heterogeneous material may be modeled as nonscattering. In the case of flames and soot-laden gases, for example, scattering can be neglected relative to absorption and emission over most of the infrared wavelengths important to radiative transfer.

In many cases, the RTE can be further simplified by assuming the medium is *nonparticipating*. Since all matter absorbs and emits thermal radiation, at first glance it may seem that this assumption only applies to a vacuum. However, the absorption and emission of common monatomic (e.g., Ar, He) and homonuclear diatomic (e.g., N_2, O_2) gases is almost always negligible, and other participating species like CO_2 and H_2O may only have a strong influence if they are present in significant amounts (e.g., in flue gases) or over long pathlengths (e.g., the Earth's atmosphere). Energy transfer across a nonscattering and nonabsorbing medium can be envisioned strictly as a surface phenomenon. Each surface emits radiation and absorbs and reflects the energy received from other surfaces. The emission and reflection characteristics of surfaces are covered in Chapters 3 and 4.

We start the analysis with the radiative intensity $I_\lambda(\theta,\phi)$ leaving surface element dA_1. To analyze the radiative exchange between two finite surfaces, we need to carry out integration over the entire area of each surface. For this, consider radiative energy leaving a small area element dA_1 and traveling in a nonparticipating medium. Assume that this energy is incident on a second small area element dA_2 on finite area A_2, at distance S_{12} from dA_1. The projected areas are formed by taking the area that the energy is passing through and projecting it normal to the direction of radiation propagation; therefore, $dA_1\cos\theta_1$ and $dA_2\cos\theta_2$ are the normal components of the infinitesimal areas along direction S_{12}, as shown in Figure 2.15a. The elemental solid angle is centered about the direction of the radiant path and has its origin at dA. Using the definition of spectral intensity $I_{\lambda,1}$ as the rate of energy passing through dA_1 per unit projected area per unit solid angle and per unit wavelength interval, the energy $dQ_{\lambda,1}$ from dA passing through dA_1 in the direction of S_{12} is:

$$dQ_{\lambda,1} = I_{\lambda,1}dA_1\cos\theta_1 d\Omega_{21}d\lambda = I_{\lambda,1}dA_1\cos\theta_1\left(\frac{dA\cos\theta}{S_{12}^2}\right)d\lambda \qquad (2.73)$$

If dA_1 is placed at a farther distance along the same direction, the rate of energy passing through would be smaller, as in Figure 2.15b. Assume another small area element $dA_2 = dA_1$ is located at a distance S_2. The energy rate at this new position is $I_{\lambda,2}dA_1\cos\theta_1 d\Omega_2 d\lambda = I_{\lambda,2}dA_1\cos\theta_1(dA_2\cos\theta_2/S_2^2)d\lambda$. If all areas are perpendicular to the direction of propagation, then all $\cos\theta$ terms equal unity, and the ratio of energy rates for distances S_1 and S_2 is given as $I_{\lambda,1}S_1^2/I_{\lambda,2}S_2^2$.

Now, consider a differential source emitting energy equally in all directions. Assume we draw two concentric spheres around this source, as in Figure 2.15c. If $dQ_{\lambda,s}d\lambda$ is the entire spectral energy leaving the differential source, the energy

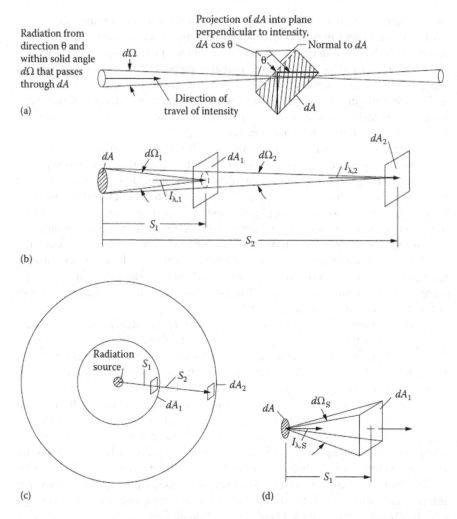

FIGURE 2.15 Demonstration that intensity is invariant along a path in a transparent medium. (a) Black element dA within a black spherical enclosure, (b) energy transfer from dA to dA_1, (c) energy transfer from dA_1 to dA, and (d) solid angle subtended by dA_1 on the sphere surface.

flux crossing the inner sphere is $dQ_{\lambda,s}d\lambda/4\pi S_1^2$ and that crossing the outer sphere is $dQ_{\lambda,s}d\lambda/4\pi S_2^2$. The ratio of the energies passing through the two elements dA_1 at S_1 and $dA_2 = dA_1$ at S_2 is then $[(dQ_{\lambda,s}d\lambda/4\pi S_1^2)dA_2]/[(dQ_{\lambda,s}d\lambda/4\pi S_2^2)dA_1] = S_2^2/S_1^2$.

This is indeed expected as the radiative energy density decreases with the square of the distance from the origin (the inverse square law). However, in the previous paragraph, this energy ratio was derived as $I_{\lambda,1}S_1^2/I_{\lambda,2}S_2^2$. For these two expressions to be equal, the intensities must remain constant along the line of sight:

$$I_{\lambda,1} = I_{\lambda,2} \tag{2.74}$$

Thus, the intensity in each direction in a nonattenuating and nonemitting (nonparticipating) medium is independent of position along that direction, including that at the starting point, that is, $I_{\lambda,s}$ originating at the surface dA_s. *This demonstrates again the invariance of intensity along a ray transecting a nonparticipating medium.* The invariance of intensity provides a convenient way to specify the magnitude of attenuation or emission when they are present in a medium, as their effects are directly shown by the change of intensity with distance along a path. This highlights why intensity is useful for describing radiative energy propagation.

Next, consider a nonabsorbing and nonscattering medium. An example of such a medium would be a kitchen oven filled with air. Since the medium is not participating, the equation will not have any gain or loss terms due to emission, absorption, or scattering.

First, consider infinitesimally small area elements; spectral radiative energy leaving dA_1 toward dA_2 is given as $dQ_{\lambda,1} = I_{\lambda,1} \cos\theta_1 dA_1 d\lambda d\Omega_2 = I_{\lambda,1} \cos\theta_1 dA_1 d\lambda \dfrac{\cos\theta_2 dA_2}{(S_2 - S_1)^2}$

where $d\Omega_2 = \dfrac{\cos\theta_2 dA_2}{(S_2 - S_1)^2}$

The radiative energy arriving at dA_2 is then found from:

$$dQ_{\lambda,2} = I_{\lambda,1} \cos\theta_1 dA_1 d\lambda d\Omega_2$$
$$= I_{\lambda,1} \cos\theta_1 dA_1 \frac{\cos\theta_2 dA_2}{(S_2 - S_1)^2} d\lambda \tag{2.75}$$

To write a similar expression for finite surfaces, Equation (2.75) is integrated over both surfaces A_1 and A_2.

$$Q_{\lambda,1\to2} = \iint_{A_1,A_2} dQ_2 = \iint_{A_1,A_2} I_{\lambda,1} \cos\theta_1 \frac{\cos\theta_2}{(S_2 - S_1)^2} d\lambda dA_1 dA_2 \tag{2.76}$$

If the intensity leaving surface A_1 is directionally uniform, then it can be taken outside the integral sign, yielding:

$$Q_{\lambda,1\to2} = I_{\lambda,1} d\lambda \iint_{A_1,A_2} \cos\theta_1 \frac{\cos\theta_2}{(S_2 - S_1)^2} dA_1 dA_2 = E_{\lambda,1} d\lambda \iint_{A_1,A_2} \frac{\cos\theta_1 \cos\theta_2}{\pi(S_2 - S_1)^2} dA_1 dA_2 \tag{2.77}$$

The double integral in this equation is a purely geometrical factor that can be evaluated independently of the spectral surface properties:

$$Q_{\lambda1\to2} = E_{\lambda,1} A_1 F_{1-2} d\lambda \tag{2.78}$$

where

$$A_1 F_{1-2} = \iint_{A_1,A_2} \frac{\cos\theta_1 \cos\theta_2}{\pi(S_2 - S_1)^2} dA_1 dA_2 \tag{2.79}$$

Here, F_{1-2} is called a *configuration factor*, *view factor*, *angle factor*, or *shape factor*, and is treated extensively in later chapters.

2.11 CONCLUDING REMARKS

In this chapter, we discussed the importance, definitions, and equations of thermal radiative transfer, and highlighted how they are related to the First and Second Laws of Thermodynamics. We derived the fundamental RTE and discussed its simplified forms for different medium properties. We also presented an intuitive relationship between the RTE formulations and the configuration factor analyses, which is used extensively for calculating radiative transfer through nonparticipating media.

While this textbook mainly focuses on engineering applications, we have shown how the development of EM theory and, especially, the concept of blackbody radiation has shaped modern physics. The development of the blackbody relations in this chapter differs from the real historical sequence. A detailed discussion of the history of the blackbody, biographies of the important contributors, and a timeline of events in developing the theory of the blackbody are given in Appendices G and H of the online appendix www.ThermalRadiation.net/appendix.html.

HOMEWORK (additional problems available in online Appendix I at www.ThermalRadiation.net/appendix.html)

2.1 What are the wave number range in a vacuum and the frequency range for the visible spectrum (0.4–0.7 μm)? What are the wave number and frequency values at the spectral boundary (25 mm) between the near- and the far-infrared regions?
Answers: 2.5×10^6 to 1.4286×10^6 m^{-1}; 4.2827×10^{14} to 7.4948×10^{14} s^{-1}; 4×10^4 m^{-1}; 1.1992×10^{13} s^{-1}

2.2 Radiation propagating within a medium is found to have a wavelength within the medium of 1.633 μm and a speed of 2.600×10^8 m/s.
(a) What is the refractive index of the medium?
(b) What is the wavelength of this radiation if it propagates into a vacuum?
Answers: (a) 1.153; (b) 1.883 μm

2.3 A material has an index of refraction $n(x)$ that varies with position x within its thickness. Obtain an expression in terms of c_o and $n(x)$ for the transit time for radiation to pass through a thickness L. If $n(x) = n_i(1 + kx)$, where n_i and k are constants, what is the relation for transit time? How does wave number (relative to that in a vacuum) vary with position within the medium?
Answers: $t = \dfrac{n_i}{c_o}\left(L + \dfrac{kL^2}{2}\right)$; $n_i(1 + kx)$

2.4 Derive Equation (2.29) by finding the maximum of the $E_{\lambda b}/T^5$ versus λT relation (Equation (2.23)).

2.5   A blackbody is at a temperature of 1350 K and is in air.
      (a) What is the spectral intensity emitted in a direction normal to the black
          surface at $\lambda = 4$ μm?
      (b) What is the spectral intensity emitted at $\theta = 50°$ with respect to the
          normal of the black surface at $\lambda = 4.00$ μm?
      (c) What is the directional spectral emissive power from the black surface
          at $\theta = 50°$ and $\lambda = 4.00$ μm?
      (d) At what λ is the maximum spectral intensity emitted from this black-
          body, and what is the value of this intensity?
      (e) What is the hemispherical total emissive power of the blackbody?
      Answers: (a) 8706 W/(m²·μm·sr); (b) 8706 W/(m²·μm·sr); (c) 5596 W/
      (m²·μm·sr); (d) 2.1465 μm, 18365 W/(m²·μm·sr); (e) 188.34 kW/m².

2.6   For a blackbody at 2450 K in air, find
      (a) The maximum emitted spectral intensity [(kW/(m²·μm·sr)]
      (b) The hemispherical total emissive power (kW/m²)
      (c) The emissive power in the spectral range between $\lambda_o = 1.5$ and 7 μm
      (d) The ratio of spectral intensity at $\lambda_o = 1.5$ μm to that at $\lambda_o = 7$ μm
      Answers: (a) 361.54 kW/m²·μm·sr; (b) 2043.1 kW/m²; (c) 1142.8 kW/m²;
      (d) 59.17

2.7   Determine the fractions of blackbody energy that lie below and above the
      peak of the blackbody curve.
      Answers: 25% and 75%

2.8   The surface of the Sun has an effective blackbody radiating temperature of
      5780 K.
      (a) What percentage of the solar radiant emission lies in the visible range
          $\lambda = 0.4$–0.7 μm?
      (b) What percentage is in the ultraviolet?
      (c) At what wavelength and frequency is the maximum energy per unit
          wavelength emitted?
      (d) What is the maximum value of the solar hemispherical spectral emis-
          sive power?
      Answers: (a) 36.7%; (b) 12.2%; (c) 0.5013 μm, 5.98×10^{14} Hz; (d) $8.301 \times$
      10^7 W/(m²·μm)

2.9   A blackbody has a hemispherical spectral emissive power of
      0.0390 W/(m²·μm) at a wavelength of 85 μm. What is the wavelength for
      the maximum emissive power of this blackbody?
      Answer: 19.71 μm

2.10 A solid copper sphere 2.5 cm in diameter has a thin black coating. Initially,
      the sphere is at 750 K, and it is then placed in a vacuum with very cold sur-
      roundings. How long will it take for the sphere to cool to 300 K? Because
      of the high thermal conductivity of copper, it is assumed that the tempera-
      ture within the sphere is uniform at any instant during the cooling process.
      (Properties of copper: density, $\rho = 8950$ kg/m³; specific heat, $c = 383$ J/kg·K.)
      Answer: 0.807 h

2.11   Spectral radiation at λ = 2.500 μm and with intensity 8.00 kW/(m²·μm·sr) enters a gas and travels through the gas along a path length of 21.5 cm. The gas is at a uniform temperature of 1200 K and has an absorption coefficient of $\kappa_{2.500\ \mu m}$ = 0.625 m^{-1}. What is the intensity of the radiation at the end of the path? Neglect scattering but include emission by the gas.
   Answer: 8.27 kW/(m²·μm·sr)

2.12   A science fiction writer uses a "slow glass" window as a plot device. The window material has such a high value of constant simple refractive index n in the visible range that light takes 5.0 minutes to traverse a 1 m thick window. Thus, the observer on one side sees a murder scene 5 minutes after it occurs on the other. He rushes outside, but the perpetrator and corpse have been gone for 5 minutes. Chaos ensues.
   What must the slow glass refractive index be for this to be possible?

   Answer: n = 9.00x10^{10} (For another approach, see Hau et al. 1999, Hau, 2001)

2.13   If the window in Problem 2.12 is 1 m thick and is made of crown glass (refractive index n = 1.50 in the visible), what will be the time delay between light entering and exiting the window in the normal direction?
   Answer: Δt = 5 ns

NOTES

1. What most people perceive as "red light" corresponds to the narrow wavelength range of roughly 620–740 nm; green light is roughly 510–530 nm, and violet is 400–430 nm.
2. A short history of important events in radiative transfer, biographies of contributors, and a timeline are in online Appendixes G and H for this text at www.ThermalRadiation.net/appendix.html.

3 Radiative Properties at Interfaces

Johann Heinrich Lambert (1728–1777) *was a self-taught mathematician, astronomer, logician, and philosopher. Aside from his work in radiation, he offered the first proof that* π *was an irrational number. In 1758, he published his first book, describing the exponential decay of light in a medium, followed in 1760 by his more complete book* Photometrie *that describes both the exponential decay and the cosine dependence of emission from a diffuse surface.*

Gustav Robert Kirchhoff (1824–1887) *proposed in 1859 and provided proof in 1861 that, in simple terms, "For an arbitrary body emitting and absorbing thermal radiation in thermodynamic equilibrium, the emissivity is equal to the absorptivity." He described the ideal radiation emitter in 1862 and called it the* schwarzer körper *(blackbody) because, as the perfect emitter, it must also be the perfect absorber and thus a zero reflector that would appear black to the eye.*

3.1 INTRODUCTION

The radiative behavior of a blackbody was presented in Chapter 2. The ideal blackbody is a thermodynamic limiting case against which the performance of real radiating bodies can be compared. The radiative behavior of a real body depends on many factors such as composition, surface finish, temperature, radiation wavelength, angle at which radiation is either emitted, intercepted, or reflected, and spectral distribution of incident radiation. We need to define the spectral, directional, or averaged emissive, absorptive, reflective, and transmissive properties to describe the radiative behavior of real materials relative to blackbody behavior.

This chapter defines the surface radiative properties of materials. The interrelationship between surface properties, and their dependence on wavelength, angle, and temperature, can sometimes be overwhelming. Consider Figure 3.1, where an infinitesimal area element dA is part of an object at temperature T and is exchanging radiation with other objects. The location of dA can be specified in any coordinate system; for convenience, consider a Cartesian system so the temperature of dA can be specified as $T(x,y,z)$. The emitted energy from dA is thus an implicit function of location through its dependence on $T(x,y,z)$ and is also almost always a function of wavelength and direction of emission.

The ability of the surface to absorb and reflect radiation depends not only on the temperature of the surface $T(x,y,z)$, but on the directional and spectral distribution of

DOI: 10.1201/9781003178996-3

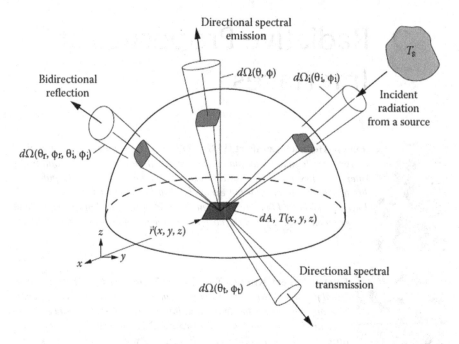

FIGURE 3.1 Radiation interactions at a surface.

the incident radiation, which, in turn, depend on the source of this radiation. A surface irradiated by a blackbody source at 6000 K will have maximum spectral incident radiation (irradiation) near 0.5 μm (see Wien's Law, Equation (2.29)), while a source at 300 K will have a spectral peak near 10 μm. The fraction of each irradiation absorbed by dA will depend on whether its absorption ability is highest near 0.5 or 10 μm.

The rates at which irradiation is reflected from, and transmitted through, dA at a given wavelength and direction also depend on the spectral characteristics of the source in addition to the spectral, directional, and temperature-dependent properties of dA. Reflected and transmitted radiation are functions of the source spectral characteristics, the direction of the source, the direction of the reflected or transmitted radiation, and the wavelength considered.

Generally, emission properties depend on $(\lambda, \theta, \phi, T)$, absorption will depend on $(\lambda, \theta_i, \phi_i, T)$, reflection on $(\lambda, \theta_i, \phi_i, \theta_r, \phi_r, T)$, and transmission on $(\lambda, \theta_i, \phi_i, \theta_t, \phi_t, T)$. To simplify the notation, the time dependence is dropped, and the wavelength dependence is shown through the presence or absence of the subscript λ.

Experimental data that provide both detailed directional and spectral data are scarce. Because of the complexities in making these measurements, tabulated property values are often averaged over all directions, all wavelengths, or both. Many engineering applications only require knowledge of spectrally or directionally averaged quantities. Furthermore, relations among averaged properties enable maximum use of available property information; for example, absorptivity data can be obtained from emissivity data if certain restrictions are observed.

Table 3.1 lists each property, its symbolic notation, the equation number of its definition, and the figure showing a physical interpretation.

Emissivity, ε, and absorptivity, α, are surface attributes. For an opaque material, the portion of incident radiation that is not reflected is transmitted through the surface and then absorbed within material that extends below the surface. This absorbing layer can be quite thin (tens of nanometers) for a material with high internal absorptance, such as a metal, or can be a few millimeters or more for a less strongly internally absorbing dielectric material. Emission comes from within a material. The amount of energy leaving the surface depends on what portion of the energy emitted by each internal volume element reaches the surface and can then be transmitted through the surface.

TABLE 3.1
Summary of Surface Radiative Property Definitions

Quantity	Symbol	Defining Equation	Descriptive Figure
Emissivity			
Directional spectral	$\varepsilon_\lambda(\theta,\phi)$	(3.3)	3.2a
Directional total	$\varepsilon(\theta,\phi)$	(3.4), (3.5)	3.2a
Hemispherical spectral	ε_λ	(3.7)	3.2b
Hemispherical total	ε	(3.8), (3.9), (3.10)	3.2b
Absorptivity			
Directional spectral	$\alpha_\lambda(\theta,\phi)$	(3.14)	3.2c
Directional total	$\alpha(\theta,\phi)$	(3.22), (3.23)	3.2c
Hemispherical spectral	α_λ	(3.29)	3.2d
Hemispherical total	α	(3.36)	3.2d
Reflectivity			
Bidirectional spectral	$\rho_\lambda(\theta_r,\phi_r,\theta_i,\phi_i)$	(3.41)	3.2e
Directional-hemispherical spectral	$\rho_\lambda(\theta_i,\phi_i)$	(3.51)	3.2f
Hemispherical-directional spectral	$\rho_\lambda(\theta_r,\phi_r)$	(3.53)	3.2g
Hemispherical spectral	ρ_λ	(3.56)	3.2h
Bidirectional total	$\rho(\theta_r,\phi_r,\theta_i,\phi_i)$	(3.64), (3.65)	3.2e
Directional-hemispherical total	$\rho(\theta_i,\phi_i)$	(3.68)	3.2f
Hemispherical-directional total	$\rho(\theta_r,\phi_r)$	(3.69)	3.2g
Hemispherical total	ρ	(3.71), (3.72)	3.2h
Transmissivity			
Bidirectional spectral	$\tau_\lambda(\theta_t,\phi_t,\theta_i,\phi_i)$	(3.73)	3.2i
Directional-hemispherical spectral	$\tau_\lambda(\theta_i,\phi_i)$	(3.75), (3.76)	3.2j
Hemispherical-directional spectral	$\tau_\lambda(\theta_i,\phi_i)$	(3.77), (3.78)	3.2k
Hemispherical spectral	τ_λ	(3.81), (3.82)	3.2l
Bidirectional total	$\tau(\theta_t,\phi_t,\theta_i,\phi_i)$	(3.83)	3.2i
Directional-hemispherical total	$\tau(\theta_i,\phi_i)$	(3.84), (3.85)	3.2j
Hemispherical-directional total	$\tau(\theta_t,\phi_t)$	(3.86), (3.87)	3.2k
Hemispherical total	τ	(3.90), (3.91)	3.2l

The properties are discussed for the surface of a medium that has enough thickness to attenuate and absorb all radiation that passes through the surface; thus, the medium is *opaque*. The radiant energy transmitted through the surface of the medium is then equal to the radiant energy absorbed by the medium. Although the ability of a surface to transmit radiation is not needed for radiative transfer among opaque surfaces, the definitions describing transmission across a surface or interface are provided for reference when nonopaque surfaces are treated (Chapter 11).

It is common to assign the *-ivity* ending to intensive properties such as electrical resistivity, thermal conductivity, or thermal diffusivity. The *-ivity* ending is used throughout this book for radiative properties of opaque materials, whether for ideal surfaces or for properties with a given surface condition. The *-ance* ending is for an extensive property such as the emittance of a partially transmitting isothermal layer of water or glass, where the emittance depends on layer thickness. The *-ance* ending is also often found in literature dealing with the experimental determination of surface properties. The term emittance is also used in some references to describe what we have called emissive power. Over the years, suggestions have been made to standardize radiation nomenclature. The US National Institute of Standards and Technology (NIST) has published a nomenclature useful in the fields of illumination and measurement (Nicodemus et al. 1977). That nomenclature is close to the set used in this text, and we follow the notation recommended by the editors of the major heat transfer journals.

The notation used here is an extension of that in Chapter 2. A functional notation is used to show explicitly the variables upon which a quantity depends. For example, $\varepsilon_\lambda(\theta,\phi,T)$ shows that the spectral emissivity is directional and can depend on four variables. The λ subscript specifies that the quantity is *spectral* and is convenient when the functional notation is omitted. A *total* quantity does not have a λ subscript. Certain quantities depend on *two* directions and have *four angles* in their functional notation. A hemispherical-directional or directional-hemispherical quantity, for example, has only two angles. Bidirectional properties require four explicit angles.

Additional notation is needed for Q, the energy rate for a *finite area*, to keep consistent mathematical forms for energy balances. The notation $Q_\lambda d\lambda$ is used to designate the spectral energy rate in the interval $d\lambda$; this is consistent with blackbody spectral energy, where terms such as $E_{\lambda b}d\lambda$ are used in Chapter 2. In $dQ_\lambda(\theta,\phi)d\lambda$, the Q has a derivative to indicate that the spectral energy is also of differential order in solid angle. The total energy $dQ(\theta,\phi)$ is a differential quantity with respect to the solid angle. If the energy is for a *differential area*, the order of the derivative is correspondingly increased.

The notation $\int_\cap d\Omega$ signifies integration over the hemispherical solid angle where $d\Omega = \sin\theta\, d\theta\, d\phi$. Hence, for any function $f(\theta,\phi)$, $\displaystyle\int_{\phi=0}^{2\pi}\int_{\theta=0}^{\pi/2} f(\theta,\phi)\sin\theta\, d\theta d\phi \equiv \int_\cap f(\theta,\phi)d\Omega.$

The four main sections of this chapter each deal with one property: emissivity, absorptivity, reflectivity, or transmissivity. In each section, the most fundamental property is presented first. Then, averaged quantities are obtained by integration. The section on absorptivity also contains various forms of Kirchhoff's Law that relate absorptivity to emissivity. The sections on reflectivity and transmissivity include reciprocity relations. Chapter 4 provides information on the predicted and observed property behavior of actual surfaces.

3.2 EMISSIVITY

Emissivity specifies the rate at which thermal radiation is emitted from a real surface as compared with emission from a blackbody at the same temperature. It can be spectral, and in general depends on surface temperature and direction. In detailed radiative exchange calculations, data for spectral-directional emissivity at the correct surface temperature are needed, but such detailed data for many materials are scarce. Most available data are for a limited spectral range and often are for the normal direction only. Because of this, averaged emissivity values are often reported over direction, wavelength, or both.

Values averaged with respect to all wavelengths are termed *total* quantities; averages with respect to all directions are termed *hemispherical* quantities. Usually, the total hemispherical quantity is implied unless otherwise specified: for example, if we simply refer to a property as "emissivity" without a modifier, we are likely referring to hemispherical total emissivity. Similarly, "spectral emissivity" usually implies hemispherical spectral emissivity.

3.2.1 DIRECTIONAL SPECTRAL EMISSIVITY, $\varepsilon_\lambda(\theta,\varphi,T)$

Consider the geometry for emitted radiation in Figure 3.2a. As in Chapter 2, the radiation intensity is the energy per unit time emitted in direction (θ,ϕ) per unit of projected area dA_p normal to this direction, per unit solid angle, and per unit wavelength interval. Unlike the intensity from a blackbody, the intensity emitted from a real body generally depends on the direction. The energy leaving a real surface dA of temperature T per unit time in the wavelength interval $d\lambda$ and within the solid angle $d\Omega = \sin\theta d\theta d\phi$ is then given by:

$$d^2Q_\lambda(\theta,\phi,T)d\lambda = I_\lambda\left(\theta,\phi,T\right)dA\cos\theta\, d\Omega d\lambda \qquad (3.1)$$

To maintain the order of the differential quantities, the d^2Q_λ indicates dependence on both dA and $d\lambda$, as noted for the order of quantities in Chapter 2.

For a blackbody, the intensity $I_{\lambda b}(T)$ is independent of direction. The T notation is introduced here to clarify when quantities are temperature-dependent, so the blackbody intensity is $I_{\lambda b}(T)$. The energy leaving a black area element per unit of time within $d\lambda$ and $d\Omega$ is:

$$d^2Q_{\lambda,b}(\theta,\phi,T)d\lambda = I_{\lambda b}\left(T\right)dA\cos\theta d\Omega d\lambda \qquad (3.2)$$

The *directional spectral emissivity* is then defined as the ratio of the rate at which a real surface radiates energy in each direction and at given wavelengths compared to a blackbody:

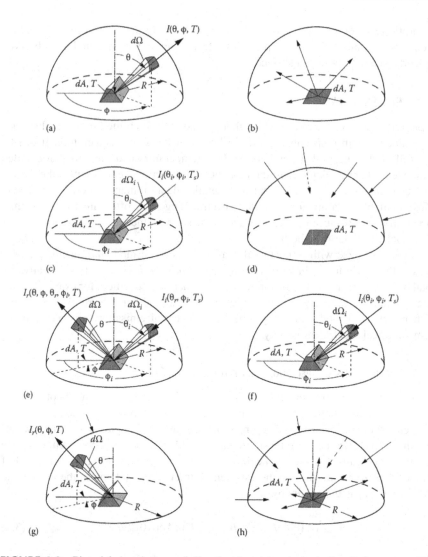

FIGURE 3.2 Pictorial descriptions of directional and hemispherical radiation properties. For spectral properties, the subscript λ is added the property definitions: (a) directional emissivity $\varepsilon(\theta,\phi,T)$; (b) hemispherical emissivity $\varepsilon(T)$; (c) directional absorptivity $\alpha(\theta_i,\phi_i,T)$; (d) hemispherical absorptivity $\alpha(T)$; (e) bidirectional reflectivity $\rho(\theta_r,\phi_r,\theta_i,\phi_i,T)$; (f) directional-hemispherical reflectivity $\rho(\theta_i,\phi_i,T)$; (g) hemispherical-directional reflectivity $\rho(\theta,\phi,T)$; (h) hemispherical reflectivity $\rho(T)$.

$$\text{Directional spectral emissivity} \equiv \varepsilon_\lambda\left(\theta,\phi,T\right) \equiv \frac{d^2Q_\lambda\left(\theta,\phi,T\right)d\lambda}{d^2Q_{\lambda,b}\left(\theta,\phi,T\right)d\lambda} = \frac{I_\lambda\left(\theta,\phi,T\right)}{I_{\lambda b}\left(T\right)} \quad (3.3)$$

This is the most fundamental emissivity expression because it includes dependencies on wavelength, direction, and surface temperature. Surface temperature influences I_λ in two ways: via the Boltzmann distribution of energy states in the surface,

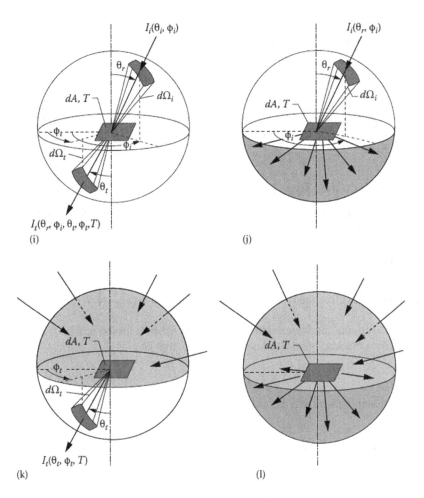

FIGURE 3.2 (Continued) Pictorial descriptions of directional and hemispherical radiation properties. For spectral properties, the subscript λ is added the property definitions: (i) bidirectional transmissivity $\tau(\theta_i,\phi_i,\theta_t,\phi_t,T)$; (j) directional-hemispherical transmissivity $\tau(\theta_i,\phi_i,T)$; (k) hemispherical-directional transmissivity $\tau(\theta_t,\phi_t,T)$; and (l) hemispherical transmissivity $\tau(T)$.

Equation (1.1), giving rise to Planck's distribution, and through the bulk electromagnetic properties of the surface material. The former effect is canceled out by the blackbody intensity in the denominator of Equation (3.3), so the temperature influences ε_λ only through the properties of the surface material.

Example 3.1

A surface heated to 1000 K in a vacuum has a directional spectral emissivity of 0.70 at a wavelength of 5 μm. The emissivity is independent of azimuthal angle φ. What is the emitted spectral intensity in this direction?

From Equation (3.3), $I_\lambda(5 \text{ μm}, 60°, 1000 \text{ K}) = \varepsilon_\lambda(5 \text{ μm}, 60°, 1000 \text{ K}) I_{\lambda b}$ (5 μm, 1000 K). Using Equation (2.20) for $I_{\lambda b}(5 \text{ μm}, 1000 \text{ K})$,

$$I_\lambda\left(5\,\mu m, 60°, 1000\,K\right) = 0.70 \times \frac{2C_1}{\lambda^5\left(e^{C_2/\lambda T}-1\right)} = 0.70 \times \frac{2 \times 0.59552 \times 10^8}{5^5\left(e^{14387.75/5000}-1\right)}$$

$$= 0.70 \times 2272.61 = 1591\,W/\left(m^2 \cdot \mu m \cdot sr\right)$$

From the directional spectral emissivity in Equation (3.3), an averaged emissivity can be derived by two approaches: averaging over all wavelengths or averaging over all directions.

3.2.2 DIRECTIONAL TOTAL EMISSIVITY, $\varepsilon(\theta,\varphi,T)$

The radiation emitted at all wavelengths into direction (θ,ϕ) is found by integrating the directional spectral intensity to obtain the *directional total intensity* (the term *total* denotes that radiation at all wavelengths is included): $I(\theta,\phi,T) \equiv \int_{\lambda=0}^{\infty} I_\lambda(\theta,\phi,T)d\lambda$.

Equation (2.32) gives the total blackbody intensity as: $I_b(T) = \int_{\lambda=0}^{\infty} I_{\lambda b}(T)d\lambda = \dfrac{\sigma T^4}{\pi}$.

The *directional total emissivity* is the ratio of $I(\theta,\phi,T)$ for the real surface to $I_b(T)$ emitted by a blackbody at the same temperature; that is:

$$\text{Directional total emissivity} \equiv \varepsilon\left(\theta,\phi,T\right) \equiv \frac{I\left(\theta,\phi,T\right)}{I_b\left(T\right)} = \frac{\pi \int_{\lambda=0}^{\infty} I_\lambda\left(\theta,\phi,T\right)d\lambda}{\sigma T^4} \qquad (3.4)$$

The $I_\lambda(\theta,\phi,T)$ in the numerator can be replaced in terms of $\varepsilon_\lambda(\theta,\phi,T)$ by using Equation (3.3) to give directional total emissivity in terms of the directional spectral emissivity:

$$\varepsilon\left(\theta,\phi,T\right) = \frac{\pi \int_{\lambda=0}^{\infty} \varepsilon_\lambda\left(\theta,\phi,T\right)I_{\lambda b}\left(T\right)d\lambda}{\sigma T^4} = \frac{\int_{\lambda=0}^{\infty} \varepsilon_\lambda\left(\theta,\phi,T\right)E_{\lambda b}\left(T\right)d\lambda}{\sigma T^4} \qquad (3.5)$$

Thus, if the wavelength dependence of $\varepsilon_\lambda(\theta,\phi,T)$ is known, $\varepsilon(\theta,\phi,T)$ can be found from Equation (3.5). The total directional emissivity can be calculated accurately only if the spectral-directional emissivity is available.

The temperature dependence of directional total emissivity arises from two distinct sources: (*i*) the spectral distribution of the emitted intensity, which affects the weighting of the spectral emissivity ε_λ through Equation (3.5); and (*ii*) the temperature dependence of ε_λ itself, which depends on the surface properties.

Example 3.2

For a surface at 700 K, the $\varepsilon_\lambda(\theta,\phi,T)$ can be approximated by 0.8 for $\lambda = 0$–5 μm, and $\varepsilon_\lambda(\theta,\phi,T) = 0.4$ for $\lambda > 5$ μm. What is the value of $\varepsilon(\theta,\phi,T)$?

From Equation (3.5),

$$\varepsilon(\theta,\phi,T) = \int_0^{5T} 0.8 \frac{E_{\lambda b}(T)}{\sigma T^5} d(\lambda T) + \int_{5T}^{\infty} 0.4 \frac{E_{\lambda b}(T)}{\sigma T^5} d(\lambda T)$$

From Equation (2.37) and by use of Equation (2.40),

$$\varepsilon(\theta,\phi,T) = 0.8 F_{0--3500} + 0.4 F_{3500--\infty} = 0.8 \times 0.38291 + 0.4 \times 0.61709 = 0.553$$

Because 61.7% of the emitted blackbody radiation at 700 K is in the region of > 5 μm, the result is heavily weighted toward the emissivity value of 0.4.

3.2.3 HEMISPHERICAL SPECTRAL EMISSIVITY, $\varepsilon_\lambda(T)$

Now, return to Equation (3.3) and consider the average obtained by integrating the directional spectral quantities over all directions of a hemisphere centered over dA (Figure 3.2b). The spectral radiation emitted by a unit surface area into all directions of the hemisphere is the *hemispherical spectral emissive power*, found by integrating the spectral energy per unit solid angle over all solid angles. This is analogous to Equation (2.18) for a blackbody and is:

$$E_\lambda(T) \equiv \int_{\phi=0}^{2\pi} \int_{\theta=0}^{\pi/2} I_\lambda(\theta,\phi,T) \cos\theta \sin\theta \, d\theta d\phi = \int_\cap I_\lambda(\theta,\phi,T) \cos\theta \, d\Omega$$

Using Equation (3.3), this becomes:

$$E_\lambda(T) = I_{\lambda b}(T) \int_\cap \varepsilon_\lambda(\theta,\phi,T) \cos\theta \, d\Omega \qquad (3.6)$$

For a blackbody, the hemispherical spectral emissive power is $E_{\lambda b}(T) = \pi I_{\lambda b}(T)$. The ratio of actual to blackbody emission from the surface is then the *hemispherical spectral emissivity*:

$$\text{Hemispherical spectral emissivity} \equiv \varepsilon_\lambda(T) \equiv \frac{E_\lambda(T)}{E_{\lambda b}(T)} = \frac{1}{\pi} \int_\cap \varepsilon_\lambda(\theta,\phi,T) \cos\theta \, d\Omega \quad (3.7)$$

3.2.4 HEMISPHERICAL TOTAL EMISSIVITY, $\varepsilon(T)$

Hemispherical total emissivity represents the average of the spectral-directional emissivity over both wavelength and direction. To derive the hemispherical total emissivity, consider that from a unit area, the spectral emissive power in any direction is derived from Equation (3.3) as $\varepsilon_\lambda(\theta,\phi,T)I_{\lambda b}(T)\cos\theta$. This is integrated over all λ and all directions to give the hemispherical total emissive power. Dividing by the hemispherical emissive power of a blackbody σT^4 gives the *hemispherical total emissivity*:

$$\text{Hemispherical total emissivity} \equiv \varepsilon(T) \equiv \frac{E(T)}{E_b(T)} = \frac{\int_{\cap}\int_{\lambda=0}^{\infty}[\varepsilon_\lambda(\theta,\phi,T)I_{\lambda b}(T)d\lambda]\cos\theta d\Omega}{\sigma T^4} \qquad (3.8)$$

Using Equation (3.5), this becomes:

$$\varepsilon(T) = \frac{1}{\pi}\int_{\cap}\varepsilon(\theta,\phi,T)\cos\theta d\Omega \qquad (3.9)$$

If the order of integrations in Equation (3.8) is interchanged and Equation (3.7) is then used, a third form is obtained:

$$\varepsilon(T) = \frac{\pi\int_{\lambda=0}^{\infty}\varepsilon_\lambda(T)I_{\lambda b}(T)d\lambda}{\sigma T^4} = \frac{\int_{\lambda=0}^{\infty}\varepsilon_\lambda(T)E_{\lambda b}(T)d\lambda}{\sigma T^4} \qquad (3.10)$$

To interpret Equation (3.10), consider Figure 3.3. The spectral profile of $\varepsilon_\lambda(T)$ at surface temperature $T = 1000$ K is plotted in Figure 3.3a. The solid curve in Figure 3.3b is the hemispherical spectral emissive power for a blackbody at T. The area under the solid curve is σT^4, which is the denominator in Equation (3.10) and equals the radiation emitted per unit area by a blackbody, including all wavelengths and directions. The dashed curve in Figure 3.3b is the product $\varepsilon_\lambda(T)E_{\lambda b}(T)$, and the area under this curve is the numerator of Equation (3.10), which is the emission from the real surface. Hence, $\varepsilon(T)$ is the ratio of the area under the dashed curve to that under the solid curve. At each λ, the $\varepsilon_\lambda(T)$ is the ordinate of the dashed curve divided by the ordinate of the solid curve. In Figure 3.3b, the hemispherical spectral emissivity at 7.5 μm is ε_λ ($\lambda = 7.5$ μm, $T = 1000$ K) $= b/a$, the ratio of the lower curve value to that of the upper curve.

Example 3.3

A surface at 1000 K has an $\varepsilon(\theta,\varphi,T_A)$ that is independent of φ, but depends on θ as in Figure 3.4. What are the hemispherical total emissivity and the hemispherical total emissive power?

The $\varepsilon(\theta, 1000$ K) is approximated by the function $0.85\cos\theta$. Then, from Equation (3.9), the total hemispherical emissivity is:

$$\varepsilon(T) = \int_{\cap}0.85\cos^2\theta d\Omega = \frac{1}{\pi}\int_{\phi=0}^{2\pi}\int_{\theta=0}^{\pi/2}0.85\cos^2\theta\sin\theta d\theta d\phi$$

$$= -1.70\frac{\cos^3\theta}{3}\Big|_0^{\pi/2} = 0.567$$

The hemispherical total emissive power is

$$E(T) = \varepsilon(T)\sigma T^4 = 0.567 \times 5.6704 \times 10^8 \times 1000^4 = 32,150 \text{ W/m}^2.$$

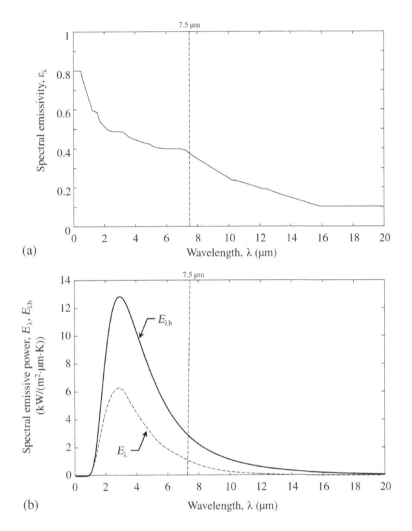

FIGURE 3.3 Physical interpretation of hemispherical spectral and total emissivities. (a) Measured emissivity values at $T = 1000$ K; (b) interpretation of emissivity as the ratio of actual emissive power (dashed curve) to blackbody emissive power (solid curve).

Generally, $\varepsilon(\theta, T)$ cannot be approximated well by a convenient analytical function, and numerical integration of Equation (3.9) is necessary.

Example 3.4

The $\varepsilon_\lambda(T)$ for a surface at $T = 650$ K is approximated, as shown in Figure 3.5. What is the hemispherical total emissivity?

Numerical integration is used for the portion where $3.5 \leq \lambda \leq 9.5$ μm. This part of the spectral emissivity is given by $\varepsilon(T) = 1.27917 - 0.10833\lambda$. The λT bounds

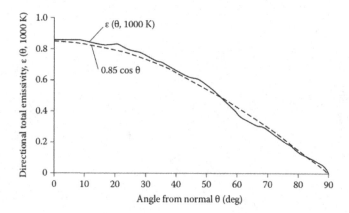

FIGURE 3.4 Directional total emissivity at 1000 K for Example 3.3.

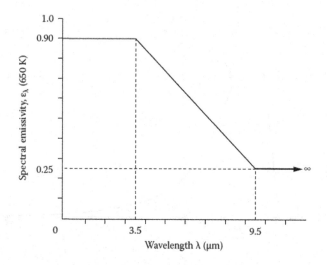

FIGURE 3.5 Spectral emissivity for Example 3.4.

are $3.5 \times 650 = 2275 \ \mu m \cdot K$ and $9.5 \times 650 = 6175 \ \mu m \cdot K$. Then, as in Example 3.3, and by use of Equations (2.36) and (2.40),

$$\varepsilon(650 \ K) = 0.90 F_{0 \to 2275} + \frac{1}{\sigma 650^4} \int_{3.5}^{9.5} (1.27917 - 0.10833\lambda) \frac{2\pi C_1}{\lambda^5 \left(e^{C_2/650\lambda} - 1\right)} d\lambda + 0.25\left(1 - F_{0 \to 6175}\right)$$

Numerically evaluating the integral gives:

$$\varepsilon(650 \ K) = 0.90 \times 0.11514 + 0.40141 + 0.25\left(1 - 0.75214\right) = 0.5670$$

3.3 ABSORPTIVITY

Absorptivity is defined as the fraction of energy incident on a body that is absorbed by the body. Compared with emissivity, absorptivity has additional complexities because directional and spectral characteristics of the incident radiation must be included along with the absorbing surface temperature. Therefore, while the total and hemispherical values of emissivity can be readily calculated from the spectral-directional emissivity at a given surface temperature, the same is not true for the corresponding absorptivity values due to its dependence on the spectral and directional character of the irradiation. For example, the total absorptivity of a steel sheet will change instantaneously as the steel is transferred from the factory surroundings at 293 K into a furnace at 1000 K due to the abrupt change in irradiation, while its total emissivity, which depends only on surface temperature, will only change gradually as the steel heats up.

It is also possible to derive relations between emissivity and absorptivity so that measured values of one may be used to calculate the other. These relations are developed in this section.

3.3.1 DIRECTIONAL SPECTRAL ABSORPTIVITY, $\alpha_\lambda(\theta_i, \varphi_i, T)$

Figure 3.2c illustrates the energy incident on a surface element dA from the (θ_i, ϕ_i) direction. The line from dA in the direction (θ_i, ϕ_i) passes normally through an area element dA_i on the surface of a hemisphere of radius R centered over dA. The incident spectral intensity passing through dA_i is $I_{\lambda,i}(\theta_i, \phi_i)$. This is the energy per unit area of the hemisphere, per unit solid angle $d\Omega d_i$, per unit time, and per unit wavelength interval. The energy within the incident solid angle $d\Omega_i$ strikes the area dA of the absorbing surface. The energy per unit time incident from the direction (θ_i, ϕ_i) in the wavelength interval $d\lambda$ is:

$$d^2Q_{\lambda,i}\left(\theta_i, \phi_i\right)d\lambda = I_{\lambda,i}\left(\theta_i, \phi_i\right)dA_i d\Omega d\lambda = I_{\lambda,i}\left(\theta_i, \phi_i\right)dA_i \frac{dA\cos\theta_i}{R^2}d\lambda \quad (3.11)$$

where $dA\cos\theta_i/R^2$ is the solid angle $d\Omega$ subtended by dA when viewed from dA_i as in Figure 3.6a.

Equation (3.11) can also be expressed in terms of the solid angle $d\Omega_i$ in Figure 3.6b. This is the solid angle subtended by dA_i when viewed from dA. Solid angle $d\Omega_i$ has its vertex at dA and hence is the convenient solid angle to use when integrating to obtain energy incident on dA from more than one direction. For a nonabsorbing medium in the region above the surface, as is being considered here, the incident intensity does not change along the path from dA_i to dA (Equation (2.74)). For these reasons, in the figures that follow, the energy incident on dA from dA_i will be pictured as that arriving in $d\Omega_i$, as shown in Figure 3.6b, rather than as the energy leaving dA in $d\Omega$ as in Figure 3.6a. To place Equation (3.11) in terms of $d\Omega_i$, note that:

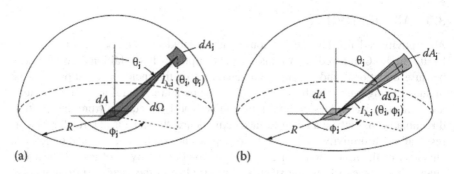

FIGURE 3.6 Equivalent ways of showing energy from dA_i that is incident upon dA. (a) Incidence within solid angle $d\Omega_i$ having an origin at dA_i; (b) incidence within solid angle $d\Omega$ having an origin at dA.

$$d\Omega dA_i \cos\theta_i = dA_i \frac{dA\cos\theta_i}{R^2} = \frac{dA_i}{R^2} dA\cos\theta_i = d\Omega_i dA\cos\theta_i \quad (3.12)$$

Equation (3.11) is then,

$$d^2 Q_{\lambda,i}\left(\theta_i, \phi_i\right)d\lambda = I_{\lambda,i}\left(\theta_i, \phi_i\right)d\Omega_i dA\cos\theta_i d\lambda \quad (3.13)$$

The fraction of the incident energy $d^2 Q_{\lambda,i}(\theta_i,\phi_i,T)d\lambda$ that is absorbed is defined as the *directional spectral absorptivity* $\alpha_\lambda(\theta_i,\phi_i,T)$. The amount of the incident energy that is absorbed is $d^2 Q_{\lambda,a}(\theta_i,\phi_i,T)d\lambda$. Then, the ratio is formed:

$$\text{Directional spectral absorptivity} \equiv \alpha_\lambda(\theta_i,\phi_i,T) = \frac{d^2 Q_{\lambda,a}(\theta_i,\phi_i,T)d\lambda}{I_{\lambda,i}(\theta_i,\phi_i)dA\cos\theta_i d\Omega_i d\lambda} \quad (3.14)$$

In addition to depending on the wavelength and direction of the incident radiation, the spectral absorptivity is also a function of the absorbing surface temperature, as it affects the electromagnetic properties of the surface, in the same way that temperature influences ε_λ.

The directional spectral absorptivity also has a probabilistic interpretation—it is the finite probability that an incident photon[1] contained within a solid angle $d\Omega_i$ centered on (θ_i,ϕ_i) and having an incident energy corresponding to a wavelength between λ and $\lambda+d\lambda$ will be absorbed by a surface. We can write this as:

$$\alpha_\lambda(\theta_i,\phi_i,T) = P\left(\text{abs}\,|\,\theta_i,\phi_i,\lambda,T\right) \quad (3.15)$$

The "*x|y*" notation highlights that this is a conditional probability, i.e., the probability that a photon will be absorbed is conditional on it having a given incident direction and wavelength, and the surface having a defined temperature, T.

3.3.2 KIRCHHOFF'S LAW

Kirchhoff's Law relates the emitting and absorbing abilities of a body and is derived from the Second Law of Thermodynamics. The law has necessary conditions, depending on whether spectral, total, directional, or hemispherical quantities are being considered. From Equations (3.1) and (3.3), the energy emitted per unit of time by an element dA in a wavelength interval $d\lambda$ and solid angle $d\Omega$ is:

$$d^2 Q_{\lambda,e} d\lambda = \varepsilon_\lambda (\theta, \phi, T) I_{\lambda b} (T) dA \cos\theta d\Omega d\lambda \tag{3.16}$$

If the element dA at temperature T is placed in an isothermal black enclosure also at temperature T, then the intensity of the energy incident on dA from the direction (θ_i, ϕ_i) (recalling the isotropy of intensity in a black enclosure) is $I_{\lambda b}(T)$. Since both dA and the emitting element on the black enclosure surface are at the same temperature, T, the Clausius statement of the Second Law requires the *net* radiation energy transfer at that wavelength to be zero. Setting these terms equal results in:

$$d^2 Q_{\lambda,e} d\lambda = \varepsilon_\lambda (\theta, \phi, T) I_{\lambda b} (T) dA \cos\theta d\Omega d\lambda$$

$$= d^2 Q_{\lambda,a} d\lambda = \alpha_\lambda (\theta_i = \theta, \phi_i = \phi, T) I_{\lambda b} (T) dA \cos(\theta_i = \theta) d\Omega d\lambda \tag{3.17}$$

$$\varepsilon_\lambda (\theta, \phi, T) = \alpha_\lambda (\theta_i = \theta, \phi_i = \phi, T) \tag{3.18}$$

This relation between the properties of the material holds without restriction. This is the most *general form of Kirchhoff's Law*. Were this condition not satisfied, one could conceive an arrangement of surfaces that would permit net energy transfer between two objects at the same temperature, which would violate the Second Law of Thermodynamics.

While Equation (3.18) was formulated by considering radiative transfer between an object within a black-walled enclosure at the same temperature, these conditions are not required for Kirchhoff's Law to apply. The relationship between $\varepsilon_\lambda(\theta, \phi, T)$ and $\alpha_\lambda(\theta, \phi, T)$ must generally apply for the Clausius statement to be satisfied under radiative equilibrium.

The equality between $\alpha_\lambda(\theta, \phi, T)$ and $\varepsilon_\lambda(\theta, \phi, T)$ implies that if the former may be interpreted as a probability, then the latter must also be a probability. But a probability of what? As described in Chapter 2, local thermodynamic equilibrium normally applies within matter, and therefore the energy levels of photons within the matter obey a probability density proportional to Planck's distribution at the surface temperature. A fraction of the RTE photons emitted by matter within a thin layer beneath the surface cross the surface boundary and issue into the surrounding medium (e.g., a vacuum or air). The spectral-directional emissivity can then be interpreted as the probability that, upon reaching the interface, a photon will cross the surface boundary and be emitted within a solid angle $d\Omega_i$ centered on (θ_i, ϕ_i):

$$\varepsilon_\lambda(\theta,\phi,T) = P\left(\text{emit}|\theta,\phi,\lambda,T\right) \tag{3.19}$$

In Chapter 2, we said that radiation may be interpreted as a transverse wave; if the wave propagates in the x-direction, the electric and magnetic field intensities fluctuate in the y–z plane, often in a complicated way. As we shall see in Chapter 4, this wave behavior may be separated into two orthogonal components, and, for the special case of blackbody radiation, the two components are equal when averaged over time. In other words, blackbody radiation is *unpolarized* radiation. In the case of incident radiation, Equation (3.18) holds only for each component of polarization, and for Equation (3.18) to be valid as written for all incident energy, the incident radiation must have equal components of polarization.

3.3.3 Directional Total Absorptivity, $\alpha(\theta_i, \varphi_i, T)$

The directional total absorptivity is the energy, including all wavelengths that is absorbed from a given direction, divided by the total energy incident from that direction. The total energy incident from the given direction is obtained by integrating the spectral incident energy [Equation (3.13)] over all wavelengths to obtain:

$$d^2 Q_i(\theta_i, \phi_i) = dA \cos\theta_i d\Omega_i \int\limits_{\lambda=0}^{\infty} I_{\lambda,i}(\theta_i, \phi_i) d\lambda \tag{3.20}$$

The radiation absorbed is found from Equation (3.20) by integrating over all λ,

$$d^2 Q_a(\theta_i, \phi_i, T_i) = dA \cos\theta_i d\Omega_i \int\limits_{\lambda=0}^{\infty} \alpha_\lambda(\theta_i, \phi_i, T) I_{\lambda,i}(\theta_i, \phi_i) d\lambda \tag{3.21}$$

The result is:

Directional total absorptivity $\equiv \alpha(\theta_i, \phi_i, T) = \dfrac{d^2 Q_a(\theta_i, \phi_i, T)}{d^2 Q_i(\theta_i, \phi_i)}$

$$= \dfrac{\displaystyle\int_{\lambda=0}^{\infty} \alpha_\lambda(\theta_i, \phi_i, T) I_{\lambda,i}(\theta_i, \phi_i) d\lambda}{\displaystyle\int_{\lambda=0}^{\infty} I_{\lambda,i}(\theta_i, \phi_i) d\lambda} \tag{3.22}$$

Using Kirchhoff's Law, Equation (3.18), an alternative form is:

$$\alpha\left(\theta_i, \phi_i, T_i\right) = \dfrac{\displaystyle\int_{\lambda=0}^{\infty} \varepsilon_\lambda\left(\theta_i, \phi_i, T\right) I_{\lambda,i}\left(\theta_i, \phi_i\right) d\lambda}{\displaystyle\int_{\lambda=0}^{\infty} I_{\lambda,i}\left(\theta_i, \phi_i\right) d\lambda} \tag{3.23}$$

The directional total absorptivity can also be interpreted as the probability that an RTE photon incident from a solid angle $d\Omega_i$ centered on (θ_i, ϕ_i) will be absorbed for any wavelength. This probability must account for the probability density of incident wavelengths,

$$p(\lambda) = I_\lambda(\lambda) \Big/ \int_{\lambda=0}^{\infty} I_\lambda(\lambda) d\lambda \qquad (3.24)$$

which, in many cases, is proportional to Planck's distribution as described in Section 2.6. Equation (3.23) can then be rewritten in terms of probabilities as:

$$\alpha(\theta_i, \phi_i, T) = P\big(\text{abs}|\theta_i, \phi_i, T\big) = \int_{\lambda=0}^{\infty} \underbrace{P\big(\text{abs}|\theta_i, \phi_i, \lambda, T\big)}_{\alpha_\lambda(\theta_i, \phi_i, T)} p(\lambda) d\lambda \qquad (3.25)$$

In the language of probability, the integral in Equation (3.25) is the *marginalization* over wavelength, and we can say that the wavelength dependence has been "marginalized out."

3.3.4 KIRCHHOFF'S LAW FOR DIRECTIONAL TOTAL PROPERTIES

The general form of Kirchhoff's Law, Equation (3.18), shows that the directional spectral properties $\varepsilon_\lambda(\theta, \phi, T)$ and $\alpha_\lambda(\theta_i, \phi_i, T)$ are equal. Now examine this equality for the *directional total* quantities. This is done by comparing a special case of Equation (3.23) with Equation (3.10). If the incident intensity originates from a blackbody source at the same temperature as the surface, it follows from the Second Law of Thermodynamics that the directional total absorptivity must equal the directional total emissivity in each direction:

$$\alpha\big(\theta_i, \phi_i, T\big) = \frac{\displaystyle\int_{\lambda=0}^{\infty} \varepsilon_\lambda\big(\theta = \theta_i, \phi = \phi_i, T\big) I_{\lambda b}(T) d\lambda}{\displaystyle\int_{\lambda=0}^{\infty} I_{\lambda b}(T) d\lambda \big(= \sigma T^4/\pi\big)} = \varepsilon\big(\theta = \theta_i, \phi = \phi_i, T\big) \qquad (3.26)$$

Hence, when $\varepsilon_\lambda(\theta_i, \phi_i, T)$ and $\alpha_\lambda(\theta_i, \phi_i, T)$ are dependent on wavelength, there is the equality $\alpha(\theta_i, \phi_i, T) = \varepsilon(\theta_i, \phi_i, T)$ *only* when the spectral distribution of incident and blackbody intensities are identical. In practice, this happens most often when a small object is at radiative equilibrium with large, isothermal surroundings, in which case the incident intensity field corresponds to that of a blackbody at the temperature of the surroundings. Otherwise, $\alpha(\theta_i, \phi_i, T) = \varepsilon(\theta_i, \phi_i, T)$ only if ε_λ and α_λ are independent of wavelength, i.e., for a *directional-gray* surface.

3.3.5 HEMISPHERICAL SPECTRAL ABSORPTIVITY, $\alpha_\lambda(T)$

Now, consider energy in a wavelength interval $d\lambda$. The hemispherical spectral absorptivity is the fraction of the spectral energy that is absorbed from the spectral energy that is incident from all directions of a surrounding hemisphere (Figure 3.2d). The spectral energy from an element dA_i on the hemisphere that is intercepted by a surface element dA is given by Equation (3.13). The incident energy on dA from all directions of the hemisphere is then given by the integral:

$$dQ_{\lambda,i}d\lambda = dAd\lambda \int_{\cap} I_{\lambda,i}(\theta_i,\phi_i)\cos\theta_i d\Omega_i \qquad (3.27)$$

The amount absorbed is given by integrating Equation (3.14) over the hemisphere:

$$dQ_{\lambda,a}(T)d\lambda = dAd\lambda \int_{\cap} \alpha_\lambda(\theta_i,\phi_i,T)I_{\lambda,i}(\theta_i,\phi_i)\cos\theta_i d\Omega_i \qquad (3.28)$$

The ratio of these quantities is:

$$\text{Hemispherical spectral absorptivity} \equiv \alpha_\lambda(T) = \frac{dQ_{\lambda,a}(T)d\lambda}{dQ_{\lambda,i}d\lambda}$$

$$= \frac{\int_{\cap} \alpha_\lambda(\theta_i,\phi_i,T)I_{\lambda,i}(\theta_i,\phi_i)\cos\theta_i d\Omega_i}{\int_{\cap} I_{\lambda,i}(\theta_i,\phi_i)\cos\theta_i d\Omega_i} \qquad (3.29)$$

or, in terms of emissivity by using Kirchhoff's Law, Equation (3.18),

$$\alpha_\lambda(T) = \frac{\int_{\cap} \varepsilon_\lambda(\theta=\theta_i,\phi=\phi_i,T)I_{\lambda,i}(\theta_i,\phi_i)\cos\theta_i d\Omega_i}{\int_{\cap} I_{\lambda,i}(\theta_i,\phi_i)\cos\theta_i d\Omega_i} \qquad (3.30)$$

The hemispherical spectral absorptivity and emissivity can now be compared by looking at Equations (3.30) and (3.7). For the general case where α_λ and ε_λ are functions of (λ,θ,ϕ,T), $\alpha_\lambda(T) = \varepsilon_\lambda(T)$ only if $I_{\lambda i}(\theta_i,\phi_i)$ is independent of θ_i and ϕ_i; that is, if the incident spectral intensity is uniform from all directions. If this is so, the $I_{\lambda i}$ can be canceled in Equation (3.29), and the denominator becomes π, and Equation (3.30) then compares with Equation (3.7).

Continuing with the probabilistic interpretation, the hemispherical spectral absorptivity can be viewed as a marginalization of the absorption probability at a given wavelength over all incident directions, so Equation (3.30) can be rewritten as:

$$\alpha_\lambda(T) = \int_{\cap} \underbrace{P\left(\text{abs}|\theta_i,\phi_i,\lambda,T\right)}_{\alpha_\lambda(\theta_i,\phi_i,T)} p\left(\Omega_i\right)d\Omega_i \qquad (3.31)$$

where $p(\Omega_i)$ is the probability density for incident angles,

$$p(\Omega_i) = \frac{I_{\lambda,i}(\theta_i,\phi_i)\cos\theta_i}{\int_\cap I_{\lambda,i}(\theta_i,\phi_i)\cos\theta_i d\Omega_i} \tag{3.32}$$

Because the incident azimuthal and polar angles are independent, this probability density can be written as the product of probability densities $p(\theta_i)$ and $p(\phi_i)$, leading to:

$$\alpha_\lambda(T) = \int_0^{2\pi}\int_0^{\pi/2} \underbrace{P\big(\mathrm{abs}\big|\theta_i,\phi_i,\lambda,T\big)}_{\alpha_\lambda(\theta_i,\phi_i,T)} p(\theta_i)p(\phi_i)d\theta_i d\phi_i \tag{3.33}$$

If the surface irradiation is diffuse (e.g., from isothermal surroundings), then $p(\theta_i) = 2\cos(\theta_i)\sin(\theta_i)$ and $p(\phi_i) = 1/(2\pi)$. This is not generally the case, however. For example, terrestrial solar irradiation typically has a lobular distribution with a peak centered on the direction of the sun, and an angular distribution caused by atmospheric scattering.

For the case where $\alpha_\lambda(\theta,\phi,T) = \varepsilon_\lambda(\theta,\phi,T) = \alpha_\lambda(T) = \varepsilon_\lambda(T)$, that is, the directional spectral properties are independent of angle, the hemispherical spectral properties are related by $\alpha_\lambda(T) = \varepsilon_\lambda(T)$ for any angular variation of incident spectral intensity. Such a surface is termed a *diffuse spectral surface*.

3.3.6 HEMISPHERICAL TOTAL ABSORPTIVITY, $\alpha(T)$

The hemispherical total absorptivity represents the fraction of energy absorbed that is incident from all directions of the enclosing hemisphere and for all wavelengths, as shown in Figure 3.2d. The total incident energy that is intercepted by a surface element dA is determined by integrating Equation (3.27) over all λ and all (θ_i,ϕ_i) of the hemisphere, giving:

$$dQ_i = dA\int_\cap\left[\int_{\lambda=0}^\infty I_{\lambda,i}(\theta_i,\phi_i)d\lambda\right]\cos\theta_i d\Omega_i \tag{3.34}$$

Similarly, by integrating Equation (3.28), the total amount of energy absorbed is:

$$dQ_a(T) = dA\int_\cap\left[\int_{\lambda=0}^\infty \alpha_\lambda(\theta_i,\phi_i,T)I_{\lambda,i}(\theta_i,\phi_i)d\lambda\right]\cos\theta_i d\Omega_i \tag{3.35}$$

The ratio of absorbed to incident energy provides the definition:

$$\textit{Hemispherical total absorptivity} \equiv \alpha(T) = \frac{dQ_a(T)}{dQ_i}$$

$$= \frac{\displaystyle\int_\cap \left[\int_{\lambda=0}^{\infty} \alpha_\lambda(\theta_i,\phi_i,T) I_{\lambda,i}(\theta_i,\phi_i) d\lambda \right] \cos\theta_i d\Omega_i}{\displaystyle\int_\cap \left[\int_{\lambda=0}^{\infty} I_{\lambda,i}(\theta_i,\phi_i) d\lambda \right] \cos\theta_i d\Omega_i}$$

(3.36)

Following the probabilistic interpretation, the hemispherical total absorptivity comes from marginalizing the photon absorption probability over both incident wavelength and direction,

$$\alpha(T) = \int_{\lambda=0}^{\infty} \int_{\phi_i=0}^{2\pi} \int_{\theta_i=0}^{\pi/2} \underbrace{P\left(\text{abs}|\theta_i,\phi_i,\lambda,T\right)}_{\alpha_\lambda(\theta_i,\phi_i,T)} p(\lambda) p(\theta_i) p(\phi_i) d\theta_i d\phi_i d\lambda \qquad (3.37)$$

which is the probability that an incident RTE photon will be absorbed for any incident wavelength and direction. Note that the marginalization has made $\alpha(T)$ independent of θ_i, ϕ_i, and λ.

Equation (3.36) can also be put in terms of emissivity by using Kirchhoff's Law, Equation (3.18),

$$\alpha(T) = \frac{\displaystyle\int_\cap \left[\int_{\lambda=0}^{\infty} \varepsilon_\lambda(\theta=\theta_i,\phi=\phi_i,T) I_{\lambda,i}(\theta_i,\phi_i) d\lambda \right] \cos\theta_i d\Omega_i}{\displaystyle\int_\cap \left[\int_{\lambda=0}^{\infty} I_{\lambda,i}(\theta_i,\phi_i) d\lambda \right] \cos\theta_i d\Omega_i}$$

(3.38)

Equation (3.38) can be compared with Equation (3.8) to find the conditions when the hemispherical total absorptivity and emissivity are equal. From Equation (3.8),

$$\varepsilon(T) = \frac{\displaystyle\int_\cap \int_{\lambda=0}^{\infty} [\varepsilon_\lambda(\theta,\phi,T) I_{\lambda b}(T) d\lambda] \cos\theta d\Omega}{\sigma T^4}$$

(3.39)

The comparison reveals that for the general case when $\varepsilon_\lambda(\theta,\phi,T)$ and $\alpha_\lambda(\theta_i,\phi_i,T)$ vary with both wavelength and angle, $\alpha(T) = \varepsilon(T)$ only when the incident intensity is independent of the incident angle and has the same spectral form as that emitted by a blackbody with temperature equal to the surface temperature T; in other words, $I_{\lambda,i}(\theta,\phi) = C(\theta,\phi) I_{\lambda b}(T)$. This only happens when an object is at radiative equilibrium with its surroundings, *in which case there can be no net energy transfer,* and thus Kirchhoff's Law, in this form, is rarely useful for analyzing radiative transfer.

TABLE 3.2
Summary of Kirchhoff's Law Relations between Absorptivity and Emissivity

Type of Quantity	Equality	Restrictions
Directional spectral	$\alpha_\lambda(\theta,\phi,T) = \varepsilon_\lambda(\theta,\phi,T)$	None
Directional total	$\alpha(\theta,\phi,T) = \varepsilon(\theta,\phi,T)$	Incident radiation must have a spectral distribution proportional to that of a blackbody at T, $I_{\lambda,i}(\theta,\phi) = C(\theta,\phi)I_{\lambda b}(T)$ or $\alpha_\lambda(\theta,\phi,T) = \varepsilon_\lambda(\theta,\phi,T)$ are independent of wavelength (directional-gray surface)
Hemispherical spectral	$\alpha_\lambda(T) = \varepsilon_\lambda(T)$	Incident radiation must be independent of angle, $I_{\lambda,i}(\lambda) = C(\lambda)$ or $\alpha_\lambda(T) = \varepsilon_\lambda(T)$ do not depend on angle (diffuse-spectral surface)
Hemispherical total	$\alpha(T) = \varepsilon(T)$	Incident radiation must be independent of angle and have a spectral distribution proportional to that of a blackbody at T, $I_{\lambda,i} = CI_{\lambda b}(T)$, or incident radiation independent of angle and $\alpha_\lambda(\theta,\phi,T) = \varepsilon_\lambda(\theta,\phi,T)$ are independent of λ (directional-gray surface), or incident radiation from each direction has spectral distribution proportional to that of a blackbody at T_A and $\alpha_\lambda(T) = \varepsilon_\lambda(T)$ are independent of angle (diffuse-spectral surface), or $\alpha_\lambda(T) = \varepsilon_\lambda(T)$ are independent of wavelength and angle (diffuse-gray surface)

The only other scenario in which $\alpha(T) = \varepsilon(T)$ is when these properties are independent of wavelength, i.e., gray, as discussed below. Some more restrictive cases are summarized in Table 3.2.

A special case of hemispherical total absorptivity is assigned in Homework Problem 3.1 at the end of this chapter. If there is a uniform incident intensity from a gray source at T_i, and if $\varepsilon_\lambda(T)$ is independent of T, then the hemispherical total absorptivity for the incident radiation is equal to the hemispherical total emissivity of the material evaluated at the source temperature T_i, $\alpha(T) = \varepsilon(T_i)$. This relation also applies if the spectral-directional absorptivity/emissivity is independent of direction (i.e., a diffuse spectral surface).

Example 3.5

The hemispherical spectral emissivity of a surface at 300 K is approximated by a step function: $\varepsilon_\lambda(\lambda, T) = 0.8$ for $0 \leq \lambda \leq 3$ μm, and 0.2 for $\lambda > 3$ μm. What is the hemispherical total absorptivity for diffuse incident radiation from a black source at $T_i = 1000$ K? What is it for diffuse incident solar radiation?

For diffuse incident radiation, $\alpha_\lambda(T) = \varepsilon_\lambda(T)$, and for a black source at T_i, Equation (3.38) becomes

$$\alpha(T) = \frac{\int_{\lambda=0}^{N} \varepsilon_\lambda(T) I_{\lambda b,i}(T_i) d\lambda}{\int_{\lambda=0}^{N} I_{\lambda b,i}(T_i) d\lambda} = \frac{\int_{\lambda=0}^{N} \varepsilon_\lambda(T) E_{\lambda b,i}(T_i) d\lambda}{\sigma T_i^4}$$

Expressing $\alpha(T)$ in terms of the two wavelength regions over which $\varepsilon_\lambda(T)$ is constant gives $\alpha_\lambda(T) = 0.8 F_{0 \to 3Ti} + 0.2(1 - F_{0 \to 3Ti})$. Using Equation (2.40), $F_{0 \to 3\,\mu m \cdot 1000\,K} = 0.27323$ so that $\alpha(T) = 0.8 F_{0 \to 3000} + 0.2(1 - F_{0 \to 3000}) = 0.364$

For incident solar radiation, $T_i = 5780$ K, $F_{0 \to 3\,\mu m \cdot 5780\,K} = 0.97880$ so that $\alpha(T) = 0.8 F_{0 \to 17,340} + 0.2(1 - F_{0 \to 17,340}) = 0.787$

Hence, for a surface with this type of spectral emissivity variation, there is a considerable increase in $\alpha(T)$ by the shift of the incident-energy spectrum toward shorter wavelengths as the source temperature is raised.

3.3.7 Diffuse-Gray Surface

As we discuss in Chapters 5 and 6, a common assumption in enclosure calculations is that surfaces are *diffuse-gray*. *Diffuse* signifies that the directional emissivity and directional absorptivity do not depend on direction. Hence, for emission, the emitted intensity is uniform over all directions as for a blackbody. The term *gray* means that the spectral emissivity and absorptivity do not depend on wavelength. They can, however, depend on temperature. Thus, at each surface temperature, for any wavelength the emitted spectral radiation is a fixed fraction of blackbody spectral radiation.

The diffuse-gray surface therefore *absorbs a fixed fraction of incident radiation from any direction and at any wavelength*. It emits intensity that is a *fixed fraction of blackbody radiation for all directions and all wavelengths* (this is the motivation for the term "gray"). The directional spectral absorptivity and emissivity are then independent of λ, θ, and ϕ so that $\alpha_\lambda(\theta_i,\phi_i,T) = \alpha(T)$ and $\varepsilon_\lambda(\theta,\phi,T) = \varepsilon(T)$, and from Kirchhoff's Law, Equation (3.18), $\alpha(T) = \varepsilon(T)$. In Equations (3.8), (3.36), and (3.38), since the $\alpha_\lambda(\theta_i,\phi_i,T)$ and $\varepsilon_\lambda(\theta,\phi,T)$ are not functions of either direction or wavelength, they can be taken out of the integrals. Then, for a diffuse-gray surface, the directional spectral and hemispherical total values of absorptivity and emissivity are *all equal*, and the hemispherical total absorptivity is *independent* of the nature of the incident radiation. This provides a great simplification for radiation transfer analysis, but rarely does it reflect the true character of the surface properties.

For diffuse-gray and other surface characteristics, the restrictions for applying Kirchhoff's Law are summarized in Table 3.2.

Example 3.6

The surface in Example 3.5 at $T = 650$ K is subject to incident radiation from a diffuse-gray source at $T_i = 925$ K. What is the hemispherical total absorptivity?

For a diffuse-gray source at T_i, $I_{\lambda,i}(\theta_i,\varphi_i) = C I_{\lambda b}(T_i)$ where C is a constant. From Table 3.2, since there is no angular dependence, $\alpha_\lambda(T) = \varepsilon_\lambda(T)$. Then from Equation (3.38),

$$\alpha(T) = \frac{\int \left[\int_{\lambda=0}^{N} \varepsilon_\lambda(T) C l_{\lambda b}(T_i) d\lambda \right] \cos\theta_i d\Omega_i}{\int \left[\int_{\lambda=0}^{N} C l_{\lambda b}(T_i) d\lambda \right] \cos\theta_i d\Omega_i} = \frac{1}{\sigma T_i^4} \int_{\lambda=0}^{N} \varepsilon_\lambda(T) E_{\lambda b}(T_i) d\lambda$$

As in Example 3.5, using λT_i values of $3.5 \times 925 = 3238$ $\mu m \cdot K$ and $9.5 \times 925 = 8788$ $\mu m \cdot K$,

$$\alpha = 0.90 F_{0\rightarrow3238} + \frac{1}{\sigma 925^4} \int_{3.5}^{9.5} (1.27917 - 0.10833\lambda) \frac{2\pi C_1}{\lambda^5(e^{C_2/925\lambda} - 1)} d\lambda + 0.25(1 - F_{0\rightarrow8788})$$

Numerical integration results in

$$\alpha = 0.90 \times 0.32650 + 0.37872 + 0.25(1 - 0.88377) = 0.7016$$

3.4 REFLECTIVITY

The reflective properties of a surface are more complicated to specify than emissivity or absorptivity because reflected energy depends not only on the angle at which the incident energy impinges on the surface, but also on the direction being considered for the reflected energy. Reflectivities are also functions of the reflective surface's temperature, T. We omit explicitly showing the temperature dependence for clarity, but it should be understood as for emissivity and absorptivity. The important reflectivity quantities are now defined.

3.4.1 SPECTRAL REFLECTIVITIES

3.4.1.1 Bidirectional Spectral Reflectivity, $\rho_\lambda(\theta_r,\phi_r,\theta_i,\phi_i)$

Consider spectral radiation incident on a surface from direction (θ_i,ϕ_i) as in Figure 3.2e. Part of this energy is reflected into the (θ_r,ϕ_r) direction. The subscript "r" denotes quantities evaluated at the reflected angle. The entire $I_{\lambda,r}(\theta_r,\phi_r)$ is the result of summing the reflected intensities produced by the incident intensities $I_{\lambda,i}(\theta_i,\phi_i)$ from *all* incident directions (θ_i,ϕ_i) of the hemisphere surrounding the surface element. The contribution to $I_{\lambda,r}(\theta_r,\phi_r)$ by the incident energy from only one direction (θ_i,ϕ_i) is designated as $I_{\lambda,r}(\theta_r,\phi_r,\theta_i,\phi_i)$.

The energy from direction (θ_i,ϕ_i) intercepted by dA per unit area and wavelength is, from Equation (3.13),

$$\frac{d^2 Q_{\lambda,i}(\theta_i,\phi_i) d\lambda}{dA d\lambda} = I_{\lambda,i}(\theta_i,\phi_i) \cos\theta_i d\Omega_i \qquad (3.40)$$

The *bidirectional spectral reflectivity* is a ratio expressing the contribution $I_{\lambda,i}(\theta_i,\phi_i)$ $\cos\theta_i d\Omega_i$ makes to the reflected spectral intensity in the (θ_r,ϕ_r) direction:

$$Bidirectional\ spectral\ reflectivity \equiv \rho_\lambda(\theta_r,\phi_r,\theta_i,\phi_i) = \frac{I_{\lambda,r}(\theta_r,\phi_r,\theta_i,\phi_i)}{I_{\lambda,i}(\theta_i,\phi_i)\cos\theta_i d\Omega_i} \quad (3.41)$$

The reflectivity also depends on surface temperature, but as noted the T notation is omitted.

The ratio in Equation (3.41) is a reflected intensity divided by the intercepted energy arriving within solid angle $d\Omega_i$. Having $\cos\theta_i d\Omega_i$ in the denominator means that when $\rho_\lambda(\theta_r,\phi_r,\theta_i,\phi_i)I_{\lambda,i}(\theta_i,\phi_i)\cos\theta_i d\Omega_i$ is integrated over all incidence angles (θ_i,ϕ_i) to provide the reflected intensity $I_{\lambda,r}(\theta_r,\phi_r)$, the reflected intensity is properly weighted by the amount of energy incident from each direction. Moreover, since $I_{\lambda,r}(\theta_r,\phi_r,\theta_i,\phi_i)$ is generally one differential order smaller than $I_{\lambda,i}(\theta_i,\phi_i)$, the $d\Omega_i$ in the denominator prevents $\rho_\lambda(\theta_r,\phi_r,\theta_i,\phi_i)$ from being a differential quantity. This also gives the bidirectional spectral reflectivity units of sr^{-1}, which is unusual for a radiative property. While emissivity and absorptivity are probabilities of RTE photon emission and absorption, and thus are bound between zero and one, the bidirectional reflectivity is the product of the probability that a photon incident from the (θ_i,ϕ_i) direction having a wavelength λ will be reflected and the probability density function that defines the direction of reflection. In other words,

$$\rho_\lambda(\theta_r,\phi_r,\theta_i,\phi_i)\cos\theta_r d\Omega_r = \underbrace{p\left(\theta_r,\phi_r\middle|ref,\theta_i,\phi_i,\lambda\right)}_{\substack{\text{probability density for}\\\text{direction on reflected direction}}} \underbrace{P\left(ref\middle|\theta_i,\phi_i,\lambda\right)}_{\substack{\text{probability of reflection}\\\text{into any direction}}} d\Omega_r \quad (3.42)$$

The probability that the photon will be reflected *exactly* into the (θ_r,ϕ_r) direction is infinitely small (0), but, by rearranging Equation (3.42), the probability that a photon will be reflected into a solid angle $d\Omega_r$ centered on the (θ_r,ϕ_r) direction is

$$P\left(ref\middle|\theta_i,\phi_i,\theta_r,\phi_r,\lambda\right) = \rho_\lambda\left(\theta_i,\phi_i,\theta_r,\phi_r\right)\cos\theta_r d\Omega_r \quad (3.43)$$

The presence of the $\cos\theta_r$ is important, because for a diffuse surface the probability of reflection depends on $\cos\theta_r$ through the projected area in the reflected direction. Therefore, a diffuse surface has a bidirectional reflectance that is independent of the reflectance angle, even though the probability of reflection depends on $\cos\theta_r$.

3.4.1.1.1 Reciprocity for Bidirectional Spectral Reflectivity

The $\rho_\lambda(\theta_r,\phi_r,\theta_i,\phi_i)$ is symmetric regarding reflection and incidence angles; the reflectivity for energy incident at (θ_i,ϕ_i) and reflected into (θ_r,ϕ_r) is equal to that for energy incident at (θ_r,ϕ_r) and reflected into (θ_i,ϕ_i). In the context of the Second Law of Thermodynamics, this is demonstrated by considering a nonblack element dA_2 in an isothermal black enclosure, as in Figure 3.7.

For the isothermal condition, the net energy exchange between black elements dA_1 and dA_3 must be zero. The energy exchange is by two paths. The direct exchange along the dashed line is between black elements at the same temperature and hence is zero. If the net exchange along this path is zero and net exchange, including all paths

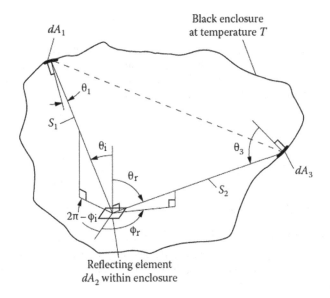

FIGURE 3.7 Enclosure used to examine reciprocity of bidirectional spectral reflectivity.

between dA_1 and dA_3, is zero, then net exchange along the path having a reflection from dA_2 must be zero. For the energy traveling along the reflected path,

$$d^3Q_{\lambda,1-2-3}d\lambda = d^3Q_{\lambda,3-2-1}d\lambda \qquad (3.44)$$

The energy reflected from dA_2 that reaches dA_3 is:

$$d^3Q_{\lambda,1-2-3}d\lambda = I_{\lambda,r}(\theta_r,\phi_r,\theta_i,\phi_i)\cos\theta_r dA_2(dA_3\cos\theta_3/S_2^2)d\lambda, \qquad (3.45)$$

or, using Equation (3.41),

$$d^3Q_{\lambda,1-2-3}d\lambda = \rho_\lambda(\theta_r,\phi_r,\theta_i,\phi_i)I_{\lambda,i}(T)\cos\theta_i\left(\frac{dA_1\cos\theta_1}{S_1^2}\right)\cos\theta_r dA_2\left(\frac{dA_3\cos\theta_3}{S_2^2}\right)d\lambda$$

$$(3.46)$$

Similarly,

$$d^3Q_{\lambda,3-2-1}d\lambda = \rho_\lambda(\theta_i,\phi_i,\theta_r,\phi_r)I_{\lambda,3}(T)\cos\theta_r\left(\frac{dA_3\cos\theta_3}{S_2^2}\right)\cos\theta_i dA_2\left(\frac{dA_1\cos\theta_1}{S_1^2}\right)d\lambda$$

$$(3.47)$$

Then from Equation (3.41) and because $I_{\lambda,1}(T) = I_{\lambda,3}(T) = I_{\lambda b}(T)$, there is the reciprocity relation:

$$\rho_\lambda(\theta_r,\phi_r,\theta_i,\phi_i) = \rho_\lambda(\theta_i,\phi_i,\theta_r,\phi_r) \tag{3.48}$$

The probabilistic interpretation based on RTE photons is particularly useful when considering microscopic reversibility, which requires any trajectory traveled by an individual photon to be reversible.[2] Since the probability density function for the direction of incident photons is proportional to $\cos\theta_i d\Omega_i$, we can write:

$$\underbrace{\rho_\lambda\left(\lambda,\theta_i,\phi_i,\theta_r,\phi_r\right)\cos\theta_r d\Omega_r\cos\theta_i d\Omega_i}_{P(ref|\theta_i,\phi_i,\theta_r,\phi_r,\lambda)} = \underbrace{\rho_\lambda\left(\lambda,\theta_r,\phi_r,\theta_i,\phi_i\right)\cos\theta_i d\Omega_i\cos\theta_r d\Omega_r}_{P(ref|\theta_i,\phi_i,\theta_r,\phi_r,\lambda)}$$

$$\tag{3.49}$$

which results in Equation (3.48).

3.4.1.2 Directional Spectral Reflectivities, $\rho_\lambda(\theta_r,\phi_r)$, $\rho_\lambda(\theta_i,\phi_i)$

Multiplying $I_{\lambda,r}(\theta_r,\phi_r,\theta_i,\phi_i)$ by $d\lambda\,\cos\theta_r dA\,d\Omega_r$ and integrating over all θ_r and ϕ_r gives the energy per unit time reflected into the entire hemisphere due to an incident intensity from a single direction (θ_i,ϕ_i):

$$d^2Q_{\lambda,r}(\theta_i,\phi_i)d\lambda = d\lambda dA\int_\cap I_{\lambda,r}(\theta_r,\phi_r,\theta_i,\phi_i)\cos\theta_r d\Omega_r$$

Using Equation (3.41), this equals:

$$d^2Q_{\lambda,r}(\theta_i,\phi_i)d\lambda = d\lambda dA I_{\lambda,i}(\theta_i,\phi_i)\cos\theta_i d\Omega_i\int_\cap \rho_\lambda(\theta_r,\phi_r,\theta_i,\phi_i)\cos\theta_r d\Omega_r \tag{3.50}$$

The *directional-hemispherical spectral reflectivity* is the energy reflected into all solid angles divided by the energy incident from direction (θ_i,ϕ_i) (Figure 3.2f). This gives Equation (3.50) divided by the incident energy from Equation (3.40):

$$\begin{matrix}\text{Directional-hemispherical spectral reflectivity} \\ \left(\text{in terms of bidirectional spectral reflectivity}\right)\end{matrix} \equiv \rho_\lambda(\theta_i,\phi_i) = \frac{d^2Q_{\lambda,r}(\theta_i,\phi_i)d\lambda}{d^2Q_{\lambda,i}(\theta_i,\phi_i)d\lambda}$$

$$= \int_\cap \rho_\lambda(\theta_r,\phi_r,\theta_i,\phi_i)\cos\theta_r d\Omega_r = P\left(ref|\theta_i,\phi_i,\lambda\right)$$

$$\tag{3.51}$$

which amounts to marginalizing the bidirectional spectral reflectivity over all possible reflection directions to obtain the probability that a photon will be reflected in any direction conditional on the incident (θ_i,ϕ_i) direction.

Likewise, by integrating Equation (3.41) over all incident directions (θ_i,ϕ_i):

$$I_{\lambda,r}(\theta_r,\phi_r) = \int_\cap \rho_\lambda(\theta_r,\phi_r,\theta_i,\phi_i)I_{\lambda,i}(\theta_i,\phi_i)\cos\theta_i d\Omega_i \tag{3.52}$$

The *hemispherical-directional spectral reflectivity* is then defined as the reflected intensity into the (θ_r, ϕ_r) direction divided by the mean incident intensity:

$$\begin{pmatrix} \text{Hemispherical-directional spectral reflectivity} \\ \left(\text{in terms of bidirectional spectral reflectivity}\right) \end{pmatrix} \equiv \rho_\lambda(\theta_r, \phi_r)$$

$$= \frac{\displaystyle\int_\cap \rho_\lambda(\theta_r, \phi_r, \theta_i, \phi_i) I_{\lambda,i}(\theta_i, \phi_i) \cos\theta_i d\Omega_i}{(1/\pi) \displaystyle\int_\cap I_{\lambda,i}(\theta_i, \phi_i) \cos\theta_i d\Omega_i} = P\left(ref \middle| \theta_r, \phi_r, \lambda\right) \tag{3.53}$$

Equation (3.53) is also a marginalization of the bidirectional spectral reflectivity, but this time over incident directions, and $\rho_\lambda(\theta_r, \phi_r)$ is the probability that a photon will be reflected from all incident directions conditional on the (θ_r, ϕ_r) direction.

3.4.1.2.1 Reciprocity for Directional Spectral Reflectivity

When the incident intensity is uniform over all incident directions (θ_i, ϕ_i), Equation (3.53) reduces to

$$\begin{pmatrix} \text{Hemispherical-directional spectral reflectivity} \\ \left(\text{for uniform incident intensity}\right) \end{pmatrix} \equiv \rho_\lambda(\theta_r, \phi_r)$$

$$= \int_\cap \rho_\lambda(\theta_r, \phi_r, \theta_i, \phi_i) \cos\theta_i d\Omega_i \tag{3.54}$$

Comparing Equations (3.51) and (3.53), and noting Equation (3.48), the reciprocal relation for ρ_λ is obtained (restricted to uniform incident intensity):

$$\rho_\lambda(\theta_i, \phi_i) = \rho_\lambda(\theta_r, \phi_r) \tag{3.55}$$

where (θ_r, ϕ_r) and (θ_i, ϕ_i) are the same angles, that is, $\theta_i = \theta_r$ and $\phi_i = \phi_r$. This means that the reflectivity of a material irradiated at a given angle of incidence (θ_i, ϕ_i), as measured by the energy collected over the entire hemisphere of reflection, is equal to the reflectivity for *uniform* irradiation from the hemisphere as measured by collecting the reflected energy at a single angle of reflection (θ_r, ϕ_r) when (θ_r, ϕ_r) are the *same* angles as (θ_i, ϕ_i). This relation is used for the design of hemispherical reflectometers for measuring radiative properties.

3.4.1.3 Hemispherical Spectral Reflectivity, ρ_λ

If the incident spectral radiation arrives from all angles of the hemisphere (Figure 3.2h), then all the radiation intercepted by the area element dA of the surface is given by Equation (3.27) as:

$$dQ_{\lambda,i} d\lambda = dA d\lambda \int_\cap I_{\lambda,i}(\theta_i, \phi_i) \cos\theta_i d\Omega_i$$

The amount of $dQ_{\lambda,i}d\lambda$ that is reflected is, by integration using Equation (3.52) to include all incident directions:

$$dQ_{\lambda,r}d\lambda = dAd\lambda \int_\cap \rho_\lambda(\theta_i,\phi_i)I_{\lambda,i}(\theta_i,\phi_i)\cos\theta_i d\Omega_i$$

The fraction of $dQ_{\lambda,i}d\lambda$ that is reflected provides the definition.

$$\text{Hemispherical spectral reflectivity} \atop \left(\text{in terms of directional-hemispherical spectral reflectivity}\right) \equiv$$

$$\rho_\lambda = \frac{dQ_{\lambda,r}d\lambda}{dQ_{\lambda,i}d\lambda} \tag{3.56}$$

$$= \frac{\displaystyle\int_\cap \rho_\lambda(\theta_i,\phi_i)I_{\lambda,i}(\theta_i,\phi_i)\cos\theta_i d\Omega_i}{\displaystyle\int_\cap I_{\lambda,i}(\theta_i,\phi_i)\cos\theta_i d\Omega_i} = P\left(ref\,|\lambda\right)$$

In a probabilistic sense, the hemispherical spectral reflectivity comes from marginalizing the bidirectional spectral reflectivity over both incident and reflected directions.

Now, consider some special cases for spectral surfaces. In the following, we use a subscript notation for clarification: "d-h" denotes directional-hemispherical, and "bi-d" is for bidirectional.

3.4.2 DIFFUSE SURFACES

For a diffuse surface, the incident energy from the direction (θ_i,ϕ_i) that is reflected produces a reflected intensity that is uniform over all (θ_r,ϕ_r) directions. When an irradiated diffuse surface is viewed, the surface is equally bright from all viewing directions, regardless of the distribution of incident intensity. The bidirectional spectral reflectivity is independent of (θ_r,ϕ_r), and Equation (3.53) simplifies to:

$$\rho_{\lambda,d-h} = \rho_{\lambda,bi-d}\int_\cap \cos\theta_r d\Omega_r = \pi\rho_{\lambda,bi-d}(\theta_i,\phi_i).$$

However, note that for an opaque surface, $\rho_{\lambda,d-h}(\theta_i,\phi_i) = 1 - \alpha_\lambda$; whereas for a diffuse surface, α_λ is not a function of incidence angle (θ_i,ϕ_i). Hence, the directional-hemispherical reflectivity must be independent of angle of incidence, so that for a diffuse surface,

$$\rho_{\lambda,d-h} = \pi\rho_{\lambda,bi-d} \tag{3.57}$$

for any incidence angle. The π arises because $\rho_{\lambda,d-h}$ accounts for energy reflected into all (θ_r,ϕ_r) directions, while $\rho_{\lambda,bi-d}$ accounts for the reflected intensity into only one direction. This is analogous to the relation $E_{\lambda b} = \pi I_{\lambda b}$.

Equation (3.52) provides the intensity in the (θ_r, ϕ_r) direction when the incident radiation is distributed over (θ_i, ϕ_i) values. If the surface is *diffuse, and if the incident intensity is uniform for all incident angles,* Equation (3.52) reduces to:

$$I_{\lambda,r} = \rho_{\lambda,\text{bi-d}} I_{\lambda,i} \int_{\cap} \cos\theta_i d\Omega_i = \pi \rho_{\lambda,\text{bi-d}} I_{\lambda,i} \qquad (3.58)$$

But, using Equation (3.57),

$$I_{\lambda,r} = \rho_{\lambda,\text{d-h}} I_{\lambda,i} = (1 - \alpha_\lambda) I_{\lambda,i} \qquad (3.59)$$

so that the reflected intensity in any direction for a diffuse surface is the uniform incident intensity multiplied by either the directional-hemispherical reflectivity or the hemispherical reflectivity.

3.4.3 Specular Surfaces

Mirror-like, or *specular,* surfaces obey the Fresnel Laws of reflection (see Chapter 4). For incident radiation from a single direction, a specular reflector, by definition, provides reflected radiation at the same magnitude of the angle from the surface normal as for the incident intensity, and in the same plane as the incident intensity and the surface normal. Hence,

$$\theta_r = \theta_i; \quad \phi_r = \phi_i + \pi \qquad (3.60)$$

At all other angles, the bidirectional reflectivity of a specular surface is zero; then,

$$\rho_\lambda(\lambda, \theta_r, \phi_r, \theta, \phi)_{\text{specular}} = \rho_\lambda(\theta_r = \theta_i, \phi_r = \phi_i + \pi, \theta_i, \phi_i) \equiv \rho_{\lambda,\text{bi-d}}(\theta_i, \phi_i) \qquad (3.61)$$

and the bidirectional reflectivity is a function of only the incident direction, as this direction also provides the reflected direction.

For the intensity of radiation reflected from a specular surface into the solid angle around (θ_r, ϕ_r), Equation (3.52) becomes, for an *arbitrary* directional distribution of incident intensity,

$$I_{\lambda,r}(\theta_r, \phi_r) = \int_{\cap} \rho_{\lambda,\text{bi-d}}(\theta_i, \phi_i) I_{\lambda,i}(\theta_i, \phi_i) \cos\theta_i d\Omega_i.$$

The integrand has a nonzero value only in the small solid angle around (θ_i, ϕ_i) because of the properties of $\rho_{\lambda,\text{bi-d}}(\theta_i, \phi_i)$ for a specular surface. When examining the radiation reflected from a specular surface into a given direction, only that radiation at the incident direction (θ_i, ϕ_i) defined by Equation (3.40) needs to be considered as contributing to the reflected intensity. Hence,

$$I_{\lambda,r}(\theta_r = \theta_i, \phi_r = \phi_i + \pi) = \rho_{\lambda,\text{bi-d}}(\theta_i, \phi_i) I_{\lambda,i}(\theta_i, \phi_i) \cos\theta_i d\Omega_i \qquad (3.62)$$

If a surface tends to be specular, the use of the bidirectional reflectivity can become less practical than for a situation with a more diffuse reflection behavior. For a specular reflection, the reflected intensity can be of the same order as the incident intensity; hence, the bidirectional reflectivity becomes very large because of the $d\Omega_i$ on the right side of Equation (3.62). In a statistical sense, the probability density that defines (θ_r,ϕ_r) in Equation (3.42) approaches a delta function. A reflectivity that equals the well-behaved quantity $\rho_{\lambda,\text{bi-d}}(\theta_i,\phi_i)\cos\theta_i d\Omega_i$ in Equation (3.62) is more useful and is now defined as $\rho_{\lambda,s}(\theta_i,\phi_i)$. From Equation (3.62), it is this quantity that equals the ratio of reflected to incident intensities. Using reciprocity, we can then define a convenient *specular* reflectivity as:

$$\frac{I_{\lambda,r}\left(\theta_r = \theta_i, \phi_r = \phi_i + \pi\right)}{I_{\lambda,i}\left(\theta_i,\phi_i\right)} = \rho_{\lambda,s}\left(\theta_i,\phi_i\right) = \rho_{\lambda,s}\left(\theta_r,\phi_r\right) \tag{3.63}$$

The angular relations of Equation (3.60) apply so that the incident radiation is reflected into one direction.

3.4.4 TOTAL REFLECTIVITIES

The previous reflectivity expressions for spectral radiation can be marginalized over all wavelengths to obtain corresponding total reflectivities.

3.4.4.1 Bidirectional Total Reflectivity, $\rho(\theta_r,\phi_r,\theta_i,\phi_i)$

This gives the contribution made by the total *energy* incident from direction (θ_i,ϕ_i) to the reflected total *intensity* into the direction (θ_r,ϕ_r). By integrating Equation (3.41),

$$\text{Bidirectional total reflectivity} \equiv \rho\left(\theta_r,\phi_r,\theta_i,\phi_i\right) = \frac{\int_{\lambda=0}^{\infty} I_{\lambda,r}\left(\theta_r,\phi_r,\theta_i,\phi_i\right)d\lambda}{\cos\theta_i d\Omega_i \int_{\lambda=0}^{\infty} I_{\lambda,i}\left(\theta_i,\phi_i\right)d\lambda}$$

$$= \frac{I_r\left(\theta_r,\phi_r,\theta_i,\phi_i\right)}{I_i\left(\theta_i,\phi_i\right)\cos\theta_i d\Omega_i} \tag{3.64}$$

Another form using Equation (3.41) is:

Bidirectional total reflectivity$\left(\text{in terms of bidirectional spectral reflectivity}\right)$

$$\equiv \rho\left(\theta_r,\phi_r,\theta_i,\phi_i\right) = \frac{\int_{\lambda=0}^{\infty} \rho_\lambda\left(\theta_r,\phi_r,\theta_i,\phi_i\right) I_{\lambda,i}\left(\theta_i,\phi_i\right)d\lambda}{\int_{\lambda=0}^{\infty} I_{\lambda,i}\left(\theta_i,\phi_i\right)d\lambda} \tag{3.65}$$

3.4.4.2 Reciprocity for Bidirectional Total Reflectivity

Rewriting Equation (3.65) for energy incident from direction (θ_r, ϕ_r) and reflected into direction (θ_i, ϕ_i) gives:

$$\rho(\theta_i, \phi_i, \theta_r, \phi_r) = \frac{\int_{\lambda=0}^{\infty} \rho_\lambda(\theta_i, \phi_i, \theta_r, \phi_r) I_{\lambda,i}(\theta_r, \phi_r) d\lambda}{\int_{\lambda=0}^{\infty} I_{\lambda,i}(\theta_r, \phi_r) d\lambda} \tag{3.66}$$

Comparing Equations (3.65) and (3.66) shows that:

$$\rho(\theta_i, \phi_i, \theta_r, \phi_r) = \rho(\theta_r, \phi_r, \theta_i, \phi_i) \tag{3.67}$$

if the spectral distribution of incident intensity is the same for all directions or, less restrictively, if $I_{\lambda,i}(\theta_i, \phi_i) = C I_{\lambda,i}(\theta_r, \phi_r)$.

3.4.4.3 Directional-Hemispherical Total Reflectivity, $\rho(\theta_r, \phi_r)$, $\rho(\theta_i, \phi_i)$

The *directional-hemispherical total reflectivity* is the fraction of the total energy incident from a single direction that is reflected into all angular directions, or, in other words, the probability that a photon having any wavelength will be reflected in any direction, conditional on an incident direction. The spectral energy from a given direction (θ_i, ϕ_i) that is intercepted by the surface is $I_{\lambda,i}(\theta_i, \phi_i)\cos\theta_i d\Omega_i d\lambda dA$. The reflected portion is $\rho_\lambda(\theta_i, \phi_i) I_{\lambda,i}(\theta_i, \phi_i)\cos\theta_i d\Omega_i d\lambda dA$. Integrating over all wavelengths and forming a ratio gives:

$$\begin{array}{l}\text{Directional-hemispherical total reflectivity} \\ \text{(in terms of directional-hemispherical spectral reflectivity)}\end{array} \equiv \rho(\theta_i, \phi_i)$$

$$= \frac{\int_{\lambda=0}^{\infty} \rho_\lambda(\theta_i, \phi_i) I_{\lambda,i}(\theta_i, \phi_i) d\lambda}{\int_{\lambda=0}^{\infty} I_{\lambda,i}(\theta_i, \phi_i) d\lambda} = P\left(ref \mid \theta_i, \phi_i\right) \tag{3.68}$$

Another directional total reflectivity specifies the fraction of radiation reflected into the angle (θ_r, ϕ_r) direction when there is diffuse irradiation. The total radiation intensity reflected into the (θ_r, ϕ_r) direction when the incident intensity is uniform over all incident directions is:

$$I_r(\theta_r, \phi_r) = \int_{\lambda=0}^{\infty} \rho_\lambda(\theta_r, \phi_r) I_{\lambda,i} d\lambda$$

where $\rho_\lambda(\theta_r, \phi_r)$ is in Equation (3.53). The total reflectivity is defined as the reflected intensity divided by the uniform incident intensity:

Hemispherical-directional total reflectivity
$$\left(\text{for diffuse irradiation}\right) \equiv \frac{I_r(\theta_r,\phi_r)}{I_i}$$

$$= \frac{\int_{\lambda=0}^{\infty} \rho_\lambda(\theta_r,\phi_r) I_{\lambda,i} d\lambda}{\int_{\lambda=0}^{\infty} I_{\lambda,i} d\lambda} \qquad (3.69)$$

3.4.4.4 Reciprocity for Directional Total Reflectivity

Compare Equations (3.68) and (3.69). Restricting to uniform incident intensity, Equation (3.55) gives:

$$\rho(\theta_r,\phi_r) = \rho(\theta_i,\phi_i) \qquad (3.70)$$

where (θ_r,ϕ_r) and (θ_i,ϕ_i) are the same angles, that is, $\theta_i = \theta_r$ and $\phi_i = \phi_r$. In addition, there is a fixed spectral distribution of the incident intensity such that $I_{\lambda,i}$ in Equation (3.68) is related to that in Equation (3.69) by $I_{\lambda,i}(\theta_i,\phi_i) = CI_{\lambda i}$.

3.4.4.5 Hemispherical Total Reflectivity, ρ

If the incident total radiation arrives from all angles of the hemisphere, the total radiation intercepted by a unit area at the surface is given by Equation (3.27). The amount of this radiation that is reflected is:

$$dQ_r = dA \int_\cap \rho(\theta_i,\phi_i) I_i(\theta_i,\phi_i) \cos\theta_i d\Omega_i.$$

The fraction of all the incident energy that is reflected into all directions of reflection, is then:

Hemispherical total reflectivity
$$\left(\text{in terms of directional-hemispherical total reflectivity}\right) \equiv \rho = \frac{dQ_r}{dQ_i}$$

$$= \frac{dA}{dQ_i} \int_\cap \rho(\theta_i,\phi_i) I_i(\theta_i,\phi_i) \cos\theta_i d\Omega_i = P\left(ref\right) \qquad (3.71)$$

which is the probability that an incident photon will be reflected, and is marginalized over all wavelengths and all incident and reflected directions. Another form is found by using the incident hemispherical spectral energy. The amount reflected is $\rho_\lambda dQ_{\lambda,i} d\lambda$ where ρ_λ is the hemispherical spectral reflectivity from Equation (3.56). Integrating yields:

Hemispherical total reflectivity
$$\left(\text{in terms of hemispherical spectral reflectivity}\right) \equiv \rho = \frac{\int_0^\infty \rho_\lambda dQ_{\lambda,i} d\lambda}{dQ_i = \int_0^\infty dQ_{\lambda,i} d\lambda} \qquad (3.72)$$

TABLE 3.3

Summary of Reciprocity Relations between Reflectivities

Type of Quantity	Equality	Restrictions
Bidirectional spectral (Equation 3.48)	$\rho_\lambda(\theta_i,\phi_i,\theta_r,\phi_r) = \rho_\lambda(\theta_r,\phi_r,\theta_i,\phi_i)$	None
Directional spectral (Equation 3.55)	$\rho_\lambda(\theta_i,\phi_i) = \rho_\lambda(\theta_r,\phi_r)$ where $\theta_i = \theta_r$, $\phi_i = \phi_r$	$\rho_\lambda(\theta_r,\phi_r)$ is for uniform incident intensity or ρ_λ is independent of θ_i,ϕ_i,θ_r, and ϕ_r.
Bidirectional total (Equation 3.67)	$\rho(\theta_i,\phi_i,\theta_r,\phi_r) = \rho(\theta_r,\phi_r,\theta_i,\phi_i)$	$I_{\lambda,i}(\theta_i,\phi_i) = CI_{\lambda,i}(\theta_r,\phi_r)$ or $\rho_\lambda(\theta_i,\phi_i,\theta_r,\phi_r)$ independent of λ.
Directional total (Equation 3.70)	$\rho(\theta_i,\phi_i) = \rho(\theta_r,\phi_r)$	For uniform incident intensity, or ρ_λ is independent of incident angle; independent of wavelength.

3.4.5 SUMMARY OF RESTRICTIONS ON REFLECTIVITY RECIPROCITY

Table 3.3 summarizes the restrictions necessary for applying the reflectivity reciprocity relations.

3.5 TRANSMISSIVITY AT AN INTERFACE

As for reflectivity, the transmissivity of radiation across a single interface depends on two sets of angles: the angle of incidence on the interface, the direction at which the radiation is transmitted after crossing the surface, and on the wavelength of the radiation. The properties defined in this section pertain only to the *fraction of incident radiation crossing the single interface*. For many real configurations, radiation that is transmitted across the interface may return to the interface by reflection from another nearby surface as is the case in a pane of glass, or by scattering from material in the region of the interface. For opaque surfaces, the transmissivity across the surface of the material is not germane to radiative transfer calculations, but the definitions are included here for later reference.

Given that the transmissivity is defined here for a single interface, it follows that the radiation incident on a surface is either reflected from the interface or transmitted through the surface, so that the transmissivity is directly related to the reflectivity. Because the definitions in this section are derived in parallel to those for reflectivity, the derivations are more terse.

3.5.1 SPECTRAL TRANSMISSIVITIES

3.5.1.1 Bidirectional Spectral Transmissivity, $\tau_\lambda(\theta_t,\phi_t,\theta_i,\phi_i)$

The ratio of intensity that is transmitted across an interface into the direction (θ_t,ϕ_t) to the incident spectral energy on the interface from an incident direction (θ_i,ϕ_i) (Figure 3.2i) is thus defined by the relation:

Bidirectional spectral transmissivity $\equiv \tau_\lambda\left(\theta_t,\phi_t,\theta_i,\phi_i\right)$

$$= \frac{I_{\lambda,t}\left(\theta_t,\phi_t,\theta_i,\phi_i\right)}{I_{\lambda,i}\left(\theta_i,\phi_i\right)\cos\theta_i d\Omega_i} \qquad (3.73)$$

which, as for the other properties, depends on the surface temperature T, and that notation is again omitted. Like the bidirectional spectral reflectivity, the bidirectional spectral transmissivity also has units of sr^{-1}, and has a similar meaning as Equation (3.42), i.e.,

$$\tau_\lambda(\theta_t,\phi_t,\theta_i,\phi_i)\cos\theta_t d\Omega_t = \underbrace{p\left(\theta_t,\phi_t\big|trans,\theta_i,\phi_i,\lambda\right)}_{\substack{\text{probability density for}\\ \text{direction of transmitted photon}}}\underbrace{P\left(trans\big|\theta_i,\phi_i,\lambda\right)}_{\substack{\text{probability of transmission}\\ \text{into any direction}}}d\Omega_t \quad (3.74)$$

3.5.1.2   Directional Spectral Transmissivities, $\tau_\lambda(\theta_t,\phi_t)$, $\tau_\lambda(\theta_i,\phi_i)$

The directional-hemispherical spectral transmissivity $\tau_\lambda(\theta_i,\phi_i)$ is the fraction of the incident spectral energy that is transmitted across the interface into all directions (Figure 3.2j). By energy conservation, it is related to the directional-hemispherical spectral reflectivity, so that:

Directional-hemispherical spectral transmissivity

$$\equiv \tau_\lambda\left(\theta_i,\phi_i\right) = \frac{\int_\cap I_{\lambda,t}\left(\theta_t,\phi_t,\theta_i,\phi_i\right)\cos\theta_t d\Omega_t}{I_{\lambda,i}\left(\lambda,\theta_i,\phi_i\right)\cos\theta_i d\Omega_i} = 1-\rho_\lambda(\theta_i,\phi_i) \qquad (3.75)$$

Using Equation (3.73), this becomes:

$$\tau_\lambda\left(\theta_i,\phi_i\right) = \frac{I_{\lambda,i}\left(\theta_i,\phi_i\right)\cos\theta_i d\Omega_i \int_\cap \tau_\lambda\left(\theta_t,\phi_t,\theta_i,\phi_i\right)\cos\theta_t d\Omega_t}{I_{\lambda,i}\left(\theta_i,\phi_i\right)\cos\theta_i d\Omega_i}$$

$$= \int_\cap \tau_\lambda\left(\theta_t,\phi_t,\theta_i,\phi_i\right)\cos\theta_t d\Omega_t \qquad (3.76)$$

Following Equations (3.52) and (3.53), the hemispherical-directional spectral transmissivity $\tau_\lambda(\theta_t,\phi_t)$ is the spectral intensity transmitted into (θ_t,ϕ_t) over the integrated average spectral intensity incident on the surface (Figure 3.2k), or

Hemispherical-directional spectral transmissivity $\equiv \tau_\lambda(\theta_t,\phi_t)$

$$= \frac{I_{\lambda,t}\left(\theta_t,\phi_t\right)}{\left(\dfrac{1}{\pi}\right)\int_\cap I_{\lambda,i}\left(\theta_i,\phi_i\right)\cos\theta_i d\Omega_i}$$

$$(3.77)$$

or, using Equation (3.75),

$$\begin{array}{c}\text{Hemispherical-directional spectral transmissivity} \\ \text{(in terms of the bidirectional spectral transmissivity)}\end{array} \equiv \tau_\lambda\left(\theta_t,\phi_t\right)$$

$$= \frac{\displaystyle\int_\cap \tau_\lambda\left(\theta_t,\phi_t,\theta_i,\phi_i\right)I_{\lambda,i}\left(\theta_i,\phi_i\right)\cos\theta_i d\Omega_i}{\displaystyle\left(\frac{1}{\pi}\right)\int_\cap I_{\lambda,i}\left(\theta_i,\phi_i\right)\cos\theta_i d\Omega_i} \quad (3.78)$$

For uniform incident spectral intensity from all directions, Equation (3.78) reduces to:

$$\begin{array}{c}\text{Hemispherical-directional spectral transmissivity} \\ \left(\text{for uniform incident intensity}\right)\end{array} \equiv \tau_\lambda\left(\theta_t,\phi_t\right)$$

$$= \int_\cap \tau_\lambda\left(\theta_t,\phi_t,\theta_i,\phi_i\right)\cos\theta_i d\Omega_i$$

$$(3.79)$$

Comparing Equation (3.79) with Equation (3.78) shows that, for uniform incident radiation,

$$\tau_\lambda\left(\theta_t,\phi_t\right) = \tau_\lambda\left(\theta_i,\phi_i\right) = 1 - \rho_\lambda(\theta,\phi) \quad (3.80)$$

3.5.1.3 Hemispherical Spectral Transmissivity, τ_λ

The fraction of spectral energy incident on the surface from all directions that is transmitted across the surface into all directions (Figure 3.2*l*) is the hemispherical spectral transmissivity. It is related through energy conservation to the hemispherical spectral reflectivity:

$$\text{Hemispherical spectral transmissivity} \equiv \tau_\lambda = \frac{\displaystyle\int_\cap I_{\lambda,t}\left(\theta_t,\phi_t\right)\cos\theta_t d\Omega_t}{\displaystyle\int_\cap I_{\lambda,i}\left(\theta_i,\phi_i\right)\cos\theta_i d\Omega_i} \quad (3.81)$$

$$= 1 - \rho_\lambda$$

or, using Equation (3.75),

$$\begin{array}{c}\text{Hemispherical-spectral transmissivity} \\ \left(\text{in terms of directional-hemispherical transmissivity}\right)\end{array} \equiv \tau_\lambda$$

$$= \frac{\displaystyle\int_\cap \tau_\lambda\left(\theta_i,\phi_i\right)I_{\lambda,i}\left(\theta_i,\phi_i\right)\cos\theta_i d\Omega_i}{\displaystyle\int_\cap I_{\lambda,i}\left(\lambda,\theta_i,\phi_i\right)\cos\theta_i d\Omega_i} \quad (3.82)$$

3.5.2 Total Transmissivities

Total transmissivities are obtained by direct analogy with the spectral property definitions.

3.5.2.1 Bidirectional Total Transmissivity, $\tau(\theta_i,\phi_i,\theta_t,\phi_t)$

The ratio of total intensity that is transmitted across an interface into the direction (θ_t,ϕ_t) to the incident total energy on the interface from direction (θ_i,ϕ_i) (Figure 3.2i) is defined by:

$$\text{Bidirectional total transmissivity} \equiv \tau\left(\theta_t,\phi_t,\theta_i,\phi_i\right) = \frac{I_t\left(\theta_t,\phi_t,\theta_i,\phi_i\right)}{I_i\left(\theta_i,\phi_i\right)\cos\theta_i d\Omega_i} \tag{3.83}$$

3.5.2.2 Directional Total Transmissivities, $\tau(\theta_i,\phi_i)$

The directional-hemispherical total transmissivity is the fraction of the incident total energy that is transmitted across the interface into all directions (Figure 3.2j). By energy conservation, it is related to the directional-hemispherical total reflectivity, so that:

Directional-hemispherical total transmissivity

$$\equiv \tau\left(\theta_i,\phi_i\right) = \frac{\int_\cap I_t\left(\theta_t,\phi_t,\theta_i,\phi_i\right)\cos\theta_t d\Omega_t}{I_i\left(\theta_i,\phi_i\right)\cos\theta_i d\Omega_i} = 1-\rho(\theta_i,\phi_i) \tag{3.84}$$

Using Equation (3.83), this becomes:

Directional-hemispherical total transmissivity
(in terms of bidirectional transmissivity) $\equiv \tau\left(\theta_i,\phi_i\right)$

$$= \frac{I_i\left(\theta_i,\phi_i\right)\cos\theta_i d\Omega_i \int_\cap \tau\left(\theta_t,\phi_t,\theta_i,\phi_i\right)\cos\theta_t d\Omega_t}{I_i\left(\theta_i,\phi_i\right)\cos\theta_i d\Omega_i} = \int_\cap \tau\left(\theta_t,\phi_t,\theta_i,\phi_i\right)\cos\theta_t d\Omega_t \tag{3.85}$$

3.5.2.3 Hemispherical-Directional Total Transmissivity, $\tau(\theta_t,\phi_t)$, $\tau(\theta_i,\phi_i)$

The hemispherical-directional total transmissivity $\tau(\theta_t,\phi_t)$ is the total intensity transmitted in the (θ_t,ϕ_t) direction over the integrated average total intensity incident on the surface (Figure 3.2k), or

Hemispherical-directional total transmissivity

$$\equiv \tau(\theta_t,\phi_t) = \frac{I_t\left(\theta_t,\phi_t\right)}{\left(\dfrac{1}{\pi}\right)\int I_i\left(\theta_i,\phi_i\right)\cos\theta_i d\Omega_i} \tag{3.86}$$

or, using Equation (3.83),

Hemispherical-directional total transmissivity
(in terms of the bidirectional total transmissivity) $\equiv \tau(\theta_t, \phi_t)$

$$= \frac{\displaystyle\int_{\cap} \tau(\theta_t, \phi_t, \theta_i, \phi_i) I_i(\theta_i, \phi_i) \cos\theta_i d\Omega_i}{\left(\dfrac{1}{\pi}\right)\displaystyle\int_{\cap} I_i(\theta_i, \phi_i) \cos\theta_i d\Omega_i} \qquad (3.87)$$

For uniform incident total intensity from all directions, Equation (3.87) reduces to

Hemispherical-directional total transmissivity
$\big($for uniform incident intensity$\big)$ $\equiv \tau(\theta_t, \phi_t)$

$$= \int_{\cap} \tau(\theta_t, \phi_t, \theta_i, \phi_i) \cos\theta_i d\Omega_i \qquad (3.88)$$

Comparing Equation (3.88) with Equation (3.85) shows that, for uniform incident radiation and because of the reciprocity of $\tau(\theta_t, \phi_t, \theta_i, \phi_i)$,

$$\tau(\theta_t, \phi_t) = \tau(\theta_i, \phi_i) \qquad (3.89)$$

3.5.2.4 Hemispherical Total Transmissivity, τ

The fraction of total energy incident on the surface from all directions that is transmitted across the surface into all directions (Figure 3.2*l*) is the hemispherical total transmissivity:

$$\text{Hemispherical total transmissivity} \equiv \tau = \frac{\displaystyle\int_{\cap} I_t(\theta_t, \phi_t) \cos\theta_t d\Omega_t}{\displaystyle\int_{\cap} I_i(\theta_i, \phi_i) \cos\theta_i d\Omega_i} \qquad (3.90)$$

or, using Equation (3.83),

Hemispherical total transmissivity
$\big($in terms of directional-hemispherical transmissivity$\big)$ $\equiv \tau$

$$= \frac{\displaystyle\int_{\cap} \tau(\theta_i, \phi_i) I_i(\theta_i, \phi_i) \cos\theta_i d\Omega_i}{\displaystyle\int_{\cap} I_i(\theta_i, \phi_i) \cos\theta_i d\Omega_i} \qquad (3.91)$$

For uniform incident intensity, this reduces to:

$$\begin{array}{l} \text{Hemispherical total transmissivity} \\ \text{(for uniform incident intensity)} \end{array} \equiv \tau = \frac{1}{\pi} \int_\cap \tau(\theta_i, \phi_i) \cos\theta_i d\Omega_i \qquad (3.92)$$

3.6 RELATIONS AMONG REFLECTIVITY, ABSORPTIVITY, EMISSIVITY, AND SURFACE TRANSMISSIVITY

From the definitions of absorptivity, reflectivity, and transmissivity as fractions of incident energy absorbed, reflected, or transmitted, it is evident that some relations exist among these properties through the First Law of Thermodynamics. For an opaque material, all energy that is transmitted across a surface is absorbed by the medium beneath the surface, so it follows that transmissivity and absorptivity are equal. (This will not be true when translucent materials of finite thickness are considered.) With the use of Kirchhoff's Law and its restrictions (Table 3.2), further relations between emissivity and reflectivity can be obtained. These relationships are summarized in Table 3.4.

The spectral energy $d^2Q_{\lambda,i}(\theta_i,\phi_i)d\lambda$ incident per unit time on an element dA of an opaque body from within a solid angle $d\Omega_i$ at (θ_i,ϕ_i) is either absorbed or reflected, so that:

$$\frac{d^2Q_{\lambda,a}(\theta_i,\phi_i)d\lambda}{d^2Q_{\lambda,i}(\theta_i,\phi_i)d\lambda} + \frac{d^2Q_{\lambda,r}(\theta_i,\phi_i)d\lambda}{d^2Q_{\lambda,i}(\theta_i,\phi_i)d\lambda} = 1 \qquad (3.93)$$

Because the radiation is incident from the angles (θ_i,ϕ_i), the two energy ratios in Equation (3.93) are the directional spectral absorptivity, Equation (3.14), (or the directional spectral transmissivity, Equation (3.75)) and the directional-hemispherical spectral reflectivity, Equation (3.51). Substituting gives:

$$\alpha_\lambda(\theta_i,\phi_i) + \rho_\lambda(\theta_i,\phi_i) = \tau_\lambda(\theta_i,\phi_i) + \rho_\lambda(\theta_i,\phi_i) = 1 \qquad (3.94)$$

Kirchhoff's Law, Equation (3.18), can be applied without restriction to give:

$$\varepsilon_\lambda(\theta_i,\phi_i) + \rho_\lambda(\theta_i,\phi_i) = \tau_\lambda(\theta_i,\phi_i) + \rho_\lambda(\theta_i,\phi_i) = 1 \qquad (3.95)$$

TABLE 3.4

Summary of Relations among Reflectivity, Transmissivity, and Absorptivity

Property Type	Equality	Equation
Directional-hemispherical spectral	$\tau_\lambda(\theta_i,\phi_i) = \alpha_\lambda(\theta_i,\phi_i) = 1 - \rho_\lambda(\theta_i,\phi_i)$	(3.94)
Hemispherical spectral	$\tau_\lambda = \alpha_\lambda = 1 - \rho_\lambda$	(3.100)
Directional-hemispherical total	$\tau(\theta_i,\phi_i) = \alpha(\theta_i,\phi_i) = 1 - \rho(\theta_i,\phi_i)$	(3.97)
Hemispherical total	$\tau = \alpha = 1 - \rho$	(3.103)

When the total energy is considered to arrive at dA in $d\Omega_i$ from the given direction (θ_i,ϕ_i), Equation (3.93) becomes:

$$\frac{d^2Q_a(\theta_i,\phi_i)}{d^2Q_i(\theta_i,\phi_i)} + \frac{d^2Q_r(\theta_i,\phi_i)}{d^2Q_i(\theta_i,\phi_i)} = 1 \tag{3.96}$$

Substituting Equation (3.22) or (3.84) and Equation (3.68) results in:

$$\alpha(\theta_i,\phi_i) + \rho(\theta_i,\phi_i) = \tau(\theta_i,\phi_i) + \rho(\theta_i,\phi_i) = 1 \tag{3.97}$$

The absorptivity (and transmissivity) are directional total values, and the reflectivity is the total directional-hemispherical value. Kirchhoff's Law for directional total properties (Section 3.3) is applied to give:

$$\varepsilon(\theta_i,\phi_i) + \rho(\theta_i,\phi_i) = \tau(\theta_i,\phi_i) + \rho(\theta_i,\phi_i) = 1 \tag{3.98}$$

under the restriction that the incident radiation obeys $I_{\lambda,i}(\theta_i,\phi_i) = C(\theta_i,\phi_i)I_{\lambda b}(T)$, which usually only happens under conditions of radiative equilibrium, or when the surface is directional-gray.

If the incident spectral intensity is arriving at dA from all directions over the hemisphere, then Equation (3.96) gives:

$$\frac{d^2Q_{\lambda,a}d\lambda}{d^2Q_{\lambda,i}d\lambda} + \frac{d^2Q_{\lambda,r}d\lambda}{d^2Q_{\lambda,i}d\lambda} = 1 \tag{3.99}$$

Equation (3.99), using the property definitions, is:

$$\alpha_\lambda + \rho_\lambda = \tau_\lambda + \rho_\lambda = 1 \tag{3.100}$$

where the properties are hemispherical spectral values from Equations (3.29), (3.56), and (3.82). Under these restrictions, Equation (3.100) becomes:

$$\varepsilon_\lambda + \rho_\lambda = \tau_\lambda + \rho_\lambda = 1 \tag{3.101}$$

If the energy incident on dA is integrated over all wavelengths and directions, Equation (3.99) results in:

$$\frac{dQ_a}{dQ_i} + \frac{dQ_r}{dQ_i} = 1 \tag{3.102}$$

or

$$\alpha + \rho = \tau + \rho = 1 \tag{3.103}$$

Note that, while $\varepsilon_\lambda(\theta_i,\phi_i) = \alpha_\lambda(\theta_i,\phi_i)$ without restriction, and $\varepsilon_\lambda = \alpha_\lambda$ if either the surface or the irradiation is diffuse, the conditions under which total emissivity can be equated with total absorptivity are far more restrictive. Specifically, this requires: (*i*) $I_{\lambda i}(T) = I_{\lambda b}(T)$, which usually implies radiative equilibrium between a surface its surroundings and therefore no net radiative transfer, or (*ii*) gray surfaces, an idealization commonly invoked to simplify the analysis, but one that rarely reflects reality.

Table 3.4 may be used to further develop the relationship of surface transmissivity to the other radiative properties of opaque surfaces (see also Table 3.2).

Example 3.7

Radiation from the sun is incident on a surface in orbit above the Earth's atmosphere. The surface is at 1000 K, and the directional total emissivity is given in Figure 3.4. If the solar energy is incident at an angle of 25° from the normal to the surface, what is the reflected energy flux?

From Figure 3.4, $\varepsilon(25°, 1000\ \text{K}) = 0.8$. Table 3.2 shows that for directional total properties $\alpha(25°, 1000\ \text{K}) = \varepsilon(25°, 1000\ \text{K})$ only when the incident spectrum is proportional to that emitted by a blackbody at 1000 K. This is not the case here, since the solar spectrum is like that of a blackbody at 5780 K. Hence $\alpha(25°, 1000\ \text{K}) \neq 0.8$, and without $\alpha(25°, 1000\ \text{K})$ we cannot determine $\rho(25°, 1000\ \text{K})$. The emissivity data given are insufficient to solve the problem. Spectral values are needed.

HOMEWORK (additional problems available in online Appendix I at www.ThermalRadiation.net/appendix.html)

3.1 A material has a hemispherical spectral emissivity that varies considerably with wavelength but is independent of surface temperature. Radiation from a gray source at T_i is incident on the surface uniformly from all directions. Show that the total absorptivity for the incident radiation is equal to the total emissivity of the material evaluated at the source temperature T_i.

3.2 Suppose that ε_λ is independent of λ (gray-body radiation). Show that $F_{0 \rightarrow \lambda T}$ represents the fraction of the total radiant emission of the gray body in the range from 0 to λT.

3.3 Find the emissivity at 450 K and the solar absorptivity of the diffuse material with the measured spectral emissivity shown in the figure.

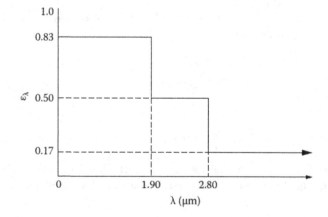

Answers: $\varepsilon(T = 450 \text{ K}) = 0.171$; $\alpha_s(T = 5780 \text{ K}) = 0.799$.

3.4 (a) Obtain the total absorptivity of a diffuse surface with properties given in the figure for incident radiation from a blackbody with a temperature of 5800 K.

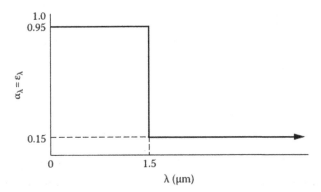

(b) What is the total emissivity of the diffuse surface with properties given in the figure if the surface temperature is 600 K?

Answers: (a) 0.855; (b) 0.15.

3.5 For the spectral properties given in the figure for a diffuse surface:
   (a) What is the solar absorptivity of the surface (assume the solar temperature is 5800 K)?
   (b) What is the total hemispherical emissivity of the surface if the surface temperature is 650 K?

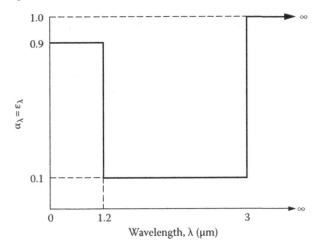

Answers: (a) 0.766; (b) 0.947.

3.6 Find the emissivity at 800 K and the solar absorptivity of the diffuse material with the measured spectral emissivity shown in the figure. This will require numerical integration.

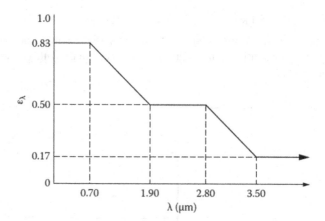

Answers: $\varepsilon(800\ K) = 0.226$; $\varepsilon(5780\ K) = \alpha_s(5780\ K) = 0.756$.

3.7   The directional total absorptivity of a gray surface is $\alpha(\theta) = 0.490\ \cos^2\theta$ where θ is the angle relative to the normal to the surface.
(a)   What is the hemispherical total emissivity of the surface?
(b)   What is the hemispherical-hemispherical total reflectivity of this surface for diffuse incident radiation (uniform incident intensity)?
(c)   What is the hemispherical-directional total reflectivity for diffuse incident radiation reflected into a direction 65° from the normal?

Answers: (a) 0.245; (b) 0.775; (c) 0.913.

3.8   A gray surface has a directional emissivity, as shown. The properties are isotropic with respect to circumferential angle ϕ.
(a)   What is the hemispherical emissivity of this surface?
(b)   If the energy from a blackbody source at 650 K is incident uniformly from all directions, what fraction of the incident energy is absorbed by this surface?
(c)   If the surface is placed in a very cold environment, at what rate must energy be added per unit area to maintain the surface temperature at 1000 K?

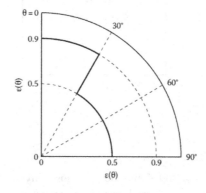

Answers: (a) 0.600; (b) 0.600; (c) 34,000 W/m².

3.9 A gray surface has a directional total emissivity that depends on angle of incidence as $\varepsilon(\theta) = 0.810 \cos\theta$. Uniform radiant energy from a single direction normal to the cylinder axis is incident on a long cylinder of radius R. What fraction of energy striking the cylinder is reflected? What is the result if the body is a sphere rather than a cylinder?

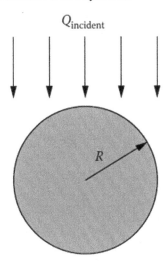

Answers: 0.364; 0.460.

3.10 A thin ceramic plate, insulated on one side, is radiating energy from its exposed side into a vacuum at very low temperature. The plate is initially at 1400 K, and is cooled to 300 K. At any instant, the plate is assumed to be at uniform temperature across its thickness and over its exposed area. The plate is 0.20 cm thick, and the surface hemispherical spectral emissivity is as shown and is independent of temperature. What is the cooling time? The density of the ceramic is 3200 kg/m³, and its specific heat is 710 J/(kg · K).

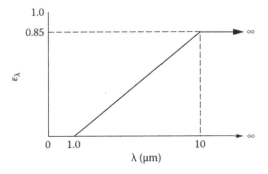

Answer: $t = 24.3$ min.

NOTES

1. While we conceptually speak of "a photon" interacting with the surface, it is important to keep in mind the limitations of the "RTE photon" interpretation discussed in Chapter 2.
2. Or, in terms of EM waves, the time-reversal invariance of Maxwell's equations.

4 Predicted and Measured Surface Properties

James Clerk Maxwell (1831–1879) *described the propagation of an electromagnetic wave using a system of 20 equations, and showed that the speed of wave propagation equals the speed of light, implying that light was itself an electromagnetic wave. Among his many accomplishments, Maxwell, in collaboration with Rudolph Clausius, used a statistical approach to find the velocity distribution in an assembly of gas molecules at a given temperature (later derived using the maximum entropy principle by Ludwig Boltzmann, and now called the Maxwell–Boltzmann distribution).*

Joseph Thomas Gier (1910–1961) *received his BS and ME degrees in Mechanical Engineering at UC Berkeley in 1933 and 1940, respectively. After serving as a lecturer and researcher, he was promoted to Associate Professor of Electrical Engineering at UC Berkeley in 1951, becoming the first tenured African-American professor in the University of California system. He formed a fruitful partnership with Robert V. Dunkle, a professor in the Mechanical Engineering department, in 1943. Together they developed instrumentation to characterize the radiative properties of surfaces and conceived of spectral-selectivity to improve the performance of solar collectors.*

4.1 PREDICTION OF RADIATIVE PROPERTIES OF OPAQUE MATERIALS

The radiative properties of surfaces are critical to analyzing a long list of engineering applications involving thermal radiation, including solar heating, thermal loading on buildings, furnace-based industrial processes, and spacecraft thermal control. However, obtaining these properties is often challenging since they depend on the composition, temperature, and roughness of the surface, and are also affected by real-world effects like surface contamination (e.g., oxidation) and surface heterogeneity. While several monographs are dedicated to cataloging the radiative properties of surfaces (e.g., Touloukian et al. 1970, 1972a, 1972b; Palik 1998), it is impossible to define a comprehensive database due to the sheer number of possible materials, finishes, and temperatures. In this context, predictions based on electromagnetic theory are quite useful; while they apply only to perfectly clean and optically smooth surfaces, they often yield useful trends and provide a unifying basis to help explain radiation phenomena. Moreover, while radiation calculations often require properties as a function of wavelength and direction, in many cases tabulated data is spectrally

DOI: 10.1201/9781003178996-4

or directionally integrated. In these cases, it is often possible to "fill in the blanks" using EM theory predictions. This chapter begins with a discussion of electromagnetic wave propagation, which underpins the theoretical predictions of radiative properties. We next compare these predictions to experimentally determined values, and discuss how real-world effects, like heterogeneity, surface roughness, and contamination, may influence radiative properties. Finally, we show how surface properties may be engineered to enhance the performance of solar collectors or to reduce the cooling load on buildings.

4.2   INTRODUCTION TO ELECTROMAGNETIC WAVE THEORY

In 1864, James Clark Maxwell provided the crowning achievement of classical physics by explaining how the propagation of electromagnetic wave energy is related to orthogonal fluctuations in electric and magnetic fields. Maxwell originally proposed a set of 20 equations, which Oliver Heaviside (Heaviside 1889; Maxwell 1890) reformulated into the four equations that we use today:

$$\nabla \times \mathbf{E} = -\mu \frac{\partial \mathbf{H}}{\partial t} \text{ (Faraday's Law)} \tag{4.1}$$

$$\nabla \times \mathbf{H} = \gamma \frac{\partial \mathbf{E}}{\partial t} + \sigma_e \mathbf{E} \quad \text{(Ampère's Law)} \tag{4.2}$$

$$\nabla \cdot \mathbf{E} = 0 \text{ (Gauss's Law)} \tag{4.3}$$

and

$$\nabla \cdot \mathbf{H} = 0 \text{ (Gauss's Law)} \tag{4.4}$$

where vectors $\mathbf{E}$ and $\mathbf{H}$ describe the electric and magnetic field intensities, σ_e is the electrical conductivity, γ is the electrical permittivity, and μ is the magnetic permeability[1] Loosely speaking, the electrical permittivity describes how easy it is for an electric field to polarize matter (e.g., to induce a dipole moment), while magnetic permeability defines how easy it is to magnetize a material in response to a magnetic field. Gauss's Law is a law of conservation: if the electric or magnetic fields strengthen in one direction, they must diminish by an equal amount in other directions. Faraday's Law and Ampère's Law define the familiar "right-hand rule" between electric and magnetic fields. Faraday's Law shows that a changing magnetic field (e.g., rotating an armature between two magnets) induces an electric current. Ampère's Law defines the operating principle of a solenoid; a time-varying current flowing through a coil surrounding a magnetized plunger induces a magnetic field that, in the case of a doorbell, pushes the plunger into the bell.

Equations (4.1) and (4.2) are *wave equations*, and the $\sigma_e\mathbf{E}$ term in Ampère's Law acts to damp the wave. Columbic forces cause charged particles within a conducting medium to move and scatter from other atoms, ions, and electrons, providing a mechanism through which the wave energy is absorbed by the medium; this is discussed in Section 4.3. The nature of wave propagation through a material and the behavior of waves at interfaces depends strongly on whether the material(s) in question are electrical conductors (large σ_e), insulators (small σ_e), or perfect dielectrics ($\sigma_e = 0$).

Maxwell's equations define a set of second-order partial differential equations that may be solved given appropriate boundary conditions. Consider a Cartesian coordinate system, and suppose an EM wave propagates in the x-direction, so the electric and magnetic field fluctuations must be confined to the y–z plane, i.e., $\mathbf{E} = [0,E_y,E_z]^T$, and $\mathbf{H} = [0,H_y,H_z]^T$. The simplest scenario is a "plane-polarized wave" in which fluctuations in electric and magnetic fields are confined to respective planes defined by the coordinate system, e.g., $E_z = 0$ and $H_y = 0$. Finally, suppose a boundary condition at $x = 0$ in which the electric field fluctuates in time according to a cosine wave:

$$E_y\left(x = 0,t\right) = E_{y0} \cos\left(\omega t\right) \tag{4.5}$$

where $\omega = 2\pi\nu$ is the angular frequency in rad/s. We choose a cosine wave for its convenient mathematical properties, and because any waveform may be represented by a superposition of cosine waves through a Fourier series. This information is sufficient to solve for the EM field as a function of position and time.

4.2.1 WAVES PROPAGATING IN AN ISOTROPIC DIELECTRIC MEDIUM

Consider first a plane-polarized EM wave that propagates at angular frequency $\omega = 2\pi\nu$ through a perfect insulator, i.e., $\sigma_e = 0$. In this case, Maxwell's Equations (4.1)–(4.4), along with the boundary condition Equation (4.5), have the solution:

$$E_y\left(x,t\right) = E_{y0} \cos\left[\omega\left(t - \sqrt{\mu\gamma}x\right)\right] = E_{y0} \cos\left[\omega\left(t - \frac{nx}{c_0}\right)\right] \tag{4.6}$$

Here c_0 is the speed of light in a vacuum and

$$n = c_0\sqrt{\gamma\mu} = \sqrt{\gamma\mu}/\sqrt{\gamma_0\mu_0} \tag{4.7}$$

is the refractive index of the material. The $t - (\gamma\mu)^{1/2}x$ term in Equation (4.6) reveals that the EM wave propagates at a speed equal to $(\gamma\mu)^{-1/2}$, and setting the electric permittivity and dielectric permeability equal to their vacuum values (see Appendix A.1) produces $c_0 = 2.998 \times 10^8$ m/s. The fact that Maxwell's equation predicted that EM waves propagate at the speed of light implied that light itself was an electromagnetic wave. It is important to note that the refractive index of most materials is a strong function of wavelength (or frequency); this behavior is further explored in Section 11.1.

The magnetic field strength is found by substituting Equation (4.6) into Equation (4.2) to give:

$$H_z\left(x,t\right) = \frac{n}{\mu c_0} E_{y0} \cos\left[\omega\left(t - \frac{nx}{c_0}\right)\right] = \sqrt{\frac{\gamma}{\mu}} E_y\left(x,t\right) \qquad (4.8)$$

The electric and magnetic fields for a plane-polarized wave propagating through a perfect dielectric are shown in Figure 4.1, revealing that these fields oscillate in phase and in directions that are mutually orthogonal to the direction of wave propagation. The energy carried by the wave is given by the magnitude of the *Poynting vector*,

$$\mathbf{S} = \mathbf{E} \times \mathbf{H} \qquad (4.9)$$

which is aligned with the direction of wave propagation. In the case of the plane-polarized wave shown in Figure 4.1, the Poynting vector has a nonzero component only in the x-direction, $S_x = E_y \times H_z$. The magnitude of the Poynting vector is the spectral intensity I_λ at the wavelength $\lambda = c/\nu = 2\pi c/\omega$.

Perfect dielectrics are also perfect insulators; consequently, the damping term in Equation (4.2) is zero, and the intensity along a ray passing through the medium is constant. While this is an idealization, it is quite a good approximation for many insulators and semiconductors over a limited portion of the spectrum. This arises from the fact the charge carriers in these materials can only absorb energy from waves of a given frequency. As an example, window glass behaves as a near-perfect dielectric over visible wavelengths, but it is opaque at wavelengths longer than approximately 2 μm. This behavior is further discussed in Section 4.3.

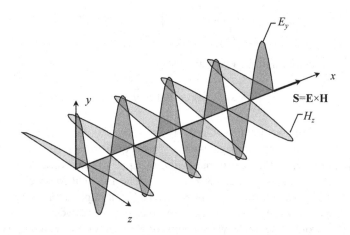

FIGURE 4.1   An EM wave propagation through a dielectric material. The electric field wave is polarized in the x–y plane, traveling in the x-direction, while the companion magnetic field wave is polarized in the x–y plane.

4.2.2 Waves Propagating in an Isotropic Conducting Medium

We next consider EM waves that propagate through a conducting medium. In this case, the damping term in Ampère's equation produces a damped wave that must be defined using complex notation. Assuming a plane-polarized wave with $E_z = H_y = 0$, and the same boundary condition as before, Equations (4.1–4.4) yield:

$$E_y(x,t) = E_{y0} \cos\left[\omega\left(t - \frac{nx}{c_0}\right)\right] \exp\left(-\frac{\omega kx}{c_0}\right) \tag{4.10}$$

where n and k are the real and imaginary components of the complex refractive index, $\bar{n} = n - ik$, defined by:

$$n^2 = \frac{\mu\gamma c_0^2}{2}\left[1 \pm \sqrt{1 + \left(\frac{\sigma_e}{\mu\omega}\right)^2}\right] \tag{4.11}$$

and

$$k^2 = \frac{\mu\gamma c_0^2}{2}\left[1 \pm \sqrt{1 - \left(\frac{\sigma_e}{\mu\omega}\right)^2}\right] \tag{4.12}$$

The positive sign is chosen in most cases for Equations (4.11) and (4.12), except for materials having negative electrical permittivity. The imaginary part, k, is called the absorption index, and is related to the absorption coefficient in the RTE. Note that n becomes Equation (4.7) and k goes to zero when $\sigma_e = 0$.

The magnetic field can then be found from Faraday's Law, Equation (4.1). After some manipulation, one arrives at:

$$H_z(x,t) = \frac{\sqrt{n^2 + k^2}}{\mu c_0} E_{y0} \cos\left[\omega\left(t - \frac{nx}{c_0} + \phi\right)\right] \exp\left(\frac{-\omega kx}{c_0}\right) \tag{4.13}$$

where $\phi = \tan^{-1}(k/n)$ is a phase shift between the electric and magnetic field oscillations. The phase shift goes to zero when $\sigma_e = 0$.

Figure 4.2 shows how the amplitudes of the electric and magnetic fields decay exponentially with path length through a conducting medium. In this case, the magnitude of the Poynting vector varies according to $\exp(-2\omega kx/c)$ or $\exp(-4\pi kx/\lambda)$. From Equation (2.49), the intensity in a homogeneous, purely absorbing medium decays according to $I = I_{\lambda,0}\exp(-\kappa_\lambda x)$, so the absorption index, k, is related to the absorption coefficient, κ_λ, according to:

$$\kappa_\lambda = \frac{4\pi k}{\lambda} \tag{4.14}$$

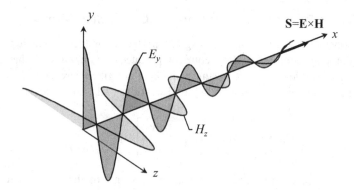

FIGURE 4.2 EM wave propagation through a weakly absorbing medium. In this case, the phase shift between the electric and magnetic waves is small.

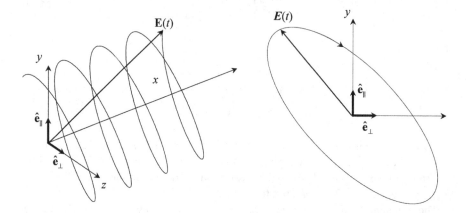

FIGURE 4.3 Trajectory of the electric field vector and vibration ellipse of an elliptically polarized wave propagating through a dielectric medium.

Therefore, the spectral absorption coefficient within a *homogenous medium* may be obtained from the extinction coefficient as a function of wavelength.

4.2.3 POLARIZATION

Until now, we have assumed that the EM waves are plane-polarized, meaning that fluctuations in the electric and magnetic field intensities are confined to mutually orthogonal planes. In most cases, however, the electric and magnetic fields will fluctuate in two dimensions normal to the direction of wave propagation. For example, the electric field vector $\mathbf{E} = [0, E_y, E_z]^T$ may follow a "corkscrew" pattern, as shown in Figure 4.3. Over time, the tip of this vector will trace an ellipse projected in the y–z plane, which is called the vibration ellipse. Many vibration ellipses are possible, depending on the source of the EM wave, and the patterns are often quite complex.

It is often convenient to represent the electric field as the superposition of two orthogonal components using Euler's formula:

$$\mathbf{E} = \mathrm{Re}\left[\mathbf{E}_0(t)e^{(ikx-i\omega t)}\right], \quad \mathbf{E}_0(t) = \mathbf{E}_\parallel(t)\hat{\mathbf{e}}_\parallel + \mathbf{E}_\perp(t)\hat{\mathbf{e}}_\perp \tag{4.15}$$

where Re denotes the real component of the argument. Orthogonal unit vectors $\hat{\mathbf{e}}_\parallel$ and $\hat{\mathbf{e}}_\perp$ may be defined in a convenient manner, e.g., relative to a surface normal vector. Due to the linearity of Maxwell's equations, any wave may be expressed this way, and each component is analyzed separately. This treatment is particularly useful for predicting how EM waves behave at interfaces, as will be discussed in Section 4.4.

The components $\mathbf{E}_\parallel(t)$ and $\mathbf{E}_\perp(t)$ may vary as a function of time. If these components are uncorrelated in time and, on average, equal in magnitude, then the EM wave is uncorrelated. This is the case for thermal emissions arising from the random processes described in Chapter 2.

4.3   ORIGINS OF OPTICAL CONSTANTS

As we have seen, the way in which a wave propagates through a medium and the way the waves behave at interfaces depend on the electromagnetic properties of the material: μ, ε, and σ_e. These properties are assembled into the complex refractive index through Equations (4.11) and (4.12). It is sometimes more convenient to work with the complex dielectric function,

$$\bar{\varepsilon} = \varepsilon_I - i\varepsilon_{II} = \frac{\gamma}{\gamma_0} - i\frac{\sigma_e}{2\omega\gamma_0} = \bar{n}^2 \tag{4.16}$$

The refractive indices and complex dielectric functions are related to each other according to:

$$\varepsilon_I = n^2 - k^2, \quad \varepsilon_{II} = 2nk \tag{4.17}$$

and

$$n^2 = \frac{1}{2}\left(\varepsilon_I + \sqrt{\varepsilon_I^2 + \varepsilon_{II}^2}\right), \quad k^2 = \frac{1}{2}\left(-\varepsilon_I + \sqrt{\varepsilon_I^2 + \varepsilon_{II}^2}\right) \tag{4.18}$$

It is usually more convenient to use $\bar{n}$ when describing wave speed and attenuation through a medium, while $\bar{\varepsilon}$ is more closely linked with how waves interact with the atoms and electrons that constitute the medium. While the wavelength dependence of these parameters is usually omitted from the notation, we shall see that they depend strongly on wavelength.

Attenuating matter can broadly be categorized according to whether the matter contains bound charges or free charges, e.g., free electrons. This distinction can be understood in a mechanistic way through classical models developed well before the advent of quantum physics. While these models grossly simplify how EM waves interact with matter, they are surprisingly robust for many (although not all) materials and are still widely used today.

4.3.1 LORENTZ MODEL (NON-CONDUCTORS)

In the case of a non-conductor, the EM wave interacts with bound charged particles of a given mass (electrons or ions) within the medium. In the Lorentz model, developed at the end of the 19th and beginning of the 20th century by Hendrick Lorentz (Lorentz 1916), these charges are attached to a lattice by a spring and damper, as shown in Figure 4.4.

The spring represents the attractive and repulsive electrostatic forces between the charged particle and other particles in the lattice, while the damper accounts for the transfer of energy between the moving charge and the lattice, through which wave energy is converted into internal energy of matter. The equation of motion for the charge is:

$$m\ddot{\mathbf{x}} + b\dot{\mathbf{x}} + K\mathbf{x} = e\mathbf{E} \qquad (4.19)$$

where $\mathbf{x}$ denotes the displacement, m is the rest mass of the charge (e.g., the rest mass of an electron), b is the damping coefficient, K is the spring constant, and e is the charge. The product $e\mathbf{E}$, which fluctuates with time, is the Columbic force that drives the motion, while the spring force and damping force oppose the motion. For simplicity, consider a plane-polarized monochromatic wave, of the type shown in Figure 4.2. The solution to Equation (4.19) is:

$$y(t) = \frac{e/m\, E_y(t)}{\omega_0^2 - \omega^2 - i\zeta\omega} \qquad (4.20)$$

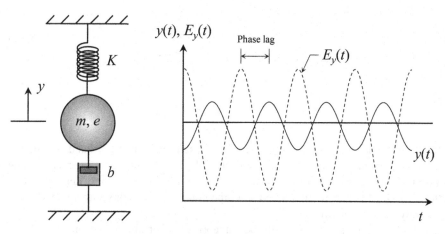

FIGURE 4.4 Schematic of the Lorentz oscillator model. The dashpot b attenuates the amplitude of the oscillator and causes a phase shift with respect to the driving EM field.

where $\omega_0 = K/m$ is the resonant frequency and $\zeta = b/m$. The dipole moment of an individual oscillator is ey, and if there are N oscillators per unit volume the dipole moment per unit volume is:

$$P(t) = Ney(t) = \frac{n^2 e^2/m E_y(t)}{\omega_0^2 - \omega^2 - i\zeta\omega} = \frac{\omega_p^2}{\omega_0^2 - \omega^2 - i\zeta\omega} \gamma_0 E_y(t) \qquad (4.21)$$

The plasma frequency, $\omega_p = [Ne^2/(m\gamma_0)]^{1/2}$, defines the frequency of motion that would occur if the charges were initially displaced and then allowed to return to their equilibrium positions in the absence of a driving force field.

Recall from Section 4.2 that the electric permittivity defines the ease with which an electric field can induce a dipole moment. Accordingly, after some manipulation Equation (4.21) becomes:

$$\bar{\varepsilon} = 1 + \frac{\omega_p^2}{\omega_0^2 - \omega^2 - i\zeta\omega} \qquad (4.22)$$

where the ϵ_0 term is added to account for the background of high-frequency oscillators, e.g., core electrons, that are not explicitly modeled by Equations (4.19)–(4.21). The electric permittivity can be separated into real and complex terms,

$$\varepsilon_I = \varepsilon_0 + \frac{\omega_p^2\left(\omega_0^2 - \omega^2\right)}{\left(\omega_0^2 - \omega^2\right)^2 + \zeta^2\omega^2} \qquad (4.23)$$

and

$$\varepsilon_{II} = \frac{\omega_p^2\zeta\omega}{\left(\omega_0^2 - \omega^2\right)^2 + \zeta^2\omega^2} \qquad (4.24)$$

which can then be used to find n and k according to Equation (4.18).

Plots of the dielectric function and refractive index in Figure 4.5(a) for a typical dielectric show that n is constant with respect to ω and k is approximately zero at very low and very high frequencies, but not in the vicinity of the resonant frequency, ω_0. The spectral variation of n causes waves to "disperse" or spatially separate according to their wavelengths, as occurs when light is shone through a prism. Hence, theories that describe this behavior are called *dispersion theories*.

At low frequencies, Lorentz theory predicts that $n = 1 + Ne^2/(2m\varepsilon_0\omega_0^2)$ and $k = 0$. In this range, the material behaves like a perfect dielectric; the wave speed depends on the number density of oscillators, their mass, and the spring constant (i.e., how tightly they are connected to the lattice). This behavior is called "normal dispersion." At frequencies slightly higher than ω_0, $n < 1$ and $k > 0$. This is called "anomalous dispersion" and coincides with the absorption of light by the material. Most insulating materials exhibit anomalous dispersion between the infrared and ultraviolet spectra, and those that absorb in the visible spectrum appear tinted. It is important to note

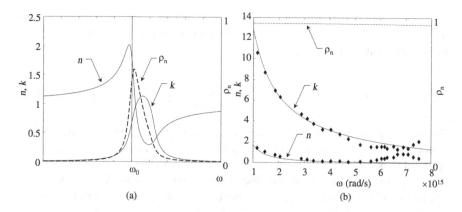

FIGURE 4.5 Refractive indices and normal spectral reflectivity for (a) a typical dielectric predicted from Lorentz theory, and (b) molten silver predicted using Drude theory. The normal spectral reflectivity is found using Equation (4.48). The symbols indicate experimentally measured values. Note that the Drude prediction misses a "bump" due to an interband transition at high frequencies.

that $n < 1$ does *not* imply a wave speed greater than c_0. A detailed discussion of this behavior is provided by Bohren and Huffman (1983).

The above analysis assumed that the medium is composed of a single type of oscillator. All real materials contain multiple oscillator types, and, due to the linearity of Maxwell's equations, the complex dielectric function can be found by summing over multiple oscillator types.

4.3.2 DRUDE MODEL (CONDUCTORS)

In the case of conductors, EM waves primarily interact with non-localized charge carriers, e.g., conduction-band electrons that are free to move against a sea of positively charged ions. In the case of metals, this is called the "nearly free electron model." In 1900, Paul Drude developed a model to describe how EM waves interact with these electrons (Drude 1900a, 1990b). Mathematically the Drude model is like the Lorentz model except that there is no "spring" that localizes the charges to the lattice. This results in:

$$\bar{\varepsilon} = 1 - \frac{\omega_p^2}{\omega^2 + i\zeta\omega} \tag{4.25}$$

where the damping constant ζ accounts for collisions between the electrons and the background ions and is equal to the inverse of the relaxation time, ζ, which is the average time interval between collisions. The plasma frequency is defined as before, but substituting the rest mass and number density of electrons, m_e and N_e, in place of m and N. The latter parameter can be found by multiplying the number density of atoms by the valence of the material, which defines the number of electrons each atom contributes to the conduction band. The relaxation time is related to the DC conductivity of the material, σ_{DC}, by:

$$\tau = \frac{1}{\zeta} = \frac{m_e \sigma_{DC}}{N_c e^2} \tag{4.26}$$

where e is the charge of an electron.

The electrical permittivity can be separated into real and complex terms,

$$\varepsilon_I = \varepsilon_0 - \frac{\omega_p^2}{\omega^2 + \zeta^2} \tag{4.27}$$

and

$$\varepsilon_{II} = \frac{\omega_p^2 \zeta}{\omega\left(\omega^2 + \zeta^2\right)} \tag{4.28}$$

which can then be used to find n and k via Equation (4.18).

The refractive index for molten silver shown in Figure 4.5(b) reveals that the magnitudes of n and k are much larger than those of an insulator, particularly at high frequencies/short wavelengths. As we will show in Section 4.4.2, this corresponds to a much higher reflectivity compared to dielectric materials.

The finite value of DC conductivity is a consequence of collisions between free electrons and background ions, and, according to Drude theory, this is the mechanism through which the wave energy is transferred to the internal (thermal) energy of the metal. This has several important implications for the radiative properties of metals; one would expect, for example, that metals having a higher electrical conductivity would absorb less incident radiation, and therefore have a higher reflectivity, compared to metals having a lower electrical conductivity. One would also expect the reflectivity to drop with increasing temperature, as thermal motion of the background ions leads to more electron scattering. As we shall see, both trends are observed in the radiative properties of metals.

While Drude theory provides accurate estimates for some materials and explains trends on how the radiative properties of metals vary with wavelength, the radiative properties of many metals differ markedly from Drude theory predictions. One of the main shortcomings of Drude's theory is that it ignores the contributions of bound electrons to radiative properties; for example, the red color of copper is associated with bound electrons. Lorentz (1905a–d,1916) extended Drude's model by adding a term to account for bound transitions:

$$\varepsilon = 1 - \underbrace{\frac{\omega_{p,e}^2}{\omega^2 + i\zeta_e\omega}}_{\text{free electons}} + \sum_j \underbrace{\frac{\omega_{p,j}^2}{\omega_{0,j}^2 - \omega^2 - i\zeta_j\omega}}_{\text{bound oscillators}} \tag{4.29}$$

This is called Drude–Lorentz theory.

4.4   EM WAVES AT INTERFACES

Maxwell's equations define a framework for obtaining the radiative properties of opaque materials and surfaces. For "optically smooth" surfaces, these equations can be used to obtain spectral directional reflectivity, transmissivity, absorptivity, and emissivity from the optical and electrical properties of the bulk material. In this context, "optically smooth" means that surface imperfections, roughness, or texture are much smaller than the wavelength of the incident radiation, and wave/surface interactions are strictly *specular* (mirror-like) in nature. (The term *specular* is not to be confused with *spectral*, which is indicates a wavelength-specific attribute.) Accordingly, the EM theory predictions described in this section work best for optics and polished surfaces, or at longer wavelengths. As an example, many unfinished metallic surfaces have roughness features on the micrometer scale, so EM theory predictions work well at wavelengths on the order of 10 μm.

The departures of real surfaces from the ideal conditions assumed in theoretical predictions (e.g., due to roughness and contamination) often result in large differences between measured and modeled properties. Despite these limitations, EM theory provides an understanding of why there are basic differences in the radiative properties of insulators and electrical conductors and reveals trends that help unify the presentation of experimental data. The theory also explains the angular dependence of directional reflectivity, absorptivity, and emissivity for smooth surfaces. Finally, since the theory applies to pure substances with ideally smooth surfaces, it also provides a means for computing limits of attainable properties, such as maximum reflectivity or minimum emissivity of a metallic surface.

4.4.1   DIELECTRIC MATERIALS

Consider an EM wave traveling through a dielectric medium with refractive index n_1, which encounters the smooth planar interface of a second medium with a refractive index n_2. If the angle of the incidence onto the surface of medium 2 is θ_i, then the wave is refracted into medium 2 at angle χ. The angles of incidence of θ_i, reflection θ_r, and refraction χ, which are all in the same plane (Figure 4.6), are related through *Snell's Law*:

$$n_1 \sin\theta_i = n_1 \sin\theta_r = n_2 \sin\chi \qquad (4.30)$$

which results in

$$\theta_i = \theta_r \qquad (4.31)$$

The angle of reflection of an EM wave from an ideal interface is thus equal to its angle of incidence rotated about the normal to the interface through a circumferential angle of $\theta = \pi$ (i.e., a specular reflection). Equation (4.30) can be rearranged into:

$$\frac{\sin\chi}{\sin\theta_i} = \frac{n_1}{n_2} = \sqrt{\frac{\varepsilon_1}{\varepsilon_2}} \qquad (4.32)$$

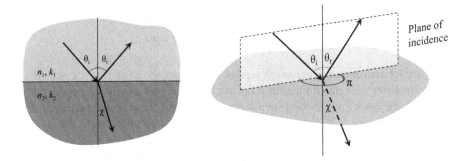

FIGURE 4.6 Geometric quantities for an EM wave at an interface between two semi-infinite media.

In the case of radiation incident from air or a vacuum, $n_1 = 1$.

As discussed in Section 4.2, we can resolve the electric and magnetic fields into two orthogonal polarization components; their vectorial sum gives the total intensity of the propagating wave. It is convenient to define the polarization components relative to the plane of incidence and the oscillating plane of the electric field. The perpendicular and parallel components oscillate perpendicular and parallel to the incidence plane, respectively, and both are perpendicular to the direction of the propagation.

4.4.1.1 Reflectivity

The specular reflectivity of a wave incident on a surface at angle θ_i and polarized parallel or perpendicular to the plane of incidence is:

$$\rho_{\lambda,\parallel}\left(\theta_i\right) = \left[\frac{\tan\left(\theta_i - \chi\right)}{\tan\left(\theta_i + \chi\right)}\right]^2 \tag{4.33}$$

and

$$\rho_{\lambda,\perp}\left(\theta_i\right) = \left[\frac{\sin(\theta_i - \chi)}{\sin(\theta_i + \chi)}\right]^2 \tag{4.34}$$

The subscript λ results because the refractive indices used to calculate χ depend on wavelength.

Useful forms containing only θ_i are obtained by eliminating χ in Equations (4.33) and (4.34) using Equation (4.32):

$$\rho_{\lambda,\parallel}\left(\theta_i\right) = \left\{\frac{(n_2/n_1)^2\cos\theta_i - [(n_2/n_1)^2 - \sin^2\theta_i]^{1/2}}{(n_2/n_1)^2\cos\theta_i + [(n_2/n_1)^2 - \sin^2\theta_i]^{1/2}}\right\}^2 \tag{4.35}$$

$$\rho_{\lambda,\perp}\left(\theta_i\right) = \left\{ \frac{\left[(n_2/n_1)^2 - \sin^2\theta_i\right]^{1/2} - \cos\theta_i}{\left[(n_2/n_1)^2 - \sin^2\theta_i\right]^{1/2} + \cos\theta_i} \right\}^2 \tag{4.36}$$

The parallel reflectance component equals zero when $\theta_i = \tan^{-1}(n_2/n_1)$, which is called *Brewster's angle*. Radiation reflected at this angle is only perpendicularly polarized (the basis for polarized sunglasses, which block the reflected perpendicular component). Figure 4.7(a) shows the spectral reflectivity components in the case of $n_2/n_1 > 1$. In general, the reflectance is lowest at $\theta_i = 0$ and highest at glancing angles, near $\theta_i = \pi/2$. This is why birds of prey like eagles and ospreys perch in high trees on the shore or circle overhead when fishing. We also experience this phenomenon when looking through window glass: during the daytime, we will see most clearly outside when looking at the window at near-normal angles, while we see a reflection at glancing angles.

When $n_2/n_1 > 1$, there is also a critical incident angle, θ_{max}, beyond which Equation (4.32) predicts $\chi > \pi/2$. As shown in Figure 4.7(b), under these conditions, all incident radiation is reflected into the original medium in a specular manner. This effect, called *total internal reflection*, is the principle that underlies light transmission along fiber optic cables, since light reflected at this angle can travel great distances without significant attenuation. The light propagates through a fiber core that has a larger refractive index than the surrounding sheath.

The reflectivity for unpolarized incident radiation is given by Fresnel's equation:

$$\rho_{\lambda}\left(\theta_i\right) = \frac{1}{2}\frac{\sin^2(\theta_i - \chi)}{\sin^2(\theta_i + \chi)}\left[1 + \frac{\cos^2(\theta_i + \chi)}{\cos^2(\theta_i - \chi)}\right] \tag{4.37}$$

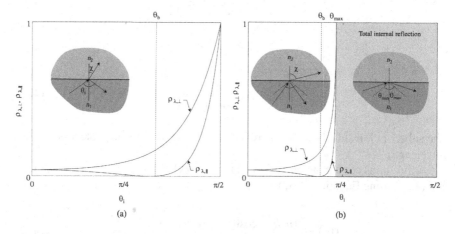

FIGURE 4.7 Spectral reflectance at the interface between two dielectrics: (a) $n_2/n_1 > 1$ and (b) $n_2/n_1 < 1$.

TABLE 4.1
Optical Property Values and Normal Spectral Reflectivity of Various Dielectric Materials at T = 300 K and λ = 0.589 μm in a Vacuum

Material	n (Refractive Index)	$\varepsilon = \varepsilon/\varepsilon o$ (Dielectric constant)	$\rho_{\lambda,n}$ Equation(4.38)
SiO$_2$ (glass)	1.458	4.42	0.035
SiO$_2$ (fused quartz)	1.544	3.75	0.046
NaCl	1.5441	5.90	0.046
KCl	1.4902	4.86	0.039
H$_2$O (liquid)	1.332	1.77	0.020
H$_2$O (ice, 0°C)	1.309	1.71	0.018
Vacuum	1	1	0.000

Source: Lide, 2008.

The average of Equations (4.35) and (4.36) is often used to obtain the angular reflectivity profile $\rho_\lambda(\theta_i)$ in terms of θ_i only. For normal incidence, $\theta_i = 0$ and the reflectivity becomes

$$\rho_{\lambda,n} = \rho_\lambda\left(\theta_i = 0\right) = \rho_{\lambda,\parallel}\left(0\right) = \rho_{\lambda,\perp}\left(0\right) = \left(\frac{n_2 - n_1}{n_2 + n_1}\right)^2 = \left[\frac{(n_2/n_1)-1}{(n_2/n_1)+1}\right]^2 \quad (4.38)$$

Electromagnetic theory requires the interface to be perfectly clean and optically smooth, which is the case for lenses, windows, and other optical components, and these can be accurately modeled using the Fresnel equations for most wavelengths of interest. Predictions of the normal reflectivity from EM theory for various dielectrics are listed in Table 4.1. These relations assume that the material is infinitely thick (i.e., opaque), so that there is no reflection from a second internal surface within medium 2. Nonopaque layers (e.g., windows and glazing) are considered in Chapter 11.

Note that the spectral reflectance of the interface increases with the refractive index, specifically relative to that of a vacuum ($n = 1$). Equations (4.32)–(4.38) show that the reflectance depends only on the ratio of the refractive indices between the two media, n_2/n_1, and the reflectance increases with this ratio. Conceptually, this can be explained by the "phase difference" undergone by the wave as it crosses the medium between the two materials. As this difference increases, a larger fraction of the light incident on the interface is reflected backward into the originating medium. Conversely, when $n_2/n_1 = 1$, there is no phase difference, and the reflectivity is equal to zero. In this scenario, the interface becomes invisible.

Example 4.1

Unpolarized radiation is incident on a dielectric surface (medium 2) in air (medium 1) at angle $\theta = 30°$ from the normal. The dielectric material has a refractive index of $n_2 = 3.0$. Find the reflectivity of each polarized component and of the unpolarized incident radiation.

Because the incident intensity is in air, $n_1 \approx 1$, and from Equation (4.32), $n_1/n_2 = 1/3.0 = \sin\chi/\sin 30°$; therefore, $\chi = 9.6°$. The reflectivity of the parallel component is, from Equation (4.33), $\rho_{\lambda,\parallel}(\theta_i = 30°) = (\tan 20.4°/\tan 39.6°)^2 = 0.202$. and that of the perpendicular component is, from Equation (4.34), $\rho_{\lambda,\perp}(\theta_i = 30°) = (\sin 20.4°/\sin 39.6°)^2 = 0.301$. The reflectivity for the unpolarized incident intensity, obtained from Equation (4.37) or from the average of the components, is $\rho_\lambda(\theta_i = 30°) = (0.202 + 0.301)/2 = 0.252$.

Example 4.2

What fraction of light is reflected for normal incidence in air on a glass surface or on the surface of water?

For glass in the visible spectrum, $n \approx 1.55$, and for water $n \approx 1.33$. Then, from Equation (4.38),

$$\rho_{\lambda,n}(\text{glass}) = \left(\frac{n-1}{n+1}\right)^2 = \left(\frac{0.55}{2.55}\right)^2 = 0.047 \ (<5\%)$$

$$\rho_{\lambda,n}(\text{water}) = \left(\frac{0.33}{2.33}\right)^2 = 0.020 \ (<2\%)$$

Note that these results are only for the given portion of the spectrum.

4.4.1.2   Emissivity

The directional emissivity is related to directional reflectivity by Kirchhoff's Law (Equation 3.18). For an opaque medium,

$$\varepsilon_\lambda(\theta_i) = 1 - \frac{\rho_{\lambda,\parallel}(\theta_i) + \rho_{\lambda,\perp}(\theta_i)}{2} \tag{4.39}$$

At $\theta_i = 0$, the normal spectral emissivity is given by:

$$\varepsilon_{\lambda,n} = 1 - \rho_{\lambda,n} = 1 - \left(\frac{n-1}{n+1}\right)^2 = \frac{4n}{(n+1)^2} \quad \left(n = \frac{n_2}{n_1} > 1\right) \tag{4.40}$$

Except for some specially tailored surfaces, emission is unpolarized. As is the case for directional reflectivity, directional emissivity also depends on the index of refraction of the emitting body and the surrounding medium. Figure 4.8 shows curves for various n_2/n_1, where $n_2 > n_1$. When $\rho(\theta)$ is computed for incident radiation in air $(n_1 \approx 1)$, the ratio n_2/n_1 reduces to the refractive index of the material on which the radiation is incident. For $(n_2/n_1 = 1)$, all radiation passes through the interface and

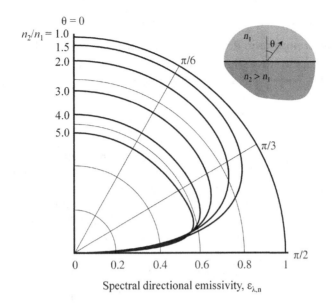

FIGURE 4.8   Directional emissivity of a dielectric predicted using EM theory.

the effective emissivity is one (as for a blackbody), and the curve in Figure 4.8 is semicircular with a radius of one. Again, the interface becomes invisible if the optical properties of the media on either side of the interface are identical.

As (n_2/n_1) increases, the curves remain approximately circular up to about $\theta = 70°$ and then decrease rapidly to zero at $\theta = 90°$. Thus, dielectric materials are good emitters except at large angles from the normal direction. Figure 4.8 also shows that the overall magnitude of emissivity drops as n_2/n_1 becomes larger. This result can be intuited from Kirchhoff's Law, because increasingly dissimilar refractive indices lead to increasing reflectivity at the interface.

Hemispherical-spectral emissivity can be computed from the directional-spectral emissivity using Equation (3.7) and expressed in terms of n_2/n_1:

$$\varepsilon_\lambda = \frac{1}{2} - \frac{(3n+1)(n-1)}{6(n+1)^2} - \frac{n^2(n^2-1)^2}{(n^2+1)^3}\ln\left(\frac{n-1}{n+1}\right) + \frac{2n^3(n^2+2n-1)}{(n^2+1)(n^4-1)}$$

$$- \frac{8n^4(n^4+1)}{(n^2+1)(n^4-1)^2}\ln n \qquad \left[n = \left(\frac{n_2}{n_1}\right) > 1\right]$$

(4.41)

Example 4.3

A dielectric medium has $n = 1.41$. What is its hemispherical emissivity into air at the wavelength at which n was measured?

From Equation (4.41), for $n_1/n_2 = n = 1.41$, $\varepsilon_\lambda = 0.92$.

4.4.2 RADIATIVE PROPERTIES OF CONDUCTORS

4.4.2.1 Electromagnetic Relations for Incidence on an Absorbing Medium

In the case of an EM wave within a dielectric medium (n_1) incident upon a conducting medium ($\bar{n}_2 = n_2 - ik_2$), Snell's Law becomes:

$$\frac{\sin \chi}{\sin \theta_i} = \frac{n_1}{\bar{n}_2} = \frac{n_1}{n_2 - ik_2} \tag{4.42}$$

Consequently, $\sin \chi$ is complex, and χ can no longer be interpreted as the angle at which the wave propagates into the material. Furthermore, except for normal incidence, n_2 is not directly related to the wave propagation velocity.

The equations for the reflected components of incident radiation are more complicated than the case of two dielectrics. Hering and Smith (1968) give an approximate form *for incidence from air or vacuum* ($n_1 = 1$, $k_1 = 0$, $n_2 = n$, $k_2 = k$):

$$\rho_{\lambda,\perp}(\theta_i) = \frac{(n\beta - \cos\theta_i)^2 + \left(n^2 + k^2\right)\alpha - n^2\beta^2}{(n\beta + \cos\theta_i)^2 + \left(n^2 + k^2\right)\alpha - n^2\beta^2} \tag{4.43}$$

and

$$\rho_{\lambda,\parallel}(\theta_i) = \frac{(n\gamma - \alpha/\cos\theta_i)^2 + \left(n^2 + k^2\right)\alpha - n^2\gamma^2}{(n\gamma + \alpha/\cos\theta_i)^2 + \left(n^2 + k^2\right)\alpha - n^2\gamma^2} \tag{4.44}$$

where

$$\alpha^2 = \left(1 + \frac{\sin^2\theta_i}{n^2 + k^2}\right) - \frac{4n^2}{n^2 + k^2}\left(\frac{\sin^2\theta_i}{n^2 + k^2}\right) \tag{4.45}$$

$$\beta^2 = \frac{n^2 + k^2}{2n^2}\left(\frac{n^2 - k^2}{n^2 + k^2} - \frac{\sin^2\theta_i}{n^2 + k^2} + \alpha\right) \tag{4.46}$$

and

$$\gamma = \frac{n^2 - k^2}{n^2 + k^2}\beta + \frac{2nk}{n^2 + k^2}\left(\frac{n^2 + k^2}{n^2}\alpha - \beta\right)^{1/2} \tag{4.47}$$

The reflectivity at normal incidence reduces to:

$$\rho_{\lambda,n} = \rho_{\lambda,\parallel}(0) = \rho_{\lambda,\perp}(0) = \left[\frac{n - ik - 1}{n - ik + 1}\right]^2 = \frac{(n-1)^2 + k^2}{(n+1)^2 + k^2} \tag{4.48}$$

The directional emissivity can be found from Kirchhoff's Law, $\varepsilon_\lambda(\theta) = 1 - [\rho_{\lambda,\perp}(\theta) + \rho_{\lambda,\parallel}(\theta)]/2$, which can then be integrated over the hemisphere to obtain the hemispherical-spectral emissivity, although the complexity of Equations (4.43)–(4.47) preclude a convenient closed-form solution.

These relations may be used in cases where n and k are not large, but k is greater than zero. This scenario occurs for many semiconductors and imperfect insulators ("lossy" dielectrics) at frequencies close to ω_0, as shown in Figure 4.5(a).

4.4.2.2 Reflectivity and Emissivity Relations for Metals (Large k)

As discussed in Sections 4.3.2 and 4.4.2.1, materials having a large σ_e (or small resistivity, $r_e = 1/\sigma_e$) are internally highly absorbing; this means that the extinction coefficient k is sufficiently large to simplify Equations (4.43) through (4.47) into more convenient results. For large k, the $\sin^2\theta$ terms in Equations (4.43) through (4.47) can be neglected relative to $n^2 + k^2$. Then $\alpha \approx \beta \approx \gamma \approx 1$ and Equations (4.43) and (4.44) become:

$$\rho_{\lambda,\parallel}(\theta_i) = \frac{(n\cos\theta_i - 1)^2 + (k\cos\theta_i)^2}{(n\cos\theta_i + 1)^2 + (k\cos\theta_i)^2} \tag{4.49}$$

and

$$\rho_{\lambda,\perp}(\theta_i) = \frac{(n-\cos\theta_i)^2 + k^2}{(n+\cos\theta_i)^2 + k^2} \tag{4.50}$$

These components are plotted in Figure 4.9(a) for a typical metal, which reveals several key differences compared to the spectral directional reflectance at the interface between two dielectrics, Figure 4.7. First, the spectral reflectance of metals is much higher than those of dielectrics, particularly at near-normal angles. The reflectance

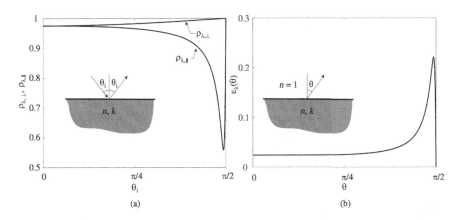

FIGURE 4.9 (a) Spectral directional reflectivity and (b) spectral emissivity at the interface between a metal and a vacuum ($n = 13$, $k = 44$).

is also very large at glancing angles and becomes unity at $\theta = \pi/2$. Finally, while the parallel component of spectral directional reflectance dips at intermediate angles, there is no angle at which it reaches zero.

For unpolarized incident radiation, $\rho(\theta_i) = [\rho_\parallel(\theta_i) + \rho_\perp(\theta_i)]/2$, and for the normal direction ρ_n reduces to Equation (4.48). Emission is unpolarized, so the directional emissivity is given by Equation (4.39), and, in the normal direction:

$$\varepsilon_{\lambda,n} = 1 - \rho_{\lambda,n} = \frac{4n}{(n+1)^2 + k^2} \tag{4.51}$$

The directional-spectral emissivity for a polished metal surface is plotted in Figure 4.9(b). The spectral emissivity is small at most angles, and gradually increases to a maximum value at a glancing angle, before dropping to zero at $\theta = \pi/2$.

Normal spectral reflectivities calculated using Equation (4.51) are shown in Table 4.2. These values agree with measured $\varepsilon_{\lambda,n}$ at $\lambda = 8$ μm (0.1 eV) but are in

---

TABLE 4.2
Optical Properties and Normal Spectral Reflectivity of Metals at $T = 298$ K and $\lambda = 0.484$ μm and 8.06 μm

Material (electrical resistivity, r_e [×10⁶ (Ω·cm)])	Wavelength, λ (μm)	Refractive index, n^1	Absorption index, k^1	Normal Spectral Reflectivity, $\rho_{\lambda,n}$ Eq.(4.48)	Eq.(4.51)
Aluminum	0.484	0.826	6.283	0.923	0.883
(2.709)	8.06	34.464	105.6	0.989	0.970
Chromium	0.484	2.75	4.46	0.676	0.826
(12.6)	8.06	11.81	29.76	0.955	0.955
Copper	0.484	1.12	2.60	0.602	0.933
(1.712)	8.06	29.69	71.57	0.980	0.983
Gold	0.484	0.50	1.86	0.650	0.924
(2.255)	8.06	8.17	82.83	0.995	0.981
Iron	0.484	2.56	3.31	0.567	0.845
(9.87)	8.06	6.41	33.07	0.978	0.960
Nickel	0.484	1.71	3.06	0.591	0.867
(7.12)	8.06	9.54	45.82	0.983	0.966
Platinum	0.484	2.03	3.54	0.626	0.839
(10.7)	8.06	13.21	44.72	0.976	0.959
Titanium	0.484	1.81	2.47	0.483	0.71
(39)	8.06	5.03	23.38	0.965	0.92
Tungsten	0.484	3.45	2.72	0.493	0.883
(5.39)	8.06	14.06	54.7	0.983	0.970

Sources: [1] Rumble, J. R., (ed.) Handbook of Chemistry and Physics, 100th ed., CRC Press, Boca Raton, 2019.
[2] Measured data is the largest reflectivity value from references cited in Touloukian (1970)

greater error at $\lambda = 0.484$ μm (2.4 eV). For the cases of poor agreement, it is difficult to ascribe the error specifically to the optical constants, to the measured emissivity, or to the theory itself, although the theory is expected to be less accurate at shorter wavelengths. The optical constants also may be in error, or the experimental samples do not meet the standards of perfection in surface preparation demanded by the theory.

Within the approximation of neglecting $\sin^2\theta_i$ relative to $n^2 + k^2$, the hemispherical-spectral emissivity for a metal in air or vacuum is found by substituting the directional-spectral emissivity found from Equations (4.49) and (4.50) into Equation (4.39). Integrating gives:

$$
\begin{aligned}
\varepsilon_\lambda = 4n - 4n^2\ln\left(\frac{1+2n+n^2+k^2}{n^2+k^2}\right) + \frac{4n(n^2-k^2)}{k}\tan^{-1}\left(\frac{k}{n+n^2+k^2}\right) + \frac{4n}{n^2+k^2} \\
- \frac{4n^2}{(n^2+k^2)^2}\ln\left(1+2n+n^2+k^2\right) - \frac{4n(k^2-n^2)}{k(n^2+k^2)^2}\tan^{-1}\left(\frac{k}{1+n}\right)
\end{aligned}
$$

$$\tag{4.52}$$

Note that, per Figure 4.7(a), the hemispherical-spectral emissivity of a metal is always larger than the normal value. Finally, it is important to remember that these results apply only to optically smooth surfaces (e.g., highly polished metals.) Real-world effects, like surface roughness and surface contamination, can cause both the magnitude and directional distribution of radiative properties to depart significantly from EM theory predictions.

4.4.2.3 Relations between Radiative Emission and Electrical Properties

We showed in Section 4.2.2 that the solutions to Maxwell's wave equations provide a means for determining n and k from the electric and magnetic properties of a material, per Equations (4.11) and (4.12). For highly conducting metals ($r_e = 1/\sigma_e$ is small) and for relatively long wavelengths, these equations simplify to:

$$
n \approx k \approx \sqrt{\frac{\lambda_0\mu_0 c_0}{4\pi r_e}} = \sqrt{\frac{0.003\lambda_0}{r_e}} \quad \left(\lambda_0 \text{ in } \mu m, \, r_e \text{ in } \Omega \cdot cm\right) \tag{4.53}
$$

where the magnetic permeability of most metals can be taken to be the vacuum value. This is the *Hagen–Rubens equation* (Hagen and Rubens 1900). This equation can also be derived from Drude theory. At long wavelengths, $\omega \ll \zeta$ and the contribution of core electrons, ϵ_0, is negligible. Accordingly, we can show that $\epsilon_I \approx 0$ and $\epsilon_{II} \approx \omega_p^2/(\omega\zeta)$. Substituting the terms for the plasma frequency and damping constant, and after some manipulation, one arrives at Equation (4.53). Given these simplifications, we only expect Hagen–Rubens theory to apply at longer wavelengths, and, indeed, it is typically only valid for $\lambda_0 > \sim 5$ μm. Predictions of n and k from this equation can be greatly in error, particularly at short wavelengths.

With the conditions in Equation (4.53) that $n \approx k$, Equation (4.51) reduces to:

$$\varepsilon_{\lambda,n} = 1 - \rho_{\lambda,n} = \frac{4n}{2n^2 + 2n + 1} \tag{4.54}$$

for a material with refractive index n radiating in the normal direction into air or a vacuum. Substituting Equation (4.53) into Equation (4.54) and expanding into a series gives the *Hagen–Rubens emissivity relation*:

$$\varepsilon_{\lambda,n} = \frac{2}{\sqrt{0.003}} \left(\frac{r_e}{\lambda_0} \right)^{1/2} - \frac{2}{0.003} \frac{r_e}{\lambda_0} + \cdots \tag{4.55}$$

Because the index of refraction of metals as predicted from Equation (4.53) is generally large at longer wavelengths so that (r_e/λ_0) is small, that is, for $\lambda_0 > \sim 5$ μm (see Table 4.2), one or two terms of the series is usually adequate. A two-term approximation is made by adjusting the coefficients of the second term to account for the remaining terms in the series:

$$\varepsilon_{\lambda,n} = 36.5 \left(\frac{r_e}{\lambda_0} \right)^{1/2} - 464 \frac{r_e}{\lambda_0} \tag{4.56}$$

Equation (4.56) affirms the key trends in spectral emissivity expected from classical Drude theory. The EM wave principally interacts with free electrons in the metal, which move due to Columbic forces imposed by the oscillating electric field. These electrons collide with ions (and other structures), and transfer some of their kinetic energy imparted by the EM wave to the internal energy of the metal. The frequency with which electrons scatter against ions in the metal is summarized by the electrical resistivity, so metals having a high electrical resistivity can absorb, and hence, emit, more radiation compared to those having a lower resistivity. A perfectly conducting material would have a reflectivity of unity and an emissivity of zero.

The normal spectral emissivity in Equation (4.56) can be integrated with respect to wavelength to yield a normal total emissivity, ε_n, per Equation (3.5). Equation (4.56) is accurate only for $\lambda_0 > \sim 5$ μm, so in performing the integration starting from a small value of λ (e.g., in the visible spectrum) assumes that the energy radiated at wavelengths shorter than 5 μm is small compared with that at wavelengths longer than 5 μm. Then, substituting into the integral, the first term of Equation (4.55) and Equation (2.20) for $I_{\lambda b}$ provides:

$$\varepsilon_n(T) \approx \pi \frac{\displaystyle\int_0^\infty 2 \left(\frac{r_e}{0.003\lambda_0} \right)^{1/2} 2C_1 \bigg/ \left[\lambda_0^5 (e^{C_2/\lambda_0 T} - 1) \right] d\lambda_0}{\sigma T^4} \tag{4.57}$$

$$= \frac{4\pi C_1 (T r_e)^{1/2}}{(0.003)^{1/2} \sigma C_2^{4.5}} \int_0^\infty \frac{\zeta^{3.5}}{e^\zeta - 1} d\zeta$$

where $\zeta = C_2/\lambda_0 T$ as was used in conjunction with Equation (2.38). The integration gives:

$$\varepsilon_n(T) \approx \frac{4\pi C_1 (Tr_e)^{1/2}}{(0.003)^{1/2} \sigma C_2^{4.5}} (12.27) = 0.575(r_e T)^{1/2} \qquad (4.58)$$

For pure metals, r_e near room temperature is approximately:

$$r_e \approx r_{e,273} \frac{T}{273} \qquad (4.59)$$

where $r_{e,273}$ is the electrical resistivity at 273 K. Substituting Equation (4.59) into Equation (4.58) gives the approximate result:

$$\varepsilon_n(T) \approx 0.0348T \sqrt{r_{e,273}} \qquad (4.60)$$

This equation predicts that the total normal emissivity of pure metals increases directly with absolute temperature, but often it applies only to temperatures below about 550 K. Again, in the context of classical Drude theory, increasing the thermal motion of the ions with increasing temperature causes more electron scattering, and more opportunities to transfer the wave energy into the internal energy of the metal. Hence, the absorptivity and emissivity of metals are expected to increase with temperature. This is demonstrated for polished nickel in Figure 4.10.

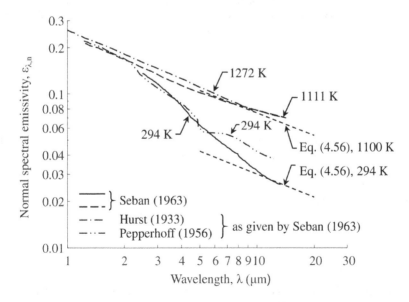

FIGURE 4.10 Comparison of measured spectral emissivity with Hagen–Rubens theory for normal spectral emissivity of polished nickel. Hagen–Rubens predictions are calculated using the room temperature resistivity of nickel in Table 4.2.

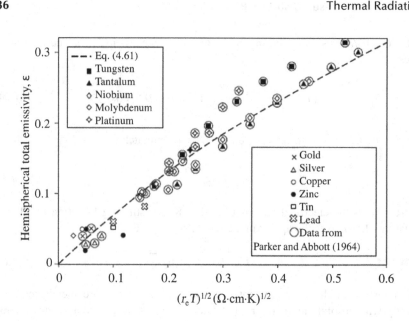

FIGURE 4.11 Comparison of measured total emissivity of various metals with predictions from Hagen–Rubens theory, Equation (4.61). This plot is of the hemispherical total emissivity of a large number of metals versus a parameter from the Hagen–Rubens equation.

The spectral directional emissivity found from the refractive index, Equation (4.39) and the Fresnel relations, Equations (4.49) and (4.50), can be integrated over all directions to obtain total hemispherical quantities. The following approximate equations for the hemispherical total emissivity fit the results in two ranges:

$$\varepsilon(T) = \begin{cases} 0.751\sqrt{r_e T} - 0.396 r_e T, & 0 < r_e T < 0.2 \\ 0.698\sqrt{r_e T} - 0.266 r_e T, & 0.2 < r_e T < 0.5 \end{cases} \tag{4.61}$$

where T and r_e are in units of K and Ω-cm, respectively. Per Equation (4.59), the resistivity r_e depends approximately on T to the first power so that the first term of each of these equations provides a T proportionality, indicating that the temperature dependence for energy emission by metals is higher than the blackbody dependence of T^4. Figure 4.11 shows predictions from Equation (4.61) compared with measured values for 11 different metals.

Example 4.4

A polished platinum surface is maintained at $T = 250$ K. Energy is incident upon the surface from a black enclosure at $T_i = 500$ K that surrounds the surface. What is the hemispherical-directional total reflectivity into the direction normal to the surface?

The directional-hemispherical total reflectivity for the normal direction can be found using:

$$\rho_n(T = 250\,\text{K}) = 1 - \alpha_n(T = 250\,\text{K})$$

where $\alpha_n(T = 250\text{ K})$ is the normal total absorptivity of a surface at 250 K for incident black radiation at 500 K:

$$\alpha_n(T = 250\,\text{K}) = \frac{\int_0^\infty \alpha_{\lambda,n}(T = 250\,\text{K}) I_{\lambda,b}(500\,\text{K}) d\lambda}{\int_0^\infty I_{\lambda,b}(500\,\text{K}) d\lambda}$$

For spectral quantities, $\alpha_{\lambda,n}(T = 250\text{ K}) = \varepsilon_{\lambda,n}(T = 250\text{ K})$. From Equation (4.59) the near-linear variation of r_e with temperature provides the approximate emissivity variation $\varepsilon_{\lambda,n}(T) \propto T^{1/2}$. Then $\varepsilon_{\lambda,n}(T = 250\text{ K}) = \varepsilon_{\lambda,n}(T = 500\text{ K})(250/500)^{1/2}$, and we obtain:

$$\alpha_n(T = 250\,K) = \frac{\sqrt{1/2}\int_0^\infty \varepsilon_{\lambda,n}(T = 250\,K) I_{\lambda,b}(500\,K) d\lambda}{\int_0^\infty I_{\lambda,b}(500\,K) d\lambda} = \frac{\varepsilon_n(T = 500\,K)}{\sqrt{2}}$$

where the last equality is obtained by examining the emissivity definition, Equation (3.5). The normal total emissivity of platinum at 500 K is given by Equation (4.60) as:

$$\varepsilon_n(T = 500\,K) = 0.348\sqrt{r_{e,273}T} = 0.348\sqrt{r_{e,293}}\sqrt{\frac{273}{293}}T$$

$$= 0.348\sqrt{10 \times 10^{-6}}\sqrt{\frac{273}{293}} \times 500 = 0.053$$

Equation (4.60) should be used only when temperatures are such that most of the emitted radiation is at wavelengths greater than 5 μm. From the blackbody functions, for $T = 500$ K about 10% of the energy is below $\lambda = 5$ μm, so a small error is introduced. The reciprocity relation of Equation (3.59) for uniform incident intensity can now be employed to give the desired result for the hemispherical-directional total reflectivity:

$$\rho_n(T = 250\,K) = 1 - \alpha_n(T = 250\,K) \approx 1 - \frac{1}{\sqrt{2}}\varepsilon_n(T = 500\,K)$$

$$= 1 - \frac{0.053}{\sqrt{2}} = 0.963$$

4.5   RADIATIVE PROPERTIES OF REAL SURFACES

The results discussed in the previous section apply to pure materials having perfectly smooth and uncontaminated interfaces. These conditions are approximately satisfied, e.g., for lenses and optical components. Most surfaces encountered in engineering practice deviate from this ideal, and their radiative properties will differ significantly from EM predictions.

4.5.1   SURFACE HETEROGENEITY AND SURFACE COATINGS

The EM predictions derived in the previous section apply to the interface between two homogeneous materials. In many practical situations, the surface may be heterogeneous, which can affect the surface properties profoundly. As an example, in the visible wavelengths, solid ice behaves as a dielectric having a refractive index of approximately 1.3, so it is nearly transparent at normal incidence (e.g., on sidewalks, much to the chagrin of pedestrians) with a normal reflectance of 0.02 according to Equation (4.38). On the other hand, freshly fallen snow has a reflectivity of approximately 0.85 or greater (to which anyone who has received a sunburn while skiing can attest.)

As shown schematically in Figure 4.12, this is because incident light waves undergo multiple scattering as it crosses the interface between the ice crystals and the air. Consequently, most incident rays are eventually reflected out of the snow into all directions.

The "snow effect" also applies to paper, fabric, chalk, ceramics, as well as some pigments and coatings: in all these scenarios, the heterogeneous nature of the material causes the reflectance to be large and the surface to be nearly diffuse. White paint often consists of dielectric particles within a dielectric binder (e.g., TiO_2 particles in a polymer resin). This paint has a high reflectivity over the visible spectrum due to the snow effect, and consequently is often applied to roofs to lower solar heating loads. Black paint, on the other hand, is made of highly absorbing particles like carbon blacks, and has a low reflectivity over all wavelengths.

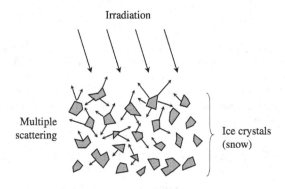

FIGURE 4.12   Heterogeneities in a dielectric can increase the reflectivity dramatically and make the surface diffuse.

While scattering from heterogeneities within a surface usually leads to diffuse reflection, this is not always the case. When the Moon is full in phase, the Sun, Earth, and Moon are nearly coincident, and the Moon is uniformly irradiated with nearly collimated solar radiation, as shown in Figure 4.13. Were the lunar surface diffuse, the reflected light would follow a Lambertian cosine distribution, and the Moon would appear brightest at its center and darker near its edges. On the contrary, however, we see the Moon as a white disk of nearly uniform intensity.

The reason for this observation can be traced to the bidirectional reflectivity measurements plotted in Figure 4.14 that increase approximately proportionally to $1/\cos \theta_i$ (shown by the dashed line in Figure 4.14). This compensates for the reduced energy incident per unit area at large angles, which is proportional to $\cos(\theta_i)$ according to the Lambertian cosine rule. The figure shows a strong back-scattering component, located at $\phi = 180°$ away from the direction of specular reflectance (Orlova 1956). This behavior, called the "opposition effect," is thought to arise from coherent backscattered light from particles and shadowing effects (Helfenstein et al. 1997.)

Dielectrics often exist as coatings on a substrate. A thin layer of varnish on an oil painting produces an optically smooth interface, and greatly enriches the colors. A particularly important case concerns the oxidation of metallic surfaces. Most metals in their "as-received" state have a thin oxide layer several nanometers in thickness, which passivates the surface. Because the oxide is a dielectric, this coating is effectively transparent, and the radiative properties of the surface are dominated by the

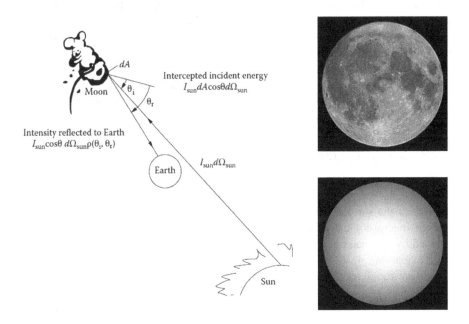

FIGURE 4.13 Were the Moon to reflect incident sunlight in a diffuse manner, it would appear darker around the edges. Instead, the Moon appears as a uniformly bright disk.

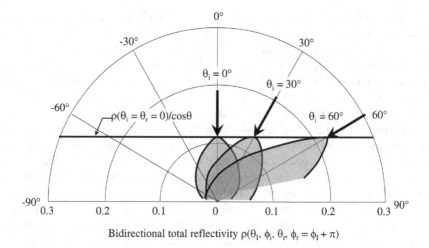

FIGURE 4.14 Bidirectional total reflectivity in the incidence plane for the mountainous regions of the lunar surface. After Orlova, N. S., *Astron. Z.* 33(1), 93, 1956.

metal substrate. If the metal surface is heated, however, the oxide layer will rapidly become thicker, and the radiative properties become dominated by those of the oxide.

In some cases, an oxide layer produces a visible "hump" or "dip" in the measured spectral reflectance or emittance with respect to wavelength. Incident waves undergo a wavelength shift as they cross the air/oxide boundary. Because the oxide behaves as a dielectric, the incident waves traverse the oxide layer without attenuation until they are reflected from the metal substrate, back through the oxide layer, and back into the air. The outgoing waves have undergone a wavelength shift that depends on both the refractive index and the thickness of the oxide layer; these shifted outgoing waves constructively and destructively interfere with the incident EM field to produce patterns in the spectral emittance or spectral reflectance. This effect is discussed in more detail in Chapter 11.

4.5.2 SURFACE ROUGHNESS

Surface topography has a profound and varied impact on the radiative properties of metals and nonmetals, depending on the bulk refractive index of the surface and the size of the surface features relative to wavelength. While surfaces may appear smooth macroscopically, microscopically they consist of asperities that may be induced by the manufacturing process, mechanical damage, or a chemical reaction (e.g., oxidation).

Surface roughness is often expressed in terms of the root-mean-square (RMS) roughness,

$$\sigma_0 = \sqrt{\left\langle \left(z - z_m \right)^2 \right\rangle} \tag{4.62}$$

where z_m is the mean surface height and $< >$ denotes the average, as shown in Figure 4.15. The roughness is readily measured using a contact profilometer, consisting of a stylus or needle that is dragged across a surface, or with an optical profilometer or digital microscope. This parameter is then used to derive the optical roughness, σ_0/λ. At small values of optical roughness, the surface may appear specular or "mirror-like," while at larger values the surface will be more diffuse. It is difficult or impossible to see your reflection in a wooden countertop or a whiteboard with the naked eye, but if you looked at one with an infrared camera, you would see a clear reflection of your image at infrared wavelengths due to the thermal radiation emitted by your body. This is because these surfaces are optically rough at visible wavelengths, but optically smooth at infrared wavelengths.

Generally, when σ_0/λ is greater than about unity, incident waves undergo multiple reflections between roughness elements; this increases the trapping of incident radiation, thereby increasing the observed surface absorptivity and consequently the emissivity. In this wavelength regime, the radiative properties of a surface can be understood through "geometric optics," in which incident rays are ray-traced through multiple interactions with the surface; at each interaction, the wave is reflected in a specular manner, and a portion of the wave energy is absorbed. At longer wavelengths and shallower glancing angles, the interactions between the EM wave and surface features become dominated by diffraction effects which are difficult to calculate.

Surface roughness may be caused by the manufacturing process or chemical reactions like oxidation. Figure 4.16 shows the hemispherical-spectral reflectivity of a cold-rolled steel alloy. At short wavelengths, the spectral reflectivity is lower due to the surface roughness, while at very long wavelengths the surface becomes optically smooth, and the experimentally measured spectral reflectance converges to the value predicted using the refractive indices of steel and Equation (4.51). Figure 4.16 also shows the spectral reflectance of a specimen that has been heated in a reducing atmosphere; the strong interference effect indicates the presence of an oxide layer.

An interesting case is how the radiative properties of aluminized steels change as they are heated in a furnace. Aluminized steels are widely used for manufacturing automotive

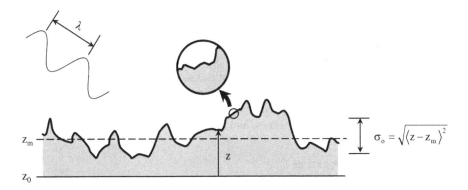

FIGURE 4.15 Definition of RMS roughness.

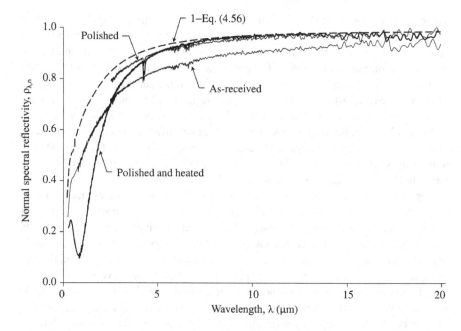

FIGURE 4.16 Spectral reflectivity of a cold-rolled steel alloy. A thin-film interference effect caused by the oxide layer is visible at very short wavelengths. The reflectivity gradually increases as the surface becomes optically smoother at longer wavelengths.

parts through "hot stamping," a process through which steel sheets are heated to approximately 950°C in a furnace and then transferred to a cooled die where they are simultaneously formed and quenched into shape. To avoid the need for atmosphere control within the furnace, the steel sheets are often equipped with a metallic 90% Al/10% Si coating that prevents the steel from oxidizing. The coating melts at about 600°C, and then reacts with iron from the substrate steel to form an intermetallic layer, which thickens and roughens with subsequent heating. The electrons in the intermetallic Al–Si–Fe layer are localized in bonds, so these materials act more like insulators (Section 4.4.1) and therefore have a larger emissivity than the as-received metallic coating.

Figure 4.17 shows the spectral emissivity of an aluminized steel sheet as it is heated in air. The spectral emissivity of the steel in its as-received state is the same as roughened aluminum; the peak at around 3 s is due to surface roughness, and the surface is optically smooth at long wavelengths. Once the coating melts, the surface has the properties of a very smooth metal; the peak associated with the roughness disappears. The spectral emissivity increases as intermetallic forms at progressively higher temperatures and heating times, and the surface becomes rougher.

4.5.3 SURFACE TEMPERATURE

4.5.3.1 Nonmetals

Theoretical predictions of the effect of temperature on the radiative properties of insulators are varied. In general, the index of refraction for non-conductors is weakly

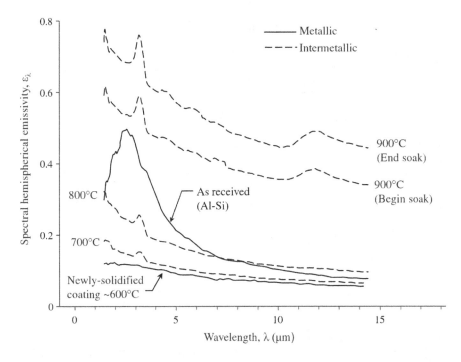

FIGURE 4.17 Spectral emissivity of aluminized steel as it is heated in a furnace. The steel initially has the radiative properties similar to solid aluminum. The emissivity is lowest when the coating melts, and then increases as the coating transforms from metallic into intermetallic.

dependent on temperature; consequently, ε_λ is insensitive to temperature, and variation of total emissivity with temperature is dominated by how the spectral shift in $E_{\lambda b}$ influences the relative weighting of ε_λ. This provides the very useful result that the ε_λ measured at one temperature can be used in the integral to calculate accurate ε_λ values over a range of nearby temperatures.

4.5.3.2 Metals

In Section 4.4.2, it was shown that the normal spectral emissivity of metals tends to decrease as wavelength increases in the IR region. Figure 4.18 shows the normal spectral emissivity for several polished metals at high temperatures. At sufficiently long wavelengths $\lambda > \sim 5$ μm, Hagen–Rubens theory predicts that the normal spectral emissivity should be proportional to $\lambda^{-1/2}$, which is the case for the data for iron, platinum, nickel, and molybdenum plotted in Figure 4.18. The curve for the copper sample illustrates an exception, as the emissivity remains relatively constant with wavelength. At very short wavelengths, the assumptions underlying Hagen–Rubens theory become invalid. Most metals exhibit a peak emissivity somewhere near the visible region, and the emissivity decreases rapidly with further increases in wavelength. This behavior is shown in Figure 4.19 for tungsten. Note that the spectral emissivity at 2 μm increases with respect to $T^{1/2}$ as predicted from Hagen–Rubens theory, even though this wavelength is shorter than the domain of applicability. Also note that the spectral emissivity becomes insensitive to temperature at a critical wavelength called

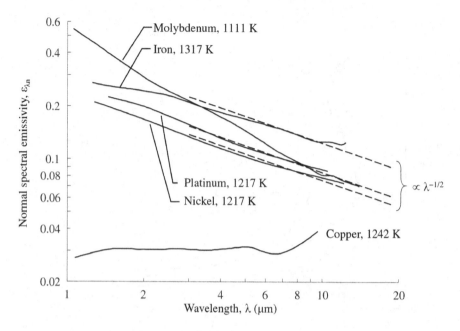

FIGURE 4.18 Variation with the wavelength of normal spectral emissivity for polished metals. (From Seban, R. A., *Thermal Radiation Properties of Materials*, pt. III, WADD-TR-60-370, University of California, Berkeley, CA, August 1963.)

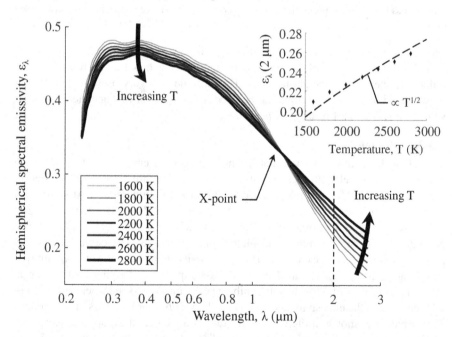

FIGURE 4.19 Effect of wavelength and surface temperature on hemispherical-spectral emissivity of tungsten. (From De Vos, J. C., *Physica*, 20, 690, 1954.)

the "X-point," which, for most metals, is in the near-infrared. At wavelengths shorter than the X-point, spectral emissivity decreases with increasing temperature.

While EM theory predicts how the emissivity of a metal surface varies with changes in wavelength and temperature, quantitative agreement between these values and experimental measurements should only be expected for polished specimens, and otherwise should be considered as a lower bound due to surface roughness. It is also important to consider the possibility of other temperature-induced changes in the surface state, such as oxidation.

4.6   SELECTIVE SURFACES FOR SOLAR APPLICATIONS

The radiative properties of surfaces are often engineered to increase or decrease their natural ability to absorb, emit, or reflect radiant energy. This is particularly so for solar-related applications: for solar thermal applications, e.g., solar hot water and air heaters, the goal is often to maximize the fraction of solar energy absorbed at short wavelengths and minimize emission losses at longer wavelengths. In other cases, it is desirable to minimize the absorption of solar radiation while maximizing the emission of radiation at long wavelengths, for example, when using "radiative-cooling" to reduce the cooling load of a building in hot climates. While so-called spectrally selective surfaces have long been used for these purposes, recent advancements in nanotechnology (Section 12.2) allow engineers to precisely "tune" surface properties according to the requirements of the specific application, leading to significant performance improvements.

To understand how these "selective surfaces" work, it is first necessary to consider the key characteristics of solar radiation.

4.6.1   CHARACTERISTICS OF SOLAR RADIATION

The key attributes of solar irradiation are the solar constant and the solar temperature. The solar constant describes the average total solar irradiation incident on a surface normal to the Sun at a distance equal to the Earth's mean radius from the Sun; it accounts for the solid angle subtended by the Sun as viewed by the Earth but excludes attenuation of solar energy by the atmosphere. The accepted value by many standards organizations, including the American Society for Standards and Measurement (ASTM), is 1366 W/m^2, although the National Oceanographic and Atmospheric Administration (NOAA) uses a value of 1376 W/m^2. The actual value fluctuates slightly with time due to changes in the solar energy output. Larger changes are due to the eccentricity of the Earth's orbit about the Sun, causing variations from 1322 W/m^2 at aphelion to 1412 W/m^2 at perihelion. In this book, we use a value of 1366 W/m^2 unless specified otherwise.

The solar temperature defines the spectral distribution of solar irradiation, which is close to that of a blackbody at a temperature of 5780 K.

4.6.2   MODIFICATION OF SURFACE SPECTRAL CHARACTERISTICS

For surfaces that collect solar energy, such as in solar distillation units, solar furnaces, or solar air and water heaters, it is desirable to maximize the rate at which

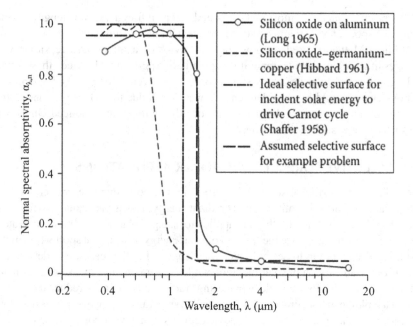

FIGURE 4.20 Characteristics of some spectrally selective surfaces.

solar irradiation is absorbed while minimizing emission losses at longer wavelengths. Some examples of this behavior are shown in Figure 4.20. In solar thermionic or thermoelectric devices, the best performance is obtained by maintaining the solar-irradiated surface at its highest possible equilibrium temperature. For photovoltaic (PV) solar cells, absorption should be maximized within the bandgap of the PV material to maximize electrical output, but absorption of incident solar radiation should be otherwise minimized to prevent the cells from heating, which degrades their photoelectric conversion efficiency.

An ideal solar-selective surface maximizes the amount of absorbed solar irradiation while minimizing losses due to emission. A good surface should therefore have a spectral absorptivity near unity over short wavelengths where the incident solar energy has a large intensity and a low spectral emissivity close to zero at longer wavelengths. The wavelength at which this transition occurs, as shown in Figure 4.20, is the *cutoff wavelength*.

Example 4.5

An ideal selective surface is exposed to a normal radiation flux equal to the average solar constant G_{solar} = 1366 W/m². The only means of energy transfer to or from the exposed surface is by radiation. Determine the maximum equilibrium temperature T_{eq} corresponding to a cutoff wavelength of λ_c = 1 μm. The solar energy can be assumed to have a spectral distribution proportional to that of a blackbody at 5780 K.

Because the only means of energy transfer is by radiation, the radiant energy absorbed must equal that emitted from the exposed side. For an ideal selective absorber, the hemispherical emissivity and absorptivity are:

$$\mu_\lambda = \alpha_\lambda = 1; \quad 0 \le \lambda \le \lambda_c$$
$$\mu_\lambda = \alpha_\lambda = 0; \quad \lambda_c < \lambda < \infty$$

The energy absorbed by the surface per unit of time is:

$$Q_a = \alpha_\lambda F_{0\to\lambda_c}(T_r)G_{solar}\,A = (1)F_{0\to\lambda_c}(T_r)G_{solar}A$$

where $F_{0\to\lambda_c}(T_r)$ is the fraction of blackbody energy in the range of the wavelengths between zero and the cutoff value, for a radiating temperature T_r. In this case, T_r is the effective solar radiating temperature of 5780 K. Similarly, the energy emitted by the selective surface i:. $Q_e = \varepsilon_\lambda F_{0\to\lambda_c}(T_{eq})\sigma T_{eq}^4 A = (1)F_{0\to\lambda_c}(T_{eq})\sigma T_{eq}^4 A$

Equating Q_e and Q_a gives:

$$T_{eq}^4 F_{0\to\lambda_c}(T_{eq}) = \frac{q_{sol}F_{0\to\lambda_c}(T_r)}{\sigma}$$

For the chosen value for λ_c all terms on the right are known, and we can solve for T_{eq} by trial and error.

For a blackbody surface ($\lambda_c \to \infty$), the equilibrium temperature is 394 K; this is the equilibrium temperature of the surface of a black object in space at the Earth's orbit when exposed to normally incident solar radiation and with all other surfaces of the object perfectly insulated. The same equilibrium temperature is reached by a gray body, since a gray emissivity cancels out of the energy-balance equation. As λ_c is decreased, the T_{eq} continues to increase even though less energy is absorbed; this is because it also becomes relatively more difficult to emit energy.

The equilibrium temperature for $\lambda_c = 1$ μm is 1337 K. Here are the values of T_{eq} for various λ_c:

Cutoff Wavelength	Equilibrium Temperature
λ_c (μm)	T_{eq}, K
0.4	2287
0.6	1817
0.8	1530
1.0	1337
1.2	1197
1.5	1048
$\to \infty$	394

A performance parameter for a solar-selective surface is the ratio of its directional total absorptivity $\alpha(\theta_i,\phi_i,T)$ for incident solar energy to its hemispherical total emissivity $\varepsilon(T)$. The ratio $\alpha(\theta_i,\phi_i,T)/\varepsilon(T)$ for the condition of incident solar energy is a measure of the theoretical maximum temperature that an otherwise insulated surface can attain when exposed to solar radiation. In general, the solar energy absorbed per unit of time by a surface element dA is:

$$dQ_a\left(\theta_i,\phi_i,T\right)=\alpha\left(\theta_i,\phi_i,T\right)I_{solar}\left(\theta_i,\phi_i\right)\Delta\Omega_{solar}dA\cos\theta_i$$
$$=\alpha\left(\theta_i,\phi_i,T\right)G_{solar}\left(\theta_i,\phi_i\right)dA\cos\theta_i \tag{4.63}$$

where I_{solar} is the incident (total) solar intensity, $\Delta\Omega_{solar}$ is the small but finite solid angle subtended by the Sun as viewed by dA, and G_{solar} is the corresponding directional irradiation (W/m²). The total energy emitted per unit of time by the surface element is:

$$dQ_e = \varepsilon(T)\sigma T^4 dA \tag{4.64}$$

If there are no other modes of energy transfer (e.g., conduction or convection losses), the emitted and absorbed energies are equated to give:

$$\frac{\alpha(\theta_i,\phi_i,T)}{\varepsilon(T_{eq})} = \frac{\sigma T_{eq}^4}{G_{solar}\left(\theta_i,\phi_i\right)\cos\theta_i} \tag{4.65}$$

where T_{eq} is the equilibrium temperature that is achieved. Thus, the ratio $\alpha(\theta_i,\phi_i,T)/\varepsilon(T)$ for $T \approx T_{eq}$ is a measure of the equilibrium temperature of the element.

For the collection and use of solar energy on Earth or in outer space applications, it is common to impose normal solar incidence so that $\alpha = \alpha_n$ and $\cos\theta_i = 1$, since this maximizes the solar radiation intercepted by the element. For the relatively low temperatures of solar collection in ground-based systems without solar concentrators, selective paints on an aluminum substrate can attain $\alpha_n = 0.92$ and $\varepsilon = 0.10$ (Moore 1985a, b). For ground-based solar collectors, convection energy transfer must be included in the energy balances, and solar collectors are often designed to minimize these losses. To attain high equilibrium temperatures for space power systems, polished metals attain α_n/ε of 5–7, and specially manufactured surfaces have α_n/ε approaching 20.

Example 4.6

The properties of a real SiO–Al selective surface are approximated by the long-dashed curve in Figure 4.20 (it is assumed that this curve can be extrapolated toward $\lambda = 0$ and $\lambda = \infty$). What is the equilibrium temperature of the surface for normally incident solar radiation at Earth orbit when energy transfer is only by radiation? What is α_n/ε for the surface? Describe the spectra of the absorbed and emitted energy at the surface. Assume normal and hemispherical emissivities are equal.

As in the derivation of Equation (4.65), equate the absorbed and emitted energies. The emissivity has nonzero constant values on both sides of the cutoff wavelength, so that:

$$Q_a = A \int_0^\infty \alpha_{\lambda,n} G_{\lambda,solar} d\lambda = \left[\varepsilon_{0 \to \lambda_c} F_{0 \to \lambda_c}(T_r) + \varepsilon_{\lambda_c \to \infty} F_{\lambda_c \to \infty}(T_r) \right] G_{\lambda,solar} A = \alpha_n G_{solar} A$$

$$Q_e = A \int_0^\infty \varepsilon_\lambda E_{\lambda b}(T_{eq}) d\lambda = \left[\varepsilon_{0 \to \lambda_c} F_{0 \to \lambda_c}(T_{eq}) + \varepsilon_{\lambda_c \to \infty} F_{\lambda_c \to \infty}(T_{eq}) \right] \sigma T_{eq}^4 A = \varepsilon \sigma T_{eq}^4 A$$

where T_r is the temperature of the radiating source. Equating Q_e and Q_a gives:

$$\left\{ 0.95 F_{0-\lambda_c}(T_r) + 0.05 \left[1 - F_{0-\lambda_c}(T_r) \right] \right\} G_{solar} = \left\{ 0.95 F_{0-\lambda_c}(T_{eq}) + 0.05 \left[1 - F_{0-\lambda_c}(T_{eq}) \right] \right\} \sigma T_{eq}^4$$

Solving by trial and error, for $\lambda_c = 15$ μm, we obtain $T_{eq} = 790$ K. The small difference in the properties of an ideal selective surface produces a significant change in T_{eq}, which in the previous example was 1048 K for an ideal selective surface with the same λ_c. The spectral curve of incident solar energy is given by $G_{\lambda,solar} \propto E_{b,\lambda}(T_r)$. It has the shape of the blackbody curve at the solar temperature, $T_R = 5780$ K, but is scaled so that the integral of E_λ over all λ is equal to G_{solar}, the total incident solar energy per unit area at Earth orbit. Multiplying this curve by the spectral absorptivity of the selective surface gives the spectrum of the absorbed energy. The spectrum of emitted energy is that of a blackbody at 790 K multiplied by the spectral emissivity of the selective surface. The integrated energies under the spectral curves of absorbed and emitted energy are equal.

The energy equation solved in Example 4.6 is a two-spectral-band approximation to the following more general energy-balance equation for a *diffuse* surface:

$$\int_{\lambda=0}^\infty \alpha_\lambda(T_{eq}) G_{\lambda,solar}(\lambda) d\lambda = \int_{\lambda=0}^\infty \varepsilon_\lambda(T_{eq}) E_{\lambda b}(T_{eq}) d\lambda$$

The $G_{\lambda,solar} d\lambda$ can have any spectral distribution, and by Kirchhoff's Law $\alpha_\lambda(T_{eq}) = \varepsilon_\lambda(T_{eq})$ for a diffuse surface.

The analysis can be extended further to consider directionally dependent properties. The emission from the surface depends on its hemispherical-spectral emissivity $\varepsilon_\lambda(T_{eq})$. The q_e is the energy flux supplied to the surface by any other means, such as convection, electrical heating, or radiation to its lower side. The energy balance then becomes:

$$\cos\theta_i \int_{\lambda=0}^\infty \alpha_\lambda(\theta_i, \phi_i, T_{eq}) G_{\lambda,solar} d\lambda + q_e = \int_{\lambda=0}^\infty \varepsilon_\lambda(T_{eq}) E_{\lambda b}(T_{eq}) d\lambda \qquad (4.66)$$

where $\alpha_\lambda(\theta_i, \phi_i, T_{eq})$ is the directional-spectral absorptivity at the incidence angle θ.

Example 4.7

A spectrally selective surface is to be used as a solar energy absorber. The surface that has the same properties as given in Example 4.6 is to be maintained at $T = 394$ K by extracting energy for a power-generating cycle. If the absorber is in orbit around the Sun at the same radius as the Earth and is normal to the solar direction, how much energy will a square meter of the surface provide? How does this energy compare with that provided by a black surface at the same temperature? Emitted energy and reflected solar energy by Earth are neglected.

The energy extracted from the surface is the difference between the absorbed and emitted radiation. The absorbed energy flux is calculated as in Example 4.6, where $T_r = 5780$ K,

$$q_a = \int_{\lambda=0}^{\infty} \varepsilon_\lambda(T) G_{\lambda,\text{solar}}(T_r) d\lambda = \left\{0.95 F_{0\to\lambda_c}(T_r) + 0.05\left[1 - F_{0\to\lambda_c}(T_r)\right]\right\} G_{\text{solar}}$$

$$= \left[0.95(0.880) + 0.05(1 - 0.880)\right]1366 = 1142.1 \ \text{W/m}^2$$

The emitted flux is:

$$q_e = \int_{\lambda=0}^{\infty} \varepsilon_\lambda E_{\lambda b}(T) d\lambda = \left\{0.95 F_{0\to\lambda_c}(T) + 0.05\left[1 - F_{0\to\lambda_c}(T)\right]\right\} \sigma T^4$$

$$= \left[0.95 \times (\sim 0) + 0.05(\sim 1)\right]5.6704 \times 10^{-8} \times 394^4 = 70.7 \ \text{W/m}^2$$

Therefore, the energy that can be used for power generation is 1142.1 − 70.7 = 1071.4 W/m². For a blackbody or gray body, the equilibrium temperature is 394 K, as obtained from Example 4.6 so, for a black or gray absorber, no useful energy can be removed for the stated conditions.

Spectrally selective surfaces can also be useful where it is desirable to cool an object exposed to incident radiation from a high-temperature source. Common situations are objects exposed to the Sun, such as a hydrocarbon storage tank, a cryogenic fuel tank in space, or the roof of a building. Equation (4.65) shows that the smaller the value of α/ε that can be reached, the lower will be the radiative equilibrium temperature. Radiative dissipation is particularly important in space applications since there are no other cooling modes.

Objects exposed to the night sky can cool by radiation to achieve temperatures below the ambient air temperature. This cooling effect can also be utilized during the day if the solar reflectivity of a surface is high (greater than about 0.95), and its emissivity is large in the IR. The use of surface nanostructures to tailor such spectral properties is discussed in Chapter 12.

4.7 CONCLUDING REMARKS

The way EM waves propagate through matter, and how they behave at interfaces, is governed by Maxwell's equations as well as the electromagnetic properties of the relevant media, usually expressed as the refractive index or dielectric constant. EM waves interact with matter principally through charges, which may either be bound (insulators) or free (conductors). While this behavior can be quite complicated, simple classical models based on harmonic oscillators can help us understand how the electromagnetic properties of the media may vary with wavelength.

Maxwell's equations may be used to predict the radiative properties of optically smooth interfaces. Extrapolating these results to real surfaces must be done with caution, however, since large property variations may result from factors that include heterogeneities, surface roughness, and contamination, as well as the assumptions used to derive simplified expressions for the radiative properties. Nevertheless, EM theory predictions are useful for determining how radiative properties may vary with wavelength and temperature. In general, both theoretical predictions and experimental measurements show that the total emissivities of dielectrics at moderate temperatures are larger than those for metals, and the spectral emissivity of metals increases with temperature over a broad range of wavelengths. By coupling analytical trends with experimental measurements, we can gain insight into what classes of surfaces will be suitable for specific applications and how surfaces may be fabricated to obtain desired radiative behavior. The latter includes spectrally selective surfaces that are of great value in practical applications such as collection of solar energy and spacecraft temperature control. Similarly, traditional building materials, such as roof tiles, paints, facades, and windows, can be spectrally optimized to decrease the energy load on the buildings.

HOMEWORK (additional problems available in online Appendix I at www.ThermalRadiation.net/appendix.html)

4.1   An electrical insulator has a refractive index of $n = 1.332$ and has a smooth surface radiating into air. What is the directional emissivity for the direction normal to the surface? What is it for the direction $60°$ from the normal? *Answers*: 0.9793; 0.9397.

4.2   A dielectric material has a refractive index of $n = 1.346$. For a smooth radiating surface, estimate:
(a)   the hemispherical emissivity of the material for emission into air.
(b)   the directional emissivity at $\theta = 60°$ into air.
(c)   the directional-hemispherical reflectivity in air for both components of polarized reflectivity. Plot both components for $n = 1.346$. Let θ_i be the angle of incidence.
Answers: (a) 0.9315, (b) 0.9379.

4.3   An inventor wants to use a light source and some polarized sunglasses to determine when the wax finish is worn from her favorite bowling alley. She reasons that the wax will reflect as a dielectric with $n = 1.35$, and that the parallel component of light from the source will be preferentially absorbed and the perpendicular component strongly reflected by the wax. When the wax is worn away, the wood will reflect diffusely. At what height should the light source be placed to maximize the ratio of perpendicular to parallel polarization from the wax as seen by the viewer?

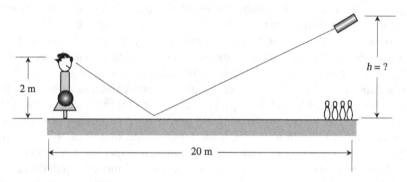

Answer: 12.8 m.

4.4 A highly polished metal disk is found to have a measured normal spectral emissivity of 0.098 at a wavelength of 12 μm. What is:
(a) the electrical resistivity of the metal (Ω-cm)?
(b) the normal spectral emissivity of the metal at $\lambda = 10$ μm?
(c) the refractive index n of the metal at $\lambda = 10$ μm?
(Note any assumptions that you make in obtaining your answers.)
Answers: (a) 9.2969 x 10^{-5} Ω-cm; (b) 0.107; (c) 17.7.

4.5 Using Equation (4.35), show that the parallel component of reflectivity becomes zero when $\theta = \tan^{-1}(n_2/n_1)$.

4.6 A smooth polished gold surface must radiate 400 W/m². What is its surface temperature as calculated from the Hagen–Rubens relations? At this temperature, what is the normal total emissivity of the gold? What is the ratio of its hemispherical total emissivity to its normal total emissivity?
Answers: 643 K; 0.0324; 1.27.

4.7 A sample of highly polished platinum has a value of normal spectral emissivity of 0.042 at a wavelength of 7.0 μm at 293 K. What value of normal spectral absorptivity do you expect the sample to have at
(a) a wavelength of 10 μm and at 293 K?
(b) a wavelength of 10 μm and at 600 K?
Answers: (a) 0.0352; (b) 0.0501.

4.8 Using Equation (4.52) with data for n and k from Table 4.2, find the hemispherical emissivity of aluminum and titanium at 0.484 and 8.06 μm, for surfaces at 298 K.
Answers: Aluminum: $\varepsilon_\lambda(0.484$ μm$) = 0.085$; $\varepsilon_\lambda(8.06$ μm$) = 0.015$. Titanium: $\varepsilon_\lambda(0.484$ μm$) = 0.678$; $\varepsilon_\lambda(8.06$ μm$) = 0.045$.

4.9 For a selective surface with $\alpha_\lambda = 0.94$ in the range $0 < \lambda < \lambda_c$, and $\alpha_\lambda = 0.095$ at $\lambda > \lambda_c$, plot the curve of equilibrium temperature versus λ_c for the range $2 \leq \lambda_c \leq 10$ μm.

4.10 Use a very simple instantaneous radiative energy balance at the Earth's surface. Assume a normal incident solar flux on the surface of 900 W/m². Calculate the equilibrium temperature of the Earth's surface when it is
   a) Covered with fine fresh snow.
   b) Plowed soil.
   Discuss how this might impact global warming in snow-covered regions as the average global air temperature increases.
   Answers: (a) –24°C; (b) 147°C.

4.11 A diffuse-spectral coating has the spectral emissivity approximated as below. The coating is placed on one face of a thin sheet of metal. The sheet is placed in an orbit around the Sun where the solar flux is 1350 W/m². The other face of the sheet is coated with a diffuse-gray coating of hemispherical total emissivity $\varepsilon = 0.530$. What is the temperature (K) of the sheet if:
   (a) The side with the spectral coating is facing normal to the Sun?
   (b) The gray side is facing normal to the Sun?
   (c) What is the normal-hemispherical total reflectivity of the diffuse-spectral coating when exposed to solar radiation?
   Take the effective solar radiating temperature to be 5780 K. Note any necessary additional assumptions.

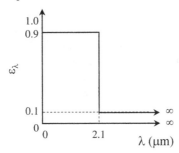

   Answers: (a) 424.4 K; (b) 376.2 K; (c) 0.143.

4.12 A spinning spherical satellite 0.85 m in diameter is in Earth orbit and is receiving solar radiation of 1353 W/m². Because of the rotation, the sphere surface temperature is assumed uniform. The sphere exterior is an SiO–Al selective surface, as in Example 4.6, with a cutoff wavelength of 2.00 μm. The surface properties do not depend on angle. Neglect energy emitted by the Earth, and solar energy reflected from the Earth.
   (a) What is the equilibrium surface temperature for energy transfer only by radiation? How does the temperature depend on the sphere diameter?
   (b) It is desired to maintain the sphere surface at 750 K. At what rate must energy be supplied to the sphere to accomplish this?
   Answers: 569 K; 1820 W.

4.13 A cylindrical concentrator (reflector) is long in the direction normal to the cross-section shown, so that end effects may be neglected. The concentrator is gray and reflects 94% of the incident solar energy onto the central tube.

The central tube receiving the energy is assumed to be at uniform temperature and is coated with a material that has the spectral properties shown. The
properties are independent of direction. Emitted energy from the tube that
is reflected by the concentrator may be neglected, as may emission from the
concentrator. The surrounding environment is at a low temperature.

If the energy exchange is only by radiation, compute the temperature of the
central tube.

If the tube is cooled to 395 K by passing a coolant through its interior, how
much energy must be removed by the coolant per meter of tube length?

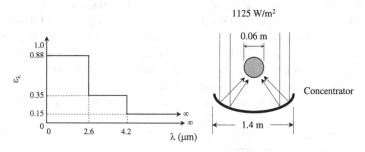

Answers: (a) 816.3 K; (b) 1239 W/m.

4.14 For a small diameter light-pipe with refractive index $n_2 = 1.4400$, plot the
reflectivity of the light-pipe end versus incident angle θ_1 for radiation incident on the end. Also, plot the angle of refraction χ versus angle of incidence, and determine the range of incident angles θ_1 that will have angles
θ_2 above the critical angle when the radiation encounters the cylindrical
light-pipe wall.

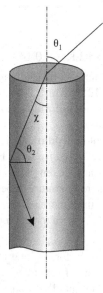

4.15 The radiative properties of silicon carbide (SiC) can be captured using a single-oscillator Lorentz model with $\omega_p = 2.716 \times 10^{14}$ rad/s, $\omega_0 = 1.439 \times 10^{14}$ rad/s, $\varepsilon_0 = 6.7$, and $\zeta = 8.966 \times 10^{11}$ s^{-1}. Calculate the dielectric constants and refractive indices for this material between 2 and 25 μm. Plot the normal reflectance as a function of wavelength and plot the perpendicular and parallel reflectance as a function of θ at 20 μm.

4.16 While silicon is a semiconductor at room temperature, when it melts its four valence electrons form a conduction band, transforming it into a metal. Given that the bulk density of molten silicon is 3600 kg/m^3 and its electrical resistivity is 8×10^{-7} Ω·, find the dielectric constants, refractive indices, and normal reflectance of this material between 300 nm and 1200 nm.

NOTE

1. These equations involve some simplifying assumptions about the nature of the material through which the waves propagate. For a more complete discussion, refer to Chapter 8 in *Thermal Radiation Heat Transfer, 7th Ed.*

5 Configuration Factors for Diffuse Surfaces

Hoyt Clarke Hottel (1903–1998) *spent his professional life as a professor at MIT from 1928 and became an emeritus professor in 1968. He developed gas emissivity charts for the important combustion gases, the crossed-string method for determining configuration factors in 2D geometries, and in 1927 established the engineering basis for treating radiation in furnaces, including the zone method. He also contributed to the fields of combustion and solar energy.*

5.1 RADIATIVE TRANSFER EQUATION FOR SURFACES SEPARATED BY A TRANSPARENT MEDIUM

Radiation interchange among surface areas must be quantified to allow calculation of energy transfer rates for effective and efficient processes, building energy loads for interior thermal comfort, illumination engineering for visual comfort, and applied optics for the development of diagnostic tools. Studies have been conducted for many years, as evidenced by the publication dates of d'Aguillon (1613) and Charle (1888). Radiative transfer studies were given impetus in the 1960s by technological advances in systems where thermal radiation is very important, including radiant heating of materials, curing, surface modification systems, satellite temperature control, devices for collection and utilization of solar energy, advanced engines with increased operating temperatures, reduction of energy leakage into cryogenic fuel storage tanks, space station power systems, thermal control during spacecraft entry into planetary atmospheres, heat islands in cities, atmospheric-ocean-land coupling to understand climate change, and many others.

The general radiative transfer equation describes the propagation of intensity along a path and was introduced in Sections 2.8 and 2.9. For the important case when the medium separating two surfaces is transparent (i.e., there is no attenuation of intensity along a path by scattering or absorption, and the medium along the path does not emit), then Equation (2.74) indicates that the intensity leaving a radiating surface is invariant along a path (Figure 5.1).

Under certain assumptions, the geometric relations between surfaces 1 and 2 can be separated from the radiative intensity, allowing the description of the geometric configuration of surface 2 relative to surface 1 to be treated independently of the thermal states of the two surfaces. The factors that contain only the geometric relations are called *configuration factors* (Equation 2.79).

DOI: 10.1201/9781003178996-5

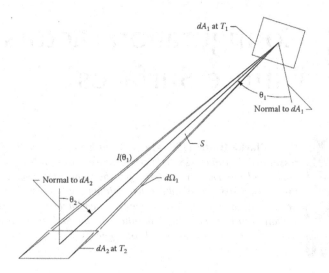

FIGURE 5.1 Intensity leaving surface 1 and transferring energy to surface 2.

The geometric configuration factors derived in this chapter are an important component for analyzing radiation exchange. In this chapter and Chapters 6 and 7, the theory is developed for radiation exchange in enclosures that are evacuated or contain radiatively nonparticipating media.

First, we must define what is meant by an *enclosure*. Any surface can be considered as surrounded by an envelope of other solid surfaces or open areas. This envelope is the enclosure for the surface, so an enclosure accounts for all directions surrounding a surface. By considering the radiation from surface 1 to all parts of the enclosure and the radiation arriving at surface 1 from all parts of the enclosure, it is assumed that all radiative contributions are included. A convenient enclosure is usually evident from the physical configuration. An opening into a large empty environment can often be treated as a bounding area of zero reflectivity. The opening may also act as a radiation source that can be diffuse or directional when radiation enters from the surrounding environment.

In Chapters 3 and 4, we examined the radiative properties of solid opaque surfaces. For some materials, the properties vary substantially with wavelength, surface temperature, and direction. For radiation computations within enclosures, the geometry of the exchange introduces another challenge in addition to the surface property variations. For simple geometries, it is possible to account in detail for property variations without the analysis becoming complex. As the geometry becomes more involved, it may be necessary to idealize the surface properties so that solutions can be obtained with reasonable effort, if the resulting decrease in accuracy is acceptable.

5.1.1 ENCLOSURES WITH DIFFUSE SURFACES

Consider an enclosure where each surface is black and isothermal. For black surfaces, there is no reflected radiation, and all emitted energy is diffuse (the intensity leaving a surface is independent of direction). The local energy balance at a surface involves the enclosure geometry, which governs how much radiation leaving a surface will reach another surface. For a black enclosure, the geometric effects are expressed in terms of the configuration factors developed in this chapter. A configuration factor is the *fraction of uniform diffuse radiation leaving a surface that directly reaches another surface.*

The computation of configuration factors involves analytical or numerical integration over the solid angles by which surfaces can view each other. Factors may be evaluated numerically, and routines have been incorporated into many of the computer programs developed for thermal analysis. It is helpful to use relations that exist among configuration factors to obtain a desired factor from factors that are already known.

The next step in increasing the reality of enclosures is to have *gray* rather than black surfaces that emit and reflect diffusely. The enclosure surfaces must have both the emitted and reflected energies uniform over each surface. The configuration factors can then be used for radiation leaving a surface by both emission and reflection. For gray surfaces, reflections among surfaces must be accounted for in the analysis; we discuss this in Chapter 6.

5.1.2 ENCLOSURES WITH DIRECTIONAL (NONDIFFUSE) AND SPECTRAL (NONGRAY) SURFACES

Sometimes the approximations of black or diffuse-gray surfaces are not sufficiently accurate, and directional and/or spectral effects must be included. If the spectral surfaces are diffuse, the configuration factors developed here remain applicable for enclosure analysis.

In some instances, surface properties have significant directional characteristics, as outlined in Chapter 4. They sometimes differ considerably from the diffuse approximation, and diffuse configuration factors cannot be used. Further information on mirrored enclosures is in online Appendix B at www.ThermalRadiation.net/appendix.html.

5.2 GEOMETRIC CONFIGURATION FACTORS BETWEEN TWO SURFACES

When calculating radiative transfer among surfaces, geometric relations are needed to determine how the surfaces view each other. In this section, relations for geometric configuration factors are developed.

5.2.1 CONFIGURATION FACTOR FOR ENERGY EXCHANGE BETWEEN DIFFUSE DIFFERENTIAL AREAS

The radiative transfer from a diffuse differential area element to another area element is used to derive relations for transfer between finite areas. Consider the two differential area elements in Figure 5.1. The dA_1 and dA_2 at T_1 and T_2, are arbitrarily oriented, and have their normal at angles θ_1 and θ_2 to the line of length S between them.

If I_1 is the total intensity leaving dA_1, the total energy per unit time leaving dA_1 and incident on dA_2 is:

$$d^2Q_{d1-d2} = I_1 dA_1 \cos\theta_1 d\Omega_1 \tag{5.1}$$

where $d\Omega_1$ is the solid angle subtended by dA_2 when viewed from dA_1 and the dash in the subscript of Q means "to."

Equation (5.1) follows directly from the definition of I_1 as the total energy leaving surface 1 per unit time, per unit area projected normal to S, and per unit solid angle. Because I_1 is diffuse, it is independent of the angle at which it leaves dA_1. It may consist of both diffusely emitted and diffusely reflected portions.

This equation can also be written for radiant energy in a wavelength interval $d\lambda$:

$$d^2Q_{\lambda,d1-d2}d\lambda = I_{\lambda,1}d\lambda dA_1 \cos\theta_1 d\Omega_1 \tag{5.2}$$

The total radiant energy is then found by integrating over all wavelengths:

$$d^2Q_{d1-d2} = dA_1 \cos\theta_1 d\Omega_1 \int_{\lambda=0}^{\infty} I_{\lambda,1}d\lambda \tag{5.3}$$

Because the geometric factors are independent of λ, they can be removed from under the integral sign, and the integration over λ is independent of geometry. Thus, the results that follow for diffuse geometric configuration factors involving finite areas apply for *both spectral and total* quantities. For simplicity in notation, the remaining development is for total quantities.

The solid angle $d\Omega_1$ is related to the projected area of dA_2 and the distance between the differential elements by:

$$d\Omega_1 = \frac{dA_2 \cos\theta_2}{S^2} \tag{5.4}$$

Substituting this into Equation (5.1) gives the relation for the total energy per unit time leaving dA_1 that is incident upon dA_2:

$$d^2Q_{d1-d2} = \frac{I_1 dA_1 \cos\theta_1 dA_2 \cos\theta_2}{S^2} \tag{5.5}$$

An analogous derivation for the radiation leaving a diffuse dA_2 that arrives at dA_1 results in:

$$d^2Q_{d2-d1} = \frac{I_2 dA_2 \cos\theta_2 dA_1 \cos\theta_1}{S^2} \tag{5.6}$$

Equations (5.5) and (5.6) contain the same geometric factors.

The *fraction* of energy leaving diffuse surface element dA_1 that arrives at element dA_2 is defined as the geometric *configuration factor* dF_{d1-d2}.

$$dF_{d1-d2} = \frac{I_1 \cos\theta_1 \cos\theta_2 dA_1 dA_2 / S^2}{\pi I_1 dA_1} = \frac{\cos\theta_1 \cos\theta_2}{\pi S^2} dA_2 \qquad (5.7)$$

where $\pi I_1 dA_1$ is the total diffuse energy leaving dA_1 within the entire 2π hemispherical solid angle over dA_1. Equation (5.7) shows that dF_{d1-d2} *depends only on the size of dA_2 and its orientation with respect to dA_1*. By substituting Equation (5.4), Equation (5.7) can also be written as:

$$dF_{d1-d2} = \frac{\cos\theta_1 d\Omega_1}{\pi} \qquad (5.8)$$

Consequently, elements dA_2 have the same configuration factor regardless of geometry and orientation *if* they subtend the same solid angle $d\Omega_1$ when viewed from dA_1 and are positioned along a path at angle θ_1 with respect to the normal of dA_1.

The factor dF_{d1-d2} has various names, being called the *view, angle, shape, interchange, exchange,* or *configuration factor*. The notation used here has subscripts designating the areas involved and a derivative consistent with the mathematical description. For the subscript notation, $d1$, $d2$, etc., indicate differential area elements, while 1, 2, etc., indicate finite areas. Thus, dF_{d1-d2} is a factor between two differential elements, and dF_{1-d2} is from finite area A_1 to differential area dA_2. The derivative dF indicates that the factor is for energy *to a differential* element; this keeps the mathematical form of equations such as Equation 5.7 consistent by having a differential quantity on both sides. The F denotes a factor to a *finite* area; thus, F_{d1-2} is from differential element dA_1 to finite area A_2.

5.2.1.1 Reciprocity for Differential Element Configuration Factors

The configuration factor for energy from dA_2 to dA_1 is found using a similar derivation to that for Equation (5.7):

$$dF_{d2-d1} = \frac{\cos\theta_1 \cos\theta_2}{\pi S^2} dA_1 \qquad (5.9)$$

Multiplying Equation (5.7) by dA_1 and Equation (5.9) by dA_2 gives the *reciprocity relation* between dF_{d1-d2} and dF_{d2-d1}:

$$dF_{d1-d2} dA_1 = dF_{d2-d1} dA_2 = \frac{\cos\theta_1 \cos\theta_2}{\pi S^2} dA_1 dA_2 \qquad (5.10)$$

5.2.1.2 Sample Configuration Factors between Differential Elements

The derivation of configuration factor expressions in terms of system geometry parameters is illustrated by a few examples.

Example 5.1

In Figure 5.2, two elemental areas are shown that are located on strips that have parallel generating lines. Derive an expression for the configuration factor between dA_1 and dA_2. The angle β is in the $y–z$ plane.

The $y–z$ plane is normal to the generating lines. The distance $S = (l^2 + x^2)^{1/2}$, and $\cos\theta_1 = (l\cos\beta)/S = (l\cos\beta)/(l^2 + x^2)^{1/2}$. The angle β is in the $y–z$ plane normal to the two strips. The solid angle subtended by dA_2, when viewed from dA_1, is:

$$d\Omega_1 = \frac{\text{Projected area of } dA_2}{S^2} = \frac{(\text{Projected width of } dA_2)(\text{Projected length of } dA_2)}{S^2}$$

$$= \frac{(ld\beta)(dx\cos\psi)}{S^2} = \frac{ld\beta dx}{S^2}\frac{l}{S}$$

Substituting into Equation (5.8) gives

$$dF_{d1-d2} = \frac{\cos\theta_1 d\Omega_1}{\pi} = \frac{l\cos\beta}{(l^2 + x^2)^{1/2}}\frac{1}{\pi}\frac{l^2 d\beta dx}{(l^2 + x^2)^{3/2}} = \frac{l^3\cos\beta d\beta dx}{\pi(l^2 + x^2)^2}$$

which is the desired configuration factor in terms of convenient parameters that specify the geometry.

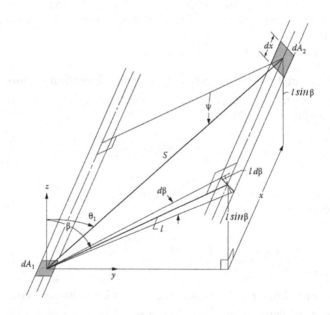

FIGURE 5.2 Geometry for configuration factor between elements on strips formed by parallel generating lines.

Example 5.2

Find the configuration factor between an elemental area and an infinitely long strip of differential width as in Figure 5.3, where the generating lines of dA_1 and $dA_{strip,2}$ are parallel.

Example 5.1 gave the configuration factor between element dA_1 and element dA_2 of length dx. To find the factor when dA_2 becomes an infinite strip, as in Figure 5.3, integrate over all x to obtain:

$$dF_{d1-strip,2} = \frac{l^3 \cos\beta d\beta}{\pi} \int_{-\infty}^{\infty} \frac{dx}{(l^2+x^2)^2} = \frac{l^3 \cos\beta d\beta}{\pi} \frac{\pi}{2l^3} = \frac{1}{2} d(\sin\beta)$$

where β is in the y–z plane normal to the strips.

Figure 5.3 also shows that since dA_1 lies on an *infinite* strip $dA_{strip,1}$ with elements parallel to $dA_{strip,2}$, the $dF_{d1-strip,2}$ is valid for dA_1 at any location along $dA_{strip,1}$. Then, since any element dA_1 on $dA_{strip,1}$ has the same fraction of its energy reaching $dA_{strip,2}$, it follows that the fraction of energy from the *entire* infinite $dA_{strip,1}$ that reaches $dA_{strip,2}$ is the same as the fraction for each element dA_1. Thus, *the configuration factor between two infinitely long strips of differential width and having parallel generating lines* must also be the same as for element dA_1 to $dA_{strip,2}$, or $(1/2)d(\sin\beta)$.

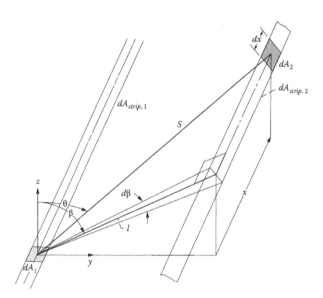

FIGURE 5.3 Geometry for configuration factor between elemental area and an infinitely long strip of differential width; area and strip are on parallel generating lines.

Example 5.3

Consider an infinitely long wedge-shaped groove, as shown in cross-section in Figure 5.4. Determine the configuration factor between the differential strips dx and $d\xi$ in terms of x, ξ, and α.

From Example 5.2, the configuration factor is $dF_{dx-d\xi} = \dfrac{1}{2}d(\sin\beta) = \dfrac{1}{2}\cos\beta\, d\beta$

From the construction in Figure 5.4b, $\cos\beta = (\xi\sin\alpha)/L$. The $d\beta$ is the angle subtended by the projection of $d\xi$ normal to L, that is, $d\beta = \dfrac{d\xi\cos(\alpha+\beta)}{L} = \dfrac{d\xi}{L}\dfrac{x\sin\alpha}{L}$. From the law of cosines, $L^2 = x^2 + \xi^2 - 2x\xi\cos\alpha$. Then,

$$dF_{dx-d\xi} = \frac{1}{2}\cos\beta\, d\beta = \frac{1}{2}\frac{x\xi\sin^2\alpha}{L^3}d\xi = \frac{1}{2}\frac{x\xi\sin^2\alpha}{(x^2+\xi^2-2x\xi\cos\alpha)^{3/2}}d\xi$$

5.2.2 Configuration Factor between a Differential Area Element and a Finite Area

Consider an isothermal diffuse element dA_1 with uniform emissivity at temperature T_1 exchanging energy with a finite area A_2 that is isothermal at temperature T_2, as in Figure 5.5. The relations for exchange between two differential elements must be extended to a finite A_2. Figure 5.5 shows that θ_2 is different for different positions on A_2 and that θ_1 and S will also vary as different differential elements on A_2 are viewed from dA_1.

Two configuration factors need to be considered. The F_{d1-2} is from the differential area dA_1 to the finite area A_2, and dF_{2-d1} is from A_2 to dA_1. Each of these is obtained by evaluating the fraction of energy leaving one diffusely emitting and reflecting area

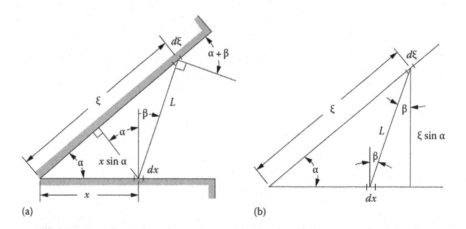

(a)   (b)

FIGURE 5.4 Configuration factor between two strips on sides of wedge groove: (a) wedge-shaped groove geometry and (b) auxiliary construction.

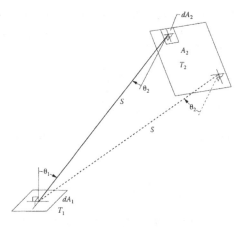

FIGURE 5.5 Radiant interchange between differential element and finite area.

that reaches the second area. The radiation leaving dA_1 is $\pi I_1 \, dA_1$. The energy reaching dA_2 on A_2 is $\pi I_1 (\cos \theta_1 \cos \theta_2 / \pi S^2) dA_1 \, dA_2$.

Then, integrating over A_2 to obtain all the energy reaching A_2 and dividing by the energy leaving dA_1 results in:

$$F_{d1-2} = \frac{\int_{A_2} I_1 \cos \theta_1 (\cos \theta_2 dA_1 / S^2) dA_2}{\pi I_1 dA_1} = \int_{A_2} \frac{\cos \theta_1 \cos \theta_2}{\pi S^2} dA_2 \qquad (5.11)$$

From Equation (5.7), the quantity inside the integral of Equation (5.11) is dF_{d1-d2}, so F_{d1-2} becomes:

$$F_{d1-2} = \int_{A_2} dF_{d1-d2} \qquad (5.12)$$

This gives the fraction of the energy reaching A_2 as the sum of the fractions reaching all the elements of A_2.

For obtaining the configuration factor from the finite area A_2 to the element dA_1, the energy leaving A_2 is $\int_{A_2} \pi I_2 dA_2$. The energy reaching dA_1 from A_2 is by integrating Equation (5.6) over A_2, $\int_{A_2} I_2 (\cos \theta_1 \cos \theta_2 / S^2) dA_2$. The configuration factor dF_{2-d1} is then,

$$dF_{2-d1} = \frac{dA_1 \int_{A_2} I_2 \dfrac{\cos \theta_1 \cos \theta_2}{S^2} dA_2}{\int_{A_2} \pi I_2 dA_2} = \frac{dA_1}{A_2} \int_{A_2} \frac{\cos \theta_1 \cos \theta_2}{\pi S^2} dA_2 \qquad (5.13)$$

The last integral on the right requires that A_2 has uniform emitted plus reflected intensity I_2 over its entire area. From Equation (5.7), the quantity under the integral sign in Equation (5.13) is dF_{d1-d2}, so the alternative form is obtained:

$$dF_{2-d1} = \frac{dA_1}{A_2} \int_{A_2} dF_{d1-d2} \tag{5.14}$$

By use of Equation (5.12) to eliminate the integral in Equation (5.14), we get the *reciprocity relation*:

$$A_2 dF_{2-d1} = dA_1 F_{d1-2} \tag{5.15}$$

5.2.2.1 Configuration Factors between Differential and Finite Areas

Certain geometries have configuration factors with closed-form algebraic expressions, while others require numerical integration of Equation (5.11). Example 5.4 illustrates how factors that have algebraic forms are obtained.

Example 5.4

An infinitely long 2D wedge has an opening angle α. Derive an expression for the configuration factor from one wall of the wedge to a strip element of width dx on the other wall at x, as in Figure 5.6a. Such configurations approximate the geometries of long fins and ribs.

Use the configuration factor between two infinitely long strip elements having parallel generating lines from Example 5.2. The angle β is measured clockwise from the normal of dx, so:

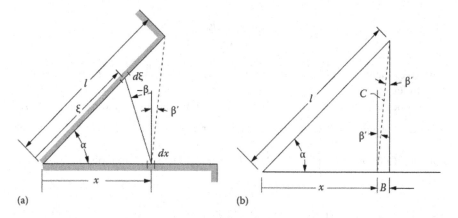

(a)   (b)

FIGURE 5.6 Configuration factor between one wall and strip on another wall of infinitely long wedge cavity: (a) wedge-cavity geometry and (b) auxiliary construction to determine sin β′.

$$F_{dx-1} = \int\limits_{\xi=0}^{l} dF_{dx-d\xi} = \int\limits_{\beta=-\pi/2}^{0} d(\sin\beta) + \int\limits_{0}^{\beta'} \frac{1}{2} d(\sin\beta) = \frac{1}{2}(1+\sin\beta')$$

The function $\sin\beta'$ is found by the auxiliary construction in Figure 5.6b to be $\sin\beta' = B/C = (l\cos\alpha - x)/(x^2 + l^2 - 2xl\cos\alpha)^{1/2}$. Then,

$$F_{dx-1} = \frac{1}{2} + \frac{l\cos\alpha - x}{2(x^2 + l^2 - 2xl\cos\alpha)^{1/2}}$$

The problem requires dF_{1-dx}. Using the reciprocal relation Equation (5.15) gives

$$dF_{1-dx} = \frac{dx}{l} F_{dx-1} = dX \left[\frac{1}{2} + \frac{\cos\alpha - X}{2(X^2 + 1 - 2X\cos\alpha)^{1/2}} \right] \text{ where } X = x/l. \text{ The only parameters}$$

are the wedge opening angle and the dimensionless position of dx from the vertex.

5.2.3 CONFIGURATION FACTOR AND RECIPROCITY FOR TWO FINITE AREAS

Consider the configuration factor for radiation with uniform intensity leaving the diffuse surface A_1 and reaching A_2, as shown in Figure 5.7.

The energy leaving A_1 is $\pi I_1 A_1$. The radiation leaving an element dA_1 that reaches dA_2 was given previously as $\pi I_1 \dfrac{\cos\theta_1 \cos\theta_2}{\pi S^2} dA_1 dA_2$. This is integrated over both A_1 and A_2 to give the energy leaving A_1 that reaches A_2. The configuration factor is then (since I_1 is constant):

$$F_{1-2} = \frac{\displaystyle\int_{A_1}\int_{A_2} \pi I_1 (\cos\theta_1 \cos\theta_2 / \pi S^2) dA_2 dA_1}{\pi I_1 A_1} = \frac{1}{A_1} \int\limits_{A_1}\int\limits_{A_2} \frac{\cos\theta_1 \cos\theta_2}{\pi S^2} dA_2 dA_1 \quad (5.16)$$

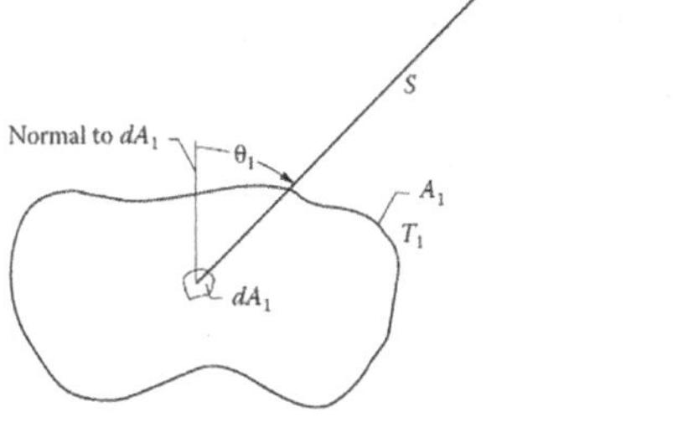

FIGURE 5.7 Geometry for energy exchange between finite areas.

This expression can be put in terms of configuration factors with differential areas:

$$F_{1-2} = \frac{1}{A_1} \int\limits_{A_1} \int\limits_{A_2} dF_{d1-d2}\, dA_1 = \frac{1}{A_1} \int\limits_{A_1} F_{d1-2} dA_1 \tag{5.17}$$

A similar derivation to Equation (5.16) gives the configuration factor from A_2 to A_1 as:

$$F_{2-1} = \frac{1}{A_2} \int\limits_{A_1} \int\limits_{A_2} \frac{\cos\theta_1 \cos\theta_2}{\pi S^2}\, dA_2 dA_1 \tag{5.18}$$

The *reciprocity relation* for configuration factors between finite areas is found from the identical double integrals in the previous equations:

$$A_1 F_{1-2} = A_2 F_{2-1} \tag{5.19}$$

Further relations among configuration factors are found by using Equation (5.17) in conjunction with the reciprocity relations Equations (5.15) and (5.19):

$$F_{2-1} = \frac{A_1}{A_2} F_{1-2} = \frac{A_1}{A_2} \frac{1}{A_1} \int\limits_{A_1} F_{d1-2} dA_1 = \frac{1}{A_2} \int\limits_{A_1} dF_{2-d1} A_2 = \int\limits_{A_1} dF_{2-d1} \tag{5.20}$$

Example 5.5

Two plates of the same finite width and of infinite length are joined along one edge at angle α as in Figure 5.6. Derive the configuration factor between the plates.

Example 5.4 gives the configuration factor dF_{l-dx} between one plate and an infinite strip on the other plate. Substituting into Equation (5.20) gives

$$F_{l-l^*} = \int\limits_{x=0}^{l^*} dF_{l-dx} = \int\limits_{0}^{l} \left[\frac{1}{2} + \frac{\cos\alpha - X}{2(X^2 + 1 - 2X\cos\alpha)^{1/2}} \right] dX \text{ where the width of the side in}$$

Figure 5.6 having element dx is designated as l^*, $X = x/l$, and $l^* = l$. Integration

yields $F_{l-l^*} = 1 - \left(\dfrac{1 - \cos\alpha}{2} \right)^{1/2} = 1 - \sin\dfrac{\alpha}{2}$.

For the present case with equal plate widths, the only parameter is α; hence, the configuration factor is the same for plates of any equal width.

Table 5.1 summarizes the integral definitions of the configuration factors and the configuration factor reciprocity relations.

TABLE 5.1

Summary of Configuration Factor and Reciprocity Relations

Geometry	Configuration Factor	Reciprocity
Elemental area to elemental area	$dF_{d1-d2} = \dfrac{\cos\theta_1 \cos\theta_2}{\pi S^2} dA_2$	$dA_1 dF_{d1-d2} = dA_2 dF_{d2-d1}$
Elemental area to finite area	$F_{d1-2} = \displaystyle\int_{A_2} \dfrac{\cos\theta_1 \cos\theta_2}{\pi S^2} dA_2$	$dA_1 F_{d1-2} = A_2 dF_{2-d1}$
Finite area to finite area	$F_{1-2} = \dfrac{1}{A_1} \displaystyle\int_{A_1}\int_{A_2} \dfrac{\cos\theta_1 \cos\theta_2}{\pi S^2} dA_2 dA_1$	$A_1 F_{1-2} = A_2 F_{2-1}$

5.3 METHODS FOR DETERMINING CONFIGURATION FACTORS

5.3.1 CONFIGURATION FACTOR ALGEBRA

Configuration factor algebra is the manipulation of various relations among configuration factors to derive new factors from those already known. The algebra is based on the reciprocal relations, the definition that the F factor is the fraction of energy that is intercepted, and energy conservation for a complete enclosure.

Consider an arbitrary isothermal area A_1 in Figure 5.8, exchanging energy with a second area A_2. The F_{1-2} is the fraction of diffuse energy leaving A_1 that is incident

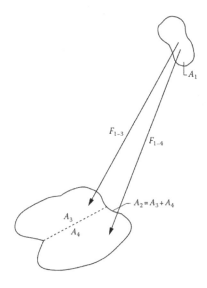

FIGURE 5.8 Energy exchange between two finite areas with one area subdivided: $F_{1-3} + F_{1-4} = F_{1-2}$.

on A_2. If A_2 is divided into A_3 and A_4, the fractions of the energy leaving A_1 that are incident on A_3 and A_4 must add to F_{1-2}:

$$F_{1-2} = F_{1-(3+4)} = F_{1-3} + F_{1-4}$$ (5.21)

$$F_{1-3} = F_{1-2} - F_{1-4}$$ (5.22)

The reciprocity relation, Equation (5.19), gives:

$$F_{3-1} = \frac{A_1}{A_3} F_{1-3} = \frac{A_1}{A_3}(F_{1-2} - F_{1-4})$$ (5.23)

As before, the dashes in the subscript mean "to" and are not to be confused with negative signs. Thus, F_{1-2} means from area A_1 to area A_2. Combined areas are grouped with parentheses, so that (3 + 4) means the sum of A_3 and A_4. Thus, the notation $F_{1-(3+4)}$ means from A_1 to the combined areas of A_3 and A_4. The reciprocity relation of Equation (5.23) is a powerful tool for obtaining new configuration factors.

Example 5.6

An elemental area dA_1 is perpendicular to a ring of outer radius r_o and inner radius r_i as in Figure 5.9. Derive an expression for the configuration factor $F_{d1-ring}$.

The configuration factor between dA_1 and the entire disk of area A_2 with outer radius r_o is (http://www.thermalradiation.net/sectionb/B-15.html):

$$F_{d1-2} = \frac{H}{2} \left\{ \frac{H^2 + R_o^2 + 1}{\left[\left(H^2 + R_o^2 + 1 \right)^2 - 4R_o^2 \right]^{1/2}} - 1 \right\}$$

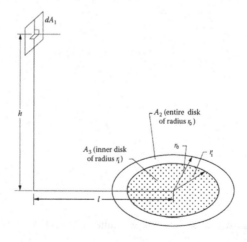

FIGURE 5.9 Energy exchange between elemental area and finite ring.

where $H = h/l$, $R_o = r_o/l$. The configuration factor to the inner disk of area A_3 with radius r_i has the same form, with $R_i = r_i/l$ substituted for R_o. Using configuration factor algebra, the desired configuration factor from dA_1 to the ring $A_2 - A_3$ is $F_{d1-2} - F_{d1-3}$, so that:

$$F_{d1-ring} = \frac{H}{2} \left\{ \frac{H^2 + R_o^2 + 1}{\left[\left(H^2 + R_o^2 + 1 \right)^2 - 4R_o^2 \right]^{1/2}} - \frac{H^2 + R_i^2 + 1}{\left[\left(H^2 + R_i^2 + 1 \right)^2 - 4R_i^2 \right]^{1/2}} \right\}$$

Example 5.7

Suppose the configuration factor is known between two parallel disks of arbitrary size whose centers lie on the same axis. From this, derive the configuration factor between the two rings A_2 and A_3 of Figure 5.10. Give the result in terms of known disk-to-disk factors from disk areas on the lower surface to disk areas on the upper surface.

The factor desired is F_{2-3}. From configuration factor algebra, F_{2-3} is equal to $F_{2-3} = F_{2-(3+4)} - F_{2-4}$. The factor $F_{2-(3+4)}$ can be found from the reciprocal relation $A_2 F_{2-(3+4)} = (A_3 + A_4) F_{(3+4)-2}$. Applying configuration factor algebra to the right-hand side results in:

$$A_2 F_{2-(3+4)} = (A_3 + A_4)[F_{(3+4)-(1+2)} - F_{(3+4)-1}]$$

$$= (A_3 + A_4) F_{(3+4)-(1+2)} - (A_3 + A_4) F_{(3+4)-1}$$

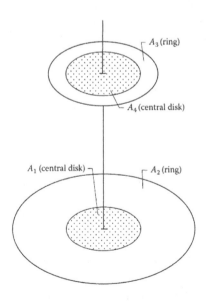

FIGURE 5.10 Energy exchange between parallel ring areas having a common axis.

Applying reciprocity to the right side gives $A_2F_{2-(3+4)} = (A_1 + A_2)F_{(1+2)-(3+4)} - A_1F_{1-(3+4)}$ where the F factors on the right are both disk-to-disk factors from the lower surface to the upper.

Next, the factor F_{2-4} is to be determined. Applying reciprocity relations and configuration factor algebra,

$$F_{2-4} = \frac{A_4}{A_2}F_{4-2} = \frac{A_4}{A_2}[F_{4-(1+2)} - F_{4-1}] = \frac{1}{A_2}[(A_1 + A_2)F_{(1+2)-4} - A_1F_{1-4}]$$

Substituting the relations for F_{2-4} and $F_{2-(3+4)}$ into the first equation gives:

$$F_{2-3} = \frac{A_1 + A_2}{A_2}[F_{(1+2)-(3+4)} - F_{(1+2)-4}] - \frac{A_1}{A_2}[F_{1-(3+4)} - F_{1-4}]$$

and all configuration factors on the right-hand side are for exchange between two disks from the lower surface to the upper surface.

Small differences in large numbers can occur in obtaining an F factor by using configuration factor algebra, as might occur on the right side of the last equation of the preceding example. Care must be taken that enough significant figures are retained.

Example 5.8

The internal surface of a hollow circular cylinder of radius R is radiating to a disk A_1 of radius r as in Figure 5.11. Express the configuration factor from the internal cylindrical side A_3, to the disk in terms of disk-to-disk factors for the case of $r \leq R$.

From any position on A_1, the solid angle subtended when viewing A_3 is the difference between the solid angle when viewing A_2, or $d\Omega_2$, and that viewing A_4, or $d\Omega_4$. This gives the F_{d1-3} factor from an area element dA_1 on A_1 to area A_3 as $F_{d1-3} = F_{d1-2} - F_{d1-4}$. By integrating over A_1 and using Equation (5.22), $F_{1-3} = F_{1-2} - F_{1-4}$. The factors on the right are between parallel disks. The result for F from the internal cylindrical side A_3 to the disk A_1 is $F_{3-1} = \frac{A_1}{A_3}(F_{1-2} - F_{1-4})$ $(r \leq R)$.

From symmetry, the configuration factor from A_1 to any sector A_s of A_1 is $(A_s/A_1)F_{3-1}$.

In formulating relations among configuration factors, it is sometimes useful to use *energy quantities* rather than fractions of energy leaving a surface that reaches another surface. For example, in Figure 5.11, the energy leaving A_2 that arrives at A_1 is proportional to A_2F_{2-1} and is equivalent to the sums of the energies from A_3 and A_4 that arrive at A_1. Thus,

$$(A_3 + A_4)F_{(3+4)-1} = A_3F_{3-1} + A_4F_{4-1} \tag{5.24}$$

This can also be proved by using reciprocity relations:

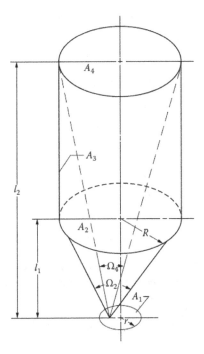

FIGURE 5.11 Geometry for a cylindrical cavity radiating to circular disk A_1 for $r \leq R$.

$$(A_3 + A_4)F_{(3+4)-1} = A_1 F_{1-(3+4)} = A_1 F_{1-3} + A_1 F_{1-4} = A_3 F_{3-1} + A_4 F_{4-1}.$$

5.3.2 CONFIGURATION FACTOR RELATIONS IN ENCLOSURES

So far, only configuration factors between two surfaces have been considered, although the subdivision of one or both surfaces into smaller areas has been examined. Consider the very important situation where the configuration factors are among surfaces that form a complete enclosure.

For an enclosure of N surfaces, such as in Figure 5.12, where for example, $N = 8$, the entire energy leaving any surface k inside the enclosure must all be incident on surfaces of the enclosure. Thus, all the fractions of energy leaving one surface and reaching the surfaces of the enclosure must total to one:

$$F_{k-1} + F_{k-2} + F_{k-3} + \cdots + F_{k-k} + \cdots + F_{k-N} = \sum_{j=1}^{N} F_{k-j} = 1 \qquad (5.25)$$

The factor F_{k-k} is included because when A_k is *concave*, it will intercept a portion of its own outgoing radiative energy.

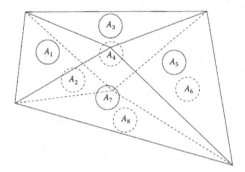

FIGURE 5.12   Geometry for an isothermal enclosure composed of black surfaces.

Example 5.9

Two diffuse isothermal concentric spheres are exchanging energy. Find all the configuration factors for this geometry if the surface area of the inner sphere is A_1 and the surface area of the outer sphere is A_2.

All energy leaving A_1 is incident upon A_2, so $F_{1-2} = 1$. Using the reciprocal relation gives $F_{2-1} = A_1F_{1-2}/A_2 = A_1/A_2$. From Equation (5.25), $F_{2-1} + F_{2-2} = 1$, giving $F_{2-2} = 1 - F_{2-1} = (A_2 - A_1)/A_2$.

Example 5.10

An isothermal cavity of internal area A_1 has a plane opening of area A_2. Derive an expression for the configuration factor of the internal surface of the cavity to itself.

Assume that a black plane surface A_2 replaces the cavity opening; this has no effect on the F_{1-1} factor, which depends only on geometry for diffuse surfaces. Then $F_{2-1} = 1$ and $F_{1-2} = A_2F_{2-1}/A_1 = A_2/A_1$, which is the configuration factor from the entire internal area to the opening. Since A_1 and A_2 form an enclosure, $F_{1-1} = 1 - F_{1-2} = (A_1 - A_2)/A_1$, which is the desired F factor.

Example 5.11

A long enclosure of triangular cross-section consists of three plane areas, each of finite width and infinite length, thus forming a hollow, infinitely long triangular prism. Derive an expression for the configuration factor between any two of the areas in terms of their widths L_1, L_2, and L_3.

For area 1, $F_{1-2} + F_{1-3} = 1$ since $F_{1-1} = 0$. Using similar relations for each area and multiplying through by the respective areas result in $A_1F_{1-2} + A_1F_{1-3} = A_1$; $A_2F_{2-1} + A_2F_{2-3} = A_2$; $A_3F_{3-1} + A_3F_{3-2} = A_3$.

By applying reciprocal relations to some terms, these equations become:

$$A_1F_{1-2} + A_1F_{1-3} = A_1; \quad A_1F_{1-2} + A_2F_{2-3} = A_2; \quad A_1F_{1-3} + A_2F_{2-3} = A_3$$

thus giving three equations for the three unknown F factors. Solving gives

$$F_{1-2} = \frac{A_1 + A_2 - A_3}{2A_1} = \frac{L_1 + L_2 - L_3}{2L_1}.$$

When $L_1 = L_2$, this reduces to the factor between infinitely long adjoint plates of equal width separated by an angle α as in Example 5.5:

$$F_{1-2} = \frac{2L_1 - L_3}{2L_1} = 1 - \frac{L_3/2}{L_1} = 1 - \sin\frac{\alpha}{2}$$

Examine the set of three simultaneous equations yielding the result in Example 5.11. The first equation involves two unknowns, F_{1-2} and F_{1-3}; the second has one additional unknown F_{2-3}; and the third has no additional unknowns. Generalizing the procedure from a three-surface enclosure to any N-sided enclosure of plane or convex surfaces shows that, of N simultaneous equations, the first involves $N-1$ unknowns, the second $N-2$ unknowns, and so forth. The total number of unknowns U is then:

$$U = (N-1) + (N-2) + \cdots 1 = \frac{N(N-1)}{2} \tag{5.26}$$

Thus, $[N(N-1)/2] - N = N(N-3)/2$ factors must be provided. For a four-sided enclosure made up of planar or convex areas, we can write four equations relating $4(4-1)/2 = 6$ unknown configuration factors. Specifying any two factors allows calculation of the rest by solving the four simultaneous equations.

If any surface k can view itself, the factor F_{k-k} must be included in each of the equations. Analyzing this situation shows that an N-sided enclosure provides N equations in $N(N+1)/2$ unknowns. Thus, $[N(N+1)/2] - N = N(N-1)/2$ factors must be specified. For a four-sided enclosure, four equations involving ten unknown F factors can be written. The specification of six factors is required; then, the simultaneous relations can be solved for the remaining four factors. If only M surfaces can view themselves, then $[N(N-3)/2] + M$ factors must be specified. For some geometries, the use of symmetry can further reduce the number of factors that are required.

5.3.3 Techniques for Evaluating Configuration Factors

Evaluating configuration factors F_{d1-2} and F_{1-2} requires integration over the finite areas involved. Various mathematical methods are useful for evaluating configuration factors when analytical integration is too cumbersome or is not possible. A few especially useful methods are discussed next.

5.3.3.1 Hottel's Crossed-String Method

Consider configurations such as long grooves in which all surfaces are assumed to extend infinitely along one coordinate. Such surfaces are generated by moving a line such that it is always parallel to its original position. A typical configuration is shown in cross-section in Figure 5.13.

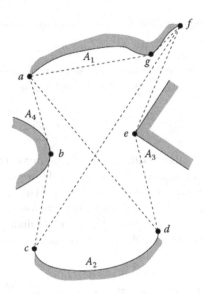

FIGURE 5.13 Geometry for determining configuration factors using Hottel's crossed-string method.

Suppose that the configuration factor is needed between A_1 and A_2 when some blockage of radiant transfer occurs because of other surfaces A_3 and A_4. To obtain F_{1-2}, first consider that A_1 may be concave. In this case, draw the dashed line agf across A_1. Then draw in the dashed lines cf and abc to complete the enclosure $abcfga$, which has three sides that are either convex or planar. The relation found in Example 5.11 for enclosures of this type is:

$$A_{agf}F_{agf-abc} = \frac{A_{agf} + A_{abc} - A_{cf}}{2} \qquad (5.27)$$

For the three-sided enclosure $adefga$, similar reasoning gives:

$$A_{agf}F_{agf-def} = \frac{A_{agf} + A_{def} - A_{ad}}{2} \qquad (5.28)$$

Further, note that:

$$F_{agf-abc} + F_{agf-cd} + F_{agf-def} = 1 \qquad (5.29)$$

Observing that $F_{agf-cd} = F_{agf-2}$ and substituting Equations (5.27) and (5.28) into Equation (5.29) results in:

$$A_{agf}F_{agf-2} = A_{agf}(1 - F_{agf-abc} - F_{agf-def}) = \frac{A_{cf} + A_{ad} - A_{abc} - A_{def}}{2} \qquad (5.30)$$

Now $F_{2-agf} = F_{2-1}$ since A_{agf} and A_1 subtend the same solid angle when viewed from A_2. Then, with the additional use of reciprocity, the left side of Equation (5.30) is:

$$A_{agf} F_{agf-2} = A_2 F_{2-agf} = A_2 F_{2-1} = A_1 F_{1-2} \qquad (5.31)$$

Substituting Equation (5.31) into Equation (5.30) results in:

$$A_1 F_{1-2} = \frac{A_{cf} + A_{ad} - A_{abc} - A_{def}}{2} \qquad (5.32)$$

If the dashed lines in Figure 5.13 are imagined as being lengths of string stretched tightly between the outer edges of the surfaces, then the term on the right of Equation (5.32) is interpreted as one-half the total quantity formed by the sum of the lengths of the crossed strings connecting the outer edges of A_1 and A_2 minus the sum of the lengths of the uncrossed strings. This is a useful way of determining configuration factors in two-dimensional (2D) geometries. It was first described by Hottel (1954).

Example 5.12

Two infinitely long semicylindrical surfaces of radius R are separated by a minimum distance D as in Figure 5.14a. Derive the configuration factor F_{1-2}.

The length of crossed-string $abcde$ is denoted as L_1 and that of uncrossed string ef as L_2. From the symmetry of the geometry, Equation (5.32) gives
$$F_{1-2} = \frac{2L_1 - 2L_2}{2\pi R} = \frac{L_1 - L_2}{\pi R}.$$
Then, $L_2 = D + 2R$. The L_1 is twice the length cde. The segment of L_1, from c to d, is found from right triangle $0cd$ to be:

$$L_{1,c-d} = \left[\left(\frac{D}{2} + R\right)^2 - R^2\right]^{1/2} = \left[D\left(\frac{D}{4} + R\right)\right]^{1/2}$$

The segment of L_1 from d to e is $L_{1,\,d-e} = R\varphi$ where, from triangle $0cd$, $\varphi = \sin^{-1}[R/(D/2 + R)]$. Then,

$$F_{1-2} = \frac{L_1 - L_2}{\pi R} = \frac{2(L_{1,c-d} + L_{1,d-e}) - L_2}{\pi R}$$

$$= \frac{[4D(D/4 + R)]^{1/2} + 2R\sin^{-1}[R/(D/2 + R)] - D - 2R}{\pi R}$$

Letting $X = 1 + D/2R$ gives:

$$F_{1-2} = \frac{2}{\pi}\left[(X^2 - 1)^{1/2} + \sin^{-1}\frac{1}{X} - X\right] \qquad (5.12.1)$$

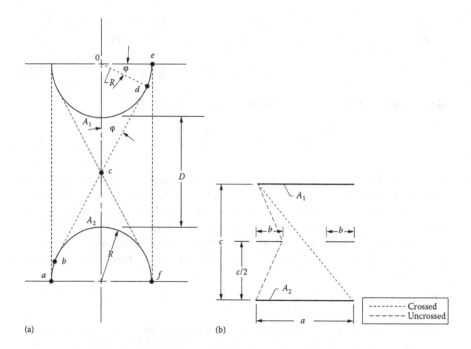

FIGURE 5.14 Examples of applying the crossed-string method: (a) configuration factor between infinitely long semicylindrical surfaces and (b) partially blocked view between parallel strips.

Example 5.13

The view between two infinitely long parallel planes of width a is partially blocked by strips of width b, as in Figure 5.14b. Obtain the factor F_{1-2}.

The length of each crossed string is $(a^2 + c^2)^{1/2}$, and the length of each uncrossed string is $2[b^2 + (c/2)^2]^{1/2}$. From the crossed-string method, the configuration factor is then:

$$F_{1-2} = \frac{\sqrt{a^2 + c^2} - 2\sqrt{b^2 + (c/2)^2}}{a} = \sqrt{1 + \left(\frac{c}{a}\right)^2} - \sqrt{\left(\frac{2b}{a}\right)^2 + \left(\frac{c}{a}\right)^2}$$

As expected, $F_{1-2} \to 0$ as b is extended inward so that $b \to a/2$. A warning: in this and many other geometries, the requirement imposed in Chapters 6 and 7 is that there is uniform emitted and reflected energy over each finite surface. This is unlikely to be the case here.

5.3.3.2 Contour Integration

A useful tool for evaluating configuration factors is to apply Stokes' theorem to reduce the multiple integrations over a surface area to a single integration around the boundary of the area. This method is used in many commercial programs for finding configuration factors and can greatly reduce the computational time. A complete exposition of the method, along with examples of its implementation is in Howell et al. (2021).

5.3.3.3 Differentiation of Known Factors

An extension of configuration factor algebra is to obtain configuration factors to differential areas by differentiating known factors to finite areas. This technique is useful in certain cases, as in the following example.

Example 5.14

As part of the determination of radiative exchange in a channel whose temperature varies longitudinally, the configuration factor dF_{d1-d2} is needed between an element dA_1 at a point along the channel and a second wall element dA_2 as in Figure 5.15c.

Configuration factor algebra plus differentiation can be used to find the required factor. Start with F_{1-2} for two parallel areas A_1 and A_2 that are cross sections of a cylindrical channel of arbitrary cross-section (Figure 5.15a). This factor

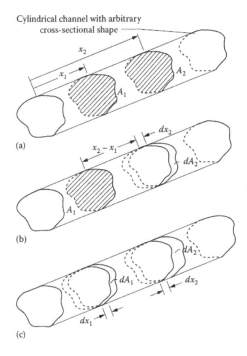

FIGURE 5.15 Geometry for the derivation of the configuration factors for differential areas starting from the factors for finite areas: (a) two finite areas, F_{1-2}; (b) finite to differential area, $dF_{1-d2} = -(\partial F_{1-2}/\partial x_2)dx_2$; and (c) two differential areas, $dF_{d1-d2} = -(A_1/dA_1)(\partial^2 F_{1-2}/\partial x_1 \partial x_2)dx_2 dx_1$.

depends on the spacing $|x_2-x_1|$ between the two areas and includes blockage due to the channel wall; that is, it is the factor by which A_2 is viewed from A_1 with the channel wall present. The factor between A_1 and dA_2 in Figure 5.15b is then given by:

$$dF_{1-d2} = -\frac{\partial F_{1-2}}{\partial x_2} dx_2 \qquad (5.14.1)$$

Equation (5.14.1) is now used to obtain dF_{d1-d2}, the configuration factor between the two differential area elements dA_1 and dA_2 in Figure 5.15c.

By reciprocity, $F_{d2-1} = (-A_1/dA_2)(\partial F_{1-2}/\partial x_2)dx_2$. Then like the derivation of Equation (5.14.1), $dF_{d2-d1} = (\partial F_{d2-1}/\partial x_1)dx_1$. Substituting F_{d2-1} results in:

$$dF_{d2-d1} = -\frac{A_1}{dA_2}\frac{\partial^2 F_{1-2}}{\partial x_1 \partial x_2} dx_2 dx_1 \qquad (5.14.2)$$

or after using reciprocity,

$$dF_{d1-d2} = -\frac{A_1}{dA_1}\frac{\partial^2 F_{1-2}}{\partial x_1 \partial x_2} dx_2 dx_1 \qquad (5.14.3)$$

Hence, by two differentiations, the factor dF_{d1-d2} can be found from F_{1-2} for the cylindrical configuration. For example, if the channel is circular with radius r, the disk-disk factor is (http://www.thermalradiation.net/sectionc/C-40.html) $F_{1-2} = \frac{1}{2}\left\{X - \left(X^2 - 4\right)^{1/2}\right\}$ where $X = (2R^2 + 1)/R^2$; $R = r/(x_2 - x_1)$. Two differentiations will give the required factor.

5.3.3.4 Unit-Sphere and Hemicube Methods

Experimental measurement of configuration factors is possible by the *unit-sphere method* introduced by Nusselt (1928). If a hemisphere of unit radius $r = 1$ is constructed over the area element dA_1 in Figure 5.16, the configuration factor from dA_1 to an area A_2 is, by Equation (5.11),

$$F_{d1-2} = \frac{1}{\pi}\int_{A_2} \cos\theta_1 \frac{\cos\theta_2 dA_2}{S^2} = \frac{1}{\pi}\int_{A_2} \cos\theta_1 d\Omega_1$$

The projection of dA_2 onto the surface of the hemisphere is $d\Omega_1$, because $d\Omega_1 = \frac{dA_s}{r^2} = dA_s = \frac{\cos\theta_2 dA_2}{S^2}$. The F_{d1-2} is then $F_{d1-2} = \frac{1}{\pi}\int_{A_s}\cos\theta_1 dA_s$. However, $dA_s\cos\theta_1$ is the projection of dA_s onto the base of the hemisphere. It follows that integrating $\cos\theta_1 dA_s$ gives the projection A_b of A_s onto the base of the hemisphere or

$$F_{d1-2} = \frac{1}{\pi}\int_{A_s} \cos\theta_1 dA_s = \frac{A_b}{\pi} \qquad (5.33)$$

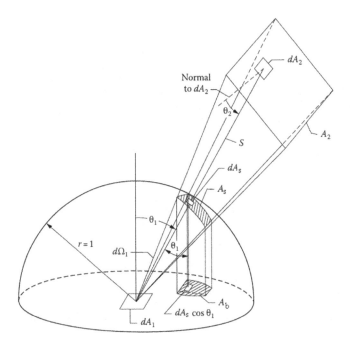

FIGURE 5.16 Geometry of unit-sphere method for obtaining configuration factors.

The relation in Equation (5.33) is the basis of several graphical and experimental methods for determining configuration factors. In one method, a spherical sector mirrored on the outside is placed over the area element dA_1. A photograph taken by a camera placed directly above the sector and normal to dA_1 then shows the projection of A_2, which is A_b. The measurement of A_b on the photograph then provides F_{d1-2} from $F_{d1-2} = A_b/\pi r_e^2$, where r_e is the radius of the experimental mirrored sector.

The *hemicube* approach to finding configuration factors is a modified unit-sphere method to meet speed requirements for applications in computer graphics. The method uses the fact that the configuration factor is the same from any object that subtends the same solid angle to the viewer, regardless of the actual shape and orientation of the object. Projection onto the face of one-half of a cube centered over the receiving element dA_1 rather than onto a hemisphere, as shown in Figure 5.17, is more convenient for some purposes. In practice, the five surfaces of the hemicube are divided into small square elements (pixels), and the configuration factors from the element dA_1 to each pixel are precomputed using element-pixel configuration factors.

5.3.3.5 Direct Numerical Integration

Configuration factors between a differential area and a finite area, or between two finite areas, can be found by direct numerical integration of Equation (5.11) or (5.16). An example of a differential-to-finite area geometry is carried out in the online Appendix C, Example C.1, at www.ThermalRadiation.net/appendix.html.

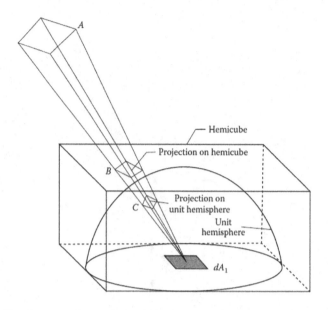

FIGURE 5.17   Geometry for hemicube analysis and relation to a unit sphere.

Commercial and freeware computer programs are available that use one or more of the methods outlined in this chapter for numerical calculation of configuration factors. These are discussed in Section E.2 of the online appendices listed at www. ThermalRadiation.net/appendix.html.

These codes provide a means to generate configuration factors for complex geometries and are invaluable for radiative analyses in such systems.

5.3.3.6   The Monte Carlo Method

The Monte Carlo method can be used to find configuration factors by following energy bundles from a diffusely emitting surface 1 to a receiving surface 2 based on the probabilities of the bundles being emitted into particular directions and from uniformly distributed locations on surface 1. The fraction of the bundles striking surface 2 is the configuration factor for the geometry under study. An exposition of the method is in Chapter 10.

5.4   CONSTRAINTS ON CONFIGURATION FACTOR ACCURACY

To obtain accurate results when configuration factors are used for calculating radiative transfer in an enclosure, the configuration factors must satisfy their physical constraints. One constraint is that factors for any pair of surfaces satisfy reciprocity. A second constraint provides that the sum of all the factors from a surface to all surrounding surfaces in the enclosure, including the view to itself, must equal 1 for energy to be conserved; this is called the "closure" constraint. These constraints may not be accurately satisfied by factors individually evaluated in complex enclosures with many surfaces. Factors may become difficult to evaluate to high precision when enclosures have complex shapes with obstructions and partial views of surfaces. Individual factors are often computed by numerical integrations that introduce approximations or by statistical methods such as Monte Carlo that may be converged into a prescribed accuracy.

It was shown in Section 5.3 that if some configuration factors are calculated in an enclosure, the physical constraints can be used to obtain the remaining factors by solving a set of simultaneous equations. Sowell and O'Brien (1972) point out that in an enclosure with many surfaces, the configuration factors specified or computed easily may not lend themselves to subsequent accurate calculation of the remaining factors. Reciprocity or the energy conservation (closure) relation, Equation (5.25), may be difficult to apply. They present a computer scheme using matrix algebra that allows calculations of all remaining factors in an N-surface planar or convex-surface enclosure once the required minimum number of configuration factors is specified.

To improve the accuracy of calculations, some methods are discussed by Larsen and Howell (1986) for smoothing sets of "direct exchange areas" for enclosures, using the constraints imposed by reciprocity and energy conservation. These methods are directly applicable to configuration factors. Van Leersum (1989) provides a method that guarantees overall energy conservation for the enclosure. An uncertainty analysis by Taylor et al. (1995) shows that strict enforcement of the reciprocity and closure constraints yields an order-of-magnitude reduction in the uncertainty in energy flux results that arises from uncertainties in the areas and configuration factors. Methods for adjusting the factors to provide enforcement are provided in Taylor and Luck (1995) and in Vercammen and Froment (1980). A method that enforces both reciprocity and conservation while also enforcing a nonnegativity constraint on all factors using least-squares minimization is given by Daun et al. (2005).

5.5 COMPILATION OF KNOWN CONFIGURATION FACTORS

Many configuration factors for specific geometries have been derived in analytical form or tabulated and are spread throughout the literature. Some factors that have convenient analytical forms are given in online Appendix L at www.ThermalRadiation.net/appendix.html for use in examples and homework problems. An open website giving extensive information on configuration factors is housed at www.ThermalRadiation.net/appendix.html, also accessible through the QR code shown. That website provides references, algebraic relations, tables, and graphical results for some 325 configurations.

HOMEWORK (additional problems available in online Appendix I at www.ThermalRadiation.net/appendix.html)

5.1   Derive the configuration factor F_{d1-2} between a differential area centered above a disk and a finite disk of unit radius parallel to the dA_1.

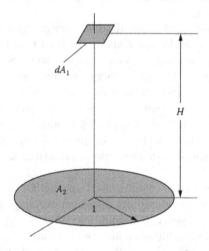

Answer: $1/(H^2 + 1)$.

5.2   Find the configuration factor F_{d1-2} from a planar element to a coaxial parallel rectangle as shown below; the dA_1 is above the center of the rectangle. Use any method.

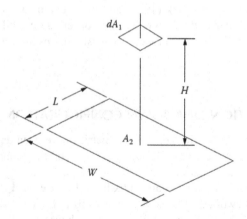

Answer:

$$F_{d1-2} = \frac{2}{\pi}\left\{\left[\frac{X}{(1+X^2)^{1/2}}\right]\tan^{-1}\left[\frac{Y}{(1+X^2)^{1/2}}\right]+\left[\frac{Y}{(1+Y^2)^{1/2}}\right]\tan^{-1}\left[\frac{X}{(1+Y^2)^{1/2}}\right],\right.$$

where $X = L/2H \; Y = W/2H$

5.3   Find the configuration factor F_{1-2} for the geometries shown below. Give a numerical value using any method:

(a)   Two plates infinitely long normal to the cross-section shown:

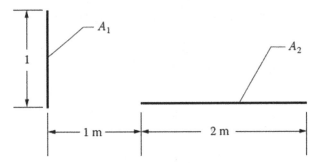

(b)   Two plates of finite length as shown:

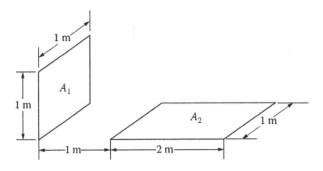

Answers: (a) 0.126; (b) 0.0417.

5.4   Find the configuration factor F_{1-2} for the geometries shown below. Give a numerical value using any method:

(a)   Disk to concentric ring

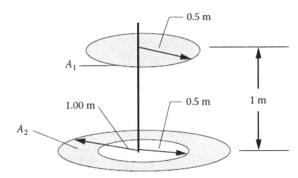

(b)   Configuration factor F_{1-2} between the interior conical surface A_1 of a frustum of a right circular cone and a hole in its top

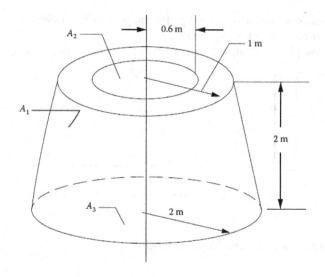

(c) Configuration factor from infinitely long rectangular shape 1 shown in cross-section to infinitely long plane 2

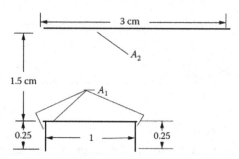

Answers: (a) 0.2973; (b) 0.02743; (c) 0.4164.

5.5   Derive a formula for F_{2-2} in terms of α. The A_1 is inside the cone, and A_2 is part of a sphere.

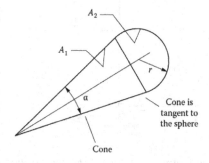

Answer: $F_{2-2} = \dfrac{1}{2}\left(1 + \sin\dfrac{\alpha}{2}\right)$.

5.6 Find the configuration factor between the two infinitely long parallel plates shown below in cross-section. Use

(a) The crossed-string method

(b) Configuration factor algebra with factors from online Appendix L at www.ThermalRadiation.net/appendix.html.

(c) Integration of differential strip-to-differential strip factors

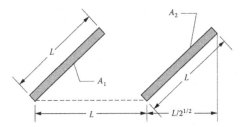

Answer: 0.3066.

5.7 For the cylindrical enclosure of diameter d and length $l = 3.3d$ shown below, find all configuration factors among the surfaces.

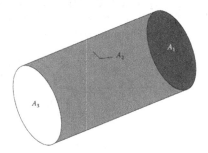

Answer: $F_{2-2} = 0.852$.

5.8 Using the factor for an "infinitely long enclosure formed of three plane areas" derived in Example 5.11, find the factor from one side of an infinitely long enclosure to its base. The enclosure cross-section is an isosceles triangle with apex angle α.

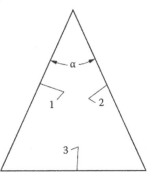

Answer: $F_{1-3} = \sin \dfrac{\alpha}{2}$.

5.9 A four-sided enclosure is formed of three mutually perpendicular isosceles triangles of short side S and hypotenuse H and an equilateral triangle of side H. Find the configuration factor F_{1-2} between two perpendicular isosceles triangles sharing a common short side.

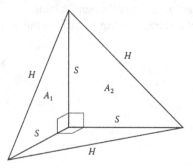

Answer: 0.2113.

5.10 Using the crossed-string method, derive the configuration factor F_{1-2} between the infinitely long plate and the parallel cylinder shown below in cross-section. Compare your result with that given for configuration 11 in online Appendix C at www.ThermalRadiation.net/text/html.

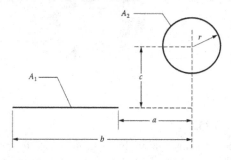

Answer: $F_{1-2} = \left[\dfrac{r}{(b-a)} \right] \left[\tan^{-1}\left(\dfrac{b}{c} \right) - \tan^{-1}\left(\dfrac{a}{c} \right) \right]$.

5.11 For the 2D geometry shown, the view between A_1 and A_2 is partially blocked by two identical cylinders. Determine the configuration factor F_{1-2}.

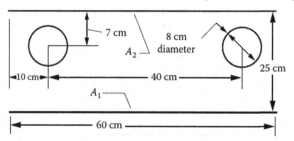

Answer: 0.667.

5.12 An infinitely long enclosure is shown below in cross-section. The outer surfaces form a square in cross-section, and the outer surfaces are parallel to the inner circular coaxial cylinder, so the geometry is 2D. Find the configuration factors F_{2-1}, F_{2-3}, and F_{2-4}.

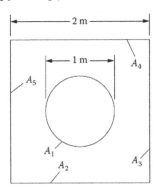

Answers: $F_{2-1} = \pi/8$; $F_{2-3} = 0.2482$; $F_{2-4} = 0.1109$.

5.13 (a) Derive the configuration factor between a sphere of radius R and a coaxial disk of radius r. (Hint: Think of the disk as being a cut through a spherical envelope concentric around the sphere.)

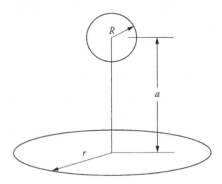

(b) What is the F from a sphere to a sector of a disk?

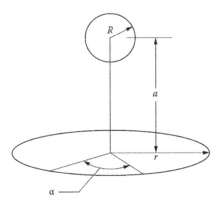

(c) What is the dF from a sphere to a portion of a ring of differential width?

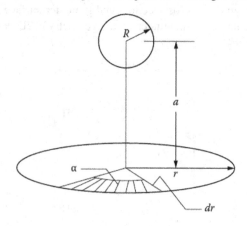

Answers:

(a) $\dfrac{1}{2}\left[1-\dfrac{a}{(a^{2}+r^{2})^{1/2}}\right]$;   (b) $\dfrac{\alpha}{4\pi}\left[1-\dfrac{a}{(a^{2}+r^{2})^{1/2}}\right]$;   (c) $\dfrac{\alpha}{4\pi}\dfrac{ar}{(a^{2}+r^{2})^{3/2}}\,dr$.

5.14 Consider the interior of a black cubical enclosure. Determine the configuration factors between (a) two adjacent walls and (b) two opposite walls. A sphere of diameter equal to one-half the length of a side of the cube is placed at the center of the cube. Determine the configuration factors between (c) the sphere and one wall of the enclosure, (d) one wall of the enclosure and the sphere, and (e) the enclosure and itself.

Answers: (a) 0.20004; (b) 0.19982; (c) 1/6; (d) 0.1309; (e) 0.8691.

6 Radiation Exchange in Enclosures Bounding Transparent Media

Leopoldo Nobili (1784–1835) (left) and **Macedonio Melloni (1798–1854)** (right) *developed a thermopile-based radiometer read by a galvanometer, and investigated radiation from various sources. They showed (1831) that different surfaces emitted differing amounts of radiation at the same temperature, and that the radiometer reacted similarly to light sources and heated surfaces.*

6.1 INTRODUCTION

Now, we first consider radiation exchange in an enclosure where all surfaces are black, so that no reflections need to be considered. These ideal surfaces emit diffusely, so the intensity leaving each surface is independent of direction. For emission from an isothermal black surface, the configuration factors discussed in Chapter 5 can then be used to calculate how much radiation will reach another surface. The relation for exchange between two surfaces is applied to multiple surfaces, each at a different uniform temperature, in an enclosure of black surfaces.

In the next step, the surfaces of the enclosure are assumed to be *diffuse* and *gray*, so the directional spectral emissivity and absorptivity of each surface do not depend on direction or wavelength but may depend on surface temperature. At any surface temperature T, the hemispherical total absorptivity and emissivity are equal and depend only on T: $[\alpha(T) = \varepsilon(T)]$. Although this behavior is approached by only a limited number of real materials, the diffuse-gray approximation greatly simplifies enclosure theory.

We then briefly consider enclosures that *do* have spectrally and/or directionally dependent surfaces.

What is meant by the individual "surfaces" or "areas" of an enclosure? Usually, the geometry tends to divide the enclosure into surface areas, such as the individual sides of a rectangular chamber. It also may be necessary to specify surface areas by their heating or cooling conditions; for example, if one side of an enclosure has portions at two different temperatures, that side would be divided into two areas so this difference in boundary condition could be included. An area also may be subdivided by surface characteristics, such as separate smooth and rough portions with different emissivities. Surface areas in the radiation analysis are then defined as each portion

DOI: 10.1201/9781003178996-6

of the enclosure boundary for which an energy balance is formed; these portions are selected based on geometry, imposed heating or temperature conditions, or surface characteristics. A further consideration is solution accuracy. If too few areas are designated, the accuracy may be poor because significant nonuniformity in reflected flux over an area is not accounted for in the analysis. Engineering judgment is required in selecting the size and shape of the enclosure areas and their number.

Enclosure surface areas can have various imposed thermal boundary conditions. A surface may have a specified temperature, a specified energy input, or be perfectly insulated from external energy transfer. The analysis in this chapter requires that each surface area for the enclosure analysis be at a uniform temperature. If the heating conditions are such that the temperature would vary markedly over an area, the area must be subdivided into smaller, more isothermal portions; if necessary, these portions can be of differential size. From this isothermal area requirement, the emitted energy is uniform over each surface.

A gray surface reflects only a portion of the incident energy. Two assumptions are made for reflected energy: (1) it is diffuse, so the reflected intensity at each position is uniform in all directions, and (2) it is uniform over each surface area. If the reflected energy is expected to vary over an area, the area again must be subdivided so that the variation is not significant over each surface area considered for the analysis. With these restrictions reasonably met, the reflected energy for each area can be assumed to be diffuse and uniform as is the emitted energy. Hence, the reflected and emitted energies can be combined into a single diffuse energy flux leaving the area. The geometric configuration factors can then be used for the enclosure analysis. The derivation of the F factors was based on a *diffuse-uniform intensity* leaving the surface; this diffuse-uniform condition must be valid for the sum of emitted and reflected energies.

An analysis assuming diffuse-gray surfaces may not yield good results. For example, if the temperatures of the individual areas of the enclosure differ considerably from each other, then an area will be emitting predominantly in the range of wavelengths characteristic of its temperature while receiving energy predominantly in different wavelength regions. If the spectral emissivity varies with wavelength, the fact that the incident radiation has a different spectral distribution from the emitted energy makes the gray assumption invalid; that is, $\varepsilon(T) \neq \alpha(T)$. When polished (specular) surfaces are present, the diffuse reflection assumption is invalid, and the directional paths of the reflected energy must be considered.

In summary, for the analysis in Sections 6.2 through 6.4, the enclosure boundary must be subdivided into areas so that *over each such area* the following restrictions are met:

1. The surface is opaque.
2. The temperature is uniform.
3. The surface properties are uniform.
4. The ε_λ, α_λ, and ρ_λ are independent of wavelength and direction so that $\varepsilon(T) = \alpha(T) = 1 - \rho(T)$, where ρ is the reflectivity.
5. All energy is emitted and reflected diffusely.
6. The incident and hence reflected energy flux is uniform over each individual area.

6.2 RADIATIVE TRANSFER AMONG BLACK SURFACES

From Equation (2.34), the blackbody total intensity is related to the blackbody hemispherical total emissive power by $I_b = E_b/\pi = \sigma T^4/\pi$, so the *net* energy per unit time $d^2Q_{d1\leftrightarrow d2}$ transferred from black element dA_1 to black element dA_2 along path S is (Figure 6.1):

$$d^2Q_{d1\leftrightarrow d2} = \sigma\left(T_1^4 - T_2^4\right)\frac{\cos\theta_1\cos\theta_2}{\pi S^2}dA_1 dA_2 \qquad (6.1)$$

Example 6.1

The Sun emits energy at a rate that can be approximated as a blackbody at $T_s = 5780$ K. A blackbody area element in orbit around the Sun at the mean radius of the Earth's orbit, 1.49×10^{11} m, is oriented normal to the line connecting the centers of the area element and the Sun. If the Sun's radius is $R_S = 6.95 \times 10^8$ m, what solar flux is incident upon the element?

To the element in orbit, the Sun appears as a diffuse isothermal black disk element of area $dA_1 = \pi R_S^2 = \pi\left(6.95 \times 10^8\right)^2 = 1.52 \times 10^{18}$ m^2. From the derivation of Equation (6.1), since $\theta_1 = \theta_2 = 0$, the incident energy flux on element dA_2 in orbit is $\left(\sigma T_s^4/\pi\right)dA_1\left(\cos\theta_1\cos\theta_2/S^2\right) = \left(\sigma T_s^4/\pi\right)\left(dA_1/S^2\right) = \left(5.6704 \times 10^{-8} \times 5780^4/\pi\right)\left[1.52 \times 10^{18}/\left(1.49 \times 10^{11}\right)^2\right] = 1379$ W/m^2. This value is consistent with measured values of the solar constant, 1353 to 1394 W/m^2 and the accepted standard value of 1366 W/m^2 (Section 4.6).

An alternative procedure is to utilize the fact that radiant energy leaves the Sun with spherical symmetry. The energy radiated from the solar sphere is

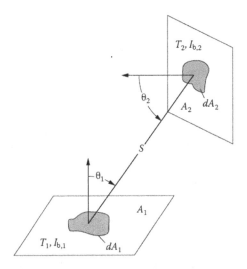

FIGURE 6.1 Radiative exchange between two black isothermal elements.

$\sigma T_s^4 4\pi R_S^2$, and the area of a sphere surrounding the Sun and having a radius equal to the Earth's orbit is $4\pi S^2$. Hence, the flux received at the Earth's orbit is $\sigma T_s^4 4\pi R_S^2 / 4\pi S^2 = \sigma T_s^4 \left(R_S/S\right)^2$, as obtained before.

Example 6.2

As shown in Figure 6.2, a black square with a side 0.25 cm is at $T_1 = 1100$ K and is near a tube 0.30 cm in diameter. The tube opening acts as a black surface at 700 K. What is the net radiative transfer from A_1 to A_2 along the connecting path S?

For this geometry, A_1 and A_2 have small dimensions compared with the length S between them. If Equation 6.1 is used, the $\cos\theta_1$, $\cos\theta_2$, and S do not vary significantly for positions along A_1 and A_2. Hence, accurate transfer results can be obtained without integrating to obtain an F factor. Then, for Equation (6.1), $\cos\theta_1$ is found from the sides of the right triangle A_2–0–A_1 as $\cos\theta_1 = 5/(8^2 + 5^2)^{1/2} = 5/89^{1/2}$. The other factors in the energy exchange equation are given, so the net radiative exchange is:

$$\sigma\left(T_1^4 - T_2^4\right)\frac{\cos\theta_1 \cos\theta_2}{\pi S^2}A_1 A_2 = 5.6704\times10^{-8}\left(1100^4 - 700^4\right)\frac{5}{89^{1/2}}\frac{\cos 20°}{\pi\left(89/10^4\right)}\frac{0.25^2}{10^4}\frac{\pi 0.30^2}{4\times10^4}$$

$$= 5.462\times10^{-5} \text{ W}$$

6.2.1 TRANSFER BETWEEN BLACK SURFACES USING CONFIGURATION FACTORS

Using Equation (5.7), the geometric factors in Equation (6.1) for black area elements are put in terms of configuration factors as:

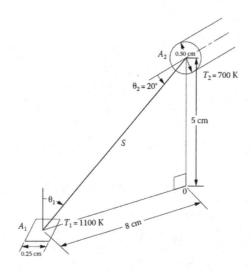

FIGURE 6.2 Radiative transfer between a small square area and a small circular-tube opening.

$$d^2Q_{d1 \leftrightarrow d2} = \sigma\left(T_1^4 - T_2^4\right)dF_{d1-d2}dA_1 = \sigma\left(T_1^4 - T_2^4\right)dF_{d2-d1}dA_2 \qquad (6.2)$$

Similarly, for radiative transfer between a black differential element and a black finite area,

$$dQ_{d1 \leftrightarrow 2} = dQ_{d1 \rightarrow 2} - dQ_{2 \rightarrow d1} = \sigma T_1^4 dA_1 \, F_{d1-2} - \sigma T_2^4 A_2 \, dF_{2-d1} \qquad (6.3)$$

or, using Equation (5.15), the net transfer is:

$$dQ_{d1 \leftrightarrow 2} = \sigma\left(T_1^4 - T_2^4\right)dA_1 \, F_{d1-2} = \sigma\left(T_1^4 - T_2^4\right)A_2 \, dF_{2-d1} \qquad (6.4)$$

For two black surfaces with finite areas:

$$Q_{1 \leftrightarrow 2} = Q_{1 \rightarrow 2} - Q_{2 \rightarrow 1} = \sigma T_1^4 A_1 F_{1-2} - \sigma T_2^4 A_2 F_{2-1} \qquad (6.5)$$

or, using the reciprocity relation, Equation (5.19),

$$Q_{1 \leftrightarrow 2} = \sigma\left(T_1^4 - T_2^4\right)A_1 \, F_{1-2} = \sigma\left(T_1^4 - T_2^4\right)A_2 \, F_{2-1} \qquad (6.6)$$

6.2.2   RADIATION EXCHANGE IN A BLACK ENCLOSURE

Black surface enclosures are seldom the case in practical applications. Also, a real surface may not be isothermal, in which case the surfaces must be subdivided into smaller ones that meet the isothermal approximation. The analysis here is for many black surfaces in an enclosure (Figure 6.3). Equations (6.5) and (6.6) can be written

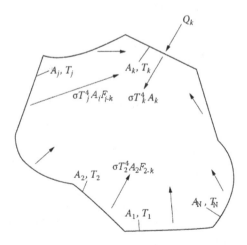

FIGURE 6.3 Enclosure composed of N black isothermal surface areas (shown in cross-section for clarity).

for black surfaces to give the total energy per unit time leaving dA_1 that is incident upon dA_2 as $I_{b,1}\,dA_1\,\cos\theta_1\,dA_2\,\cos\theta_2/S^2$ and the total radiation leaving dA_2 that arrives at dA_1 is $I_{b,2}\,dA_2\,\cos\theta_2\,dA_1\,\cos\theta_1/S^2$. For a black receiving element, all incident energy is absorbed.

An energy balance is formed on an enclosure surface A_k, as in Figure 6.3. The Q_k is the combined energy supplied to A_k by all sources other than by radiation inside the enclosure; this maintains A_k at T_k while A_k exchanges radiation with the enclosure surfaces. The Q_k could be composed of convection energy transfer to the inside of the wall, and/or conduction energy transfer through the wall from an energy source outside, and/or an energy source in the wall itself, such as an electrical heater. For wall cooling, such as by cooling channels within the wall, the contribution to Q_k is negative. The emission from A_k into the enclosure is $\sigma T_k^4 A_k$. The radiant energy received by A_k from another surface A_j is $\sigma T_j^4 A_j F_{j-k}$. The radiation energy balance is:

$$Q_k = \sigma T_k^4 A_k - \sum_{j=1}^{N} \sigma T_j^4 A_j F_{j-k} \tag{6.7}$$

where the summation includes energy arriving from all surfaces inside the enclosure, including A_k itself if A_k is concave. Equation (6.7) can be put in alternative forms. Applying reciprocity to the terms in the summation results in:

$$Q_k = A_k\left(\sigma T_k^4 - \sum_{j=1}^{N} \sigma T_j^4 F_{k-j}\right) \tag{6.8}$$

For a complete enclosure, from Equation (5.25), $\sum_{j=1}^{N} F_{k-j} = 1$, so that:

$$Q_k = A_k\left(\sigma T_k^4 \sum_{j=1}^{N} F_{k-j} - \sigma \sum_{j=1}^{N} T_j^4 F_{k-j}\right) = \sigma A_k \sum_{j=1}^{N}\left(T_k^4 - T_j^4\right) F_{k-j} \tag{6.9}$$

This is the sum of the net radiative energy transferred from A_k to each surface k within the enclosure.

Example 6.3

The three-sided black enclosure of Example 5.11 has its black surfaces maintained at T_1, T_2, and T_3. To maintain each temperature, determine the rate of energy that must be supplied to each surface by means other than radiation inside the enclosure.

This energy rate equals the net radiative loss from each surface by radiative exchange within the enclosure. Equation (6.9) is written for each surface as:

$$Q_1 = A_1\left[F_{1-2}\sigma\left(T_1^4 - T_2^4\right) + F_{1-3}\sigma\left(T_1^4 - T_3^4\right)\right]$$

$$Q_2 = A_2\left[F_{2-1}\sigma\left(T_2^4 - T_1^4\right) + F_{2-3}\sigma\left(T_2^4 - T_3^4\right)\right]$$

$$Q_3 = A_3\left[F_{3-1}\sigma\left(T_3^4 - T_1^4\right) + F_{3-2}\sigma\left(T_3^4 - T_2^4\right)\right]$$

The configuration factors are in Example 5.11. Thus, all factors on the right sides are known, and the Q_k values can be computed. A check on the numerical results is that, from overall energy conservation, the net Q_k added to the entire enclosure, $\sum_{k=1}^{N} Q_k$, must be zero to maintain steady temperatures. This is also shown by adding all the terms on the right sides of the three equations and using reciprocity, such as $A_2 F_{2-1} = A_1 F_{1-2}$. (Try it!)

Example 6.4

The enclosure of Example 5.11 has two sides maintained at T_1 and T_2. The third side is insulated on the outside, and there is only radiative transfer on the inside, so that $Q_3 = 0$. Determine Q_1, Q_2, and T_3.
    Equation (6.9) for each surface is:

$$Q_1 = A_1\left[F_{1-2}\sigma\left(T_1^4 - T_2^4\right) + F_{1-3}\sigma\left(T_1^4 - T_3^4\right)\right]$$

$$Q_2 = A_2\left[F_{2-1}\sigma\left(T_2^4 - T_1^4\right) + F_{2-3}\sigma\left(T_2^4 - T_3^4\right)\right]$$

$$0 = A_3\left[F_{3-1}\sigma\left(T_3^4 - T_1^4\right) + F_{3-2}\sigma\left(T_3^4 - T_2^4\right)\right]$$

The final equation is solved for T_3^4. This is inserted into the first two equations to obtain Q_1 and Q_2.

6.3   RADIATION AMONG FINITE DIFFUSE-GRAY AREAS

6.3.1   NET-RADIATION METHOD FOR ENCLOSURES

Consider an enclosure of N discrete internal surface areas as in Figure 6.4. To analyze the radiation exchange between the surface areas within the enclosure, two common problems are to find: (1) the required energy supplied to a surface when the surface temperature is specified, and (2) the temperature that a surface will achieve when a known energy input is imposed. Some more general boundary conditions are considered later.

A complex radiative exchange occurs inside the enclosure as radiation leaves a surface, travels to other surfaces, is partially reflected, and is then re-reflected many times within the enclosure with partial absorption at each contact with a surface. It is complicated to follow the radiation as it undergoes this process; fortunately, this is not always

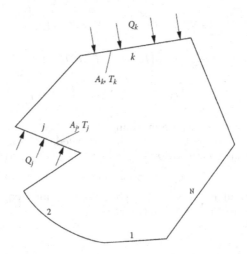

FIGURE 6.4 Enclosure composed of N discrete surface areas with typical surfaces j and k (shown in cross-section for clarity).

necessary. A convenient analysis can be formulated by using the *net-radiation method*. This method was first devised by Hottel and later developed in a different manner by Poljak (1935). An alternative approach was given by Gebhart (1961, 1971). All the methods are basically equivalent. The Hottel–Poljak approach is now given.

Consider the kth *inside* surface area A_k of the enclosure in Figures 6.4 and 6.5.

The G_k and J_k are the rates of incoming and outgoing radiant energy per unit area of A_k. The quantity q_k is the energy flux supplied to A_k by some means other than the radiation inside the enclosure, to make up for the net radiative gain or loss and thereby maintain the specified inside surface temperature. For example, if A_k is the inside surface of a wall of finite thickness, Q_k could be the energy conducted through the wall from the outside to A_k. An energy balance for A_k provides the relation:

$$Q_k = q_k A_k = \left(J_k - G_k\right) A_k \tag{6.10}$$

A second equation results from the energy flux leaving the surface being composed of emitted plus reflected energy. This gives, noting $\alpha_k = \varepsilon_k$ for a gray diffuse surface,

$$J_k = \varepsilon_k \sigma T_k^4 + \rho_k G_k = \varepsilon_k \sigma T_k^4 + \left(1 - \alpha_k\right) G_k = \varepsilon_k \sigma T_k^4 + \left(1 - \varepsilon_k\right) G_k \tag{6.11}$$

where $\rho_k = 1 - \alpha_k = 1 - \varepsilon_k$ has been used for opaque gray surfaces. The term *radiosity* is often used for J_k. The incident flux or *irradiation*, G_k, is derived from the portions of radiant energy leaving all surfaces inside the enclosure that arrive at the kth surface. If the kth surface can view itself (is concave), a portion of its outgoing flux will contribute to its incident flux. The incident energy is then equal to:

$$A_k G_k = A_1 J_1 F_{1-k} + A_2 J_2 F_{2-k} + \cdots + A_j J_j F_{j-k} + \cdots + A_k J_k F_{k-k} + \cdots + A_N J_N F_{N-k} \tag{6.12}$$

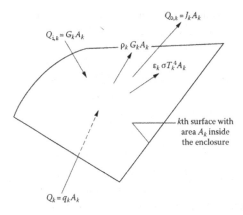

FIGURE 6.5 Energy quantities incident upon and leaving a typical surface inside enclosure.

From configuration-factor reciprocity, Equation (5.19),

$$A_1 F_{1-k} = A_k F_{k-1}; \quad A_2 F_{2-k} = A_k F_{k-2}; \quad \cdots \quad A_N F_{N-k} = A_k F_{k-N} \tag{6.13}$$

Then Equation (6.12) can be written so the only area appearing is A_k:

$$A_k G_k = A_k J_1 F_{k-1} + A_k J_2 F_{k-2} + \cdots + A_k J_j F_{k-j} + \cdots + A_k J_k F_{k-k} + \cdots + A_k J_N F_{k-N} \tag{6.14}$$

so that the incident flux is:

$$G_k = \sum_{j=1}^{N} J_j F_{k-j} \tag{6.15}$$

Equations (6.10), (6.11), and (6.15) are simultaneous relations among q_k, T_k, G_k, and J_k for each surface, and, through Equation (6.15), to the other surfaces. One solution procedure is to note that Equations (6.11) and (6.15) provide two different expressions for G_k. These are each substituted into Equation (6.10) to eliminate G_k and provide two energy-balance equations for q_k in terms of T_k and J_k:

$$\frac{Q_k}{A_k} = q_k = \frac{\varepsilon_k}{1 - \varepsilon_k} \left(\sigma T_k^4 - J_k \right) \tag{6.16}$$

$$\frac{Q_k}{A_k} = q_k = J_k - \sum_{j=1}^{N} J_j F_{k-j} = \sum_{j=1}^{N} \left(J_k - J_j \right) F_{k-j} \tag{6.17}$$

The q_k can be regarded as either the energy flux supplied to surface k by means other than internal radiation (such as by convection or conduction to A_k) or as the net radiative loss from A_k by radiation exchange inside the enclosure. Equation (6.16) or

Equation (6.17) is the balance between net radiative energy loss and energy supplied by means other than the radiation inside the enclosure.

Equations (6.16) and (6.17) are written for each of the N surfaces. This provides $2N$ equations in $2N$ unknowns. The J_k are N of the unknowns, and the remaining unknowns consist of q and T, depending on what boundary quantities are specified. Later, the J_k are eliminated to give N equations relating the N unknown q and T. Other forms of Equation (6.16) are:

$$J_k = \sigma T_k^4 - \frac{1-\varepsilon_k}{\varepsilon_k} q_k \quad \text{or} \quad \sigma T_k^4 = \frac{1-\varepsilon_k}{\varepsilon_k} q_k + J_k \tag{6.18}$$

Equations (6.16) and (6.18) were obtained by eliminating G_k from Equations (6.10) and (6.11). If J_k is eliminated instead, then the results are:

$$q_k = \varepsilon_k \sigma T_k^4 - \varepsilon_k G_k \quad \text{or} \quad G_k = \sigma T_k^4 - \frac{q_k}{\varepsilon_k} \tag{6.19}$$

Equation (6.19) shows that the net energy leaving the surface by radiation is the emitted energy minus the absorbed incident energy $\varepsilon_k G_k = \alpha_k G$. Hence, if q_k and T_k^4 are available from a solution, the incident energy that is absorbed by a gray surface can be found from:

$$\alpha_k G_k = \varepsilon_k G_k = \varepsilon_k \sigma T_k^4 - q_k \tag{6.20}$$

The absorbed energy added to the energy supplied by other means is equal to the emitted energy, $\alpha_k G_k + q_k = \varepsilon_k \sigma T_k^4$.

Before continuing with the development, examples are given to illustrate using Equations (6.16) and (6.17) as simultaneous equations for each enclosure surface.

Example 6.5

Derive the expression for the net radiative energy exchange between two infinite parallel flat plates in terms of their temperatures T_1 and T_2 (Figure 6.6).

Since for infinite plates, all radiation leaving one plate will arrive at the other plate, the $F_{1-2} = F_{2-1} = 1$. Equations (6.16) and (6.17) for each plate are then:

$$q_1 = \frac{\varepsilon_1}{1-\varepsilon_1}\left(\sigma T_1^4 - J_1\right), \quad q_1 = J_1 - J_2 \tag{6.5.1}$$

$$q_2 = \frac{\varepsilon_2}{1-\varepsilon_2}\left(\sigma T_2^4 - J_2\right), \quad q_2 = J_2 - J_1 \tag{6.5.2}$$

Comparing Equations (6.5.1) and (6.5.2), $q_1 = -q_2$, so the energy added to surface 1 is removed from surface 2. The q_1 is thus the net energy transferred from 1 to 2, as requested in the problem statement. Equations (6.5.1) and (6.5.2) yield:

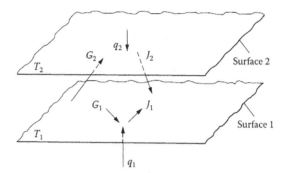

FIGURE 6.6 Energy fluxes for radiant interchange between infinite parallel flat plates.

$$J_1 = \sigma T_1^4 - \frac{1-\varepsilon_1}{\varepsilon_1} q_1, \quad J_2 = \sigma T_2^4 - \frac{1-\varepsilon_2}{\varepsilon_2} q_2 = \sigma T_2^4 + \frac{1-\varepsilon_2}{\varepsilon_2} q_1$$

These are substituted into Equation (6.5.1), and the result solved for q_1:

$$q_1 = -q_2 = \frac{\sigma\left(T_1^4 - T_2^4\right)}{1/\varepsilon_1\left(T_1\right) + 1/\varepsilon_2\left(T_2\right) - 1} \tag{6.5.3}$$

The functional notation $\varepsilon(T)$ is used to emphasize that ε_1 and ε_2 can be functions of temperature. Since T_1 and T_2 are specified in this case, ε_1 and ε_2 are substituted at their proper temperatures, and q_1 is directly calculated.

Example 6.6

For the parallel-plate geometry of the previous example, what temperature will surface 1 reach for a given energy flux input q_1 while T_2 is held at a specified value? Equation (6.5.3) applies and, when solved for T_1, gives:

$$T_1 = \left\{ \frac{q_1}{\sigma} \left[\frac{1}{\varepsilon_1\left(T_1\right)} + \frac{1}{\varepsilon_2\left(T_2\right)} - 1 \right] + T_2^4 \right\}^{1/4} \tag{6.6.1}$$

Since $\varepsilon_1(T_1)$ is a function of T_1, which is unknown, an iterative solution is necessary. A trial T_1 is selected, and ε_1 is chosen at this value. Equation (6.6.1) is solved for T_1, and this value is used to select ε_1 for the next approximation. The process is continued until $\varepsilon_1(T_1)$ and T_1 do not change with further iterations.

Example 6.7

Derive an expression for the net-radiation exchange between two uniform-temperature concentric diffuse-gray spheres as in Figure 6.7.

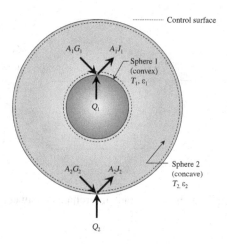

FIGURE 6.7 Energy quantities for radiant interchange between two concentric spheres.

This situation is more complicated than for infinite parallel plates, as the two surfaces have unequal areas, and surface 2 can partially view itself. The configuration factors were derived in Example 5.9 as $F_{1-2} = 1$, $F_{2-1} = A_1/A_2$, and $F_{2-2} = 1 - (A_1/A_2)$. Equations (6.16) and (6.17) are written for each of the two sphere surfaces as:

$$Q_1 = A_1 \frac{\varepsilon_1}{1-\varepsilon_1}\left(\sigma T_1^4 - J_1\right), \quad Q_1 = A_1\left(J_1 - J_2\right) \tag{6.7.1}$$

$$Q_2 = A_2 \frac{\varepsilon_2}{1-\varepsilon_2}\left(\sigma T_2^4 - J_2\right) \tag{6.7.2}$$

$$Q_2 = A_2\left[J_2 - \frac{A_1}{A_2}J_1 - \left(1 - \frac{A_1}{A_2}\right)J_2\right] = A_1\left(J_2 - J_1\right) \tag{6.7.3}$$

Comparing Equations (6.7.1) and (6.7.3) reveals that $Q_1 = -Q_2$, as expected from an overall energy balance. The four equations (Equations (6.7.1), (6.7.2), and (6.7.3)) can be solved for the four unknowns J_1, J_2, Q_1, and Q_2. This yields the net energy transfer (supplied to surface 1 and removed at surface 2):

$$Q_1 = \frac{A_1\sigma\left(T_1^4 - T_2^4\right)}{1/\varepsilon_1\left(T_1\right) + \left(A_1/A_2\right)\left[1/\varepsilon_2\left(T_2\right) - 1\right]} \tag{6.7.4}$$

If the spheres in Example 6.7 are not concentric, all the radiation leaving surface 1 is still incident on surface 2. The configuration factor F_{1-2} is again 1 and, with the use of the same assumptions, the analysis follows as before, leading to Equation

(6.7.4). However, when sphere 1 is relatively small (say, one-half the diameter of sphere 2) and the eccentricity is large, the geometry is so different from the concentric case that using Equation (6.7.4) seems intuitively incorrect. The error in using Equation (6.7.4) is that it was derived on the basis that q, G, and J are uniform over each of A_1 and A_2. These conditions are exactly met only for the concentric case. For the eccentric case, the A_1 and A_2 need to be subdivided to improve accuracy.

Example 6.8

A gray isothermal body with area A_1 and temperature T_1 and without any concave indentations is completely enclosed by a much larger gray isothermal enclosure having area A_2. How much energy is being transferred by radiation from A_1 to A_2? Area A_1 cannot see any part of itself, and A_1 is not near A_2.

Since A_1 is completely enclosed and $F_{1-1} = 0$, the configuration factors and analysis are the same as in Example 6.7, which results in Q_1 given by Equation (6.7.4). This is valid, as A_1 is specified as rather centrally located within A_2 and hence the energy fluxes tend to be uniform over A_1. For the present situation, $A_1 \ll A_2$, and Equation (6.7.4) reduces to (unless ε_2 is very small):

$$Q_1 = A_1 \varepsilon_1 (T_1) \sigma (T_1^4 - T_2^4) \tag{6.8.1}$$

Note that this result is independent of the emissivity ε_2 of the enclosure (the enclosure acts like a black cavity unless ε_2 is very small so that A_2 is highly reflective).

Example 6.9

Consider a long enclosure made up of three surfaces, as in Figure 6.8. The enclosure has a uniform cross-section and is long enough that its ends can be neglected in the radiative energy balances. How much energy must be supplied to each surface (equal to the net radiative energy loss from each surface resulting from exchange within the enclosure) to maintain the surfaces at temperatures T_1, T_2, and T_3?

Write Equations (6.16) and (6.17) for each surface:

$$\frac{Q_1}{A_1} = \frac{\varepsilon_1}{1 - \varepsilon_1} (\sigma T_1^4 - J_1) \tag{6.9.1}$$

$$\frac{Q_1}{A_1} = J_1 - F_{1-1}J_1 - F_{1-2}J_2 - F_{1-3}J_3 \tag{6.9.2}$$

$$\frac{Q_2}{A_2} = \frac{\varepsilon_2}{1 - \varepsilon_2} (\sigma T_2^4 - J_2) \tag{6.9.3}$$

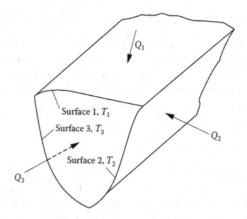

FIGURE 6.8 Long enclosure composed of three surfaces (ends neglected); cross-section is uniform.

$$\frac{Q_2}{A_2} = J_2 - F_{2-1}J_1 - F_{2-2}J_2 - F_{2-3}J_3 \qquad (6.9.4)$$

$$\frac{Q_3}{A_3} = \frac{\varepsilon_3}{1-\varepsilon_3}\left(\sigma T_3^4 - J_3\right) \qquad (6.9.5)$$

$$\frac{Q_3}{A_3} = J_3 - F_{3-1}J_1 - F_{3-2}J_2 - F_{3-3}J_3 \qquad (6.9.6)$$

The first of each of these three pairs is solved for J, and the J is substituted into the second equation of each pair to obtain:

$$\frac{Q_1}{A_1}\left(\frac{1}{\varepsilon_1} - F_{1-1}\frac{1-\varepsilon_1}{\varepsilon_1}\right) - \frac{Q_2}{A_2}F_{1-2}\frac{1-\varepsilon_2}{\varepsilon_2} - \frac{Q_3}{A_3}F_{1-3}\frac{1-\varepsilon_3}{\varepsilon_3}$$
$$= (1-F_{1-1})\sigma T_1^4 - F_{1-2}\sigma T_2^4 - F_{1-3}\sigma T_3^4 \qquad (6.9.7)$$
$$= F_{1-2}\sigma\left(T_1^4 - T_2^4\right) + F_{1-3}\sigma\left(T_1^4 - T_3^4\right)$$

$$-\frac{Q_1}{A_1}F_{2-1}\frac{1-\varepsilon_1}{\varepsilon_1} + \frac{Q_2}{A_2}\left(\frac{1}{\varepsilon_2} - F_{2-2}\frac{1-\varepsilon_2}{\varepsilon_2}\right) - \frac{Q_3}{A_3}F_{2-3}\frac{1-\varepsilon_3}{\varepsilon_3}$$
$$= -F_{2-1}\sigma T_1^4 + (1-F_{2-2})\sigma T_2^4 - F_{2-3}\sigma T_3^4 \qquad (6.9.8)$$
$$= F_{2-1}\sigma\left(T_2^4 - T_1^4\right) + F_{2-3}\sigma\left(T_2^4 - T_3^4\right)$$

$$-\frac{Q_1}{A_1}F_{3-1}\frac{1-\varepsilon_1}{\varepsilon_1}-\frac{Q_2}{A_2}F_{3-2}\frac{1-\varepsilon_2}{\varepsilon_2}+\frac{Q_3}{A_3}\left(\frac{1}{\varepsilon_3}-F_{3-3}\frac{1-\varepsilon_3}{\varepsilon_3}\right)$$

$$=-F_{3-1}\sigma T_1^4-F_{3-2}\sigma T_2^4+(1-F_{3-3})\sigma T_3^4 \qquad (6.9.9)$$

$$=F_{3-1}\sigma\left(T_3^4-T_1^4\right)+F_{3-2}\sigma\left(T_3^4-T_2^4\right)$$

Because the T values are known, the ε can be specified from surface-property data at their appropriate T values, and the three simultaneous equations are solved for the Q supplied to each surface. The results are approximations, because the radiosity leaving each surface is not uniform as assumed by using Equations (6.16) and (6.17). This is because the reflected flux is not uniform due to the enclosure geometry. Greater accuracy could be obtained by dividing each of the three sides into additional surface areas to yield a larger set of simultaneous equations.

6.3.1.1 System of Equations Relating Surface Heating Rate Q and Surface Temperature T

The form of Equations (6.9.7) through (6.9.9) shows that the Q and T for an enclosure of N surfaces can be related by a system of N equations. It is evident from these equations that the general form for the kth surface is:

$$\sum_{j=1}^{N}\left(\frac{\delta_{kj}}{\varepsilon_j}-F_{k-j}\frac{1-\varepsilon_j}{\varepsilon_j}\right)\frac{Q_j}{A_j}=\sum_{j=1}^{N}\left(\delta_{kj}-F_{k-j}\right)\sigma T_j^4=\sum_{j=1}^{N}F_{k-j}\sigma\left(T_k^4-T_j^4\right) \qquad (6.21)$$

Corresponding to each surface, k takes on one of the values 1, 2,...N, and δ_{kj} is Kronecker delta, defined as:

$$\delta_{kj}=\begin{cases}1 & \text{when } k=j \\ 0 & \text{when } k\neq j\end{cases}$$

When all surface temperatures are specified, the right side of Equation (6.21) is known and there are N simultaneous equations for the unknown radiation energy rate Q. In general, the energy inputs to some of the surfaces may be specified and the temperatures of these surfaces are to be determined. There is a total of N unknown Q and T, and Equation (6.21) provides the necessary number of relations. If the values of ε depend on temperature, it is necessary initially to guess the unknown T. Then, the ε (T) values can be chosen and the system of equations solved. The T values are used to select new $\varepsilon(T)$, and the process is repeated until the T and ε (T) values no longer change. Again, note that the results may be approximate because the uniform radiosity assumption is not perfectly fulfilled over each finite area and subdivision into smaller areas is required to improve accuracy.

At least one enclosure surface must have a specified temperature as a boundary condition. If all surfaces have a specified radiative energy flux (even if they meet the energy conservation requirement that $\sum_k q_k A_k=0$), then the left-hand side of Equation (6.21) is a known constant, and the surface temperatures are indeterminate.

Example 6.10

Consider an enclosure of three sides, as in Figure 6.8. Side 1 is maintained at T_1, side 2 is uniformly heated with flux q_2, and the third side is perfectly insulated on the outside. What are the equations to determine Q_1, T_2, and T_3?

The conditions give $Q_2/A_2 = q_2$ and $Q_3 = 0$. Then Equation (6.21) yields three equations, where the unknowns have been gathered on the left sides:

$$\frac{Q_1}{A_1}\left(\frac{1}{\varepsilon_1} - F_{1-1}\frac{1-\varepsilon_1}{\varepsilon_1}\right) + F_{1-2}\sigma T_2^4 + F_{1-3}\sigma T_3^4 = \left(1-F_{1-1}\right)\sigma T_1^4 + q_2 F_{1-2}\frac{1-\varepsilon_2}{\varepsilon_2} \quad (6.10.1)$$

$$-\frac{Q_1}{A_1}F_{2-1}\frac{1-\varepsilon_1}{\varepsilon_1} - \left(1-F_{2-2}\right)\sigma T_2^4 + F_{2-3}\sigma T_3^4 = -F_{2-1}\sigma T_1^4 - q_2\left(\frac{1}{\varepsilon_2} - F_{2-2}\frac{1-\varepsilon_2}{\varepsilon_2}\right) \quad (6.10.2)$$

$$-\frac{Q_1}{A_1}F_{3-1}\frac{1-\varepsilon_1}{\varepsilon_1} + F_{3-2}\sigma T_2^4 - \left(1-F_{3-3}\right)\sigma T_3^4 = -F_{3-1}\sigma T_1^4 + q_2 F_{3-2}\frac{1-\varepsilon_2}{\varepsilon_2} \quad (6.10.3)$$

For this simple situation, the solution could be shortened by using overall energy conservation to give $Q_1 = -Q_2$. However, it is generally a good idea to solve directly for all the unknowns and use the overall energy balance $\sum_{k=1}^{N} Q_k = 0$ as a check.

Example 6.11

A hollow cylinder is heated on the outside so that the cylinder is maintained at a uniform temperature. The outside of the cylinder is otherwise insulated, so the energy must be transferred by radiation from the inside of the cylinder through the cylinder ends. The system is in a vacuum, so radiation is the only mode of energy transfer. As shown in Figure 6.9, there is a disk centered on the cylinder axis and facing normal to the ends of the cylinder. The disk is exposed on both

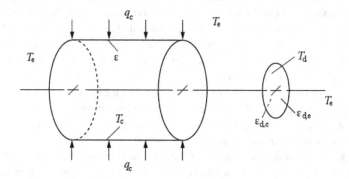

FIGURE 6.9 Hollow cylinder and disk configuration for Example 6.11.

sides, so the side facing away from the cylinder radiates to the surroundings at T_e. The other side receives radiation from the inside of the cylinder. Provide equations to compute the disk temperature, T_d, if the inside of the cylinder is at a uniform temperature T_c.

The energy loss from the surface of the disk facing away from the cylinder can be considered as radiation into a large environment at T_e. Hence, from Equation (6.8.1), the net radiant energy flux leaving this surface is $q_{d,e} = \varepsilon_{d,e}\sigma(T_d^4 - T_e^4)$. The exchange with the cylinder is analyzed as a three-surface enclosure consisting of the inside of the cylinder, the side of the disk facing the cylinder, and the open area between the disk and the cylinder (a frustum of a cone) combined with the open area at the left side of the cylinder (this is a first approximation without subdividing the cylinder and disk areas to achieve greater accuracy). In the configuration-factor designations, the subscript e designates the combination of the two open areas at T_e. Since the energy lost to the environment is not asked for, only two equations are written, one for the inside of the cylinder and one for the surface of the disk facing the cylinder. The $q_{d,c} = -\varepsilon_{d,e}\sigma\left(T_d^4 - T_e^4\right)$ is the radiative heat flux supplied to the left surface of the disk. From Equation (6.21), the two equations are:

$$q_c\left(\frac{1}{\varepsilon_c} - F_{c-c}\frac{1-\varepsilon_c}{\varepsilon_c}\right) - q_{d,c}F_{c-d}\frac{1-\varepsilon_{d,c}}{\varepsilon_{d,c}} = F_{c-d}\sigma\left(T_c^4 - T_d^4\right) + F_{c-e}\sigma\left(T_c^4 - T_e^4\right) \qquad (6.11.1)$$

$$-q_cF_{d-c}\frac{1-\varepsilon_c}{\varepsilon_c} + q_d\frac{1}{\varepsilon_{d,c}} = F_{d-c}\sigma\left(T_d^4 - T_c^4\right) + F_{d-e}\sigma\left(T_d^4 - T_e^4\right) \qquad (6.11.2)$$

The q_c is eliminated from Equations (6.11.1) and (6.11.2); this yields an analytical relation for T_d.

6.3.2 MATRIX INVERSION FOR ENCLOSURE EQUATIONS

When many surfaces are present in an enclosure, a large set of simultaneous equations results, such as Equations (6.21). The set can be written in a shorter form. Let the right side be C_k and the quantities in parentheses on the left be a_{kj}. Then, the k equations can be written as:

$$\sum_{j=1}^{N} a_{kj}q_j = C_k \qquad (6.22)$$

where

$$a_{kj} = \frac{\delta_{kj}}{\varepsilon_j} - F_{k-j}\frac{1-\varepsilon_j}{\varepsilon_j}; \quad C_k = \sum_{j=1}^{N} F_{k-j}\sigma\left(T_k^4 - T_j^4\right) \qquad (6.23)$$

For an enclosure of N surfaces, the set of equations then has the form:

$$a_{11}q_1 + a_{12}q_2 + \cdots + a_{1j}q_j + \cdots + a_{1N}q_N = C_1$$

$$a_{21}q_1 + a_{22}q_2 + \cdots + a_{2j}q_j + \cdots + a_{2N}q_N = C_2$$

$$\vdots \qquad \vdots \qquad \vdots \qquad \vdots \qquad \vdots$$

$$a_{k1}q_1 + a_{k2}q_2 + \cdots + a_{kj}q_j + \cdots + a_{kN}q_N = C_k \qquad (6.24)$$

$$\vdots \qquad \vdots \qquad \vdots \qquad \vdots \qquad \vdots$$

$$a_{N1}q_1 + a_{N2}q_2 + \cdots + a_{Nj}q_j + \cdots + a_{NN}q_N = C_N$$

The array of a_{kj} coefficients is the matrix of coefficients **a**:

$$\mathbf{a} \equiv \left[a_{kj} \right] \equiv \begin{bmatrix} a_{11} & a_{12} & \cdots & a_{1j} & \cdots & a_{1N} \\ a_{21} & a_{22} & \cdots & a_{2j} & \cdots & a_{2N} \\ \vdots & \vdots & \vdots & \vdots & \vdots & \vdots \\ a_{k1} & a_{k2} & \cdots & a_{kj} & \cdots & a_{kN} \\ \vdots & \vdots & \vdots & \vdots & \vdots & \vdots \\ a_{N1} & a_{N2} & \cdots & a_{Nj} & \cdots & a_{NN} \end{bmatrix} \qquad (6.25)$$

To solve a set of equations such as Equation (6.22), obtain the *inverse* of matrix **a**, denoted as $\mathbf{a}^{-1}$. Most computer mathematics packages such as MATLAB® can be used to obtain the inverse coefficients A_{kj} from the matrix of a_{kj} values.

After the inverse matrix is obtained, the q_k values in Equation (6.22) are found as the sum of products of A and C:

$$q_k = \sum_{j=1}^{N} A_{kj}C_j \qquad (6.26)$$

Thus, the solution for each q_k is in the form of a weighted sum of $\left(T_k^4 - T_j^4 \right)$.

For a given enclosure, the configuration factors F_{k-j} are fixed. If in addition the ε_k are constant, then the elements a_{kj}, and hence the inverse elements A_{kj}, are fixed for the enclosure. The fact that the A_{kj} remain fixed has utility when it is desired to compute radiation quantities within an enclosure for many different T values. The matrix is inverted only once, and then Equation (6.26) is applied for different values of C.

For any appreciable number of simultaneous equations, the solution would be found with a computer subroutine or solver. The following example illustrates the matrix operations.

Note that in this section, it is required that either Q_k or T_k is specified for every enclosure surface. In engineering design problems, it may be that *both* Q_k and T_k are specified for some surfaces, and either or both T_k and Q_k are to be found for others. Such problems are *ill-posed* and cannot be directly solved by the methods in this section. Treatment of such problems is discussed in Section 12.3.

Example 6.12

Figure 6.10 is a two-dimensional (2D) rectangular enclosure that is long in the direction normal to the cross-section shown.

All the surfaces are diffuse-gray and their temperatures and emissivities are given. For simplicity, the four sides are not subdivided. The configuration factors are obtained by the crossed-string method as $F_{1-3} = F_{3-1} = 0.2770$, $F_{2-4} = F_{4-2} = 0.5662$, $F_{1-2} = F_{1-4} = F_{3-2} = F_{3-4} = 0.3615$, and $F_{4-1} = F_{4-3} = F_{2-1} = 0.2169$. In what follows, an abbreviated notation is used: $E = (1 - \varepsilon)/\varepsilon$, $\vartheta = T^4$, and $F_{kj} = F_{k-j}$. Then, the four equations from Equation (6.21) are:

$$\frac{q_1}{\varepsilon_1} - q_2 F_{12} E_2 - q_3 F_{13} E_3 - q_4 F_{14} E_4 = \sigma\left(\vartheta_1 - F_{12}\vartheta_2 - F_{13}\vartheta_3 - F_{14}\vartheta_4\right)$$

$$-q_1 F_{21} E_1 + \frac{q_2}{\varepsilon_2} - q_3 F_{23} E_3 - q_4 F_{24} E_4 = \sigma\left(-F_{21}\vartheta_1 + \vartheta_2 - F_{23}\vartheta_3 - F_{24}\vartheta_4\right)$$

$$-q_1 F_{31} E_1 - q_2 F_{32} E_2 + \frac{q_3}{\varepsilon_3} - q_4 F_{34} E_4 = \sigma\left(-F_{31}\vartheta_1 - F_{32}\vartheta_2 + \vartheta_3 - F_{34}\vartheta_4\right)$$

$$-q_1 F_{41} E_1 - q_2 F_{42} E_2 - q_3 F_{43} E_3 + \frac{q_4}{\varepsilon_4} = \sigma\left(-F_{41}\vartheta_1 - F_{42}\vartheta_2 - F_{44}\vartheta_3 + \vartheta_4\right)$$

All quantities except the q are known. Substitution and solution by a computer software package give $q_1 = -2877$, $q_2 = 1613$, $q_3 = 1509$, and $q_4 = -792.0$ W/m^2. (Try it! You might also discuss how many significant figures are appropriate in the answers.)

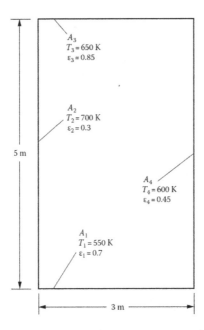

A_3
$T_3 = 650$ K
$\varepsilon_3 = 0.85$

A_2
$T_2 = 700$ K
$\varepsilon_2 = 0.3$

5 m

A_4
$T_4 = 600$ K
$\varepsilon_4 = 0.45$

A_1
$T_1 = 550$ K
$\varepsilon_1 = 0.7$

3 m

FIGURE 6.10   Rectangular geometry for Example 6.12.

6.4 RADIATION ANALYSIS USING INFINITESIMAL AREAS

6.4.1 GENERALIZED NET-RADIATION METHOD USING INFINITESIMAL AREAS

In the previous section, the enclosure was divided into finite areas. The accuracy of
the results is limited by the assumptions that the temperature and the energy incident
on and leaving each surface are uniform over that surface. If these quantities are not
uniform over part of the enclosure boundary, then that boundary must be subdivided,
possibly into infinitesimal area elements.

Formulation in terms of infinitesimal areas leads to energy balances in the form
of integral equations. Usually, an analytical solution is not possible, and the integral
equations must be solved numerically. Analytical solution methods for evaluating
integral equations are addressed in online Appendix C.2 at www.ThermalRadiation.
net/appendix.html.

Consider an enclosure of N finite areas. The areas would usually be the major geometric divisions of the enclosure or the areas on which a specified boundary condition
is held constant. Some or all areas are further subdivided into differential area elements,
as in Figure 6.11. The surfaces are diffuse-gray, and for simplicity here, the additional
restriction is made that the radiative properties are independent of temperature.

An energy balance on element dA_k at $\mathbf{r}_k$ gives:

$$q_k\left(\mathbf{r}_k\right) = J_k\left(\mathbf{r}_k\right) - G_k\left(\mathbf{r}_k\right) \tag{6.27}$$

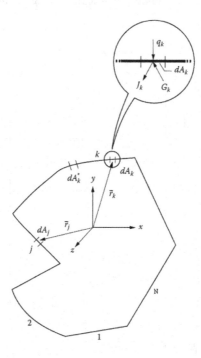

FIGURE 6.11 Enclosure composed of N discrete surface areas with areas subdivided into
infinitesimal elements.

The outgoing flux is composed of emitted and reflected energy:

$$J_k\left(\mathbf{r}_k\right) = \varepsilon_k \sigma T_k^4\left(\mathbf{r}_k\right) + \left(1 - \varepsilon_k\right)G_k\left(\mathbf{r}_k\right) \tag{6.28}$$

The incoming flux in Equation (6.27) is composed of portions of the outgoing fluxes from the other area elements of the enclosure. This is a generalization of Equation (6.12) in that an integration is performed over each finite surface to determine the contribution that the local flux leaving that surface makes to G_k:

$$dA_k G_k\left(\mathbf{r}_k\right) = \int_{A_1} J_1\left(\mathbf{r}_1\right)dF_{d1-dk}\left(\mathbf{r}_1,\mathbf{r}_k\right)dA_1 + \cdots + \int_{A_k} J_k\left(\mathbf{r}_k^*\right)dF_{dk^*-dk}\left(\mathbf{r}_k^*,\mathbf{r}_k\right)dA_k^* +$$
$$\cdots + \int_{A_N} J_N\left(\mathbf{r}_N\right)dF_{dN-dk}\left(\mathbf{r}_N,\mathbf{r}_k\right)dA_N \tag{6.29}$$

The second integral is the contribution that other differential elements dA_k^* on surface A_k make to the incident energy at dA_k. By using reciprocity, a typical integral can be transformed to obtain:

$$\int_{A_j} J_j\left(\mathbf{r}_j\right)dF_{dj-dk}\left(\mathbf{r}_j,\mathbf{r}_k\right)dA_j = dA_k\int_{A_j} J_j\left(\mathbf{r}_j\right)dF_{dk-dj}\left(\mathbf{r}_j,\mathbf{r}_k\right)$$

Operating on all the integrals in Equation (6.29) in this way results in:

$$G_k\left(\mathbf{r}_k\right) = \sum_{j=1}^{N}\int_{A_j} J_j\left(\mathbf{r}_j\right)dF_{dk-dj}\left(\mathbf{r}_j,\mathbf{r}_k\right) \tag{6.30}$$

Equations (6.29) and (6.30) provide two expressions for $G_k(\mathbf{r}_k)$. These are each substituted into Equation (6.28) to provide two expressions for $q_k(\mathbf{r}_k)$ comparable to Equations (6.16) and (6.17):

$$q_k\left(\mathbf{r}_k\right) = \frac{\varepsilon_k}{1-\varepsilon_k}\left[\sigma T_k^4\left(\mathbf{r}_k\right) - J_k\left(\mathbf{r}_k\right)\right] \tag{6.31}$$

$$q_k\left(\mathbf{r}_k\right) = J_k\left(\mathbf{r}_k\right) - \sum_{j=1}^{N}\int_{A_j} J_j\left(\mathbf{r}_j\right)dF_{dk-dj}\left(\mathbf{r}_j,\mathbf{r}_k\right) \tag{6.32}$$

As shown by Equation (5.9), the differential configuration factor dF_{dk-dj} contains the differential area dA_j. To place Equation (6.32) in a form where the variable of integration is explicitly shown, define a quantity $K(\mathbf{r}_j,\mathbf{r}_k)$ as:

$$K\left(\mathbf{r}_j,\mathbf{r}_k\right) \equiv \frac{dF_{dk-dj}\left(\mathbf{r}_j,\mathbf{r}_k\right)}{dA_j} \tag{6.33}$$

Then, Equation (6.32) becomes the integral equation:

$$q_k\left(\mathbf{r}_k\right) = J_k\left(\mathbf{r}_k\right) - \sum_{j=1}^{N} \int_{A_j} J_j\left(\mathbf{r}_j\right) K\left(\mathbf{r}_j, \mathbf{r}_k\right) dA_j \tag{6.34}$$

The $K(\mathbf{r}_j, \mathbf{r}_k)$ is the *kernel* of the integral equation.

As in the previous discussion for finite areas, there are two paths that can be followed: (1) When the temperatures and imposed energy fluxes are important, Equations (6.31) and (6.34) can be combined to eliminate the J. This gives a set of simultaneous relations relating the surface temperatures T and the energy fluxes q, and the problem is determinant if half of all the T and q are specified. Along each surface area, T or q can be specified, and the unknown T and q are found by solving the simultaneous relations. (2) Alternatively, when J is an important quantity (as in remote radiation measurement), the unknown q can be eliminated by combining Equations (6.31) and (6.34) for each surface that does not have its q specified. For a surface where q is known, Equation (6.34) can be used to directly relate the J_k to each other. This yields a set of simultaneous relations for the J in terms of the specified q and T. After solving for the J, Equations (6.31) can be used, if desired, to relate the q and T, where either the q or the T is known at each surface from the boundary conditions. Each of these procedures is now examined.

6.4.1.1 Relations between Surface Temperature T and Surface Radiation Energy Flux q

To eliminate the J in the first solution method, Equation (6.31) is solved for $J_k(\mathbf{r}_k)$, giving:

$$J_k\left(\mathbf{r}_k\right) = \sigma T_k^4\left(\mathbf{r}_k\right) - \frac{1-\varepsilon_k}{\varepsilon_k} q_k\left(\mathbf{r}_k\right) \tag{6.35}$$

Equation (6.35) in the form shown and with k changed to j is then substituted into Equation (6.31) to eliminate J_k and J_j. This gives:

$$\begin{aligned}
\frac{q_k\left(\mathbf{r}_k\right)}{\varepsilon_k} &- \sum_{j=1}^{N} \frac{1-\varepsilon_j}{\varepsilon_j} \int_{A_j} q_j\left(\mathbf{r}_j\right) dF_{dk-dj}\left(\mathbf{r}_j, \mathbf{r}_k\right) \\
&= \sigma T_k^4\left(\mathbf{r}_k\right) - \sum_{j=1}^{N} \int_{A_j} \sigma T_j^4\left(\mathbf{r}_j\right) dF_{dk-dj}\left(\mathbf{r}_j, \mathbf{r}_k\right) \\
&= \sum_{j=1}^{N} \int_{A_j} \sigma \left[T_k^4\left(\mathbf{r}_k\right) - T_j^4\left(\mathbf{r}_j\right) \right] dF_{dk-dj}\left(\mathbf{r}_j, \mathbf{r}_k\right)
\end{aligned} \tag{6.36}$$

Equation (6.36) relates the surface temperatures to the energy fluxes supplied to the surfaces. It corresponds to Equation (6.21) in the formulation for finite uniform surfaces.

Example 6.13

An enclosure of the general type in Figure 6.8 consists of three plane surfaces. Surface 1 is heated uniformly, and surface 2 has a uniform temperature. Surface 3 is black and at $T_3 \approx 0$. What are the equations needed to determine the temperature distribution along surface 1?

With $T_3 = 0$, $\varepsilon_3 = 1$, and the self-view factors $dF_{dj-dj} = 0$, Equation (6.36) is written for the two plane surfaces 1 and 2 having uniform q_1 and T_2:

$$\frac{q_1}{\varepsilon_1} - \frac{1-\varepsilon_2}{\varepsilon_2} \int_{A_2} q_2\left(r_2\right) dF_{d1-d2}\left(r_2, r_1\right) = \sigma T_1^4\left(r_1\right) - \sigma T_2^4 \int_{A_2} dF_{d1-d2}\left(r_2, r_1\right) \quad (6.13.1)$$

$$\frac{q_2\left(r_2\right)}{\varepsilon_2} - q_1 \frac{1-\varepsilon_1}{\varepsilon_1} \int_{A_1} dF_{d2-d1}\left(r_1, r_2\right) = \sigma T_2^4 - \int_{A_1} \sigma T_1^4\left(r_1\right) dF_{d2-d1}\left(r_1, r_2\right) \quad (6.13.2)$$

An equation for surface 3 is not needed since Equations (6.13.1) and (6.13.2) do not involve the unknown $q_3(r_3)$ because $\varepsilon_3 = 1$ and $T_3 = 0$. From the definitions of F factors, $\int_{A_2} dF_{d1-d2} = F_{d1-2}$ and $\int_{A_1} dF_{d2-d1} = F_{d2-1}$. Equations (6.13.1) and (6.13.2) simplify to the following relations where the unknowns are on the left:

$$\sigma T_1^4\left(r_1\right) + \frac{1-\varepsilon_2}{\varepsilon_2} \int_{A_2} q_2\left(r_2\right) dF_{d1-d2}\left(r_2, r_1\right) = \sigma T_2^4 F_{d1-2}\left(r_1\right) + \frac{q_1}{\varepsilon_1} \quad (6.13.3)$$

$$\int_{A_1} \sigma T_1^4\left(r_1\right) dF_{d2-d1}\left(r_1, r_2\right) + \frac{q_2\left(r_2\right)}{\varepsilon_2} = \sigma T_2^4 + q_1 \frac{1-\varepsilon_1}{\varepsilon_1} F_{d2-1}\left(r_2\right) \quad (6.13.4)$$

Equations (6.13.3) and (6.13.4) can be solved simultaneously for the distributions $T_1(r_1)$ and $q_2(r_2)$.

6.4.1.2 Solution Method in Terms of Outgoing Radiative Flux J

Another method results from eliminating the $q_k(\mathbf{r}_k)$ from Equations (6.35) and (6.36) for the surfaces where $q_k(\mathbf{r}_k)$ is unknown. This provides a relation between J and the T specified along a surface:

$$J_k\left(\mathbf{r}_k\right) = \varepsilon_k \sigma T_k^4\left(\mathbf{r}_k\right) + \left(1 - \varepsilon_k\right) \sum_{j=1}^{N} \int_{A_j} J_j\left(\mathbf{r}_j\right) dF_{dk-dj}\left(\mathbf{r}_j, \mathbf{r}_k\right) \quad (6.37)$$

When $q_k(\mathbf{r}_k)$, the energy rate supplied to surface k, is known, Equation (6.35) can be used directly to relate T_k and J. The combination of Equations (6.35) and (6.37) provides a complete set of relations for the unknown J in terms of known T and q.

This set of equations for the J is now formulated more explicitly. In general, an enclosure can have surfaces 1, 2,..., m with specified temperature distributions; for these surfaces, Equation (6.37) is used. The remaining N–m surfaces $m + 1, m + 2,...,$

N have an imposed energy flux distribution specified; for these surfaces, Equation (6.32) is applied. This results in N equations for the unknown J distributions:

$$J_k(\mathbf{r}_k) - (1-\varepsilon_k) \sum_{j=1}^{N} \int_{A_j} J_j(\mathbf{r}_j) dF_{dk-dj}(\mathbf{r}_j, \mathbf{r}_k) = \varepsilon_k \sigma T_k^4(\mathbf{r}_k) \quad 1 \le k \le m \quad (6.38)$$

$$J_k(\mathbf{r}_k) - \sum_{j=1}^{N} \int_{A_j} J_j(\mathbf{r}_j) dF_{dk-dj}(\mathbf{r}_j, \mathbf{r}_k) = q_k(\mathbf{r}_k) \quad m+1 \le k \le N \quad (6.39)$$

After the J is found from these simultaneous integral equations, Equations (6.38) and (6.39) are applied to determine the unknown q or T distributions:

$$q_k(\mathbf{r}_k) = \frac{\varepsilon_k}{1-\varepsilon_k} \left[\sigma T_k^4(\mathbf{r}_k) - J_k(\mathbf{r}_k) \right] \quad 1 \le k \le m \quad (6.40)$$

$$\sigma T_k^4(\mathbf{r}_k) = \frac{1-\varepsilon_k}{\varepsilon_k} q_k(\mathbf{r}_k) + J_k(\mathbf{r}_k) \quad m+1 \le k \le N \quad (6.41)$$

6.4.1.3   Special Case When Imposed Heating Is Specified for All Surfaces

An interesting special case is when the imposed energy flux q is specified for all surfaces of the enclosure except one and it is desired to determine the surface temperature distributions. Note that all but one q can be specified independently since $\sum_k q_k A_k = 0$ for a steady state; however, *at least one enclosure surface tempera-ture must be given*, or the temperatures become related but indeterminate. Often, the environmental temperature provides this anchor. For this case, the use of the method of the previous section, where the J is first determined, has an advantage over the method given by Equation (6.36), where the T are directly determined from the specified q. This advantage arises from Equation (6.39) being independent of the radiative surface properties. For a given set of q, the J need to be determined only once from Equation (6.39). Then the temperature distributions are found from Equation (6.41), which introduces the emissivity dependence. This is helpful when it is desired to examine temperature variations for various emissivity values when there is a fixed set of q. A useful relation is obtained by first considering the case in which the surfaces are all black, $\varepsilon_k = 1$. Equation (6.41) shows that $J_k(\mathbf{r}_k) = \sigma T_k^4(\mathbf{r}_k)_{\text{black}}$. Because the J_k are independent of the emissivities, the solution in Equation (6.41) for $\varepsilon_k \ne 1$ is:

$$\sigma T_k^4(\mathbf{r}_k) = \frac{1-\varepsilon_k}{\varepsilon_k} q_k(\mathbf{r}_k) + \sigma T_k^4(\mathbf{r}_k)_{\text{black}} \quad (6.42)$$

This relates the temperature distributions in an enclosure for $\varepsilon_k \ne 1$ to the temperature distributions in a black enclosure having the same imposed energy fluxes, $q_k(\mathbf{r}_k)$.

Thus, once the temperature distributions have been found for the black enclosure, the $\sigma T_k^4(\mathbf{r}_k)$ for gray surfaces are found simply by adding $[(1 - \varepsilon_k)/\varepsilon_k]q_k(\mathbf{r}_k)$.

Example 6.14

The circular tube in Figure 6.12 is open at both ends and insulated on the outside surface (Usiskin and Siegel 1960; Buckley 1928). For a uniform energy addition (such as by electrical heating in the tube wall) to the inside surface of the tube wall and a surrounding environment at 0 K, what is the temperature distribution along the tube? If the surroundings are at T_e, how does this influence the temperature distribution?

Since the open ends of the tube are nonreflecting, they can be assumed to act as black disks at the surrounding temperature 0 K. Then with $\varepsilon_1 = \varepsilon_3 = 1$, Equation (6.41) gives $J_1 = J_3 = \sigma T_1^4 = \sigma T_3^4 = 0$. Consequently, the summation in Equation (6.39) provides only radiation from surface 2 to itself. Since the tube is axisymmetric, the two differential areas dA_k and dA_k^* can be rings located at x and y. Then, Equation (6.39) yields:

$$J_2(\xi) - \int_{\eta=0}^{\eta=l} J_2(\eta) dF_{d\xi-d\eta}(|\eta - \xi|) = q_2 \tag{6.14.1}$$

where $\xi = x/D$, $\eta = y/D$, $l = L/D$, and $dF_{d\xi-d\eta}(|\eta - \xi|)$ is the configuration factor for two rings a distance $|\eta - \xi|$ apart and is given by factor 15 in online Appendix L at www.ThermalRadiation.net/appendix.html as:

$$dF_{d\xi-d\eta}(|\eta - \xi|) = \left\{ 1 - \frac{|\eta - \xi|^3 + \frac{3}{2}|\eta - \xi|}{\left[(\eta - \xi)^2 + 1\right]^{3/2}} \right\} d\eta \tag{6.14.2}$$

Absolute-value signs are used because the configuration factor depends only on the magnitude of the distance between the rings.

When $|\eta - \xi| = 0$, $dF = d\eta$, and this is the configuration factor from a differential ring to itself. Equation (6.14.1) can be divided by the constant q_2 and the dimensionless quantity $J_2(\xi)/q_2$ found by numerical or approximate solution methods for linear integral equations. The $J_2(\xi)/q_2$ distribution is in Figure 6.12b for a tube four diameters in length. From Equation (6.41), the $T_2^4(\xi)$ along the tube is given by:

$$\sigma T_2^4(\xi) = \frac{1 - \varepsilon_2}{\varepsilon_2} q_2 + J_2(\xi)$$

Since q_2 is constant, the $T_2^4(\xi)$ has the same shape as $J_2(\xi)$. The wall temperature is high in the central region of the tube and low near the ends where energy is radiated more readily to the low-temperature environment.

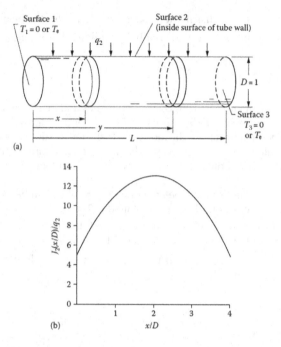

(a)

(b)

FIGURE 6.12 Uniformly heated tube insulated on the outside and open to the environment at both ends, (a) geometry and coordinate system and (b) distribution of J on the inside of the tube for $L/D = 4$.

Now, consider the environment at $T_e \neq 0$. The open ends of the tube can be regarded as perfectly absorbing disks at T_e, and Equation (6.39) yields:

$$J_2(\xi) - \int_{\eta=0}^{l} J_2(\eta) dF_{d\xi-d\eta}(|\eta-\xi|) - \sigma T_e^4 F_{d\xi-1}(\xi) - \sigma T_e^4 F_{d\xi-3}(l-\xi) = q_2$$

where $F_{d\xi-1}(\xi)$ is the configuration factor from a ring element at ξ to disk 1 at $\xi = 0$, $F_{d\xi-1}(\xi) = \left\{ \left(\xi^2 + \tfrac{1}{2}\right) \middle/ \left(\xi^2+1\right)^{1/2} \right\} - \xi$. Since the integral equation is linear in $J_2(\xi)$, let a trial solution be the sum of two parts where, for each part, either $T_e = 0$ or $q_2 = 0$:

$$J_2(\xi) = J_2(\xi)\big|_{T_e=0} + J_2(\xi)\big|_{q_2=0}$$

Substitute the trial solution into the integral equation to get:

$$J_2(\xi)\big|_{T_e=0} + J_2(\xi)\big|_{q_2=0} - \int_{\eta=0}^{l} J_2(\eta)\big|_{T_e=0} dF_{d\xi-d\eta}(|\eta-\xi|)$$

$$- \int_{\eta=0}^{l} J_2(\eta)\big|_{q_2=0} dF_{d\xi-d\eta}(|\eta-\xi|) - \sigma T_e^4 F_{d\xi-1}(\xi) - \sigma T_e^4 F_{d\xi-3}(l-\xi) = q_2$$

For $T_e = 0$, Equation (6.14.1) applies; subtract this equation to obtain:

$$J_2(\xi)\Big|_{q_2=0} - \int_{\eta=0}^{l} J_2(\eta)\Big|_{q_2=0} \, dF_{d\xi-d\eta}\big(|\eta-\xi|\big) - \sigma T_e^4 F_{d\xi-1}(\xi) - \sigma T_e^4 F_{d\xi-3}(l-\xi) = 0$$

The solution is $J_2|_{q_2=0} = \sigma T_e^4$, and this would be expected physically for an unheated surface in a uniform-temperature environment. The tube temperature distribution is found in Equation (6.41) as:

$$\sigma T_2^4(\xi) = \frac{1-\varepsilon_2}{\varepsilon_2} q_2 + J_2(\xi)\Big|_{T_e=0} + J_2(\xi)\Big|_{q_2=0} = \frac{1-\varepsilon_2}{\varepsilon_2} q_2 + J_2(\xi)\Big|_{T_e=0} + \sigma T_e^4$$

where $J_2(\xi)\Big|_{T_e=0}$ was found in the first part of this example. The environment has thus added σT_e^4 to the solution for $\sigma T_2^4(\xi)$ found previously for $T_e = 0$.

To arrive at a general conclusion for the effect of T_e in this type of problem, the result of Example 6.14 is rewritten as:

$$\sigma\left[T_2^4(\xi) - T_e^4\right] = \frac{1-\varepsilon_2}{\varepsilon_2} q_2 + J_2(\xi)\Big|_{T_e=0} \tag{6.43}$$

Thus, for nonzero T_e, the quantity $T_2^4(\xi) - T_e^4$ equals $T_2^4(\xi)$ for the case when $T_e = 0$. This illustrates a general way of accounting for a finite environment temperature. For energy transfer only by radiation, the governing equations are linear in T^4. As a result, in a cavity, the wall temperature $T_w|_{T_e=0}$ can be calculated for a zero-temperature environment. Then, by superposition, the wall temperature for any finite T_e is $T_w^4|_{T_e\neq0} = T_w^4|_{T_e=0} + T_e^4$. Hence, the thermal characteristics of a cavity having a wall temperature variation T_w and an external environment at T_e are the same as a cavity with wall temperature variation $\left(T_w^4 - T_e^4\right)^{1/4}$ and a zero-temperature environment.

Example 6.15

Consider emission from a long cylindrical hole drilled into a material at uniform temperature T_w (Figure 6.13a). The hole is long, so the portion of its internal surface at its bottom end can be neglected in the radiative energy balances. The outside environment is at T_e.

As in the previous discussion, the solution is obtained by using the reduced temperature $T_r = \left(T_w^4 - T_e^4\right)^{1/4}$. If a position is viewed at x on the cylindrical side wall, the energy leaving is $J(x)$. An apparent emissivity is defined as $\varepsilon_a(x) = J(x)/\sigma T_r^4$. The analysis determines how $\varepsilon_a(x)$ is related to the surface emissivity ε, where ε is constant over the cylindrical side of the hole. The integral equation governing the radiation exchange within a hole was first derived by Buckley (1927, 1928) and later by Eckert (1935); both obtained approximate analytical solutions. Results were evaluated numerically by Sparrow and Albers (1960).

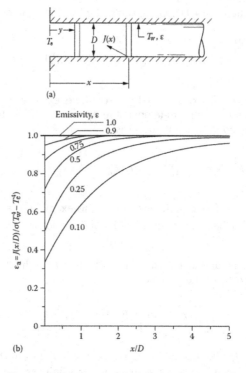

(a)

(b)

FIGURE 6.13   Radiant emission from cylindrical hole at a uniform temperature, (a) geometry and coordinate system and (b) apparent emissivity of cylinder wall.

Using the reduced temperature, the opening of the hole is approximated by a perfectly absorbing disk at zero reduced temperature. Then, from Equation (6.41) (because $\varepsilon = 1$ and $T_r = 0$ for the opening area), $J = 0$ from the opening into the cavity. Hence, the governing equation for the enclosure is Equation (6.38), written for the cylindrical side wall and including only radiation from the cylindrical wall to itself. As in Example 6.14, the configuration factor is for a ring of differential length on the cylindrical enclosure exchanging radiation with a ring at a different axial location. Equation (6.38) then yields along the surface of the hole:

$$J(\xi)-(1-\varepsilon)\int_{\eta=0}^{\infty}J(\eta)dF_{d\xi-d\eta}\left(\left|\eta-\xi\right|\right)=\varepsilon\sigma T^{4} \qquad (6.15.1)$$

where $\xi = x/D$, $\eta = y/D$, and $dF_{d\xi-d\eta}(|\eta-\xi|)$ is in Equation (6.14.2). After division by σT_r^4, the apparent emissivity $\varepsilon_a(\xi)$ is governed by the integral equation:

$$\varepsilon_{a}(\xi)-(1-\varepsilon)\int_{\eta=0}^{\infty}\varepsilon_{a}(\eta)dF_{d\xi-d\eta}\left(\left|\eta-\xi\right|\right)=\varepsilon \qquad (6.15.2)$$

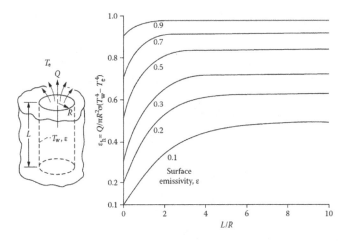

FIGURE 6.14 Apparent emissivity of cavity opening for a cylindrical cavity of finite length with diffuse reflecting walls at a constant temperature. (From Lin, S.H. and Sparrow, E.M., *J. Heat Trans.*, 87(2), 299, 1965.)

The solution of Equation (6.15.2) was carried out for various surface emissivities ε, and the results are in Figure 6.13b. The radiation leaving the surface approaches that of a blackbody as the wall position is at greater depths into the hole. At the mouth of the hole $\varepsilon_a \approx \sqrt{\varepsilon}$, as shown in Buckley (1927, 1928).

Radiation from a hole of finite depth was analyzed by Lin and Sparrow (1965), and their results are in Figure 6.14. The effective hemispherical emissivity ε_h in Figure 6.14 is a different quantity from that in Figure 6.13b; it gives the total energy leaving the mouth of the cavity ratioed to that emitted from a black-walled cavity. The latter is the same as the energy emitted from a black area across the mouth of the cavity. For each surface emissivity ε, the ε_h increases to a limiting value as the cavity depth increases. Unless ε is small, a cavity more than only a few diameters deep radiates the same amount as an infinitely deep cavity.

Example 6.16

What are the integral equations for radiation exchange between two parallel opposed plates finite in 1D and infinite in the other, as in Figure 6.15? Each plate has a specified temperature variation that depends only on the x- or y-coordinate, and the environment is at T_e.

From the discussion in Example 5.2, the configuration factors between the infinitely long parallel strips dA_1 and dA_2 are:

$$dF_{d1-d2} = \frac{1}{2}d(\sin\beta) = \frac{1}{2}\frac{a^2}{\left[(y-x)^2 + a^2\right]^{3/2}}dy; \quad dF_{d2-d1} = \frac{1}{2}\frac{a^2}{\left[(y-x)^2 + a^2\right]^{3/2}}dx$$

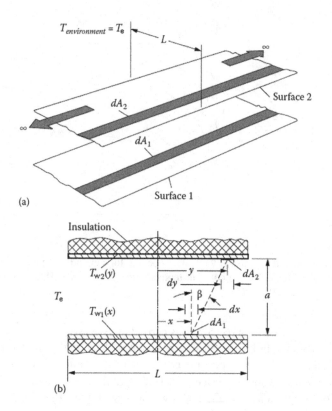

FIGURE 6.15 Geometry for radiation between two parallel plates infinitely long in one direction and of finite width, (a) parallel plates of width L and infinite length and (b) coordinates in cross-section of the gap between parallel plates.

The distribution of energy flux added to each plate is found by applying Equation (6.36) to each plate. As discussed before in Example 6.15, reduced temperatures are used to account for T_e. With $T_1 = \left(T_{w1}^4 - T_e^4\right)^{1/4}$ and $T_2 = \left(T_{w2}^4 - T_e^4\right)^{1/4}$, the governing equations are:

$$\frac{q_1(x)}{\varepsilon_1} - \frac{1-\varepsilon_2}{\varepsilon_2} \int_{-L/2}^{L/2} q_2(y)\frac{1}{2}\frac{a^2}{\left[(y-x)^2 + a^2\right]^{3/2}}\,dy = \sigma T_1^4(x) - \int_{-L/2}^{L/2} \sigma T_2^4(y)\frac{1}{2}\frac{a^2}{\left[(y-x)^2 + a^2\right]^{3/2}}\,dy$$

$$(6.16.1)$$

$$\frac{q_2(y)}{\varepsilon_2} - \frac{1-\varepsilon_1}{\varepsilon_1} \int_{-L/2}^{L/2} q_1(x)\frac{1}{2}\frac{a^2}{\left[(y-x)^2 + a^2\right]^{3/2}}\,dx = \sigma T_2^4(y) - \int_{-L/2}^{L/2} \sigma T_1^4(x)\frac{1}{2}\frac{a^2}{\left[(y-x)^2 + a^2\right]^{3/2}}\,dx$$

$$(6.16.2)$$

An alternative formulation using Equation (6.37) yields two equations for $J_1(x)$ and $J_2(y)$:

$$J_1(x) - (1-\varepsilon_1) \int_{-L/2}^{L/2} J_2(y) \frac{1}{2} \frac{a^2}{\left[(y-x)^2 + a^2\right]^{3/2}} dy = \varepsilon_1 \sigma T_1^4(x) \qquad (6.16.3)$$

$$J_2(y) - (1-\varepsilon_2) \int_{-L/2}^{L/2} J_1(x) \frac{1}{2} \frac{a^2}{\left[(y-x)^2 + a^2\right]^{3/2}} dx = J_2 \sigma T_2^4(y) \qquad (6.16.4)$$

After the $J_1(x)$ and $J_2(y)$ are found, the desired $q_1(x)$ and $q_1(y)$ are obtained from Equation (6.40) as:

$$q_1(x) = \frac{\varepsilon_1}{1-\varepsilon_1} \left[\sigma T_1^4(x) - J_1(x)\right] \qquad (6.16.5)$$

$$q_2(y) = \frac{\varepsilon_2}{1-\varepsilon_2} \left[\sigma T_2^4(y) - J_2(y)\right] \qquad (6.16.6)$$

6.5   COMPUTER PROGRAMS FOR ENCLOSURE ANALYSIS

So far, we have addressed radiative energy exchange within enclosures having black and/or diffuse-gray surfaces. The surface areas can be of finite or infinitesimal size. Each enclosure surface can have a specified net energy flux added to it by some external means, such as conduction or convection, or it can have a specified surface temperature. Various methods were presented for solving the array of simultaneous linear equations or the linear integral equations that result from the formulation of these interchange problems. It was pointed out that most practical problems become so complex that numerical techniques are required for the solution. To solve these problems, several computer programs have been developed. For example, the program Thermal Synthesizer System (TSS) (Panczak et al. 1991; Chin et al. 1992), developed under NASA sponsorship, incorporates solvers for enclosures with a very large number of surfaces as used for satellite and space vehicle design. Some other commercially available computer programs for thermal analysis also incorporate solvers for radiative enclosures in addition to including computation of configuration factors (see Sections 5.3 and 5.5 and online Appendix E at www.ThermalRadiation. net/appendix.html). Computer subroutines such as matrix solvers and integration subroutines are available in mathematics software packages.

6.6   NONGRAY AND/OR NONDIFFUSE SURFACES

As discussed in Chapter 4, most materials deviate from the idealizations of being black, gray, diffuse, or having temperature-independent radiative properties. The assumption of idealized surfaces is made to greatly simplify the computations. This can be reasonable because the radiative properties may not be known with high accuracy, especially their detailed dependence on wavelength and direction. Detailed computations are not meaningful if only very approximate property data are available. Also, the many reflections and re-reflections in an enclosure tend to average out radiative nonuniformities. For example, emitted plus reflected radiation J leaving a directionally emitting surface may be nearly diffuse if it is chiefly composed of reflected energy from radiation incident from many directions.

In some applications, more precision is required, and the gray and/or diffuse assumption cannot be trusted to be sufficiently accurate. It is then necessary to carry out exchange computations using as exact a procedure as reasonably possible. To provide techniques for more accurate computations, some methods are examined here for radiative exchange between realistic surfaces. Such analyses are inherently more difficult than for idealized surfaces, and a treatment of real surfaces including all types of variations, while possible in principle, is not usually attempted or justified, but rapidly increasing computational capacity and speed are making such analyses more feasible. For a detailed solution, the directional spectral properties must be available. Property variations with wavelength for the normal direction are available for some materials, but for many materials there is little detailed information. Certain solutions obviously must include spectral property variations, such as for spectrally selective coatings used for temperature control of space vehicles exposed to solar radiation and for solar concentrating power systems. Chapter 4 gives some means for extrapolating limited data.

Directional variations for some materials with optically smooth surfaces can be computed using electromagnetic theory, but real materials can deviate widely from ideal behavior.

As noted in Chapter 3, both directional and spectral surface properties can be interpreted in terms of probabilities that radiation will be emitted, absorbed, or reflected, including directional and spectral variations. This means that analysis based on a probabilistic interpretation can be carried out for the cases examined in this chapter. This is the basis of the Monte Carlo method presented in Section 10.7, which can be used to treat such problems in detail.

6.7 ENCLOSURE THEORY FOR DIFFUSE NONGRAY SURFACES

By assuming diffusely emitting and reflecting surfaces, directional surface effects are eliminated. If a surface is diffuse, spectral effects can be separated from directional ones, and one can understand how spectral property variations need to be accounted for. Under the diffuse assumption, emissivity, absorptivity, and reflectivity are independent of direction, but properties must be available as functions of λ and T to evaluate the radiative exchange.

For diffuse spectral surfaces, configuration factors are valid because they involve only geometry and are for diffuse radiation leaving a surface. The energy-balance equations and methods in Sections 6.2 through 6.4 remain valid if they are written for energy in each wavelength interval $d\lambda$. However, the boundary conditions usually involve *total* (including all wavelengths) energy fluxes. Total energy flux boundary conditions cannot be applied to the spectral energies. To illustrate this, consider the surface of Figure 6.16 with locally incident radiation energy flux G, and total radiosity J leaving by combined emission and reflection. If the surface is perfectly insulated, the G and J must be equal,

$$J - G = q_{\text{adiabatic}} = 0 \tag{6.44}$$

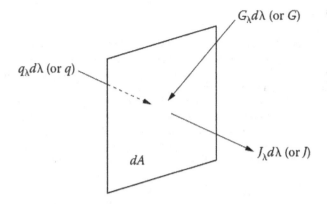

FIGURE 6.16 Spectral (or total) energy fluxes at a surface element.

However, for an insulated surface with $q = 0$, in each $d\lambda$ the incident and outgoing spectral fluxes are not generally equal so that:

$$J_\lambda d\lambda - G_\lambda d\lambda = q_\lambda d\lambda \neq 0 \tag{6.45}$$

An adiabatic surface only has a *total* radiation gain of zero, or, with Equation (6.45) restated in terms of spectral quantities,

$$q_{adiabatic} = \int_{\lambda=0}^{\infty} q_\lambda d\lambda = \int_{\lambda=0}^{\infty} \left(J_\lambda - G_\lambda\right) d\lambda = 0 \tag{6.46}$$

The $q_\lambda d\lambda$ is the net energy flux supplied in $d\lambda$ at λ by the radiative exchange process. For an adiabatic surface, the $q_\lambda d\lambda$ can vary substantially with λ, depending on the spectral property variations and the spectral distribution of incident energy.

Now consider a nonadiabatic surface where a total energy flux q is supplied by some means, such as energy conduction or convection. Then, from Figure 6.16,

$$q = \int_{\lambda=0}^{\infty} q_\lambda d\lambda = \int_{\lambda=0}^{\infty} \left(J_\lambda - G_\lambda\right) d\lambda \tag{6.47}$$

The q may be specified or may be a quantity to be determined if the surface is at a specified temperature. In any $d\lambda$, the net spectral energy $(J_\lambda - G_\lambda)d\lambda$ is unknown and may be positive or negative. The usual boundary conditions state only that the integral of all such spectral energy values must locally equal the total q.

6.7.1 Parallel Plates

To build familiarity while avoiding geometric complexity, consider a geometry of two infinite parallel plates.

Example 6.17

Two infinite parallel plates of tungsten at specified temperatures T_1 and T_2 $(T_1 > T_2)$ are exchanging radiation. Branstetter (1961) determined the temperature-dependent hemispherical spectral emissivity of tungsten by using electromagnetic theory relations to extrapolate limited experimental data, and some of his results are in Figure 6.17. Using these data, compare the net energy exchange between the tungsten plates with that for gray parallel plates using total emissivities.

The solution for gray plates is in Example 6.5, and the present analysis follows that example by writing the equations spectrally and solving for the desired quantities. The spectral result is Equation (6.5.3) written for a wavelength interval $d\lambda$,

$$q_{\lambda,1}d\lambda = -q_{\lambda,2}d\lambda = \frac{E_{\lambda b,1}(T_1) - E_{\lambda b,2}(T_2)}{\dfrac{1}{\varepsilon_{\lambda,1}(T_1)} + \dfrac{1}{\varepsilon_{\lambda,2}(T_2)} - 1}\,d\lambda \tag{6.17.1}$$

The total energy flux exchanged (supplied to surface 1 and removed from surface 2) is found by substituting the property data of Figure 6.17 into Equation (6.17.1) and integrating over all wavelengths:

$$q_1 = -q_2 = \int_{\lambda=0}^{\infty} q_{\lambda,1}d\lambda = \int_{\lambda=0}^{\infty} \frac{E_{\lambda b,1}(T_1) - E_{\lambda b,2}(T_2)}{\dfrac{1}{\varepsilon_{\lambda,1}(T_1)} + \dfrac{1}{\varepsilon_{\lambda,2}(T_2)} - 1}\,d\lambda \tag{6.17.2}$$

The integration is performed numerically for each set of specified plate temperatures T_1 and T_2.

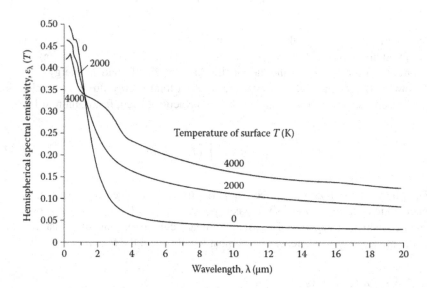

FIGURE 6.17 Hemispherical spectral emissivity of tungsten. (Branstetter, J. R.: *Radiant Heat Transfer between Nongray Parallel Plates of Tungsten*, NASA TN D-1088, Washington, DC, 1961.)

The examples illustrate the additional complication of including spectrally dependent surface properties. The enclosure calculations must be carried out in wavelength intervals, and the total energy quantities must then be obtained by integration of the spectral energy values. The integration is done rather easily for parallel plates since the formulation can be algebraically reduced to only one exchange equation. However, when many surfaces are present in an enclosure, the computations increase substantially since there are multiple exchange equations that need to be evaluated for each wavelength interval.

6.7.1.1 Spectral and Finite Spectral Band Relations for an Enclosure

For an enclosure of N diffuse surfaces, as in Figure 6.4, the relation in Equation (6.21) is written spectrally for $d\lambda$ for each of the k surfaces:

$$\sum_{j=1}^{N}\left[\frac{\delta_{kj}}{\varepsilon_{\lambda,j}(T_j)}-F_{k-j}\frac{1-\varepsilon_{\lambda,j}(T_j)}{\varepsilon_{\lambda,j}(T_j)}\right]q_{\lambda,j}d\lambda=\sum_{j=1}^{N}\left(\delta_{kj}-F_{k-j}\right)E_{\lambda b,j}(T_j)d\lambda \quad (k=1,2,\ldots,N)$$

(6.48)

If the N surface temperatures are all specified, the right sides of these simultaneous equations are known, as are the quantities in the square brackets on the left. The solution yields the $q_{\lambda,k}$ at λ for each of the surfaces. As a check on the calculations, the overall spectral energy balance within each $d\lambda$ yields the summation $\sum_{k=1}^{N}A_k q_{\lambda,k}=0$. The total energy supplied to each surface by other means is then $Q_k=A_k\int_{\lambda=0}^{\infty}q_{\lambda,k}d\lambda$. If some of the q_k are specified, and hence some of the surface temperatures are unknown, an iterative procedure may be required, as discussed following Equation (6.21). This can be especially difficult for a spectrally dependent case because the specified q_k is a *total* energy quantity and represents an integral of the values found from Equation (6.48). The boundary condition does not provide the spectral values of $q_{\lambda,k}$.

Equation (6.48) is usually applied in finite spectral bands. The bands are selected so that the properties for all surfaces are approximated as constant within each band. Equation (6.48) is then integrated over a band $\Delta\lambda$ to yield $\left(\text{after noting } q_{\Delta\lambda,j}\equiv\int_{\Delta\lambda}q_{\lambda,j}d\lambda\right)$:

$$\sum_{j=1}^{N}\left[\frac{\delta_{kj}}{\varepsilon_{\Delta\lambda,j}(T_j)}-F_{k-j}\frac{1-\varepsilon_{\Delta\lambda,j}(T_j)}{\varepsilon_{\Delta\lambda,j}(T_j)}\right]q_{\Delta\lambda,j}d\lambda$$

$$=\sum_{j=1}^{N}\left(\delta_{kj}-F_{k-j}\right)F_{\lambda T_j\to(\lambda+\Delta\lambda)T_j}E_{\lambda b,j}(T_j)\Delta\lambda \quad (k=1,2,\ldots,N)$$

(6.49)

If the values of T are specified, Equation (6.49) provides N equations for the N unknown $q_{\Delta\lambda,k}$ for each $\Delta\lambda$ band. The solution is carried out for *each* $\Delta\lambda$. Then for each of the k surfaces, the total q_k is found by summing over all wavelength bands:

$$q_k = \sum_{\Delta\lambda} q_{\Delta\lambda,k} \tag{6.50}$$

If for some surfaces neither q nor T is specified, and for others both are specified, the problem may become indeterminate (ill-posed), and the methods of this chapter will fail (Section 12.3).

Example 6.18

An enclosure consists of three diffuse surfaces of finite width and infinite length normal to the cross-section shown in Figure 6.18. One surface is concave. The radiative properties of each surface depend on wavelength and temperature, and the specified surface temperatures are T_1, T_2, and T_3. Provide a set of spectral band equations for the radiative energy exchange among the surfaces.

From Equation (6.49) for $k = 1, 2,$ and 3, and for wavelength band $\Delta\lambda$, the following equations are written. As a more compact notation, let $\bar{E}_{\Delta\lambda,j}(T_j) \equiv \left[1 - \varepsilon_{\Delta\lambda,j}(T_j)\right]/\varepsilon_{\Delta\lambda,j}(T_j)$ and $F_{\Delta\lambda T_j} \equiv F_{\lambda T_j \to (\lambda + \Delta\lambda)T_j}$ to yield:

$$\left[\frac{1}{\varepsilon_{\Delta\lambda,1}(T_1)} - F_{1-1}\bar{E}_{\Delta\lambda,1}(T_1)\right]q_{\Delta\lambda,1} - F_{1-2}\bar{E}_{\Delta\lambda,2}(T_2)q_{\Delta\lambda,2} - F_{1-3}\bar{E}_{\Delta\lambda,3}(T_3)q_{\Delta\lambda,3}$$
$$= (1 - F_{1-1})F_{\Delta\lambda T_1}\sigma T_1^4 - F_{1-2}F_{\Delta\lambda T_2}\sigma T_2^4 - F_{1-3}F_{\Delta\lambda T_3}\sigma T_3^4 \tag{6.18.1}$$

$$-F_{2-1}\bar{E}_{\Delta\lambda,1}(T_1)q_{\Delta\lambda,1} + \frac{1}{\varepsilon_{\Delta\lambda,2}(T_2)}q_{\Delta\lambda,2} - F_{2-3}\bar{E}_{\Delta\lambda,3}(T_3)q_{\Delta\lambda,3}$$
$$= -F_{2-1}F_{\Delta\lambda T_1}\sigma T_1^4 + F_{\Delta\lambda T_2}\sigma T_2^4 - F_{2-3}F_{\Delta\lambda T_3}\sigma T_3^4 \tag{6.18.2}$$

$$-F_{3-1}\bar{E}_{\Delta\lambda,1}(T_1)q_{\Delta\lambda,1} - F_{3-2}\bar{E}_{\Delta\lambda,2}(T_2)q_{\Delta\lambda,2} + \frac{1}{\varepsilon_{\Delta\lambda,3}(T_3)}q_{\Delta\lambda,3}$$
$$= -F_{3-1}F_{\Delta\lambda T_1}\sigma T_1^4 - F_{3-2}F_{\Delta\lambda T_2}\sigma T_2^4 + F_{\Delta\lambda T_3}\sigma T_3^4 \tag{6.18.3}$$

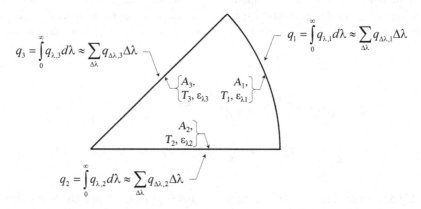

FIGURE 6.18 Radiant interchange in an enclosure with surfaces having spectrally varying radiation properties.

These are three simultaneous equations for $q_{\Delta\lambda,1}$, $q_{\Delta\lambda,2}$, and $q_{\Delta\lambda,3}$. The solution is carried out for the $q_{\Delta\lambda,k}$ values in each band $\Delta\lambda$. The $q_{\Delta\lambda,k}$ is the energy supplied to surface k in wavelength interval $\Delta\lambda$ because of external energy addition to the surface (e.g., conduction and/or convection) and energy transferred in from other wavelength bands by radiation exchange within the enclosure. Such boundary conditions are discussed in more detail in Chapter 7 for cases including conduction, convection, or imposed heating. Finally, q_k at each surface is found by summing $q_{\Delta\lambda,k}$ for that surface over all wavelength bands as in Equation (6.50). This is the energy flux that must be supplied to surface k by some means other than internal radiation, to maintain its specified surface temperature.

Example 6.19

Consider the geometry of Figure 6.18. Total energy flux is supplied to the three surfaces at rates q_1, q_2, and q_3. Determine the surface temperatures.

The equations are the same as in Example 6.18. Now, however, the prescribed boundary conditions have made the solution much more difficult. Because the surface temperatures are unknown, the $F_{\Delta\lambda T}$ and $\varepsilon_{\Delta\lambda}$ are unknown because they depend on temperature. The solution is carried out as follows: A temperature is assumed for each surface, and $q_{\Delta\lambda,k}(T)$ for each surface is computed. This is done for each of the wavelength bands. The $q_{\Delta\lambda,k}(T)$ values are then summed over all the $\Delta\lambda$ to find q_1, q_2, and q_3, which are compared to the specified boundary values. New temperatures are chosen, and the process is repeated until the computed q_k values agree with the specified values. The new temperatures for successive iterations are guessed based on the $F_{\Delta\lambda T}$ and $\varepsilon_{\Delta\lambda}$ variations and the trends of how changes in T_k produce changes in q_k throughout the system. As noted, the temperature must be specified on at least one surface of the enclosure. Thus, there may be many solutions to this problem.

6.8 DIRECTIONAL-GRAY SURFACES

Radiant exchange between gray surfaces with directional properties can be treated. However, complete data for directional surface properties are seldom available, so the treatment if needed is relegated to the online Appendix B at www.ThermalRadiation. net/appendix.html.

6.9 RADIATION EXCHANGE IN ENCLOSURES WITH SPECULARLY REFLECTING SURFACES

In Chapter 5 and so far in this chapter, the enclosure surfaces are all diffuse emitters and reflectors. This section considers a particular class of reflections. Except for one special example, the surfaces are still assumed to emit diffusely. Some of the surfaces in an enclosure are assumed to reflect diffusely; the remaining surfaces are assumed to be specular, that is, to reflect in a mirror-like manner. From the discussion of roughness in Section 4.5.2, an important surface parameter for radiative properties is the optical roughness (the ratio of the root-mean-square roughness height to the radiation wavelength.) As the radiation wavelength increases, a smooth surface tends toward being optically smooth, and reflections at larger wavelengths tend to become more specular. Thus, although a surface may not appear mirror-like to the eye for the short wavelengths of the visible spectrum, it may be specular for the longer wavelengths in the infrared.

When reflection is diffuse, the directional history of the incident radiation is lost upon reflection, and the reflected energy has the same directional distribution as if it had been absorbed and diffusely re-emitted. For specular reflection, the reflection angle relative to the surface normal is equal in magnitude to the angle of incidence, and the directional history of the incident radiation is *not* lost upon reflection. When dealing with specular surfaces, it is necessary to consider the specific directional paths that the reflected radiation follows between surfaces.

The specular reflectivity here is assumed independent of the incidence angle; the same fraction of the incident energy is reflected for any angle of incidence. In addition, all the surfaces are assumed gray (properties do not depend on wavelength).

6.9.1 CASES WITH SIMPLE GEOMETRIES

Radiation exchange between infinite parallel plates, concentric cylinders, and concentric spheres, as shown in Figure 6.19, is of practical importance for predicting the energy transfer performance of radiation shields, Dewar vessels, and cryogenic insulation applications, many of which employ specular surfaces. Specular exchange for these geometries is well understood, having been discussed and analyzed long ago by Christiansen (1883) and Saunders (1929).

Consider radiation between two infinite, gray, parallel specularly reflecting surfaces (Figure 6.19a). Although the reflections are specular, the emission is diffuse. All emitted and reflected radiation leaving surface 1 directly reaches surface 2; similarly, all emitted and reflected radiation leaving surface 2 directly reaches surface 1. This is true whether the surfaces are specular or diffuse reflectors. For diffuse emission, the surface absorptivity is independent of direction, so there are no directional considerations. Hence, for specular reflections and diffuse emission, Equation (6.5.3) applies, and the net radiative energy transfer from surface 1 to surface 2 is:

$$Q_1 = -Q_2 = \frac{A_1\sigma\left(T_1^4 - T_2^4\right)}{\left[1/\varepsilon_1\left(T_1\right)\right] + \left[1/\varepsilon_2\left(T_2\right)\right] - 1} \tag{6.51}$$

Now consider radiation between the concentric cylinders or spheres in Figure 6.19b and c, assuming that the emission is diffuse. Typical radiation paths for specular reflections are in Figure 6.19d. As shown by path *a*, all radiation emitted by surface 1 will directly reach surface 2. A portion will be reflected from surface 2 back to surface 1, and a portion of this will be re-reflected from 1. This sequence of reflections continues until insignificant energy remains because radiation is partially absorbed on each contact with a surface. From the symmetry of the concentric geometry and the equal incidence and reflection angles for specular reflections, no part of the radiation following path *a* can ever be reflected directly from surface 2 to another location on surface 2. Thus, the exchange process for radiation emitted

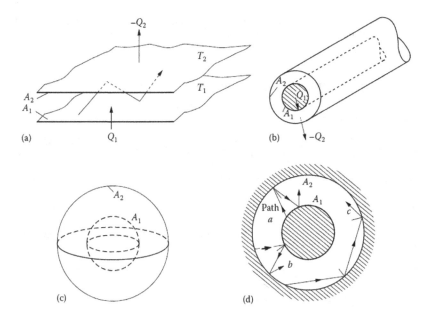

FIGURE 6.19 Radiation exchange for specular surfaces having simple geometries: (a) infinite parallel plates; (b) gap between infinitely long concentric cylinders; (c) gap between concentric spheres; and (d) paths for specular radiation in a gap between concentric cylinders or spheres.

from surface 1 is the same as though the two concentric surfaces were infinite parallel plates. However, the radiation emitted from the outer surface 2 can travel along either of two types of paths, b or c, as shown in Figure 6.19d. Since emission is assumed diffuse, the fraction F_{2-2} will follow paths of type c (F is a diffuse configuration factor). From the geometry of specular reflections, these rays will always be reflected along surface 2, with none reaching surface 1. The fraction $F_{2-1} = A_1/A_2$ is reflected back and forth between the surfaces along path b in the same fashion as radiation emitted from surface 1. The amount of radiation following this type of path is $A_2\varepsilon_2 F_{2-1}\sigma T_2^4 = A_2\varepsilon_2\left(A_1/A_2\right)\sigma T_2^4 = A_1\varepsilon_2\sigma T_2^4$. The fraction of the radiation leaving surface 2 that impinges on surface 1 thus depends on A_1 and not on A_2. Hence, for specular surfaces the exchange behaves as if both surfaces were equal portions of infinite parallel plates with sizes equal to the area of the inner body. The net radiative energy transfer from surface 1 to surface 2 is thus given by Equation (6.5.3).

Example 6.20

A spherical vacuum bottle consists of two silvered, concentric glass spheres, the inner being 15 cm in diameter and the evacuated gap between the spheres being 0.65 cm. The emissivity of the silver coating is 0.02. If hot coffee at 368 K is in the bottle and the outside temperature is 294 K, what is the initial radiative energy leakage rate from the bottle?

Equation (6.51) applies for concentric specular spheres. For the small rate of energy leakage expected, it is assumed that the surfaces will be close to 368 and 294 K. This gives:

$$Q_1 = \frac{\pi(0.15)^2 5.6704 \times 10^{-8}(368^4 - 294^4)}{(1/0.02) + (1/0.02) - 1} = 0.440 \text{ W}$$

If, instead of using the specular formulation, both surfaces are assumed diffuse reflectors with ε still 0.02, then Equation (6.5.3) applies. The denominator of the above

equation for Q_1 becomes: $\dfrac{1}{\varepsilon_1} + \dfrac{A_1}{A_2}\left(\dfrac{1}{\varepsilon_1} - 1\right) = \dfrac{1}{0.02} + \left(\dfrac{15}{16.3}\right)^2\left(\dfrac{1}{0.02} - 1\right) = 91.5$

instead   of   for   the   specular   case   where   the   denominator   is

$\dfrac{1}{\varepsilon_1} + \left(\dfrac{1}{\varepsilon_1} - 1\right) = \dfrac{1}{0.02} + \left(\dfrac{1}{0.02} - 1\right) = 99.0$. For diffuse surfaces, the energy loss

increases to 0.476 W.

Example 6.21

For the previous example, to illustrate the insulating ability of a vacuum bottle, how long will it take for the coffee to cool from 368 to 322 K if the only energy loss is by radiation?

The cooling rate of the coffee is $\rho_M Vc dT_1/dt$. Assuming the coffee is always mixed well enough that it is at a uniform temperature, the cooling rate is equal to the *instantaneous* radiation loss. The energy loss by radiation at any time t, given by Equation (6.51), is related to the change of internal energy of the coffee by:

$$-\rho_M Vc\frac{dT_1}{dt} = \frac{A_1\sigma\left[T_1^4(t) - T_2^4\right]}{1/\varepsilon_1 + 1/\varepsilon_2 - 1}$$

where it is assumed that surface 1 is at the coffee temperature and surface 2 is at the outside environment temperature. Then:

$$-\int_{T_1=T_i}^{T_1=T_F} \frac{dT_1}{T_1^4 - T_2^4} = \frac{A_1\sigma}{\rho_M Vc(1/\varepsilon_1 + 1/\varepsilon_2 - 1)}\int_0^\tau dt$$

where T_1 and T_F are the initial and final temperatures, respectively, of the coffee and ε_1 and ε_2 are assumed independent of temperature. Carrying out the integration and solving for t gives the cooling time from T_1 to T_F as:

$$t = \frac{\rho_M Vc(1/\varepsilon_1 + 1/\varepsilon_2 - 1)}{A_1\sigma}\left[\frac{1}{4T_2^3}\ln\left|\frac{(T_F + T_2)}{(T_1 + T_2)}\right| + \frac{1}{2T_2^3}\left(\tan^{-1}\frac{T_F}{T_2} - \tan^{-1}\frac{T_1}{T_2}\right)\right]$$

Substituting $\rho_M = 975$ kg/m³, $V = \frac{1}{6}\pi(0.15)^3$ m³, $c = 4195$ J/(kg · K), $\varepsilon_1 = \varepsilon_2 = 0.020$, $A_1 = \pi \cdot (0.15)^2$ m², $\sigma = 5.6704 \times 10^{-8}$ W/(m²·K⁴), $T_2 = 294$ K, $T_1 = 368$ K, and $T_f = 322$ K gives $t = 1.65$ h to cool if energy losses were only by radiation. Conduction losses through the glass wall of the bottle neck and free molecular transfer by the low-density near-vacuum gas between the cylinders usually cause the cooling to be faster.

Table 6.1 summarizes exchange between two surfaces that both have diffuse emission. When both surfaces are specular reflectors, Equation (6.51) applies for infinite parallel plates, infinitely long concentric cylinders, and concentric spheres. For infinite parallel plates, Equation (6.51) also applies when both surfaces are diffuse reflectors, or when one surface is diffuse, and the other is specular. For cylinders and spheres, Equation (6.5.3) will apply when the surface of the inner body (surface 1) is a diffuse reflector if the outer body (surface 2) is specular.

Enclosures with some specular and some diffuse surfaces can be treated by extension of the configuration factors for diffuse surfaces. It is also possible to treat surface exchange when surfaces have a reflectivity that is made up of a diffuse and a specular component. These are treated in detail in online www.ThermalRadiation. net/appendix.html.

6.10 MULTIPLE RADIATION SHIELDS

The results in Table 6.1 can be extended to obtain the performance of multiple radiation shields, as shown in Figures 6.20 and 6.21. The shields are thin, parallel, highly reflecting sheets, placed between radiating surfaces to reduce energy transfer between them. Highly effective insulation can be obtained by using many sheets separated by a vacuum to provide a series of alternate radiation and conduction barriers. One approach is depositing highly reflecting metallic films such as aluminum or silver on both sides of plastic films spaced apart by placing between them a cloth net having a large open area between its fibers. A stacking of 20 radiation shields per centimeter of thickness can be obtained. An important use of this multilayer insulation is in providing emergency protective blankets for forest-fire fighters and in low-temperature applications such as insulation of cryogenic storage tanks. It is used in satellites and other space vehicles (Lin et al. 1996).

For the results here, the spaces between the shields are evacuated so that energy transfer is only by radiation, and all emission is assumed diffuse. To analyze shield performance, consider N radiation shields between two surfaces at temperatures T_1 and T_2 with emissivities ε_1 and ε_2. As a general case, let a typical shield n have emissivity ε_{n1} on one side and ε_{n2} on the other, as in Figure 6.20.

As a result of the energy flow, the nth shield will be at temperature T_{sn}. Because for steady state, the same Q passes through the entire series of shields, Equation (6.51) for each pair of adjacent surfaces is:

TABLE 6.1
Radiant Interchange between Some Simply Arranged Surfaces (Emission Is Diffuse)

Geometry	Configuration	Type of Surface Reflection	Energy Transfer Rate, Q_1
Infinite parallel plates		A_1 or A_2, either specular or diffuse	$\dfrac{A_1\sigma\left(T_1^4 - T_2^4\right)}{1/\varepsilon_1 + 1/\varepsilon_2 - 1}$
Infinitely long concentric cylinders	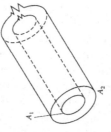	A_1, specular or diffuse; A_2, diffuse A_1 specular or diffuse; A_2 specular	$\dfrac{A_1\sigma\left(T_1^4 - T_2^4\right)}{1/\varepsilon_1 + \left(A_1/A_2\right)\left(1/\varepsilon_2 - 1\right)}$ $\dfrac{A_1\sigma\left(T_1^4 - T_2^4\right)}{1/\varepsilon_1 + 1/\varepsilon_2 - 1}$
Concentric spheres	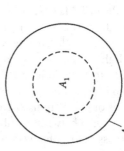	A_1, specular or diffuse; A_2, diffuse A_1, specular or diffuse; A_2 specular	$\dfrac{A_1\sigma\left(T_1^4 - T_2^4\right)}{1/\varepsilon_1 + \left(A_1/A_2\right)\left(1/\varepsilon_2 - 1\right)}$ $\dfrac{A_1\sigma\left(T_1^4 - T_2^4\right)}{1/\varepsilon_1 + 1/\varepsilon_2 - 1}$

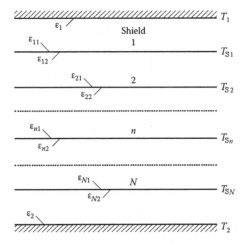

FIGURE 6.20   Parallel walls separated by N radiation shields.

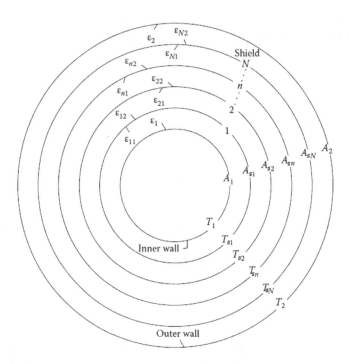

FIGURE 6.21   Radiation shields between concentric cylinders or spheres.

$$q\left(\frac{1}{\varepsilon_1}+\frac{1}{\varepsilon_{11}}-1\right)=\sigma\left(T_1^4-T_{s1}^4\right)$$

$$q\left(\frac{1}{\varepsilon_{12}}+\frac{1}{\varepsilon_{21}}-1\right)=\sigma\left(T_{s1}^4-T_{s2}^4\right)$$

$$\vdots$$

$$q\left(\frac{1}{\varepsilon_{(N-1)2}}+\frac{1}{\varepsilon_{N1}}-1\right)=\sigma\left(T_{s(N-1)}^4-T_{sN}^4\right)$$

$$q\left(\frac{1}{\varepsilon_{N2}}+\frac{1}{\varepsilon_2}-1\right)=\sigma\left(T_{sN}^4-T_2^4\right)$$

Adding these equations and dividing by the resulting factor multiplying q on the left-hand side gives, after some rearrangement,

$$q=\frac{\sigma\left(T_1^4-T_2^4\right)}{1/\varepsilon_1+1/\varepsilon_2-1+\sum_{n=1}^{N}\left(1/\varepsilon_{n1}+1/\varepsilon_{n2}-1\right)} \tag{6.52}$$

In most instances, ε is the same on both sides of each shield, and all the shields have the same ε. Let all the shield emissivities be ε_s; then, q becomes:

$$q=\frac{\sigma\left(T_1^4-T_2^4\right)}{1/\varepsilon_1+1/\varepsilon_2-1+N\left(2/\varepsilon_s-1\right)} \tag{6.53}$$

If the wall emissivities are the same as the shield emissivities, $\varepsilon_1=\varepsilon_2=\varepsilon_s$, Equation (6.53) reduces to:

$$q=\frac{\sigma\left(T_1^4-T_2^4\right)}{\left(N+1\right)\left(2/\varepsilon_s-1\right)} \tag{6.54}$$

In this instance, q decreases as $1/(N+1)$ as the number of shields N increases. When there are no shields, $N=0$, and Equation (6.53) reduces to $q=\sigma\left(T_1^4-T_2^4\right)/\left[\left(1/\varepsilon_1\right)+\left(1/\varepsilon_2\right)-1\right]$, as in Table 6.1. To illustrate shield performance, if in Equation (6.53) $\varepsilon_1=\varepsilon_2=0.8$ and $\varepsilon_s=0.05$, then $q=\sigma\left(T_1^4-T_2^4\right)/\left(1.5+39N\right)$ and the ratio $q(N$ shields$)/q($no shields$)=1.5/(1.5+39N)$.

In Table 6.2, the factor $1/(N+1)$ is the q ratio (independent of ε) when the walls have the same ε value as for the shields, as in Equation (6.54).

The expressions in Table 6.1 can be used to derive the energy transfer through a series of concentric cylindrical or spherical radiation shields, as in Figure 6.21. If the walls A_1 and A_2 and all the shields A_{sn} are diffuse reflectors, the energy transfer is (emission is diffuse):

TABLE 6.2
Effect of N Shields on Radiant Energy Transfer between Surfaces

	$N = 0$	$N = 1$	$N = 10$	$N = 100$
$q(N)/q(N = 0)$	1	0.0370	0.00383	0.00038
$1/(N + 1)$	1	0.5	0.0909	0.0099

$$Q = \frac{A_1\sigma\left(T_1^4 - T_2^4\right)}{1/\varepsilon_1 + \left(A_1/A_2\right)\left(1/\varepsilon_2 - 1\right) + \sum_{n=1}^{N}\left(A_1/A_{sn}\right)\left(1/\varepsilon_{n1} + 1/\varepsilon_{n2} - 1\right)} \tag{6.55}$$

If the walls are diffuse reflectors and all the shields are specular, then:

$$Q = \frac{A_1\sigma\left(T_1^4 - T_2^4\right)}{\left\{1/\varepsilon_1 + 1/\varepsilon_{11} - 1 + \sum_{n=1}^{N-1}\left(A_1/A_{sn}\right)\left(1/\varepsilon_{n2} + 1/\varepsilon_{(n+1)1} - 1\right) + \left(A_1/A_{sN}\right)\left[1/\varepsilon_{N2} + \left(A_{sN}/A_2\right)\left(1/\varepsilon_2 - 1\right)\right]\right\}} \tag{6.56}$$

If all the walls and shields are specular, Equation (6.56) applies if A_{sN}/A_2 is replaced by 1 in the last term in the denominator. In this instance, if all shield emissivities are the same and equal to ε_s, the result is:

$$Q = \frac{A_1\sigma\left(T_1^4 - T_2^4\right)}{1/\varepsilon_1 + 1/\varepsilon_s - 1 + \sum_{n=1}^{N-1}\left[\left(A_1/A_{sn}\right)\left(2/\varepsilon_s - 1\right)\right] + \left(A_1/A_{sN}\right)\left(1/\varepsilon_s + 1/\varepsilon_{n2} - 1\right)} \tag{6.57}$$

The sum in Equations (6.56) and (6.57) is zero if $N = 1$.

6.11 CONCLUDING REMARKS

This chapter outlines the treatment of radiative transfer among surfaces. Most analyses consider only diffuse surfaces, and in many instances, the radiative exchange of energy in enclosures is modified only slightly by considering specular in place of diffuse reflecting surfaces. In other applications such as solar concentrators and solar furnaces, and radiative transmission through highly reflecting tubes to guide radiative energy to a detector, and fiber optics, specular reflection is a dominant requirement. The design for these types of devices is often done by ray tracing; this has not been treated here, where the emphasis is on analysis of enclosures using configuration factors. A ray-tracing technique by the Monte Carlo method is presented in Section 10.7, and can embrace directionally- and spectrally-dependent surfaces and complex geometries.

The multiple specular reflections during radiative transmission within a highly reflecting circular tube or similar device can lead to significant polarization effects. As discussed in connection with the Fresnel reflection equations, the reflectivity differs for the two components of polarization, as in Equations (4.33) and (4.34). This provides different absorption for each of the components, and hence after reflection an initially unpolarized intensity becomes polarized. After multiple reflections at the same incidence angle during transit through a specular tube, for example, the radiation can become strongly polarized. In contrast, for diffuse reflections, a multiple-reflection process involves many different incidence angles, and polarization effects are not a concern.

Figure 6.22(a) shows a black surface irradiating a real surface. It might be inferred that the energy transfer between two real surfaces can be bracketed by calculating two limiting magnitudes: (1) exchange between diffuse surfaces of the same total hemispherical emissivities as the real surfaces, and (2) interchange between specularly reflecting surfaces of the same total hemispherical emissivities as the real surfaces. This is *not* generally correct. Consider, for example, a surface with strong reflection in the near-specular direction, like Figure 6.22(b). This is typical of many real surfaces with some surface irregularities.

If surface 2 is specular, it will not return any reflected energy to the black surface (Figure 6.22I). If 2 is diffuse, it will return a portion of the incident energy by

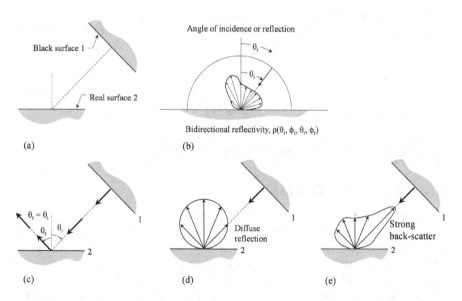

(a)

(b)

(c)   (d)   (e)

FIGURE 6.22 Effect of different surface reflectivity distributions on radiation transfer between surfaces.

reflection (Figure 6.22(d)). For a rough surface, as typified by the Lunar surface (Figure 4.13), however, the surface may reflect more energy to the black surface than for *either* of the ideal surfaces (Figure 6.22(e)). The ideal directional surfaces (specular or diffuse) therefore do not constitute limiting cases for energy transfer in general.

HOMEWORK (additional problems available in the online Appendix I at www.ThermalRadiation.net/appendix.html)

6.1 What is the net radiative energy transfer from black surface dA_1 to black surface A_2?

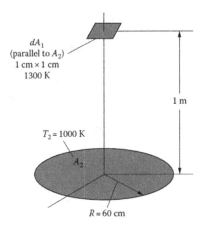

Answer: $dQ_{1-2} = 2.79$ W.

6.2 A person holds her open hand (approximated by a circular disk 8 cm in diameter) 6 cm directly above and parallel to a black heater element in the form of a circular disk 20 cm in diameter (such as an electric range element). The heater element is at 750 K. How much radiant energy from the heater element is incident on the hand?
Answer: 54.3 W.

6.3 A black electrically heated rod is in a black vacuum jacket. The rod must dissipate 110 W without exceeding 950 K. Calculate the maximum allowable jacket temperature (neglect end effects).

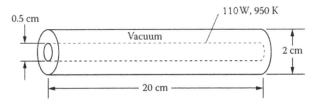

Answer: 666 K.

6.4   A regular tetrahedron of side length 0.5 m has black internal surfaces with
      the following characteristics:

Surface	Q(W)	T(K)
1	$Q_1 = 0$	T_1
2	Q_2	650
3	Q_3	1000
4	400	T_4

What are the values of T_1, Q_2, Q_3, and T_4?

Answers: $T_1 = 88\ 5$ K; $Q_2 = -356\ 2$ W; $Q_3 = 316\ 2$ W; $T_4 = 902$ K.

6.5   A rectangular solar collector resting on the ground has a black upper surface
      and dimensions of 1.5×3 m. The lower surface is very well insulated. The
      collector is tilted at an angle of 40° from the horizontal. Fluid at 95°C is
      passed through the collector at night. The night sky acts as a blackbody with
      an effective temperature of 210 K and the ground around the collector is at
      5°C. Neglecting conduction and convection to the surroundings, at night,
      (a) At what rate is energy lost by the collector (W)?
      (b) If flow through the collector is stopped, what temperature (K) will the
           collector attain?

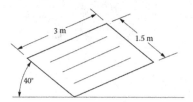

Answers: (a) 4070 W; (b) 222 K.

6.6   (a) A circular cylindrical enclosure has black interior surfaces, each main-
          tained at uniform interior temperature as shown. The outside of the
          entire cylinder is insulated so that the outside does not radiate to the
          surroundings. How much Q (W) is supplied to each area because of the
          interior radiative exchange? (Perform the calculation without subdivid-
          ing any surface areas.)

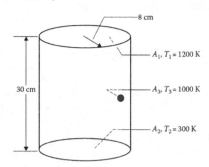

(b) For the same enclosure and the same surface temperatures, divide A_3 into two equal areas A_4 and A_5. What is the Q to each of these two areas, and how do they and their sum compare with Q_3 from part (a)?

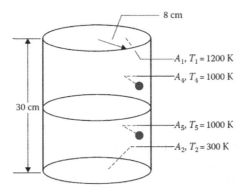

8 cm

$-A_1, T_1 = 1200$ K

$-A_4, T_4 = 1000$ K

30 cm

$-A_5, T_5 = 1000$ K

$-A_2, T_2 = 300$ K

(c) What are Q_6/A_6 and $\dot{Q}_7/A_7$ for the same enclosure and surface temperatures? How do they compare with Q_3/A_3 from part (a) and Q_4/A_4 and Q_5/A_5 from part (b)?

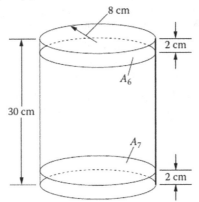

8 cm

2 cm

A_6

30 cm

A_7

2 cm

Answers:
(a) $Q_1 = 1295$ W; $Q_2 = -1207$ W; $Q_3 = -87.3$ W.
(b) $Q_1 = 1295$ W; $Q^2 = -1207$ W; $Q_4 = -853$ W; $Q_5 = 765$ W.
(c) $Q_1 = 1295$ W; $Q_2 = -1207$ W; $Q_6 = -261$ W; $Q_7 = 240$ W.

6.7   A 3 cm diameter, very long heater rod at the center of a long box furnace of 10 cm × 10 cm cross-section is heated in a vacuum to 1000°C. The top and bottom walls of the box furnace are thermally insulated on the back. The two side walls of the box furnace are covered by metallic panels with the temperature maintained at 700°C. All surfaces are diffuse and gray. The emissivity is 0.8 for the heater rod, 0.5 for the top and bottom walls of the box furnace, and 0.2 for the two metallic panels. Assuming uniform irradiation and temperature on each surface, find the rate of electrical heating applied to the heater rod and the temperature of the top and bottom walls.

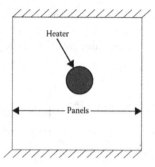

Answers: $q_{heater} = 29.7$ kW/m²; $q_{wall} = -14.0$ kW/m²; $T_{top} = 1200$ K (925°C).

6.8   Consider a diffuse-gray right circular cylindrical enclosure. The diameter is the same as the height of the cylinder, 3.00 m. The top is removed. If the remaining internal surfaces are maintained at 900 K and have an emissivity of 0.7, determine the radiative energy escaping through the open end. Assume uniform irradiation on each surface, so that subdivision of the base and curved wall is not included. The outside environment is at $T_e = 0$ K. How do the results compare with Figure 6.14? Explain any difference. If $T_e = 500$ K, what percent reduction occurs in radiative energy loss?
Answer: −204 k W; 90.5% reduction in energy rate.

6.9   A 2D diffuse-gray enclosure (infinitely long normal to the cross-section shown) has each surface at a uniform temperature. Compute the energy added per meter of enclosure length at each surface to account for the radiative exchange within the enclosure, Q_1, Q_2, and Q_3. (Assume for simplicity that it is not necessary to subdivide the three areas.) The conditions are $T_1 = 1100$ K, $\varepsilon_1 = 0.6$, $T_2 = 300$ K, $\varepsilon_2 = 0.5$, $T_3 = 800$ K, $\varepsilon_3 = 0.7$.

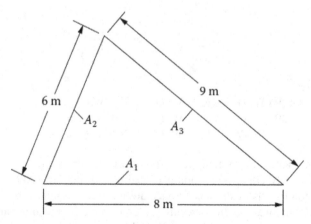

Answers: 267.4 kW/m; −126.9 kW/m; −140.5 kW/m.

6.10 Two infinite parallel black plates are at temperatures T_1 and T_2. A perforated thin sheet of gray material of emissivity ε_s is inserted between the plates.

SF, the shading factor, is the ratio of the solid area of the perforated sheet to the total area of the sheet, A_g/A. Determine the temperature of the perforated plate and find the ratio of the energy transfer Q_1 between the black plates when $SF = 1$ (i.e., with a single solid gray radiation shield) to the energy transfer with a perforated shield, Q_{SF}.

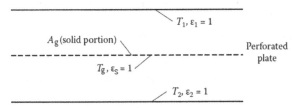

$$\text{Answer: } T_g^4 = \frac{T_1^4 + T_2^4}{2}; \quad Q_{1,\text{shield}} / Q_{1,\text{perforated}} = \frac{\varepsilon_s}{2} \frac{1}{\left[1 + \left(\varepsilon_s SF/2 \right) - SF \right]}.$$

6.11 A frustum of a cone has its base heated as shown. The top is held at 800 K, while the side is perfectly insulated. All surfaces are diffuse-gray. What is the temperature attained by surface 1 because of radiative exchange within the enclosure? (For simplicity, do not subdivide the areas.)

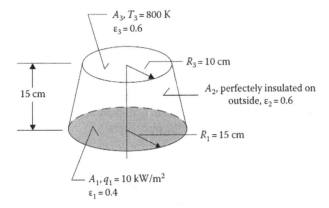

Answer: 1087 K.

6.12 An enclosure has four sides that are all equilateral triangles of the same size (i.e., it is an equilateral tetrahedron). The sides are of length $L = 4.5$ m and have the following conditions:

Side 1 is black and is at uniform temperature $T_1 = 1100$ K.

Side 2 is diffuse-gray and is perfectly insulated on the outside.

Side 3 is black and has a uniform energy flux of 8 kW/m² supplied to it.

Side 4 is black and is at $T_4 = 0$ K.

Find q_1, T_2, and T_3. For simplicity, do not subdivide the surface areas.
Answers: 51.4 kW/m²; 941 K; 972 K.

6.13 A rod 2 cm in diameter and 20 cm long is at temperature $T_1 = 1200$ K and has a hemispherical total emissivity of $\varepsilon_1 = 0.30$. It is within a thin-walled concentric cylinder of the same length having a diameter of 8 cm. The emissivity on the inside of the cylinder is $\varepsilon_2 = 0.21$, and on the outside is $\varepsilon_o = 0.15$. All surfaces are diffuse-gray. The entire assembly is suspended in a large vacuum chamber at $T_e = 300$ K. What is the temperature T_2 of the cylindrical shell? For simplicity, do not subdivide the surface areas. (Hint: $F_{2-1} = 0.225$, $F_{2-2} = 0.617$.)

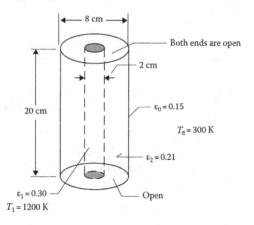

Answer: 722 K.

6.14 Three parallel plates of finite width are shown in cross-section. The plates are maintained at $T_1 = 700$ K and $T_2 = 400$ K. The surroundings are at $T_e = 300$ K. The plates are very long in the direction normal to the cross-section shown. What are the values of q_1 and q_2? All plate surfaces are diffuse-gray. (For simplicity, do not subdivide the plate areas.)

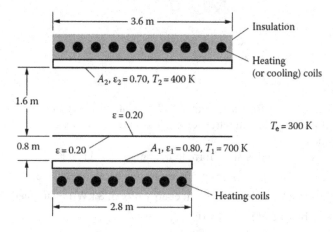

Answers: $q_1 = 5941.2$ W/m²; $q_2 = 97.5$ W/m².

6.15 Two infinitely long, directly opposed, parallel diffuse-gray plates of width
      W have the same uniform energy flux q supplied to them. The environment
      is at temperature T_e. Set up the governing equation for determining the
      temperature distribution $T(x)$ on the bottom plate. There is no energy loss
      from the top side of the upper plate or from the bottom side of the lower
      plate.

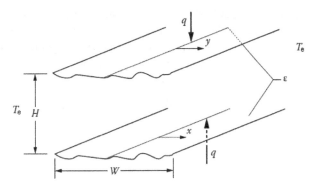

6.16 For the geometry specified in Homework 6.15, solve for the temperature
      distribution $T(x) = T(y)$ on either plate if $q = 3000$ W/m², $\varepsilon = 0.25$, $H =$
      0.25 m, and $W = 1$ m. A numerical solution is required. Present your results
      graphically. The environment temperature is $T_e = 300$ K.
      Answer: Peak $T(x/W) = 1/2$ of 796 K and a temperature at the plate edge of
      712 K.

6.17 Two diffusely emitting and reflecting parallel plates are of infinite length.
      The lower plate has a uniform temperature of $T_1 = 1000$ K, and the upper
      plate has a uniform temperature of $T_2 = 500$ K. The surroundings have a
      temperature of $T_e = 300$ K. Surface 1 has gray emissivity $\varepsilon_1 = 0.8$ and sur-
      face 2 has gray emissivity $\varepsilon_2 = 0.2$.

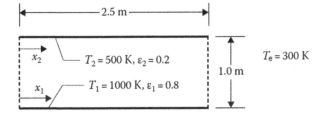

      (a) Find the average net radiative energy fluxes q_1 and q_2 on surfaces 1 and
           2, assuming uniform irradiation over each surface.
      (b) Partition each of the two parallel plates into four equal segments (upper:
           1, 2, 3, 4; lower: 5, 6, 7, 8). Find the net radiative energy flux for each
           segment, assuming uniform irradiation on each segment individually.
           Compare the results with those of part (a).

(c) Do not assume uniform irradiation on either of the parallel plates and find the distribution of net radiative energy flux $q_1(x_1)$ and $q_2(x_2)$ on each surface. Plot these results and the results of parts (a) and (b) on the same graph.

Answers: (a) 30,390 W/m, −5970 W/m; (b) $q_1 = q_3 = 32,190$ W/m, $q_2 = 28,120$ W/m, $q_5 = q_7 = -5200$ W/m, $q_6 = -6770$ W/m.

6.18 Two plates are joined at 90° and are very long normal to the cross-section shown. The vertical plate (plate 1) is heated uniformly with an energy flux of 300 W/m². The horizontal plate has a uniform temperature of 400 K. Both plates have an emissivity of 0.6. The environment is at $T_e = 300$ K. The dimensions are shown in the figure.

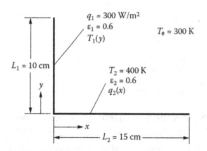

Derive the equations necessary for finding the temperature distribution on surface 1 and the net radiative energy flux on surface 2.

6.19 A three-surface enclosure has the properties shown in the Table. Assuming each surface can be treated as having uniform radiosity, find the unknown temperatures and energy fluxes.

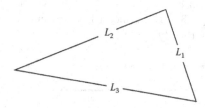

Answers: $T_2 = 1116$ K; $q_1 = 37.5$ kW/m²; $q_3 = -25.0$ kW/m².

	L (m)	ε	*T* (K)	*q* (W/m²)
L_1	0.8	1.0	1200	
L_2	1.0	0.75		0
L_3	1.2	0.25	300	

6.20 A bakery oven will burn biomass for energy to produce organic nonfat sugarless gluten-free whole-grain cookies. The geometry and conditions are shown in the following. The oven is quite long in the dimension of the paper.

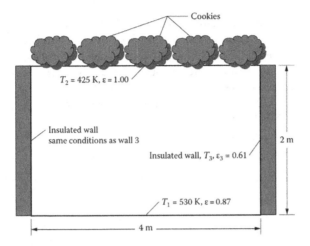

(a) What radiative energy flux q_1 must the biomass fuel produce on the heated wall at $T_1 = 530$ K if the cookie tray is to be heated to 425 K?
(b) What will the temperature of the sidewalls (T_3) be under these conditions?

Answers: (a) 1895 W/m; (b) 480 K.

6.21 For the geometry shown with nonconducting walls,
  (a) Provide the final governing equations necessary for finding $T_1(x_1)$ and $q_2(x_2)$ in two forms (1) explicit in the required variables and (2) in non-dimensional form using appropriate nondimensional variables.
  (b) Find the temperature distribution $T_1(x_1)$ and the energy flux distribution $q_2(x_2)$ and show them on appropriate graphs.

Boundary conditions are $q_1(x_1) = \left[100x_1 - 50x_1^2\right]$ (kW/m) where x_1 is in meters, $T_2 = 475$ K, and $T_3 = T_4 = 300$ K. Properties for gray (or black) diffuse surfaces are shown in the figure.

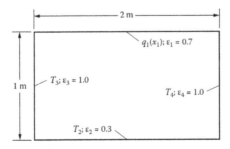

Show that your solution is grid independent and meets overall energy conservation.

Answers: $T_{1, max} = 1117$ K; $q_{2, max} = 10.4$ kW/m.

6.22 Solar radiation with a flux of $q_{solar} = 1353$ W/m^2 is normally incident on a front surface of a parallel radiation shield system. The front surface has solar absorptivity $\alpha_{solar} = 0.10$, and with IR emissivity of 0.86. The back of the surface has an emissivity of 0.015 and faces a four-layer shield system; each surface of the shields has an emissivity of $\varepsilon_{IR} = 0.015$. The outer face of the last layer of the shield faces space that can be taken as having a temperature of 4 K. What is the temperature of each layer (i.e., the front surface and each of the shields)?

Answers: $T_1 = 228.2$ K; $T_2 = 214.4$ K; $T_3 = 197.4$ K; $T_4 = 174.2$ K; $T_5 = 134$ K; $q = 2.74$ W/m^2.

7 Radiation Combined with Boundary Conduction and Convection

Ernst Rudolph George (E.R.G.) Eckert (1904–2004) *researched radiation from solids and gases, and published measurements of directional emissivity from various materials as well as directional reflectivity from blackbody irradiation. He also developed optical methods for obtaining configuration factors. In 1937, he turned to measurement of the emissivity of CO_2–N_2 mixtures as well as water vapor at various temperatures and partial pressures.*

Raymond Viskanta (1931–2021) *was born in Lithuania, and migrated to Germany with his family in 1944 and to the US in 1949. He received his PhD from Purdue University in 1960 and joined the Purdue faculty in 1962. He made contributions to engineering understanding and applications to highly nonlinear problems involving radiation coupled with conduction and/or convection, radiating systems with transients, radiation transfer in combustion systems and for glass manufacturing, and radiation effects on melting and solidification, porous media, and buoyancy-driven systems.*

7.1 INTRODUCTION

In the preceding chapters, an enclosure theory was formulated to find radiative exchange among surfaces. The local net-radiation loss at a surface was balanced by energy supplied by "some other means" that were not explicitly described. This chapter considers this energy balance at the surface as affected either by conduction from within the volume interior to the surface (such as from within a wall of an enclosure) or by convection or conduction at the surface from a surrounding medium. At each location along the surface, the radiation, convection, and conduction need to be combined to set the required boundary conditions. The solution of the energy equation subject to these boundary conditions supplies the surface temperature and energy flux distributions. The radiation analysis has the same restrictions as before: we assume the surfaces are opaque, and the medium between the radiating surfaces is transparent. The medium between the radiating surfaces may be conducting or convecting energy, but it does not interact with radiation.

DOI: 10.1201/9781003178996-7

One example of combined-mode energy transfer is a space radiator that is part of a vapor-cycle power plant operating in outer space. Waste energy must be rejected by radiation transfer to outer space. In a space radiator (see Figure 7.1a), the vapor of the working fluid in a thermodynamic cycle is condensed, releasing its latent heat. This energy is conducted through the condenser wall and into fins that radiate the energy into outer space. The temperature distribution in the fins and their radiating efficiency depend on combined radiation and conduction energy transfer.

A fin-tube geometry is commonly used for the absorber in a flat-plate solar collector. Solar energy is incident on the absorber plate through one or more transparent cover glasses that reduce convective losses to the atmosphere. A fluid is heated as it flows through tubes attached to the absorber plate. The collector design requires analysis of radiation, conduction, and convection.

In one type of steel-strip cooler in a steel mill (Figure 7.1b), a sheet of hot metal radiates energy to a bank of cold tubes as it moves past them. At the same time, cooling gas is blown across the sheet. A combined radiation and convection analysis is needed to find the temperature distribution along the steel strip. In a concept for a nuclear rocket engine, illustrated by Figure 7.1c, transparent hydrogen gas is heated

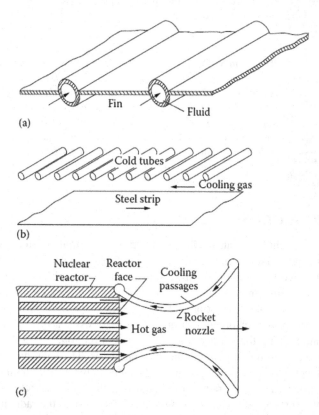

(a)

(b)

(c)

FIGURE 7.1 Energy transfer devices involving combined radiation, conduction, and convection: (a) space radiator or absorber plate of flat-plate solar collector, (b) steel-strip cooler, and (c) nuclear rocket.

by flowing through a high-temperature nuclear reactor. The hot gas then passes out through the rocket nozzle. The interior surface of the rocket nozzle receives radiant energy from the exit face of the reactor core and by convection from the flowing gas. Cooling the nozzle is done by conducting this energy through the nozzle wall and removing it to a flowing coolant. (NASA has conducted tests on such an engine; see Robbins and Finger 1991.)

These examples involve energy transfer by two or more modes. The modes can be in series, such as conduction through a wall followed by radiation from its surface. Energy transfer can also be by parallel modes, such as simultaneous conduction and radiation through a transparent material such as glass, or simultaneous radiation and convection from a hot surface. In some cases, the energies transferred from a surface by radiation and convection are independent, and they can be computed separately and added. In other instances, the interaction can be complex, such as when surface radiation is coupled with natural convection.

The various energy transfer modes depend on temperature and/or temperature differences to different powers. When radiation exchange between black surfaces is considered, the energy fluxes depend on surface temperatures to the fourth power. For nonblack surfaces, the temperature dependence may differ somewhat because of surface property dependence on temperature. Energy conduction depends on the local temperature gradient. Convection from a surface to a fluid depends approximately on the first power of their temperature difference. The exact power depends on the type of flow; for example, natural convection depends on temperature difference to a power from 1.25 to 1.4. Physical properties that vary with temperature introduce added temperature dependencies. The various powers and dependencies of temperature produce nonlinear energy transfer relations, and it is usually necessary to use numerical solution techniques. This chapter examines the governing energy-balance relations and presents some common solution methods.

7.2 ENERGY RELATIONS AND BOUNDARY CONDITIONS

7.2.1 GENERAL RELATIONS

In the analyses developed for enclosures, the net radiative energy flux at any position on the boundary was balanced by the energy flux q supplied by "some other means." The means considered here are conduction, convection, or wall internal energy sources such as electric heaters or nuclear reactions. Since we assume opaque enclosure walls, the absorption of radiation is at the surface, and the energy balance supplies a boundary condition. Although there can be conduction or convection in a medium between radiating surfaces, the medium is assumed here to be perfectly transparent, so radiation passes through with undiminished intensity. The radiation exchange relations developed previously for an enclosure are unchanged. If convection is expressed in terms of an energy transfer coefficient h, the energy balance at a boundary location "bo" is:

$$q = h(T_g - T_{bo}) - k \frac{\partial T}{\partial n}\bigg|_{bo} = J - G \qquad (7.1)$$

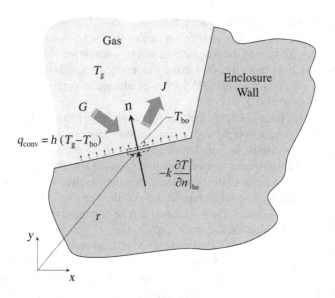

FIGURE 7.2 Boundary condition at location r on the surface of an opaque-walled enclosure.

where all quantities are at r on the surface of the enclosure wall in Figure 7.2.

The J is the total outgoing radiation (radiosity), and G is the total incident radiation. The earlier enclosure relations are valid as they are written in terms of q. For example, Equation (6.36) relates T and q along the enclosure boundaries. If the T are given, Equation (6.36) is solved for the q; then Equation (7.1) yields $(\partial T/\partial n)_{bo}$. Either T or the temperature gradient $\partial T/\partial n$ at the wall surface may be the boundary conditions for the energy equation at the interior of the wall:

$$\rho c \frac{\partial T}{\partial t} = \nabla \cdot (k \nabla T) + \dot{q} \tag{7.2}$$

The form of the energy equation inside the enclosure in the space between the radiating surfaces depends on the type of convection energy transfer mode present. It may be forced convection in a channel, a boundary layer flow, or natural convection. If the convection depends significantly on the boundary temperatures or energy flux distributions, the solution may require a simultaneous solution of the radiation exchange, energy conduction in the wall, and convection energy transfer in the medium.

In some practical problems, the net energy added to the surface by external means is specified directly rather than by the temperature gradient shown in Equation (7.1). For example, if electric heating is used, we can express it as the energy flux q_e in the wall. If there is insulation on one boundary (or, if there is negligible energy conduction along the wall, as shown in Figure 7.3), then the q_e appearing at the radiating boundary can be used in the governing equation. Then, Equation (7.1) becomes,

$$q = h(T_g - T_w) + q_e = J - G \tag{7.3}$$

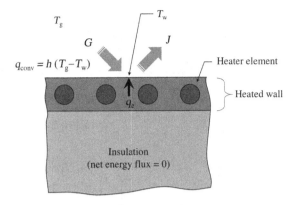

FIGURE 7.3   Boundary condition at a surface of opaque wall with specified energy flux.

The q_e may be uniform over the surface area, or it can have a specified variation with location. The heating might be by a passage of electric current within the wall as for an electrically heated wire.

7.2.2   UNCOUPLED AND COUPLED ENERGY TRANSFER MODES

In general, coupled problems of heat transfer are quite complex. However, if radiation, conduction, and convection contributions are independent of each other, then the problem becomes much simpler. This is called an *uncoupled* problem; for this case, the separate contributions are found, and the results are combined for the final temperature and energy flux profiles.

Example 7.1

Consider the region between two large gray parallel walls with a transparent gas between them (Figure 7.4). The internal surface temperatures T_1 and T_2 are specified. There is free convection in the gas, and the free-convection energy transfer coefficient h_{fc} depends on T_1 and T_2. What is the steady-state energy transfer from wall 1 to wall 2?

The energy transfer is the net radiative exchange and the transfer by free convection. It is equal to the flux q_1 that must be added to wall 1 to keep it at its specified temperature. Since T_1 and T_2 are given, the h_{fc} can be computed from free-convection correlations and the net energy transfer is, by use of Equation (6.5.3) plus the convection term,

$$q_1 = \frac{\sigma\left(T_1^4 - T_2^4\right)}{1/\varepsilon_1\left(T_1\right) + 1/\varepsilon_2\left(T_2\right) - 1} + h_{fc}\left(T_1, T_2\right) \cdot \left(T_1 - T_2\right)$$

The radiative and convective components are *uncoupled*. The q for each mode is computed independently, and the contributions are added. The methods of radiative computation developed earlier can be applied without modification.

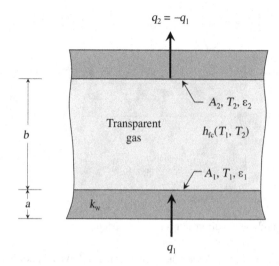

FIGURE 7.4 Parallel wall geometry for Examples 7.1 and 7.2 ($T_1 > T_2$).

Coupled problems are more common in the real-world rather than uncoupled problems. In coupled problems, the desired unknown quantity cannot be found by adding separate solutions; the energy relations must be solved with the transfer modes simultaneously included, usually resulting in highly nonlinear equations. They need to be solved in an iterative fashion.

Example 7.2

Let T_0 and T_2 in Figure 7.4 be specified. Since energy must be conserved in crossing surface 1 of the lower wall, the conduction through the lower wall must equal the transfer from surface 1 to surface 2 by combined radiation and free convection. Then, for constant thermal conductivity k_w,

$$q_1 = \frac{k_w}{a}(T_o - T_1) = \frac{\sigma\left(T_1^4 - T_2^4\right)}{1/\varepsilon_1\left(T_1\right) + 1/\varepsilon_2\left(T_2\right) - 1} + h_{fc}\left(T_1, T_2\right)\cdot\left(T_1 - T_2\right)$$

The problem is *coupled* since the unknown T_1 must be found from an equation that simultaneously incorporates all energy transfer processes. The equation for T_1 is highly nonlinear. T_1 can be obtained by iteration or by a computer math package root solver.

The type of prescribed boundary condition determines whether a problem is to be solved in coupled or uncoupled fashion. If all temperatures are specified, the energy fluxes can often be found using uncoupled analysis. If one or more energy fluxes are specified, the problem is coupled due to the nonlinear dependence of the unknown temperatures.

Combined mode problems with radiation are inherently nonlinear. For example, the equation in Example 7.2 includes both T and T^4. Numerical solution of such equations can lead to quite unsuspected difficulties (Howell, 2017).

7.2.3 CONDUCTION ALONG THIN WALLS

If the radiating wall is thin, then the principal temperature variations are along the length and width of the wall rather than across its thickness. One example is energy dissipation by radiating fins in devices that operate in outer space. Energy is conducted along the fin and radiated from the fin surface. The determination of the fin temperature distribution and performance requires a coupled conduction–radiation solution. The analysis is simplified by assuming a uniform temperature across the thin fin thickness at each location.

For this problem, a control volume across the thickness is used to derive the energy-balance equation. A volume element of area dx/dy and thickness a is shown in Figure 7.5. The thickness is small, so $T(x,y)$ is considered uniform within the z-axis of the element. Transparent fluids at $T_{m,1}$ and $T_{m,2}$ are assumed to flow across the upper and lower surfaces, with convective energy transfer coefficients h_1 and h_2. The temperature can change with time, and there can be internal energy generation within the element, such as by electric heating. A general energy balance showing that the change with time of internal energy of the element equals the energy gains by radiation exchange, conduction, convection, and internal energy sources is given by Equation (7.4), which we use in examples later:

$$\rho c_p a \frac{\partial T}{\partial t} = G_1 - J_1 + G_2 - J_2 + \frac{\partial}{\partial x}\left(ka \frac{\partial T}{\partial x} \right) + \frac{\partial}{\partial y}\left(ka \frac{\partial T}{\partial y} \right)$$

$$+ h_1\left(T_{m,1} - T\right) + h_2\left(T_{m,2} - T\right) + \dot{q}a$$

(7.4)

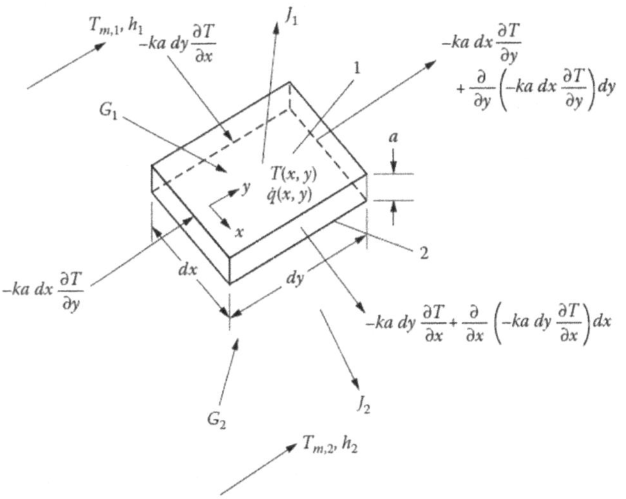

FIGURE 7.5   Element of thin plate for the derivation of the energy equation in a control volume.

7.3 RADIATION TRANSFER WITH CONDUCTION BOUNDARY CONDITIONS

Combined conduction and radiation transfer analysis allows us to find the energy losses from radiating fins in outer space, energy transfer through the walls of a vacuum Dewar, energy transfer through multilayer insulation made of highly reflective material, and temperature distributions in satellite and spacecraft structures. If radiation or conduction dominates, more approximations can be made in the analysis. They are demonstrated with a few examples.

7.3.1 THIN FINS WITH 1D OR 2D CONDUCTION

7.3.1.1 1D Energy Flow

Consider the energy transfer performance of a thin circular fin. From circular symmetry, the energy flow is one-dimensional (1D) in the radial direction.

Example 7.3

A thin annular fin in a vacuum is embedded in insulation, so it is insulated on one face and around its outside edge (Figure 7.6). The disk has thickness a, inner radius r_i, outer radius r_o, and thermal conductivity k. Energy is supplied to the inner edge from a solid rod of radius r_i that fits the central hole and keeps the inner edge at T_i. The exposed annular surface, which is diffuse-gray with emissivity ε,

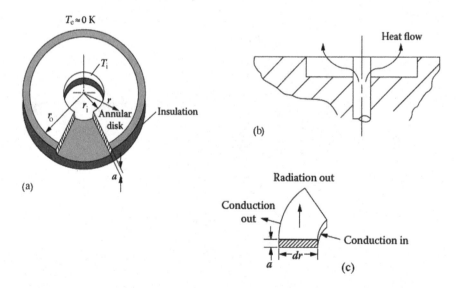

FIGURE 7.6 Geometry for finding temperature distribution in thin radiating annular plate insulated on one side and around outside edge: (a) disk geometry, (b) energy flow path through the fin, (c) part of ring element on annular disk.

radiates to the environment at $T_e \sim 0$ K. Find the temperature distribution as a function of radial position across the disk.

Assume the disk is thin enough that the local temperature can be considered constant across the disk thickness a, which is the usual thin fin assumption. For surroundings at zero temperature, there is no incoming radiation. If a and k are constant, the control volume equation for a ring element of width dr (Figure 7.6c) gives:

$$ka\frac{1}{r}\frac{d}{dr}\left(r\frac{dT}{dr}\right) - \varepsilon\sigma T^4 = 0 \tag{7.3.1}$$

This is to be solved for $T(r)$ subject to two boundary conditions: at the inner edge, $T = T_i$ at $r = r_i$ and at the insulated outer edge where there is no energy flow $dT/dr = 0$ at $r = r_o$. Using dimensionless variables $\vartheta = T/T_i$ and $R = (r - r_i)/(r_o - r_i)$ and two parameters $\delta = r_o/r_i$ and $\gamma = (r_o - r_i)^2 \varepsilon\sigma T_i^3/ka$ results in:

$$\frac{d^2\vartheta}{dR^2} + \frac{1}{R + 1/(\delta - 1)}\frac{d\vartheta}{dR} - \gamma\vartheta^4 = 0 \tag{7.3.2}$$

with the boundary conditions $\vartheta = 1$ at $R = 0$ and $d\vartheta/dR = 0$ at $R = 1$. Equation (7.3.2) is a second-order highly nonlinear differential equation where $\vartheta(R)$ depends on the two parameters, δ and γ. Solutions can be obtained by numerical methods, and solvers are available in computer mathematics software packages.

A design parameter for cooling fins is the *fin efficiency*, η. This is the energy radiated by the fin divided by the energy that would be radiated if the entire fin were at the maximum temperature T_i, which would occur for a fin with infinite thermal conductivity. The fin efficiency for the circular radiating fin is then:

$$\eta = \frac{2\pi\varepsilon\sigma\int_{r_i}^{r_o} rT^4(r)dr}{\pi(r_o^2 - r_i^2)\varepsilon\sigma T_i^4} = \frac{2\int_0^1 [R(\delta - 1) + 1]\vartheta^4(R)dR}{\delta + 1}$$

and is evaluated after $\vartheta(R)$ has been found from the differential Equation (7.3.2). The η is in Figure 7.7.

In a more general situation, if the environment is at T_e and the fin is nongray with a total absorptivity α for the incoming radiation spectrum (such as for incident solar radiation in a space application), the energy balance in Equation (7.3.1) becomes:

$$ka\frac{1}{r}\frac{d}{dr}\left(r\frac{dT}{dr}\right) - \sigma\left(\varepsilon T^4 - \alpha T_e^4\right) = ka\frac{1}{r}\frac{d}{dr}\left(r\frac{dT}{dr}\right) - \varepsilon\sigma\left(T^4 - \frac{\alpha}{\varepsilon}T_e^4\right) = 0 \tag{7.5}$$

where ε is the total emissivity for the spectrum emitted by the fin and $[(\alpha/\varepsilon)T_e^4]$ is an added parameter. For a gray fin, $\alpha = \varepsilon$; hence, a nongray fin acts like a gray fin in an effective radiating environment of $(\alpha/\varepsilon)T_e^4$. By using this effective environment, results for gray fins can be applied to nongray fins.

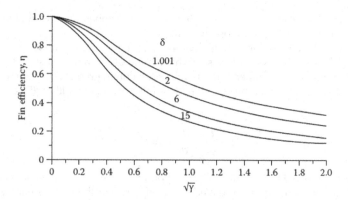

FIGURE 7.7 Radiation fin efficiency for the fin of Example 7.3. (From Chambers, R.L. and Somers, E.V., *JHT*, 81(4), 327, 1959.)

For transient fin temperature distributions, the energy storage term in Equation (7.4) must be included. The partial differential equation for $T(r,\tau)$ is then:

$$\rho c a \frac{\partial T}{\partial \tau} = ka \frac{1}{r} \frac{\partial}{\partial r}\left(r \frac{\partial T}{\partial r}\right) - \varepsilon\sigma\left(T^4 - \frac{\alpha}{\varepsilon}T_e^4\right) \tag{7.6}$$

Example 7.4

A thin plate of thickness a and length $2L$ is between two tubes in a radiator used to dissipate energy in orbit (Figure 7.8). The fin is long in the direction normal to the cross-section shown. Both sides of the plate have the same emissivity and are losing energy by radiation to space.

Radiation from the surroundings, such as from the Sun, Earth, or a planet, is incident on the plate surfaces, and the fluxes absorbed on the top and bottom sides are $q_{abs,t}$ and $q_{abs,b}$. The plate is diffuse-gray with emissivity ε on both sides and has constant thermal conductivity. Find an expression for the plate temperature distribution in the x direction. Neglect radiative interaction with the tube surfaces.

From the control volume relation Equation (7.4), the energy equation for a plate element of width dx is:

$$-ka\frac{d^2T}{dx^2} + 2\varepsilon\sigma T^4 = q_{abs,t} + q_{abs,b} \tag{7.4.1}$$

The boundary conditions for the thin plate are $T = T_{tube}$ specified at $x = 0$ and, from symmetry, $dT/dx = 0$ at $x = L$. To find $T(x)$, multiply Equation (7.4.1) by dT/dx and integrate to give:

$$-\frac{ka}{2}\left(\frac{dT}{dx}\right)^2 + \frac{2}{5}\varepsilon\sigma\left[T^5 - T^5(L)\right] = \left(q_{abs,t} + q_{abs,b}\right)\left[T - T(L)\right]$$

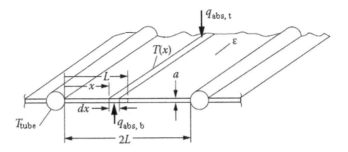

FIGURE 7.8   Flat-plate fin geometry for Example 7.4.

where $T(L)$ (which is unknown) is inserted to satisfy the boundary condition at $x = L$. Solve for dT/dx to give:

$$\frac{dT}{dx} = -\left(\frac{4\varepsilon\sigma}{5ka}\right)^{1/2} \left\{T^5 - T^5(L) - \frac{5}{2\varepsilon\sigma}(q_{abs,t} + q_{abs,b})\left[T - T(L)\right]\right\}^{1/2}$$

The minus sign was chosen for the square root because $T(x)$ must be decreasing with x. Separate variables and integrate them again to obtain:

$$x = \left(\frac{5ka}{4\varepsilon\sigma}\right)^{1/2} \int_{T}^{T_{tube}} \frac{dT}{\left\{T^5 - T^5(L) - (5/2\varepsilon\sigma)(q_{abs,t} + q_{abs,b})\left[T - T(L)\right]\right\}^{1/2}} \qquad (7.4.2)$$

which satisfies $T = T_{tube}$ at $x = 0$. To obtain $T(L)$, the relation is used that at the known length L,

$$L = \left(\frac{5ka}{4\varepsilon\sigma}\right)^{1/2} \int_{T(L)}^{T_{tube}} \frac{dT}{\left\{T^5 - T^5(L) - (5/2\varepsilon\sigma)(q_{abs,t} + q_{abs,b})\left[T - T(L)\right]\right\}^{1/2}} \qquad (7.4.3)$$

A mathematics solver can be used to obtain $T(L)$ from Equation (7.4.3). The temperature distribution is then found by numerically evaluating the integral in Equation (7.4.2) to find x for various T values (in the lower limit of the integral) between T_{tube} and $T(L)$.

Examples 7.3 and 7.4 considered a single radiating fin. When there are multiple fins that have radiative exchange among them, integral terms are introduced into the energy equations, as shown in the next example.

Example 7.5

An infinite array of identical thin fins of thickness a, width W in the x direction, and infinite length in the z direction is attached to a black base kept at a constant

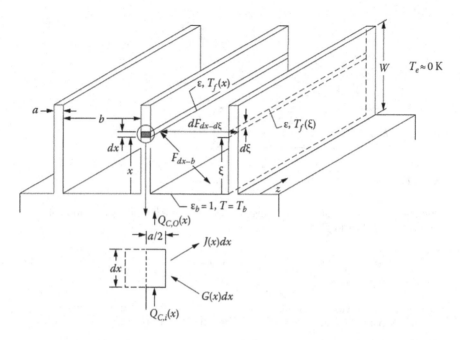

FIGURE 7.9 Geometry for determination of local temperatures on parallel fins.

temperature T_b as in Figure 7.9. The fin surfaces are diffuse-gray and are in a vacuum. Set up the equation describing the local fin temperature, assuming the environment is at $T_e \approx 0$ K.

Because the fins are thin, their local temperature is assumed constant across the thickness a, and the control volume Equation (7.4) is used for the circled differential element in Figure 7.9. Since there is an infinite array of fins, the surroundings are identical for each fin and are the same on both sides of each fin. From symmetry, only half the fin thickness needs to be considered. All the fins are the same, so the energy balance needs to be considered for only one fin, and the fin temperature distributions are all the same.

The net conduction into the element dx per unit time and *per unit length of fin* in the z direction is, for constant thermal conductivity, $(ka/2)(d^2T_f/dx^2)dx$. The radiation relations are formulated from the net-radiation method as given by the enclosure Equation (6.36). Writing this for an element dx along the fin gives:

$$\frac{q(x)}{\varepsilon} - \frac{1-\varepsilon}{\varepsilon}\int_{\xi=0}^{W} q(\xi)dF_{dx-d\xi} = \sigma T_f^4(x) - \sigma T_b^4\int_0^b dF_{dx-db} - \int_{\xi=0}^{W}\sigma T_f^4(\xi)dF_{dx-d\xi} \qquad (7.5.1)$$

Using the relations $\int_0^b dF_{dx-db} = F_{dx-b}$, $q(x) = \dfrac{ka}{2}\dfrac{d^2T_f(x)}{dx^2}$, $q(\xi) = \dfrac{ka}{2}\dfrac{d^2T_f(\xi)}{d\xi^2}$

gives, after rearrangement:

$$-\frac{ka}{2\varepsilon}\frac{d^2T_f(x)}{dx^2}+\sigma T_f^4(x)=\sigma T_b^4 F_{dx\text{-}b}+\int\limits_{\xi=0}^{W}\left[-\frac{1-\varepsilon}{\varepsilon}\frac{ka}{2}\frac{d^2T_f(\xi)}{d\xi^2}+\sigma T_f^4(\xi)\right]dF_{dx\text{-}d\xi}\quad(7.5.2)$$

In dimensionless form, this becomes:

$$-\mu\frac{d^2\vartheta(X)}{dX^2}+\vartheta^4(X)=F_{dX\text{-}B}+\int\limits_{Z=0}^{1}\left[-\mu(1-\varepsilon)\frac{d^2\vartheta(Z)}{dZ^2}+\vartheta^4(Z)\right]dF_{dX\text{-}dZ}\quad(7.5.3)$$

where $\vartheta(X)=\dfrac{T_f(X)}{T_b}$, $B=\dfrac{b}{W}$, $\mu=\dfrac{ka}{2\varepsilon\sigma T_b^3 W^2}$, $X=\dfrac{x}{W}$, $Z=\dfrac{\xi}{W}$

Equation (7.5.3) is a nonlinear integrodifferential equation and can be solved numerically. Two boundary conditions are needed. At the base of the fin, $T_f(x = 0) = T_b$, so:

$$\vartheta=1\quad\text{at }X=0\qquad\qquad(7.5.4)$$

At the tip of the fin, $x = W$, the conduction to the tip boundary must equal the energy radiated: $-k\partial T_f/\partial x\big|_{x=W}=\varepsilon\sigma T_f^4(W)$. In terms of ϑ,

$$-\frac{d\vartheta}{dX}=\frac{\varepsilon\sigma T_b^3 W}{k}\vartheta^4=\frac{1}{2\mu}\frac{a}{W}\vartheta^4\quad\text{at }X=1\qquad(7.5.5)$$

and the fin thickness-to-width ratio a/W is another parameter. If $(a/W)/2\mu$ is small, $(d\vartheta/dX)_{X=1}$ can be approximated as zero. The configuration factors needed in Equation (7.5.3) are found by the methods of Examples 5.2 and 5.4.

A black base surface without fins provides the maximum radiative emission. Adding an infinite array of fins to a plane nonblack surface produces a series of radiating cavities that can approach the performance of a black surface. The fins provide added weight and complexity, and hence it is better to simply use a planar surface with a high emissivity than to use a planar surface with a large array of fins. However, a finned surface can supply directional emission or absorption characteristics that make it attractive for some applications, as discussed in Section 3.4.3 of Howell et al. (2021).

7.3.1.2 2D Energy Flow

Some applications use 2D fins, such as for removing excess energy from electronic equipment in satellites and other space devices. The 2D fin is typically a thin metal plate, such as aluminum, with energy-generating equipment in good thermal contact with a part of the plate area. The energy transferred to the plate is conducted away in 2D and is dissipated by radiation to cooler surroundings.

Consider the diffuse-gray fin in Figure 7.10 with uniform thickness and constant thermal properties. Energy to be dissipated to cool equipment is transferred to the

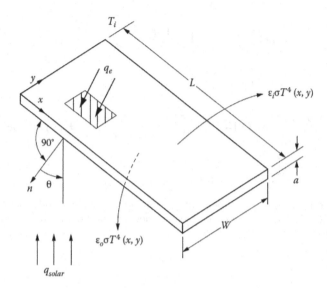

FIGURE 7.10 A 2D radiating fin with energy flux addition q_e to an area on one side.

shaded area on one side and is radiated away from both sides. The radiating plate is exposed to the Sun on the outside and to surroundings at T_i on the inside. The energy equation within the zone (shown shaded) over which energy from the equipment is being transferred to the plate is given by Equation (7.4) as:

$$ka\left(\frac{\partial^2 T}{\partial x^2} + \frac{\partial^2 T}{\partial y^2}\right) + \alpha_s q_{solar} \cos\theta + q_e - \varepsilon_o \sigma T^4(x,y) = 0 \tag{7.7}$$

net conduction + absorbed solar energy + internal generation − emission loss = 0

In this region, an element of the plate receives energy by 2D conduction within the plate, by absorption of solar radiation, and by energy addition from electronic equipment and/or other sources.

The plate loses energy by radiation from its outside surface. For the other portions of the plate, where $q_e = 0$, the energy equation becomes:

$$ka\left(\frac{\partial^2 T}{\partial x^2} + \frac{\partial^2 T}{\partial y^2}\right) + \alpha_s q_{solar} \cos\theta + \sigma\varepsilon_i\left[T_i^4 - T^4(x,y)\right] - \varepsilon_o \sigma T^4(x,y) = 0 \tag{7.8}$$

where there is a term for the net-radiation loss from the surface that is inside the enclosure. Energy losses from the end edges of the fin were assumed small, so insulated edge boundary conditions are used:

$$\frac{\partial T}{\partial x} = 0 \quad \text{at } x = (0,L); \quad \frac{\partial T}{\partial y} = 0 \quad \text{at } y = (0,W)$$

Along the boundary between the area of energy addition (shaded area) and the remaining area of the plate, there is continuity of fin temperature and energy flow in the x and y directions.

7.3.1.3 Multidimensional and Transient Energy Conduction with Radiation

For a thin radiating fin, the local fin temperature can usually be assumed uniform across the fin thickness. Then, the temperature variations are only in directions along the radiating surfaces. If the conducting solid is thick, however, the temperature will also vary normal to the radiating surface and can be steady or transient. The surfaces can usually be assumed opaque, so that without external convection, the net radiation at the surface is just the boundary condition for conduction within the solid.

If n is the outward normal from the surface, the conduction energy flux at the surface, flowing outward from within the solid, is $-k(\partial T/\partial n)$. This is the local flux supplied to the surface to balance the net radiative loss. Hence, in the absence of convection, there is the boundary condition, Equation (7.1), $-k(\partial T/\partial n)|_{wall} = (J - G) = q_r$. The governing partial differential equation in the solid is Equation (7.2). The distribution of $(J - G)$ along the conducting surfaces is found from the radiative enclosure methods described previously. For a nongray surface, the radiative flux q_r is the integral of the spectral fluxes as developed in Chapter 6. Subject to these boundary conditions, the multidimensional energy conduction equations can be solved by finite-difference or finite-element methods (FEM), as shown in Section 7.5.

If the temperature distributions within the solid are transient as well as spatially dependent, the problem can be complex, and the analytical solutions are difficult. Transient solutions are generally simpler if the geometry is 1D. An example is the electric heating of a thin wire where the transient temperature distribution varies only along the wire length. From the control volume approach, the 1D energy equation for the wire with radius r is:

$$r^2 \rho c \frac{\partial T}{\partial t} = kr^2 \frac{\partial^2 T}{\partial x^2} - 2r \left[J(x,t) - G(x,t) \right] + r^2 q_e \qquad (7.9)$$

If the properties are assumed constant, the solution can be simpler. The J and G depend on the radiative exchange with the surroundings and are functions of axial position x and time t. The J value varies considerably as the wire temperature changes with time and position. Solution for $T(x,t)$ requires an initial temperature distribution, an assumption that q_e takes on a value at $t = 0$, and two boundary conditions on x, which could be fixed electrode temperatures at the ends of the wire.

7.4 RADIATION WITH CONVECTION AND CONDUCTION

Interactions of radiation, convection, and conduction are found in many engineering applications, such as convective and radiative cooling of high-temperature components, cooling of hypersonic and reentry vehicles, and interactions of solar radiation with the Earth's surface. In many other cases, complex convection patterns may be seen, including in convection cells and free-convection patterns in oceans with the absorption of solar energy, among others. For these complex cases, results must be found using numerical techniques. If a medium is transparent, the governing equations for a flowing medium (momentum and energy equations) are unaffected by the presence of radiation, which only enters through the boundary conditions.

7.4.1 THIN RADIATING FINS WITH CONVECTION

Thin fins with multiple energy transfer modes are used extensively for supplying effective energy dissipation. We analyze one such problem with an example.

Example 7.6

Examine the performance of the fin in Figure 7.11; it depends on joint conduction, convection, and radiation. A gas at T_e is flowing over the fin and removing energy by convection. The environment to which the fin radiates is also assumed to be at T_e. The fin cross-section has area A and perimeter P.

The fin is nongray with total absorptivity α for radiation incident from the environment. Using a control volume, an energy balance on a fin element of length dx gives:

$$kA\frac{d^2T}{dx^2}dx = \sigma\left[\varepsilon T^4(x) - \alpha T_e^4\right]Pdx + hPdx\left[T(x) - T_e\right] \tag{7.6.1}$$

The term on the left is the net conduction into the element, and on the right are the radiative and convective losses. The radiative exchange between the fin and its base is neglected here. This equation is to be solved for $T(x)$, which can then be used to obtain energy dissipation. Multiply by $[1/(kA\,dx)]\,dT/dx$, and integrate once to yield:

$$\frac{1}{2}\left(\frac{dT}{dx}\right)^2 = \frac{\varepsilon\sigma P}{kA}\left(\frac{T^5}{5} - \frac{\alpha}{\varepsilon}TT_e^4\right) + \frac{hP}{kA}\left(\frac{T^2}{2} - TT_e\right) + C \tag{7.6.2}$$

where C is a constant of integration.

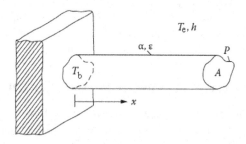

FIGURE 7.11 Fin of constant cross-sectional area transferring energy by radiation and convection. (Flowing gas and the environment are both at T_e.)

For convection and radiation from the end surface of the fin, or for the end of the fin assumed insulated, the dT/dx in Equation (7.6.2) is expressed in terms of the proper end conditions. For a case that allows an analytical solution, let $T_e \approx 0$ and let the fin be long. For large x, $T(x) \to 0$ and $dT/dx \to 0$, and from Equation (7.6.2), $C = 0$. Solving for dT/dx gives:

$$\frac{dT}{dx} = -\left(\frac{2}{5} \frac{P\varepsilon\sigma}{kA} T^5 + \frac{hP}{kA} T^2 \right)^{1/2}$$
(7.6.3)

The minus sign is used for the square root since T decreases as x increases. The variables in Equation (7.6.3) are separated and the equation integrated with the condition that $T(0) = T_h$,

$$\int_0^x dx = -\int_{T_b}^T \frac{dT}{T \left[\frac{2}{5}(P\varepsilon\sigma/kA)T^3 + hP/kA \right]^{1/2}}$$
(7.6.4)

$$x = \frac{1}{3} M^{-1/2} \left[\ln \frac{\left(GT_b^3 + M \right)^{1/2} - M^{1/2}}{\left(GT_b^3 + M \right)^{1/2} + M^{1/2}} - \ln \frac{\left(GT^3 + M \right)^{1/2} - M^{1/2}}{\left(GT^3 + M \right)^{1/2} + M^{1/2}} \right]$$

where $G = \frac{2}{3} P\varepsilon\sigma/kA$, $M = hP/kA$.

For this simplified limiting case, an analytical relation for $T(x)$ is obtained. The solution can be carried somewhat further, as considered in Homework Problem (7.2). A detailed treatment of this type of fin is in Shouman (1965, 1968).

7.4.2 Channel Flows

Next, we consider the flow of a transparent gas through a heated radiating tube. The enclosure energy-balance equations, such as Equations (6.31) and (6.32), can be used as before, and the q_k at the wall surface will have convective energy addition to the wall.

Example 7.7

A transparent gas flows through a black circular tube (Figure 7.12). The tube wall is thin, and its outer surface is perfectly insulated. The wall is heated electrically to supply uniform energy input q_e per unit area and time. The wall temperature along the tube length is to be found. The convective energy transfer coefficient h between the gas and the inside of the tube is assumed constant. The gas has a mean velocity u_m, specific heat c_p, and density ρ_f. Axial conduction in the thin tube wall can be neglected.

If radiation were not considered, the local energy addition to the gas would equal the local electric heating (since the outside of the tube is insulated) and hence would be invariant with x along the tube. The gas temperature and wall

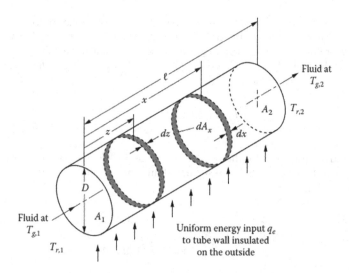

FIGURE 7.12 Flow through tube with uniform internal energy input to the wall and outer surface insulated.

temperature would both rise linearly with x. If convection were not considered, the only means for energy removal would be by radiation out of the tube ends, as in Example 6.14. In this instance, for equal environment temperatures at the tube ends, the wall temperature is a maximum at the center of the tube and decreases toward each end. The solution for joint radiation and convection is expected to show trends imposed by both limiting solutions.

Consider a ring element dA_x of length dx on the interior of the tube wall at x, as in Figure 7.12. The energy supplied per unit time is composed of electric heating, energy radiated to dA_x by other wall elements of the tube interior (see Example 6.14), and energy radiated to dA_x through the tube inlet and exit:

$$q_e \pi D dx + \int_{z=0}^{l} \sigma T_w^4(z) dF_{dz-dx}\left(|z-x|\right) \pi D dz + \sigma T_{r,1}^4 \frac{\pi D^2}{4} dF_{1-dx}(x) + \sigma T_{r,2}^4 \frac{\pi D^2}{4} dF_{2-dx}(l-x)$$

The tube ends are assumed to act as black disks at the inlet and outlet reservoir temperatures, which are taken equal to the inlet and outlet gas temperatures. The energy leaving the ring element at x by convection and radiation is $\left\{h\left[T_w(x) - T_g(x)\right] + \sigma T_w^4(x)\right\} \pi D dx$. Neglecting axial energy conduction in the wall, the energy quantities are equated to give (reciprocity was used on the F factors so that dx could be divided out):

$$h\left[T_w(x) - T_g(x)\right] + \sigma T_w^4(x) = q_e + \int_{z=0}^{l} \sigma T_w^4(z) dF_{dx-dz}\left(|x-z|\right) + \sigma T_{r,1}^4 F_{dx-1}(x) + \sigma T_{r,2}^4 F_{dx-2}(l-x) \quad (7.7.1)$$

Equation (7.7.1) has two unknowns, $T_w(x)$ and $T_g(x)$; a second equation is needed before a solution can be found. This is obtained from an energy balance on a volume element of length dx in the tube. The energy carried into this volume by the gas is $u_m \rho_f C_p T_g(x)(\pi D^2/4)$ and that added by convection from the wall is $\pi D h[T_w(x) - T_g(x)]dx$. The energy carried out by the gas is $u_m \rho_f C_p (\pi D^2/4)\{T_g(x) + [dT_g(x)/dx]dx\}$. The energy balance is:

$$u_m \rho_f C_p \frac{D}{4} \frac{dT_g(x)}{dx} = h[T_w(x) - T_g(x)] \qquad (7.7.2)$$

Defining the dimensionless quantities,

$$St = \frac{4h}{u_m \rho_f C_p} = \frac{4Nu}{Re\,Pr}; \quad H = \frac{h}{q_e}\left(\frac{q_e}{\sigma}\right)^{1/4}; \quad \vartheta = T\left(\frac{\sigma}{q_e}\right)^{1/4}$$

and $X = x/D$, $Z = z/D$, and $L = l/D$, the energy balances on the wall and fluid are:

$$\vartheta_w^4(x) + H[\vartheta_w(x) - \vartheta_g(x)] = 1 + \int_0^X \vartheta_w^4(Z)dF_{dX-dZ}(X-Z) + \int_X^L \vartheta_w^4(Z)dF_{dX-dZ}(Z-X) \qquad (7.7.3)$$

$$+ \vartheta_{r,1}^4 F_{dX-1}(X) + \vartheta_{r,2}^4 F_{dX-2}(L-X)$$

$$\frac{d\vartheta_g(X)}{dX} = St[\vartheta_w(X) - \vartheta_g(X)] \qquad (7.7.4)$$

The two equations have the unknowns $\vartheta_w(X)$ and $\vartheta_g(X)$ and five parameters: St, H, L, $\vartheta_{r,1}$, and $\vartheta_{r,2}$.

Equation (7.7.4) can be solved by using an integrating factor. The boundary condition is that $\vartheta_g(X)$ has a specified value $\vartheta_{g,1}$ at $X = 0$. The solution is:

$$\vartheta_g(X) = St\,e^{-StX} \int_0^X e^{StZ} \vartheta_w(Z)dZ + \vartheta_{g,1}e^{-StX} \qquad (7.7.5)$$

This is substituted into Equation (7.7.3) to yield an integral equation for $\vartheta_w(X)$:

$$\vartheta_w^4(X) + H\vartheta_w(X) - HSt\,e^{-StX} \int_{Z=0}^X e^{StZ} \vartheta_w(Z)dZ - H\vartheta_{g,1}e^{-StX}$$

$$\qquad\qquad\qquad\qquad\qquad\qquad\qquad\qquad\qquad\qquad (7.7.6)$$

$$= 1 + \int_{Z=0}^X \vartheta_w^4(Z)dF_{dX-dZ}(Z-X) + \int_{Z=X}^L \vartheta_w^4(Z)dF_{dX-dZ}(Z-X) + \vartheta_{r,1}^4 F_{dX-1}(X) + \vartheta_{r,2}^4 F_{dX-2}(L-X)$$

Solutions to Equation (7.7.6) were obtained by Perlmutter and Siegel (1962) and some results are in Figure 7.13. The predicted temperatures for joint radiation and convection fall below the temperatures predicted for either convection or radiation acting independently. For a short tube, radiation effects are significant over the entire tube length, and for the parameters shown, the combined-mode temperature distribution is like that for radiation alone. For a long tube, the combined-mode distribution is close to that for convection alone over the central portion of the tube. The energy transfer resulting from combined convection–radiation is more efficient than

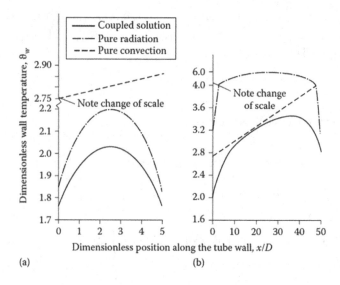

FIGURE 7.13 Tube wall temperatures resulting from combined radiation and convection for transparent gas flowing in a uniformly heated black tube for St = 0.02, $H = 0.8$, $\vartheta_{r,1} = \vartheta_{g,1} = 1.5$, and $\vartheta_{r,2} = \vartheta_{g,2}$. (a) Tube length, $l/D = 5$; (b) tube length, $l/D = 50$.

by either mode alone. Hence, the wall temperature distribution in the combined problem is below the distributions predicted by using either mode alone.

Example 7.8

What are the governing energy equations if the tube interior in Example 7.7 is diffuse-gray with emissivity ε rather than being black?

A convenient derivation is by use of the enclosure Equation (6.36). The energy flux added to the interior surface of the wall by means other than internal radiative exchange is $q_w(x) = h[T_g(x) - T_w(x)] + q_e$. The enclosure equation yields:

$$\frac{q_w(x)}{\varepsilon} - \frac{1-\varepsilon}{\varepsilon} \int_{z=0}^{l} q_w(z) dF_{dx-dz}(z,x) = \sigma T_w^4(x)$$

$$\hspace{4cm} (7.8.1)$$

$$- \int_{z=0}^{l} \sigma T_w^4(z) dF_{dx-dz}(z,x) - \sigma T_{r,1}^4 F_{dx-1} - \sigma T_{r,2}^4 F_{dx-2}$$

where $q_w(x)$ can be substituted to yield an equation with $T_w(x)$ and $T_g(x)$. Equation (7.7.5) is unchanged by having the wall gray. Thus, Equations (7.8.1) and (7.7.5) are two equations for the unknowns $T_w(x)$ and $T_g(x)$.

7.4.3 RADIATION COUPLED TO NATURAL CONVECTION

At moderate temperatures, radiative fluxes are small, but in conjunction with natural convection in air, the radiative transfer may be comparable to that due to natural convection. If a single vertical plate in air is internally heated, there can be an interaction of radiation with free convection, depending on the heating conditions of the plate. Electric heating of a wire, for example, may result in a steady-state temperature profile along the wire. For that to happen, input energy from the electricity must be dissipated to the environment by radiation, convection, and conduction. Also, in electronic circuit boards, there are isolated energy sources (from chips) which need to be cooled by combined radiation and free convection.

Natural convection instabilities can be produced or changed by radiation exchange. The stability of confined plane horizontal fluid layers has application to the design of flat-plate solar collectors. The radiation exchange between surfaces can have significant effects on crystals being grown from the vapor phase within an enclosure.

Because of the bifurcation/chaos characteristics of natural convection analysis, care should be taken in assuming that 2D solutions are valid, as 3D cell structures can form in apparently 2D geometries (e.g., long channels with asymmetrically heated walls). These structures can be augmented or retarded by the presence of radiation.

7.5 NUMERICAL SOLUTION METHODS

Numerical solution techniques are needed to solve the governing equations of radiation coupled with conduction and/or convection energy transfer. In earlier sections, energy equations were derived from an energy balance on each element of the system used to model the real configuration. With conduction and convection included, finer gridding is often needed because these transfer modes depend on local temperature derivatives; exact and detailed temperature distributions must be obtained so that derivatives can be evaluated accurately. Sometimes the difference between large incoming and outgoing radiation fluxes is needed to find a relatively small amount of conduction and/or convection, and this leads to difficulties with convergence and/or accuracy.

A local energy balance on each element of the system involves the net radiation that is absorbed at the surface, and this depends on the summation of all the contributions from the surroundings. In a detailed formulation, the radiative part may be set up as integrals involving the temperature distributions and configuration factors from the surrounding surfaces.

The benefits of nondimensionalization are discussed for multimode analyses. The relative sizes of the dimensionless parameters may supply insight into the best numerical approach.

The Monte Carlo method discussed in Section 10.7 may be used for the solution of the radiative transfer equation coupled with the other heat transfer modes. This statistical method follows many small quantities of radiant energy along their paths during radiative transfer. The method is easy to set up for complex problems that involve spectral effects and/or directional surfaces. The solutions may require long computer times (although this is rapidly becoming immaterial with increasing computer speeds and the solving capability of GPUs), and Monte Carlo may be the best way to attack some complex problems.

7.5.1 Numerical Integration Methods for Use with Enclosure Equations

The numerical solution of the radiation transfer equation coupled with conduction and convection requires integration of the governing equations subjected to the prescribed boundary conditions.

There are many ways to numerically approximate an integral. Because the integrands in radiative enclosure formulations are usually well-behaved at the end points, *closed* numerical integration forms are often used that include the end points. *Open* methods do not include the end points and can be used when end-point values are indeterminate, such as for improper integrals that yield finite values when integrated. In analyses including convection and/or conduction, the numerical integration will often use the grid spacing imposed by the differential terms.

The standard numerical integration methods are discussed in detail in online Appendix C at www.ThermalRadiation.net/appendix.html. Most methods are included in the standard mathematical packages such as MATLAB®.

7.5.2 Numerical Formulations for Combined-Mode Energy Transfer

Figures 7.2 and 7.3 illustrate the boundary condition for an opaque surface where the net radiation $(G - J)$ supplies the local addition of radiative energy flux. The $(G - J)$ is found by analyzing the radiative enclosure surrounding the surface area. This boundary condition can be applied to obtain the 3D conduction solution within the wall. As a simplified case, Figure 7.5 shows a control volume derivation for a thin wall in which the temperature distribution is 2D, as it is assumed not to vary significantly across the wall thickness. The control volume approach is further developed here to illustrate combined-mode solutions by using thin-walled enclosures as an example.

Consider an enclosure as in Figure 6.11 and let one or more of the walls be thin and energy conducting. There is uniform energy generation within each wall and convection at the inside surfaces, as in Figure 7.14. The outside of the enclosure is assumed well insulated for simplicity, but external energy flows can be added in a comparable way to those considered here.

A local rectangular coordinate system is positioned along a typical wall A_k (Figure 7.14). For a wall element at $\mathbf{r}_k$, an energy balance is:

$$J_k(\mathbf{r}_k) - G_k(\mathbf{r}_k) = q_k(\mathbf{r}_k) = \left[-\rho c a \frac{\partial T_k}{\partial t} + ka\left(\frac{\partial^2 T_k}{\partial x^2} + \frac{\partial^2 T_k}{\partial y^2} \right) + h\left(T_\mathrm{m} - T_k\right) + \dot{q}a \right]_{r_k} \quad (7.10)$$

The local radiative energy loss $q_k(\mathbf{r}_k)$ is found from the enclosure Equation (6.36):

$$\frac{q_k(\mathbf{r}_k)}{\varepsilon_k} - \sum_{j=1}^{N} \frac{1-\varepsilon_j}{\varepsilon_j} \int_{A_j} q_j(\mathbf{r}_j) dF_{dk-dj}(\mathbf{r}_j, \mathbf{r}_k) = \sigma T_k^4(\mathbf{r}_k) -$$

$$\sum_{j=1}^{N} \int_{A_j} \sigma T_j^4(\mathbf{r}_j) dF_{dk-dj}(\mathbf{r}_j, \mathbf{r}_k)$$

$$(7.11)$$

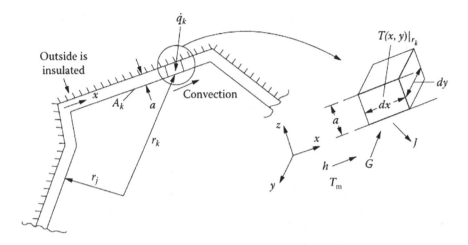

FIGURE 7.14 Radiative enclosure with thin walls in which there is 2D energy conduction.

The local temperature of the convecting medium, $T_m(\mathbf{r}_k)$, is obtained from other convective energy transfer relations. The equations are placed in dimensionless form to give:

$$\tilde{q}_k(\mathbf{R}_k) = \left[-\frac{\partial \vartheta_k}{\partial \tilde{t}} + N_{CR} \left(\frac{\partial^2 \vartheta_k}{\partial X^2} + \frac{\partial^2 \vartheta_k}{\partial Y^2} \right) + H(\vartheta_m - \vartheta_k) + \dot{S} \right]_{\mathbf{R}_k} \tag{7.12}$$

$$\frac{\tilde{q}_k(\mathbf{R}_k)}{\varepsilon_k} - \sum_{j=1}^{N} \frac{1-\varepsilon_j}{\varepsilon_j} \int_{A_j} \tilde{q}_j(\mathbf{R}_j) \, dF_{dk-dj}(\mathbf{R}_j, \mathbf{R}_k) = \vartheta_k^4(\mathbf{R}_k)$$

$$- \sum_{j=1}^{N} \int_{A_j} \vartheta_j^4(\mathbf{R}_j) \, dF_{dk-dj}(\mathbf{R}_j, \mathbf{R}_k) \tag{7.13}$$

where $\tilde{q} = q/\sigma T_{\text{ref}}^4$, $\quad N_{CR} = k/a\sigma T_{\text{ref}}^3$, $\quad H = h/\sigma T_{\text{ref}}^3$, $\quad \dot{S} = \dot{q}a/\sigma T_{\text{ref}}^4$

$$\tilde{t} = \sigma T_{\text{ref}}^3/\rho c a, \quad \mathbf{R} = \mathbf{r}/a, \quad X = x/a, \quad Y = y/a, \quad \vartheta = T/T_{\text{ref}}$$

In Equations (7.12) and (7.13), the dimensionless temperature ϑ and the dimensionless net radiative energy flux $\tilde{q}$ are the dependent variables, and X, Y, and $\tilde{t}$ are independent variables. The dimensionless parameters N_{CR}, $\dot{S}$, and H (or slight modifications of them) often appear in combined-mode problems involving radiative transfer. They supply a measure of the importance, compared to radiation, of conduction, internal energy generation, and convection, respectively. Their relative magnitudes can help to choose the best solution method for a given problem. If H is large, solution of convective energy transfer becomes more critical, with a small effect of radiation. The solution probably converges best if ϑ is chosen as the dependent variable. For small H, the problem might best be solved as a radiative transfer problem using ϑ^4 as the dependent variable. In transient cases, the variation in temperature level may shift the relative importance of the modes.

7.5.2.1 Numerical Solution Techniques

Examples 7.7 and 7.8 result in a matrix of nonlinear algebraic equations of the form:

$$\left[A_{ij}\right]\left[\vartheta_j\right]+\left[B_{ij}\right]\left[\vartheta_j^4\right]=\left[C_i\right] \tag{7.14}$$

It is important to examine the relative values of the elements A_{ij} and B_{ij}. If the A_{ij} are comparatively large, the problem can be treated as linear in ϑ_j; conversely, for large B_{ij}, the problem can be treated as linear in ϑ_j^4. When the coefficients A and B are approximately equal, other treatments are in order.

If we define $A_{ij}^* = A_{ij} + B_{ij}\vartheta_j^3$, Equation (7.14) becomes:

$$\left[A_{ij}\right]\left[\vartheta_j\right]+\left[B_{ij}\right]\left[\vartheta_j^4\right]=\left[A_{ij}+B_{ij}\vartheta_j^3\right]\left[\vartheta_j\right]=\left[A_{ij}^*\right]\left[\vartheta_j\right]=\left[C_i\right] \tag{7.15}$$

This is a set of linear algebraic equations with coefficients A_{ij}^* that are variable and nonlinear.

These equations cannot be solved by elimination or direct matrix inversion, because A_{ij}^* are temperature dependent and thus they are not initially known. Many techniques are available for the numerical solution of the nonlinear equations typical of combined-mode problems with radiation. Some of these are gathered and explored in online Appendix C.3 at www.ThermalRadiation.net/appendix.html.

7.5.3 Verification, Validation, and Uncertainty Quantification

Computer solutions to radiative transfer problems, particularly multimode problems, can be challenging. To find the quality of a numerical code, first define what it is you really want to know, called the quantity of interest (QoI). Is it the energy flux at a boundary? The temperature at a location on the boundary? Secondary factors might be allowed to have imprecise predictions if the prediction of the QoI is accurate.

In all computational problems, it is necessary to set up *a priori* conditions for the convergence and insensitivity to grid resolution of the QoI. But these are not enough by themselves. It should be recognized that any computer model or code ideally should be subjected to three categories of tests: *verification*, *validation*, and *uncertainty quantification* (UQ). This applies to any code, not just those for radiative transfer.

7.5.3.1 Verification

A software engineer needs to ask the following questions to verify his or her code: Is the code bug-free, and is it supplying results that reflect the governing equations within the assumptions of the analysis of the physical problem being addressed? To ensure that a code is bug-free and gives correct results within the physical assumptions, one possibility is the comparison of code predictions with known benchmark solutions. Benchmarks can be well-accepted solutions from the literature or can be generated. A radiation code can be tested against known simple limiting cases:

1. If all surfaces are isothermal at the same T, is the predicted energy transfer among all elements = 0?
2. If $\varepsilon = 0$ on a surface, is the predicted radiative energy flux = 0 at that surface?
3. For a combined-mode problem, does the code give correct results for each mode independently when the other modes are set to zero?

7.5.3.2   Validation

Even if verified, the numerical code may not reflect reality, since assumptions are usually implicit in the code (gray and/or diffuse surfaces, temperature-independent properties, etc.) The code should therefore be compared with "reality." Usually, that is a comparison with available experimental data. For comparison of predictions with experiment, we must know both the error limits on the experimental data and the limits on the accuracy in the code predictions to find whether the predictions lie within the range of the experimental measurements. Because perfect agreement between measurement and prediction is seldom obtained, *a priori* limits on what constitutes acceptable validation in the prediction of the QoI should be made. Must the prediction be within 1%, 5%, or 20% before the code predictions are acceptable? The answer will depend on the particular problem and the QoI being predicted.

7.5.3.3   Uncertainty Quantification

What is the uncertainty in the results of the code (the QoI)? A code prediction usually has at least three types of error. First, input data to the code, such as physical properties, dimensions, and energy source values, will have some degree of uncertainty (see Hively et al. 2013). Second, the model itself has uncertainties due to the assumptions within the model (e.g., angular and spectral discretization errors, gray and/or diffuse assumptions, and assumptions as to whether a 1D, 2D, or 3D analysis is acceptable). Finally, errors due to increment size in space and/or time and machine accuracy (number of allowed significant figures) can introduce numerical uncertainty. Having defined possible sources of uncertainty in input data, the model itself, and numerical uncertainties, how do we quantify how these uncertainties propagate through the code so that the uncertainty in the QoI prediction can be known? This is quite a tricky question to answer and is an ongoing research area. It is, however, an important question, because unless some bounds are available on the uncertainty of the QoI, the code results are not useful. For example, if the code is predicting global warming effects, but the uncertainty allows predictions that range from a new ice age to the immediate loss of the polar ice caps, then the code is not of practical benefit.

If we find that the uncertainty in the QoI is so large that the code results are not useful, how do we continue to reduce prediction uncertainty? It is possible to find the sensitivity of the code to the input data, the modeling uncertainties, and machine errors and to then improve the data or code to reduce the uncertainty in the most sensitive factors. This is a daunting task when there are many parameters and much data to be considered. The field of verification, validation, and UQ for complex codes is an active research field.

HOMEWORK (additional problems available in online Appendix I at www.ThermalRadiation.net/appendix.html)

7.1   A thin 2D fin in a vacuum radiates to outer space, which is assumed at $T_e \approx 0$ K. The base of the fin is at T_b, and the energy loss from the end edge of the fin is negligible. The fin surface is gray with emissivity ε. Write the differential equation and boundary conditions in dimensionless form to find the temperature distribution $T(x)$ of the fin. (Neglect any radiant interaction with the fin base.) Can you separate variables and show the integration necessary to obtain the temperature distribution? (Hint: $\int (d^2\theta/dx^2)\,(d\theta/dx) = (1/2)\,(d\theta/dx)^2 + $ constant.)

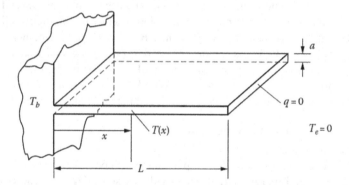

7.2   Consider the fin in Figure 7.11 as analyzed in Example 7.6. The energy transfer coefficient at the tip of the fin is h_L, and the emissivity of the end area is ε as for the rest of the fin surface. Formulate the boundary condition for the end face of the fin and apply this condition to the general solution of the fin energy equation. Formulate the analytical relations and describe how you would obtain the fin efficiency.

7.3   A very small-diameter pipe is at $T_{pipe} = 650$ K. The pipe is thin-walled polished copper, has a diameter $D = 0.2$ cm, and is in a large room at $T_e = 300$ K. The radiative emissivity of the copper is $\varepsilon_c = 0.04$. A cylindrical opaque insulation layer with thickness t and thermal conductivity $k = 0.07$ W/m·K is added to the surface of the pipe. The emissivity of the outer insulation surface is $\varepsilon_i = 0.85$. The natural convective energy transfer coefficient on the surface of the insulation is $h = 15$ W/m²·K. For simplicity, h is assumed to be independent of insulation diameter and surface temperature, but a more precise analysis should include these effects. It is found that adding the insulation increases the rate of energy loss from the pipe. Find the thickness of insulation $t = t_{max}$ that *maximizes* the energy loss from the pipe. *Answer*: $t_{max} = 3.4$ mm

7.4   A space radiator is composed of a series of plane fins of thickness $2t$ between tubes of radius R. The tubes are at uniform temperature T_b. The tubes are black, and the fins are gray with emissivity ε. The radiator operates in a vacuum with an environment temperature of $T_e \approx 0$ K. Formulate the differential equation (including the analytical expressions for the configuration factors) and boundary conditions to obtain the temperature distribution $T(x)$ along the fin. Include the interaction between the fin and the tubes.

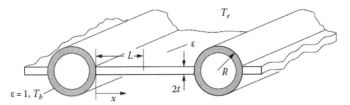

7.5   Steam is condensing inside a thin-walled tube of radius r_i. The tube has a coating of emissivity $\varepsilon_t = 1$ on the outer surface. The saturation temperature of the steam is T_b. Identical annular fins of outer radius r_o and emissivity ε_f are evenly spaced a distance L (between fin faces) along the tube. The fins are of thickness δ and thermal conductivity k. The environment surrounding the fin-tube assembly is at $T_e \approx 0$ K. Convection can be ignored. The configuration factor from a ring element on the tube to a ring element on fin 1 is $dF_{dt-d1}(x,\rho_1)$ and from a ring element on fin 2 to a ring element on fin 1 is $dF_{d2-d1}(L,\rho_1,\rho_2)$.

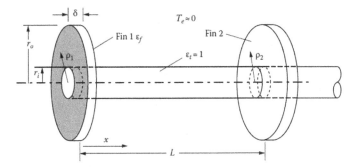

Set up the governing equation for the temperature distribution of the fin, $T(\rho)$.

7.6   Two directly opposed parallel diffuse-gray plates of finite width W have a uniform energy flux q_e supplied to each of them. The plates are infinitely long in the direction normal to the cross-section shown. They are separated by a distance H. The plates are each of thickness t ($t \ll W$) and thermal conductivity k, and both have emissivity ε. The plates are in a vacuum, and the surroundings are at T_e. Set up the governing integral equation for finding $T(x)$, the temperature distribution across both plates. The outer surface of each plate is insulated, so all radiative energy exchange is from the inner surfaces.

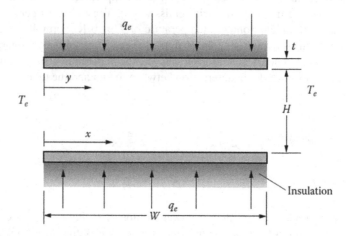

7.7   A copper–constantan thermocouple ($\varepsilon = 0.15$) is in a transparent gas stream at 300 K next to a large blackbody surface at 900 K. The energy transfer coefficient from the gas to the thermocouple is $h_{tc} = 35$ W/(m^2·K). Estimate the thermocouple temperature if it is (a) bare or (b) surrounded by a single polished aluminum radiation shield with $\varepsilon_{shield} = 0.075$ in the form of a cylinder open at both ends. The energy transfer coefficient from the gas to both sides of the shield is $h_s = 15$ W/(m^2·K).

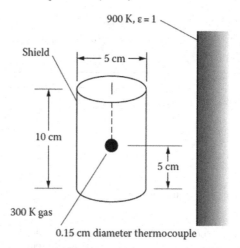

Answers: (a) 376 K; (b) 358 K.

7.8 A 1 cm diameter thin cryogenic electronic device is glued by epoxy
to the bottom surface of a 0.5 cm deep, 1 cm diameter cavity of a high
thermal conductivity ceramic package. The ceramic package is kept at
a liquid nitrogen temperature of 77 K inside a relatively large vacuum
enclosure at the ambient temperature of 300 K. The epoxy provides a
thermal interface conductance of 1×10^4 W/m$^2 \cdot$K between the electronic
device and the ceramic package. All surfaces are gray with diffuse emission. The emissivity is 0.6 on the device surface and 0.3 on the ceramic
surface. Calculate the temperature of the electronic device when the
device is turned off.

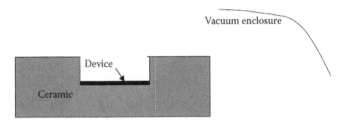

Answer: $T_d = 77$ K.

7.9 A rod of circular cross-section extends out from a slender space vehicle in
Earth orbit into surroundings at T_e. The rod axis is normal to the direction
of the Sun. The rod is coated so that its infrared emissivity is ε_{IR} and its
solar absorptivity is α_s. The base of the rod is at $T_b > T_e$. Derive a differential equation to predict the rod temperature distribution, $T(x)$. State the
boundary conditions, including radiation at the circular end face. Neglect
temperature variations within the rod cross-section at each x. Neglect radiation to the rod from the slender vehicle surface and neglect any emitted or
reflected radiation from the Earth.

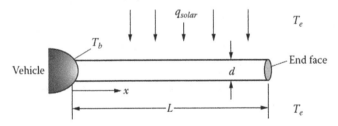

7.10 A_1 and A_2 are diffuse concentric spheres. The inner surface of the inner
sphere is heated with nonradiating combustion products at $T_{comb} =$
1100 K with a convective energy transfer coefficient to the surface of
50 W/(m$^2 \cdot$K). Calculate the temperature T_1 of the inner sphere for the
surface spectral emissivities shown, where $\varepsilon_{\lambda,1}$ is assumed independent of
temperature.

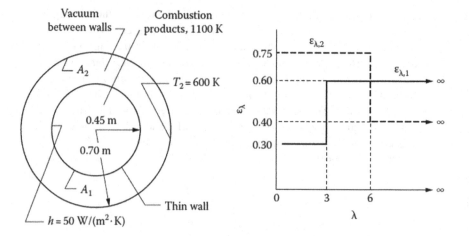

Answer: 869 K.

7.11 A wire between two electrodes is heated electrically with a total of Q_e. The wire resistivity r_e is constant, and the current is I. One end of the wire is at T_1 and the other is at T_2. The immediate surroundings are a vacuum, and the surroundings have a radiating temperature of T_0. The wire is gray with emissivity ε, diameter D, thermal conductivity k, and length L.

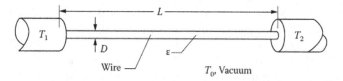

(a) Set up the differential equation for the steady-state temperature distribution along the wire, neglecting radial temperature variations within the wire. Integrate the equation, and find an expression (in the form of an integral) for the wire temperature as a function of x. The results should hold only the quantities given and should not have dT/dx. Explain how you would evaluate the expression to find $T(x)$.

(b) Let both the emissivity and the electrical resistivity be proportional to T (see Equations (4.59) and (4.60)). Derive the solution for this case and describe how it can be evaluated for a fixed value of Q_e.

7.12 A spherical temperature sensor, 0.15 cm in diameter, is on the axis of a short pipe, halfway between the open ends. Air at 400 K is flowing through the pipe, and the convective energy transfer coefficient on the sensor is 23 $W/(m^2 \cdot K)$. Calculate the sensor temperature. All surfaces are gray. Neglect blockage (shadowing) by the sensor when computing configuration factors between the boundaries of the pipe. Do not subdivide surfaces.

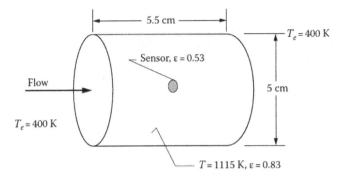

Answer: 909 K.

7.13 Two plates are joined at 90° and are very long, normal to the cross-section shown. The vertical plate (plate 1) is heated uniformly with an energy flux of 300 W/m². The horizontal plate has a uniform temperature of 400 K. Both plates have an emissivity of 0.6. The environment is at $T_e = 300$ K. Both plates are made of tungsten, with thermal conductivity $k = 174$ W/m·K. The plates are 0.056 cm in thickness. At the edge where the plates join, a thin layer of ideal insulation provides an infinite contact resistance. The dimensions are shown in the figure.

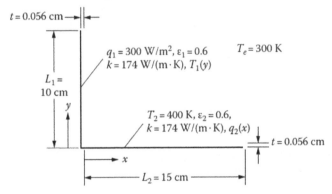

Derive the equations necessary for finding the temperature distribution on surface 1 and the net radiative energy flux on surface 2. Find reasonable boundary conditions for the plate ends; justify your choice. (This is a continuation of Homework Problem 6.18.)

7.14 An electrically heated nickel wire is suspended in a vacuum between two water-cooled electrodes kept at 810 K. The wire diameter is 0.15 cm, and the wire length is 4.15 cm. The surroundings are at a uniform temperature of $T_e = 900$ K and act as a black environment. Air at a temperature of T_e flows over the wire, causing an energy transfer coefficient between the wire and the air of $h = 35$ W/m²·K. Set up the combined radiation and conduction relations to determine the wire temperature $T(x)$, assuming that the radial temperature distribution within the wire is uniform at each x. Determine how many Watts must be generated within the wire for its center temperature to be 1050 K. The wire thermal conductivity is constant with a value of 85 W/(m·K), and the wire emissivity is constant with the value $\varepsilon_w = 0.13$. What is the required wattage if the nickel becomes oxidized so that $\varepsilon_w = 0.47$?

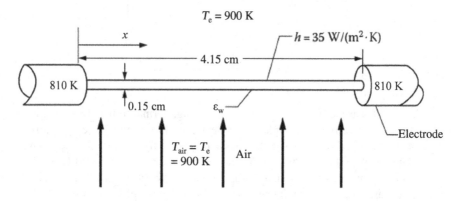

$T_e = 900$ K

$h = 35$ W/(m²·K)

4.15 cm

810 K   810 K

0.15 cm ε_w

—Electrode

$T_{air} = T_e$ Air
$= 900$ K

Answers:

ε_w	Q (W)
0.13	14.5
0.47	16.0

7.15 A long solid rectangular region in a vacuum has the cross-section shown and thermal conductivity k_w. One-half of one of the long sides is heated by contact with an opaque source of uniform flux q_e. The surroundings, which act as a black environment, are at a uniform temperature T_e. The exposed surfaces of the region are gray and have emissivity ε_w. Using the grid shown (for simplicity), set up the finite-difference relations to be solved for the steady temperature distribution in the rectangular solid.

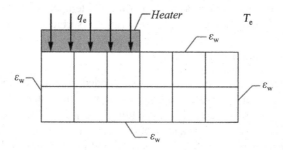

q_e —Heater T_e

ε_w

ε_w ε_w

ε_w

7.16 A long stainless-steel tube with a thick wall is filled with a highly insulating material and is near an infinite black hot wall. Divide the circumference
of the tube symmetrically about $\theta = 0$ into eight increments and obtain an
expression for the radiant energy from the hot wall to each tube increment.
Then, using a finite-difference approximation, obtain an approximate temperature distribution around the tube wall. Neglect radial temperature distributions in the tube wall. The tube wall material has thermal conductivity k_w
= 34 W/(m·K), and its outer surface is gray with emissivity $\varepsilon_w = 0.187$. The
surroundings are at a low temperature that can be neglected in the analysis,
$T_e \approx 0$ K.

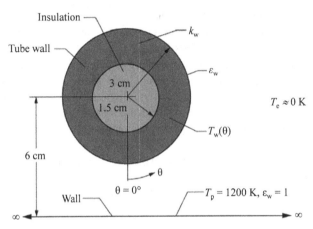

Answers: $T_1 = 1022$ K; $T_2 = 1014$ K; $T_3 = 1004$ K, $T_4 = 996$ K.

7.17 A tube has a length of $L = 2.00$ m and has an inside diameter of $D_i = 0.5$
m. The tube has a wall thickness of $b = 1$ cm, and the tube wall material
has a thermal conductivity of $k = 300$ W/m·K. An electric heating tape
is wrapped around the outside of the tube and is uniformly and carefully
insulated on its outer surface. The tape heater imposes a uniform energy
flux of $q_e = 6000$ W/m² at the inner tube surface. A transparent fluid at
$T_{f,in} = 300$ K, enters the tube from a large plenum at the same temperature. The fluid flows through the tube at a mass flow rate of 0.2 kg/s. The
fluid has specific heat $c_p = 4100$ J/(kg·K) and density $\rho = 1000$ kg/m³.
At the given flow rate, the energy transfer coefficient between the fluid
and the tube surface is $h_x = 1000/(x + 0.01)^{1/2}$ W/(m²·K), where x is the
distance from the tube entrance in meters. Four diffuse-gray coatings
are available to cover the inside of the tube. These have emissivities of
$\varepsilon = 0.05, 0.35, 0.65,$ and 0.95.

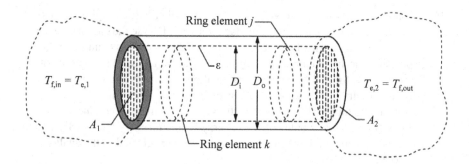

(a) Derive the energy equations that govern the energy transfer behavior to find the tube wall and fluid temperatures along the tube length and place them in dimensionless form.

(b) Find the maximum tube surface temperature, the position of that temperature, $x(T_{max})$, and the mean fluid temperature at the tube exit, $T_f(x = L)$ for each of the three possible emissivities. Show a plot of the tube's inner surface and fluid temperature distributions versus x for each emissivity. These may be in dimensionless form.

(c) Discuss the temperature results, giving some physical discussion of the relative effects of the various energy transfer mechanisms on the shapes of both the tube wall and fluid temperature profiles.

Discuss the numerical accuracy of the results. Are they converged, and are they within acceptable accuracy? Estimate the possible error in your solutions.

7.18 A high-temperature nuclear reactor is cooled by a transparent gas. The gas flows through tubular cylindrical fuel elements. The fuel elements are of length L m and inside diameter D m, and the gas has specific heat c_p kJ/(kg·K), density ρ kg/m^3 (both temperature independent), and a mass flow rate of $\dot{m}$ kg/s through each fuel element. The tube has wall thickness b. Energy is generated in a sinusoidal distribution along the length x of the tube at a rate $q(x) = q_{max} \sin\left(\dfrac{\pi x}{L}\right)$ W/m^2 based on the inside tube area.

The energy transfer coefficient h between the gas and the tube surface is assumed constant with x and has units of W/(m^2·K). The gas enters the tube at temperature $T_{g,i}$ from a large chamber at temperature $T_{r,i}$. The gas leaves the tube at $T_{g,e}$ and enters a mixing plenum that is at temperature $T_{r,e}$. The tube interior surface is diffuse-gray with emissivity ε. The tube exterior surface is perfectly insulated. The tube wall material has thermal conductivity k W/(m·K).

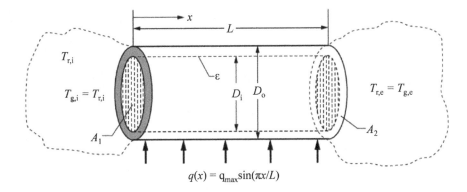

$$q(x) = q_{max}\sin(\pi x/L)$$

(a) Set up the governing equations for finding the wall and gas temperature distributions along the tube length (see Examples 7.7 and 7.8 for some help).

(b) Using the dimensionless groups given in Example 7.7 or modifications appropriate for this problem, put the equations in dimensionless form.

(c) For values of the dimensionless parameters $St = 0.02$, $N = 5$, H based on $q_{max} = 0.08$, $\vartheta_{r,i} = \vartheta_{g,i} = 1.5$, and $\vartheta_{r,e} = \vartheta_{g,e} = 5$, solve for the wall temperature distribution along the tube for the cases $\varepsilon = 1$ and $\varepsilon = 0.5$. You may wish to obtain the pure radiation and pure convection results as limiting cases.

8 Properties of Participating Media

Margaret Lindsay Huggins (1848–1915) *was a pioneer in measuring stellar spectra. Along with her husband William Huggins, she was the first to show that the stars were indeed distant suns, based on the similarity of their emission spectra (especially lines of magnesium and calcium) with that of the Sun. Their instrumentation was based on a spectroscope model proposed by Robert Bunsen. Observations of Sirius showed a slight Doppler shift in the measured spectra, which indicated that it was moving away from the Earth. This eventually led to the discovery of the expanding universe.*

Gustav Adolf Feodor Wilhelm Ludwig Mie (1868–1957) *researched colloids at the University of Greifswald in Germany. One of his students investigated the scattering and attenuation of light by gold colloids. Mie used his knowledge of the Maxwell equations and solutions of similar problems to treat the theoretical problem of scattering and absorption of light by a small absorbing sphere. These calculations were done by hand and limited his results to three term expansions and treatment of particles smaller than 200 nm. He developed the Mie system of units in 1910 with the basic units Volt, Ampere, Coulomb and second (VACS-system).*

8.1 INTRODUCTION

A participating medium absorbs, emits, and scatters electromagnetic (EM) waves propagating through it. In such media, radiation exchange takes place not only between bounding surfaces but between all volume and surface elements, and all interactions need to be accounted for in the full solution of the problem. Thus, the spectral properties of all gases and particles and the directional properties of particles need to be known before modeling radiative transfer problem in such systems. In this chapter, we discuss these properties and outline different models and approximations to determine them.

In atmospheric studies or for high-temperature combustion applications, several different gases need to be considered if radiation transfer is to be accounted for in detail. Among them, water vapor (H_2O), carbon dioxide (CO_2), and carbon monoxide (CO) are the most important ones; however, depending on the application, nitrogen (N_2), nitrogen oxides (NO_2, NO, NO_x), methane (CH_4), or others may need to be considered.

Radiation transfer through gases has been studied for over 150 years. Early work was focused on the absorption and transmission of solar radiation by the Earth's atmosphere. This interfered with observations of light from the Sun and distant stars. The solar spectrum received on Earth was recorded in detail by Samuel Langley

DOI: 10.1201/9781003178996-8

(1883) and has been updated many times since then. Figure 8.1 depicts our current knowledge of the solar spectrum, where the dashed curve is for a scaled blackbody emission spectrum at the effective solar temperature of 5780 K. The intermediate curve is the solar spectrum outside the Earth's atmosphere. The lowest solid curve with several sharp dips shows the spectrum received at ground level after the solar radiation has followed a path through the Earth's atmosphere. The dips in the curve show where radiation has been absorbed by various atmospheric constituents, mainly water vapor and carbon dioxide. Absorption occurs in specific wavelength regions, illustrating that gas radiation properties vary considerably with wavelength.

The observed spectrum caused by the emission or absorption of radiation by a gas is characteristic of a specific gas, and it can be used as a diagnostic tool to determine the gas temperature and concentration. The importance of gas radiation in industry was recognized in the 1920s for energy transfer in furnaces. Combustion products, mainly carbon dioxide and water vapor, were found to be significant emitters and absorbers of radiant energy. The energy emitted from flames arises not only from the gases but also from hot carbon (soot) particles within the flame and from other particulate material, including pulverized coal, char, fly-ash, soot agglomerates, and others. Radiation can be significant in engines and combustion chambers, where temperatures can reach a few thousand Kelvins.

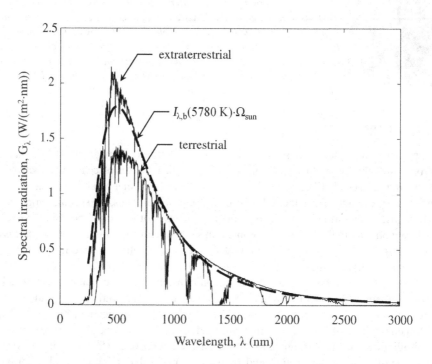

FIGURE 8.1 Attenuation by Earth's atmosphere of incident solar spectral irradiation and comparison with a 5780 K scaled blackbody spectrum. (Data extracted from NREL/ASTM G173-03 Tables of Reference Solar Spectral Irradiance, 2012.)

The details of gas radiation spectra are extremely important to understand climate change and to be able to determine the effects of different physical mechanisms on solar and terrestrial absorption and emission rates. Using these spectral gas properties, models of stellar atmospheres and the energy transfer processes within them have been constructed and compared with observed stellar behavior.

The spectral properties of solids and liquids are also of interest for predicting radiation transfer in complex industrial systems or for climate change. Gardon (1958, 1961) observed that the temperature distribution measured in a deep tank of molten glass was more uniform than expected from energy conduction alone and in the late 1940s, it became evident that radiative transfer by absorption and re-emission within the glass was a significant means of energy transfer and caused the effect. Water, on the other hand, is the most important liquid on Earth, and has a peculiar absorption spectrum, as shown in Figure 1.2. Note that water absorbs the least amount of solar energy at the visible wavelength band, from about 400 nm to 1000 nm.

The study of radiation in absorbing, emitting, and scattering media is quite challenging for three reasons. The first challenge is to account for the spatial variation of radiative properties throughout the medium of interest. Absorption and emission by gases and particles, and directional scattering by different shapes and structures of particles and agglomerates, can occur at all locations within a medium, depending on their concentration distributions and local temperature variations. A complete solution for radiative energy exchange requires knowing the radiation intensity, temperature, and physical properties throughout the medium. The underlying physics of this problem is quite complex, and, consequently, its solution is not trivial.

The second difficulty is that spectral variations of gases, particles and the surfaces need to be considered. These variations are often much more pronounced in gases, translucent solids, and translucent liquids than for solid surfaces. The radiative properties of most absorbing–emitting media, particularly polyatomic gases, are strongly wavelength-dependent. Therefore, radiative transfer calculations must consider the effect of these variations, and in the extreme must be repeated at thousands of important wavelength intervals. For this reason, past and present efforts seek to provide useful correlations of the properties that can give accurate results without the extreme computational requirements imposed by detailed spectral calculations. This is an ongoing research field and new methods, and improvements of existing methods, continue to appear in the literature.

The third challenge is the directional distribution of radiative energy in a medium due to scattering by particles and agglomerates, and reflection by surfaces. To account for such directional information, particle and agglomerate shapes and structures are needed along with their fundamental material properties, including the spectral complex index of refraction data. After that, different models can be employed, such as the Lorenz–Mie theory, discrete dipole approximation, and T-matrix method, among many others, to determine the required particle absorption and scattering cross-sections, as discussed below.

In this chapter, we first define the required absorption, emission, and scattering properties of participating media, and then discuss how this information can be used in radiative transfer calculations. Models, approximations, and correlations that are useful in engineering radiative transfer are then presented.

8.1.1 PROPAGATION OF RADIATION IN ABSORBING MEDIA

Radiative intensity (Section 2.3) is attenuated in media according to the relation $I_\lambda(S)/I_\lambda(0) = \exp(-4\pi k(\lambda)S/\lambda)$. Here, $k(\lambda)$ is the *absorption index*, which is related to the complex index of refraction, $\bar{n}(\lambda)$, expressed as $\bar{n}(\lambda) = n(\lambda) - ik(\lambda)$. Note that $n(\lambda)$ is the spectral index of refraction of the matter, and i is the imaginary number.

The electromagnetic absorption index $k(\lambda)$ should not be confused with the κ_λ in Equation (2.43), which is the volumetric *absorption coefficient*. The index $k(\lambda)$ is related to the magnetic permeability, electrical resistivity, and electrical permittivity of the medium (see Equations (4.16)–(4.18)). The *absorption coefficient* κ_λ is related to $k(\lambda)$ by:

$$\kappa_\lambda = \frac{4\pi k(\lambda)}{\lambda} \tag{8.1}$$

Note that κ_λ is a function of λ, which is related to the wavelength in a vacuum through n_λ.

These relations provide a theoretical basis for Bouguer's Law (Equation (2.45)), which was originally based on experimental observations. Equation (8.1) provides a simple expression for obtaining the spectral absorption coefficient from data using the electromagnetic absorption index $k(\lambda)$ for homogeneous materials.

The absorption coefficient of gases often varies strongly with wavelength, temperature, and pressure. Thus, it can explicitly be denoted as a function of temperature T and pressure P, as $\kappa_\lambda(T,P)$. It can be expressed as a function of location, $\kappa_\lambda(S)$, as S is related to the local T and P. For simplicity, we denote it as κ_λ. Analytical determinations of κ_λ require detailed quantum-mechanical calculations. Considerable analytical and experimental effort has been expended to determine κ_λ for various gases, liquids, and solids, as we discuss below.

As discussed in Section 2.2, the spectral dependence may be expressed either in terms of wavelength, wave number, or frequency. Frequency has the advantage that it does not change when radiation passes from one medium into another with a different refractive index. Wavelength does change because of the change in propagation velocity.

8.2 SPECTRAL LINES AND BANDS

8.2.1 PHYSICAL MECHANISMS

As shown in Chapter 4, the property variations with wavelength for optically dense solids, which can be treated as opaque, range from smooth to somewhat irregular. However, gas properties exhibit very irregular wavelength dependencies. Absorption or emission by gases is significant only in certain wavelength regions, especially when the gas temperature is below a few thousand Kelvins. Note that here "absorptance" refers to what is not transmitted through a cloud of a given gas species. (It is defined by Equation (8.11)). The absorptance spectrum for CO_2 is shown in Figure 8.2 at a particular P, T, and pathlength.

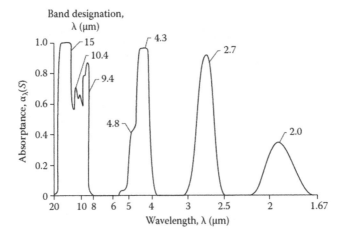

FIGURE 8.2 Low-resolution spectrum of absorption bands for carbon dioxide gas at 830 K, 10 atm, and for path length through a gas of 0.388 m. (From Edwards, 1976.)

A radiating gas can be composed of molecules, atoms, ions, and free electrons that are at various energy levels. In a molecule, the atoms form a dynamic system with vibrational and rotational modes that have specific quantized energy levels. A schematic diagram of the energy levels for an atom or ion is in Figure 8.3. Zero energy is assigned to the ground state e_1 (lowest-energy bound state), with the higher bound states being at positive energy levels. The energy e_I is the ionization potential, which is the minimum energy required to produce ionization from the ground state. For energies above e_I, ionization takes place and free electrons are produced.

As discussed before, light, or any type of EM radiation, is a wave. However, sometimes it is convenient to discuss the radiation process by utilizing a photon or quantum point of view (radiative transfer equation (RTE) photons). In this book, we will use the photon concept developed in the early 20th century to outline the absorption and emission by gases. The photon is then taken as the basic unit of radiative energy. Radiative emission releases photons, and absorption is the capture of photons. When a photon is emitted or absorbed, the energy of the emitting or absorbing particle is correspondingly decreased or increased. Figure 8.3 indicates three types of transitions that can occur in a gas molecule: bound–bound, bound–free, and free–free. A photon can also transfer part of its energy in certain inelastic scattering processes that are of minor engineering importance.

The magnitude of a radiative energy transition is related to the frequency of the emitted or absorbed radiation. The energy of a photon is $h\nu$, where h is Planck's constant and ν is the frequency of the photon. This shows how the electromagnetic wave approach (i.e., the frequency of waves) is tied to the description of energy propagation by photons (quanta of energy). For an energy transition from bound state e_3 down to bound state e_2 in Figure 8.3, a photon is emitted with energy $e_3 - e_2 = h\nu$. The frequency of the emitted energy is then $\nu = (e_3 - e_2)/h$, so a *fixed frequency* is associated with the transition from one specific energy level to another. Thus, in the

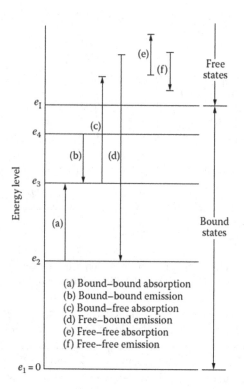

FIGURE 8.3 Schematic diagram of energy states and transitions for atom or ion. e_1 is the ground state and e_I is the ionization energy for the gas.

absence of other effects, the spectrum of the emitted radiation is a discrete spectral line at that frequency. Conversely, in a transition between two bound states when a molecule absorbs energy, the absorption is such that the particle can only go to one of the discrete higher energy levels. Consequently, for a photon to be absorbed, the frequency of the photon energy must have one of certain discrete values. For example, a molecule in the ground state in Figure 8.3 can absorb photons with frequencies $(e_2 - e_1)/h$, $(e_3 - e_1)/h$, or $(e_4 - e_1)/h$ and undergo a transition to a higher bound energy level. Photons with other frequencies in the range $0 < \nu < e_1/h$ cannot be absorbed.

When a photon is absorbed or emitted by an atom or molecule, and there is no ionization or recombination of ions and electrons, the process is a *bound–bound* absorption or emission (processes (a) and (b) in Figure 8.3). An atom or molecule goes from one quantized bound energy state to another. These states can be rotational, vibrational, or electronic in molecules, and electronic in atoms. Since bound–bound energy changes are associated with specific energy levels, the absorption and emission coefficients are sharply peaked functions of frequency in the form of a series of spectral lines. The lines have a finite spectral width resulting from various line-broadening effects discussed in Section 8.2.3.

Vibrational energy modes are always coupled with rotational modes. The rotational spectral lines superimposed about a vibrational line give a band of closely spaced spectral lines. If these lines overlap to form a continuous region, a *vibration-rotation band* is formed. Rotational transitions within a given vibrational state are associated with low energies (long wavelengths, from ~8–1000 μm). Vibration-rotation transitions are at infrared energies corresponding to wavelengths at about 1.5–20 μm. Electronic transitions usually occur at short wavelengths in the visible region 0.4–0.7 μm and at portions of the ultraviolet and infrared spectra near the visible region. At the temperatures of many industrial processes, radiation is principally from vibrational and rotational transitions; electronic transitions become important at temperatures above several thousand Kelvins.

Process (c) in Figure 8.3 is a *bound–free* absorption (photoionization). An atom absorbs a photon with energy sufficient to cause ionization. The resulting ion and electron are free to take on any kinetic energy; hence, the bound–free absorption coefficient is a continuous function of photon frequency ν if the photon energy $h\nu$ is large enough to cause ionization. The reverse (process (d) in Figure 8.3) is free–bound emission (photorecombination). Here an ion and free electron combine, a photon is released, and the energy of the resulting atom drops to that of a discrete bound state. Free–bound emission produces a continuous spectrum, as the combining particles can have any initial energy.

In an ionized gas, a free electron can pass near an ion and interact with its electric field. This can produce a *free–free* transition that is often called *bremsstrahlung*, meaning brake radiation. The electron can absorb a photon (process (e) in Figure 8.3), thereby going to a higher kinetic energy, or it can emit a photon (process (f)) and drop to a lower free energy. Since the initial and final free energies can have any values, a continuous absorption or emission spectrum is produced.

8.2.2   LOCAL THERMODYNAMIC EQUILIBRIUM (LTE)

In earlier chapters we assumed that opaque solids emit energy based solely on their temperature and physical properties. The spectrum of emitted energy was assumed unaffected by the characteristics of any incident radiation. This is usually true because all the absorbed incident energy is quickly redistributed into an equilibrium distribution of internal energy states at the temperature of the solid. However, there are exceptions. In a gas, the redistribution of absorbed energy occurs by interactions among the atoms, molecules, electrons, and ions that constitute the gas. Under most engineering conditions, this redistribution occurs rapidly, and the energy states of the gas are populated in equilibrium distributions described by the local temperature. When this is true, the Planck spectral distribution, along with the spectral absorption coefficient, Equation (2.62), describes the emission from a gas volume element. The assumption that a gas emits according to Equation (2.62) regardless of the spectral distribution of intensity passing through and being absorbed within a small volume element dV is called "local thermodynamic equilibrium."

It is assumed here that LTE exists, the properties defined in this chapter can be computed based on LTE states, and that the spontaneous emission from dV is governed by Equation (2.62). Non-LTE problems are not within the general scope of this book.

8.2.3 SPECTRAL LINE BROADENING

If a gas is not dissociated or ionized, its internal energy is expressed in discrete vibrational, rotational, and electronic energy states of its atoms or molecules. If the energies of the upper and lower discrete states are e_j and e_i, only photons of energy e_p can cause a transition, where:

$$e_p = e_j - e_i = h\nu_{ij} = hc\eta_{ij} \tag{8.2}$$

The discrete transitions result in the absorption of photons of definite frequencies or wave numbers, causing the appearance of dark lines in the transmission spectrum; this is *line absorption*. Equation (8.2) predicts that very little energy could be absorbed from the entire incident spectrum by an absorption line, because only those photons having *exactly* the single wave number given by Equation (8.2) could be absorbed. However, other effects cause the line to be broadened and consequently to have a finite wave number span around the transition wave number η_{ij}. The wave number span of the broadened line, and the variation within it of its absorption ability, depends on the physical mechanism causing the broadening. Some of the important mechanisms are natural, Doppler, collision, and Stark broadening. Collision broadening is the most important mechanism for engineering applications involving infrared radiation.

The variation of the absorption coefficient with wave number within a broadened line is the line *shape*. The shape is important as it is related to the trends of gas absorption with temperature, pressure, and path length through the gas. This shape determines the amount of energy absorbed or emitted by a gas near a spectral absorption wavelength or frequency. The shape of a typical line is illustrated in Figure 8.4.

The *line intensity* S_{ij} is the integral under the $\kappa_{\eta,ij}$ versus wave number curve,

$$S_{ij} = \int_{0}^{\infty} \kappa_{\eta,ij} d\eta = \int_{-\infty}^{\infty} \kappa_{\eta,ij} d(\eta - \eta_{ij}) \tag{8.3}$$

The $\kappa_{\eta,ij}$ is small except for η close to η_{ij}. The regions away from η_{ij}, where $\kappa_{\eta,ij}$ is small, are the "wings" of the line. The magnitudes of $\kappa_{\eta,ij}$ and S_{ij} depend on the number of molecules in energy level i and hence depend on gas density. Taking the ratio $\kappa_{\eta,ij}/S_{ij}$ tends to cancel the effect of density on the magnitude and shows the effect of density in changing the line shape.

One characteristic of the line shape is the *line half-width*, γ. This is the spectral line width (in units of wave number for the present discussion) at half the maximum line height (Figure 8.4). It provides a definite width to help describe the line. Since $\kappa_{\eta,ij}$ goes to zero asymptotically as $|\eta - \eta_{ij}|$ increases, it is not possible to define a line width in terms of a wave number where $\kappa_{\eta,ij}$ becomes zero, so the half-width is used instead. Next, we discuss the four mechanisms that cause line broadening.

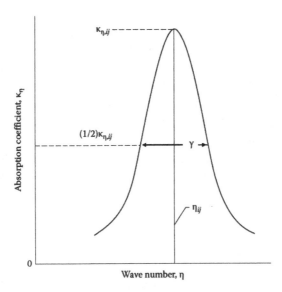

FIGURE 8.4 Absorption coefficient for a symmetric broadened spectral line for transition between energy levels i and j.

8.2.3.1 Natural Broadening

A stationary emitter unperturbed by any external effects emits energy over a finite spectral interval about each transition wave number. This *natural line broadening* arises from the uncertainty in the exact levels e_i and e_j of the transition energy states, which is related to the Heisenberg uncertainty principle. The effect of natural broadening is usually quite small compared with that of other line-broadening mechanisms and is usually neglected.

8.2.3.2 Doppler Broadening

The atoms or molecules of an absorbing or emitting gas are not stationary but have a distribution of velocities associated with their thermal energy. If an atom or molecule is emitting at wave number η_{ij} and at the same time is moving at velocity v toward an observer, the waves arrive at the observer at an increased η given by $\eta = \eta_{ij}[1 + (v/c)]$. If the emitter is moving away from the observer, v has a negative value, and the observed wave number is less than η_{ij}. In thermal equilibrium, the gas molecules have a Maxwell–Boltzmann distribution of velocities. This velocity distribution results in a spectral line shape with a Gaussian distribution:

$$\frac{\kappa_{\eta,ij}}{S_{ij}} = \frac{2}{\gamma_D}\sqrt{\frac{\ln 2}{\pi}} \exp\left[-4\left(\eta - \eta_{ij}\right)^2 \frac{\ln 2}{\gamma_D^2}\right] \tag{8.4}$$

The γ_D is the line half-width for Doppler broadening:

$$\gamma_D = \frac{2\eta_{ij}}{c}\left(\frac{2kT}{M}\ln 2\right)^{1/2} \tag{8.5}$$

The dependence of γ_D on $T^{1/2}$ shows that Doppler broadening becomes important at high temperatures.

8.2.3.3 Collision Broadening

As the pressure of a gas increases, the collision rate experienced by an atom or molecule with surrounding particles is increased. Collisions can perturb the energy states of the atoms or molecules, resulting in collision broadening of the spectral lines. For noncharged particles, the line has a Lorentz profile:

$$\frac{\kappa_{\eta,ij}}{S_{ij}} = \frac{\gamma_c/2\pi}{\gamma_c^2/4 + (\eta - \eta_{ij})^2} \tag{8.6}$$

The collision half-width γ_c is determined by the collision rate. An approximate value for a monatomic or monomolecular gas from kinetic theory is:

$$\gamma_c = \frac{1}{c}\frac{4D^2P}{(\pi MkT)^{1/2}} \tag{8.7}$$

where,
  D is the diameter of the atoms or molecules
  P is the gas pressure for the single-component gas
  M is the mass of an individual molecule

Equation (8.7) shows that collision broadening becomes important at high pressures and low temperatures or, from the perfect gas law, at high pressures and densities. Collision broadening is often the main contributor to line broadening for engineering infrared conditions, and the other line-broadening mechanisms can usually be neglected.

8.2.3.4 Stark Broadening

When strong electric fields are present as in partially ionized gases, the energy levels of the radiating gas particles can be greatly perturbed. This is the *Stark effect*, which can produce very large line broadening and splitting (Stark 1905). These line shapes are unsymmetrical and complicated, and the related calculations must be performed using the principles of quantum mechanics.

Broadening mechanisms discussed above are under the assumption that only one atomic or molecular species is present. If the gas has more than one component, collision broadening is caused by collisions among like molecules (self-broadening) and with other species. Both collision processes must be included in calculating the line shapes.

8.2.3.5 Absorption or Emission at a Single Spectral Value

Property Definitions along a Path in a Uniform Absorbing and Emitting Medium

The variation of intensity along a path is described by the RTE, Equation (2.72). In the following discussion, we consider a simplified form of the RTE and neglect scattering effects. This means that the extinction coefficient is equal to the absorption coefficient, i.e., $\beta_\lambda = \kappa_\lambda$. Starting from Equation (2.72), the spectral intensity attenuated along a path by absorption and augmented by emission is given by the following radiative transfer equation:

$$\frac{\partial I_\lambda}{\partial S} = -\kappa_\lambda(S)I_\lambda(S) + \kappa_\lambda(S)I_{\lambda b}(S) \tag{8.8}$$

If the local refractive index $n_\lambda \neq 1$, the $I_{\lambda b}$ contains an n^2 factor as in Equation (2.21).

Consider a gas with uniform temperature and uniform composition, such as in a well-mixed furnace or combustion chamber. The κ_λ and $I_{\lambda b}$ are then constant throughout the volume, and Equation (8.8) is integrated from $S = 0$ to S starting from $I_\lambda(0)$ at $S = 0$ to give:

$$I_\lambda(S) = I_\lambda(0)e^{-\kappa_\lambda S} + I_{\lambda b}\left[1 - e^{-\kappa_\lambda S}\right] \tag{8.9}$$

The $e^{-\kappa_\lambda S}$ is the spectral *transmittance* (fraction transmitted), $t_\lambda(S)$, of the initial intensity. Then $1 - e^{-\kappa_\lambda S}$ is the fraction of $I_\lambda(0)$ that was absorbed; this is the spectral *absorptance* $\alpha_\lambda(S)$ along the path. By virtue of Kirchhoff's Law, this quantity appears in the spectral emission along the path as given by the last term of Equation (8.9), which becomes:

$$I_\lambda(S) = I_\lambda(0)t_\lambda(S) + I_{\lambda b}\alpha_\lambda(S) \tag{8.10}$$

and $\alpha_\lambda(S) = 1 - t_\lambda(S) = 1 - e^{-\kappa_\lambda S}$.

By integrating over all λ, the *total absorptance* along a path in a uniform gas (uniform composition, temperature, and pressure) is:

$$\alpha(S) = \frac{\int_0^\infty I_\lambda(0)\alpha_\lambda(S)d\lambda}{\int_0^\infty I_\lambda(0)d\lambda} = \frac{\int_0^\infty I_\lambda(0)[1 - e^{-\kappa_\lambda S}]d\lambda}{\int_0^\infty I_\lambda(0)d\lambda} \tag{8.11}$$

Similarly, the total emittance along a path in a uniform gas is:

$$\varepsilon(S) = \frac{\int_0^\infty I_{\lambda b}\alpha_\lambda(S)d\lambda}{\int_0^\infty I_{\lambda b}d\lambda} = \frac{\pi\int_0^\infty I_{\lambda b}[1 - e^{-\kappa_\lambda S}]d\lambda}{\sigma T^4} \tag{8.12}$$

where T is the uniform gas temperature. The *total transmittance* along the gas path is:

$$t(S) = \frac{\int_0^\infty I_\lambda(0)t_\lambda(S)d\lambda}{\int_0^\infty I_\lambda(0)d\lambda} = \frac{\int_0^\infty I_\lambda(0)e^{-\kappa_\lambda S}d\lambda}{\int_0^\infty I_\lambda(0)d\lambda} \qquad (8.13)$$

which leads to:

$$1 - \alpha(S) = t(S) \qquad (8.14)$$

The Lorentz line shape for collision broadening (Equation (8.6)) is the most important broadening shape for engineering applications in the infrared spectrum. Increasing gas density or pressure have direct effects on the Lorentz line. This is due to the increase in absorption because of the presence of more molecules along the radiation path, and/or due to the increased width of the spectral line because of greater collisional line broadening.

Databases for the Line Absorption Properties of Molecular Gases

Compilations of the line characteristics of common molecular gases are available for download from published data bases, mainly from the *Journal of Quantitative Spectroscopy and Radiative Transfer (JQSRT)*. These line-by-line data sources generally provide the wave number of the line center, Lorentz air-broadened line half-width, and the line intensity for all important lines of common molecular gases. This is the basic data required for performing detailed line-by-line spectral analyses. Their high resolution makes them impractical for engineering calculations, but they can be used to provide benchmark calculations for determining the accuracy of less detailed engineering models.

The HITRAN compilation provides spectral lines for 49 different molecules at 297 K, and details of how the data are compiled is in Rothman (1996), Rothman et al. (2010, 2013), and Gordon et al. (2017, 2022). A special issue of *JQSRT* is devoted to the updated HITRAN 2016 (Gordon et al. 2017, Bernath and Rothman 2017). Karman et al. (2019) have extended the HITRAN database by correcting some and adding additional collision pairs to account for collision-induced line absorption with particular attention to pairs found in planetary atmospheres. Hargreaves et al. (2020) added extensive line data for methane to HITRAN. The most recent version of HITRAN (2020) is outlined in Gordon et al. (2022).

At higher temperatures, certain lines ("hot lines") that are unimportant and are neglected near room temperature become much more prominent, and the low-temperature database is not adequate for determining spectral properties. The HITEMP database extends the HITRAN database to include hot lines (Varanasi 2001, Rothman et al. 2010, 2013, Gordon et al. 2007). The HITEMP 2010 version extends the HITRAN data for CO_2, H_2O, CO, NO, and OH. The HITEMP 2010 data for water vapor has been validated up to 4000 K, and data for some species is being extended.

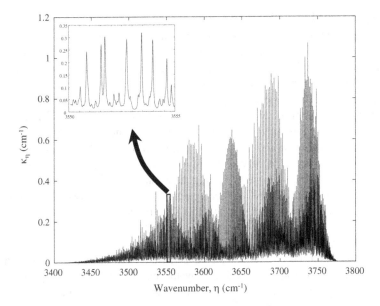

FIGURE 8.5 Example spectral absorption coefficient data for lines in the 2.7 μm band of homogeneous CO_2 with a uniform temperature of $T = 1000$ K, 1 atm. (Generated using a computer program based on HITRAN 2020.)

HITRAN 2020 absorption coefficient data for a CO_2 band are shown in Figure 8.5. The HITRAN database is published (Rothman et al. 2013, Gordon et al. 2022) within special issues of *JQSRT*.[1] More specialized line-by-line databases are also available, but access may require security clearance and/or payment.

8.3 BAND MODELS

8.3.1 Probability Density Function-Based Band Correlations

Recent approaches to band correlations have led to helpful simplifications. Most of these methods involve transforming the spectral distribution of the absorption coefficient into a probability density versus absorption coefficient space for a given narrow or wide band, and in some cases integrating the probability density function (PDF) to give a cumulative distribution function (CDF) of the absorption coefficient. The smoothly varying PDF or CDF is then used to provide spectral property dependence in the radiative transfer relations. The PDFs and CDFs must be developed from fundamental data on the spectral line absorption behavior, such as from the HITRAN database for low temperatures (Rothman et al. 2010, 2012; Gordon 2017, 2022), and in HITEMP for higher temperature data (Rothman et al. 2010). A comprehensive discussion of the various models is in Solovjev et al. (2017).

8.3.1.1 *k*-Distribution Method

To find solutions for total radiative energy transfer, the radiative transfer equation described in Chapter 2 can be solved in many spectral intervals and integration is then carried out to determine the spectral energy. Since κ_η varies sharply with η, this integration requires solving the transfer equations at many very closely spaced spectral intervals, followed by integration across the spectrum to obtain the total flux or flux divergence. For complex line spectra, such a line-by-line solution and integration is time-consuming and, in most cases, impractical for engineering calculations.

The *k-distribution method* uses a transformation of variables to reduce the extent of the spectral calculations. It was introduced for astrophysical applications (Lacis 1973) and was originally limited to homogeneous absorbing media. Many approaches to improving the method for engineering applications have been proposed. Modest (2013) provided a review of developments of the method.

8.3.1.2 Full-Spectrum *k*-Distribution Methods

Full-spectrum methods provide *k*-distributions valid over the entire wavelength spectrum. The *k*-distributions are weighted by the blackbody spectrum evaluated at a reference temperature. For a homogeneous gas, the CDF for the full-spectrum *k*-distribution (FSK) is defined by:

$$g(\kappa) = \frac{\pi}{\sigma T^4} \int_{\lambda \text{ s.t. } \kappa_\lambda \leq \kappa} I_{\lambda,b} \, d\lambda \qquad (8.15)$$

This is also known as the absorption-line blackbody distribution function (ALBDF).

Figure 8.6 illustrates the regions over which the CDF is generated for a particular value of κ for a region of the entire spectrum. As noted earlier, for an actual molecular spectrum, the solution of the RTE at each spectral interval followed by integration of intensity over the spectrum requires an extremely small increment size for accurate integration (e.g., see Figure 8.5). Generation of the CDF at a particular temperature is also time-consuming and requires extremely small increments in wavenumber. Because the calculations are done prior to the solution of the RTE, the RTE then needs to be solved over only a relatively few increments of the smoothly varying CDF. The presence of the HITRAN and HITEMP databases for molecular lines makes possible an accurate generation of the FSK CDF across the entire spectrum.

Figure 8.6 shows the FSK CDF generated from HITRAN2020 data for CO_2 at $T = 1000$ K. Rather than absorption coefficient κ, the CDF is shown in terms of the absorption cross-section C, where $C = \kappa/N$ and N is the number density (molecules/volume) of the CO_2 found for an ideal gas from $N = p/RT$. Here, p is the partial pressure of the CO_2. The range of wavenumber from 0 to 14,000 encompasses over 99.9% of the energy in the blackbody spectrum at 1000 K (Table A.3).

When the CDF $g(C)$ is generated, the RTE for a nonscattering medium (Equation (2.72) for steady state and $\sigma_s = 0$) for each value of $g(C)$ becomes:

$$\frac{dI_g}{ds} = \kappa_g (I_b - I_g) \qquad (8.16)$$

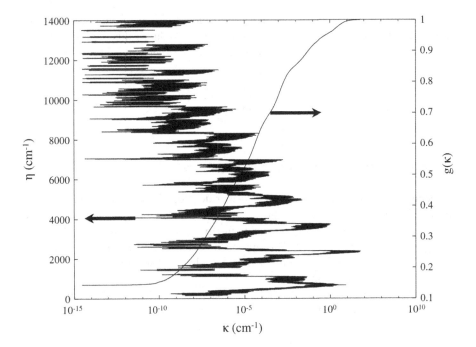

FIGURE 8.6   CDF of k-distribution for 10% CO_2 in air at 1000 K, 1 atm. (Data calculated from a computer program based on HITRAN 2020.)

After solving for I_g, the total intensity is found by integrating over g as:

$$I = \int_{g=0}^{1} I_g \, dg \qquad (8.17)$$

Because g is a monotonic smoothly increasing function, it generally suffices to solve for I_g in a relatively few g intervals and then to use a simple quadrature to carry out the integration. This is much simpler as compared with a line-by-line integration over κ_η; that is the great advantage of the full-spectrum cumulative k-distribution or FSCK approach.

8.3.1.3   Effect of Temperature and Concentration Gradients in a Medium

As described in Equation (8.17), the $g(C)$ was derived based on a single reference temperature. However, for a nonisothermal or nonhomogeneous medium, the $g(C)$ will vary across the medium, introducing error into the solution. Approximations for handling this problem are available.

8.3.2   Weighted-Sum-of-Gray-Gases Approach

The weighted-sum-of-gray-gases (WSGG) approach seeks to replace the integration of spectral properties with a summation over a small set of J gray gases to simulate the properties of the nongray gas and is thus closely related to the FSK method. Hottel (1954) was one of the first engineers who introduced this method and proposed that the emittance of an isothermal gas could be approximated by:

$$\varepsilon(S) = \sum_{j=1}^{J} w_j (1 - e^{-\kappa_j S}) \tag{8.18}$$

where

κ_j is the absorption coefficient of gas j

w_j is the corresponding weight for the jth gray gas

Denison and Webb (1996a, 1996b) further developed this approach, using line-by-line data weighted by the blackbody spectral distribution and integrated across narrow bands. The number, weights, and effective absorption coefficients of a set of gray gases were then obtained. This approach is called the "spectral-line-weighted sum-of-gray-gases" (SLWSGG) method, and it allows for windows in the spectrum, as one of the gray gases can be nonabsorbing.

The RTE for use with the SLWSGG approach in a nonscattering medium is that proposed by Modest (1991):

$$\frac{dI_j(S)}{dS} = \kappa_j \left(w_j \frac{\sigma T^4(S)}{\pi} - I_j(S) \right) \tag{8.19}$$

This equation is solved for I_j for each gray gas, and the results are summed over the J gray gases to obtain the total intensity gradient.

The weights and absorption coefficients are obtained using the SLWSGG method as follows (Denison and Webb 1996a, 1996b; Pearson et al. 2014): An absorption-line blackbody distribution function (equivalent to the k-distribution) is defined as *the fraction of the blackbody energy in the portions of the total spectrum where the high-resolution spectral molar absorption cross-section of the gas $\bar{C}_{abs,n}$ is less than a prescribed value $\bar{C}_{abs}$*. This distribution function is expressed for an arbitrary absorbing species (e.g., H_2O, CO_2) as:

$$F(\bar{C}_{abs}, T_b, T_g, P, Y_s) = \frac{\pi}{\sigma T_b^4} \int_{(\eta, \bar{C}_{abs}, T_g, P, Y_s)} I_{b\eta}(T_b, \eta) d\eta \tag{8.20}$$

The equivalence to the FSK method becomes obvious by comparing Equation (8.20) with Equation (8.15).

The fractional function F in Equation (8.20) has a monotonic increase between 0 and 1 with an increased absorption cross-section. The function is dependent on the molar absorption cross-section $\bar{C}_{abs}$, blackbody source temperature T_b, gas temperature T_g, total pressure P, and mole fraction of broadening species Y_s. The dependence of the function on the spectrum is through the spectral interval of integration, η.

For total emissivity calculations of isothermal gaseous media, the temperatures T_b and T_g, are equal. However, for gas total absorptivity calculations, the gas and the source temperatures may differ.

After the function F is found for a particular gas at known conditions, the weights w_j and absorption coefficients κ_j for use in the WSGG Equation (8.19) are found by segmenting the spectrum into J regions. The weights are given for any region j by choosing values for the absorption cross-sections at the limits of each region, $\bar{C}_j$. Then

$$w_j = F(\bar{C}_{j+1}, T_b, T_g, Y) - F(\bar{C}_j, T_b, T_g, Y) \tag{8.21}$$

and the corresponding absorption coefficients $\kappa_j = N\bar{C}_j^*$ are then found from appropriate mean values of $\bar{C}_j^*$ within each interval (these will generally not be equal to $\bar{C}_j$) and from N, the molar concentration of the absorbing–emitting gas in g·mol/m³. For $\bar{C}_j^*$, a suggested mean is given by:

$$\bar{C}_j^* = \exp\left[\frac{\ln(\bar{C}_j) + \ln(\bar{C}_{j+1})}{2}\right] \tag{8.22}$$

The weights and absorption coefficients are used in Equation (8.19), and the resulting values for intensity are then summed over the J spectral regions to obtain total intensity values.

The research on these approximations is continuing. Zheng et al. (2021) compared the accuracy of the various models against LBL calculation for homogeneous and nonhomogeneous 2D and 3D systems. André et al. (2022) have reported further improvement in the method.

8.4   GAS TOTAL EMITTANCE CHARTS

The total emittance was defined in Equation (8.12), and when it is obtained from κ_λ it involves an integration of the significant radiation contributions over the spectrum. Graphical presentations of the total emittance for the important radiating gases have been developed from total radiation measurements, by integration using spectral measurements of absorption lines and bands, and from the theories of line and band absorption. They are useful in simplified calculations of radiation transfer through absorbing–emitting media.

At the temperatures in industrial furnaces and combustion chambers, only heteropolar gases such as CO_2, H_2O, CO, SO_2, NO, and CH_4 are significant absorbers and emitters. Gases with symmetric diatomic molecules, such as N_2, O_2, and H_2, are transparent to infrared radiation and do not contribute to radiation exchange; however, they can become important absorbing/emitting contributors at very high temperatures. Charts for the total ε_g values were originally developed by Hottel (1954) based on experimental measurements. The thickness of the gas cloud enters through the parameter L_e, which is an average path length through a uniform gas volume. Values of L_e are in given Sections 9.4 and 9.5 for various volume shapes. The gas pressure also enters as a parameter because κ_λ depends on gas density.

If the gas is a mixture, then both the mixture pressure and the partial pressures of the radiating constituent need to be considered in the calculations. Additional values for sulfur dioxide, ammonia, carbon monoxide, methane, and a few other gases were also provided by Hottel and Sarofim (1967) and updated charts for H_2O–CO_2–CO and their mixtures at extended P and T ranges are reported in Alberti et al. (2018). (The full Alberti charts are available online in Appendix A at www.ThermalRadiation.net/appendix.html). Tam and Yuen (2019) provide an open-source tool for narrow-band-based emittance and absorptance of CO_2–H_2O–N_2–O_2–soot mixtures for radiative transfer calculations in combustion systems.

The discussion in this chapter is limited to CO_2 and water vapor and their mixtures. Before proceeding to describe the use of total emittance in radiative transfer calculations, two examples are given below to show how the total properties are developed by integration of the spectral properties. They outline how the transparent spectral regions between the absorbing parts of the spectrum are important in limiting the size of total emittance values and illustrate the limit for a thick layer of absorbing gas.

Example 8.1

Calculate the absorptance of CO_2 at $T_g = 830$ K and 10 atm from Figure 8.2. Outline an approximation that consists of four bands having vertical boundaries between the wavelengths of $\lambda = 1.8$ and 2.2, 2.6 and 2.8, 4.0 and 4.6, and 9 and 19 μm. What is the total emittance of a thick layer of gas at this temperature?

For a thick layer (large L_e), Equation (8.12) shows that $\varepsilon_\lambda = 1 - e^{-\kappa_\lambda L_e} = 1$ in the absorbing regions. Hence, the thick gas emits like a blackbody in the four absorption bands. In the nonabsorbing regions between the bands, ε_λ is small and is neglected in this simplified model. The total emittance becomes:

$$\varepsilon(T_g, P, S) = \frac{\int_0^\infty \varepsilon_\lambda(T_g, P, S) E_{\lambda b, g} d\lambda}{\sigma T_g^4} = \frac{\int_{\substack{absorbing \\ bands}} E_{\lambda b, g} d\lambda}{\sigma T_g^4}$$

The emittance is thus the fractional emission of a blackbody in the wavelength intervals of the absorbing bands. The required values are:

λ (μm)	λT_g (μm·K)	$F_{0 \to \lambda T_g}$	λ (μm)	λT_g (μm·K)	$F_{0 \to \lambda T_g}$
1.8	1494	0.01250	4.0	3320	0.34446
2.2	1826	0.04248	4.6	3818	0.44684
2.6	2158	0.09323	9	7470	0.83292
2.8	2324	0.12479	19	15,770	0.97275

Then, the emittance is:

$$\varepsilon(T_g, P, S) = \sum_{\substack{\text{absorbing} \\ \text{bands}}} \left(F_{0 \to (\lambda T_g)_{upper}} - F_{0 \to (\lambda T_g)_{lower}} \right)_{band}$$

$$\varepsilon = (0.04248 - 0.01250) + (0.12479 - 0.09323) + (0.44684 - 0.34446)$$
$$+ (0.97275 - 0.83292) = 0.304$$

Even for an optically thick CO_2 volume, the emittance is much less than that for a blackbody.

Example 8.2

What fraction of incident solar radiation is absorbed by a thick layer of CO_2 at 10 atm and 830 K? Use the approximate absorption bands of Example 8.1.
    The effective radiating temperature of the Sun is $T_s = 5780$ K. The desired result is the fraction of the solar spectrum that lies within the four CO_2 bands, as this is the only portion of the incident radiation that will be absorbed. Using the $F_{0 \to \lambda T_s}$ factors obtained using T_s gives:

λ (μm)	λT_g (μm·K)	$F_{0 \to \lambda T_s}$	λ (μm)	λT_s (μm·K)	$F_{0 \to \lambda T_s}$
1.8	10,400	0.92195	4.0	23,120	0.99028
2.2	12,720	0.95251	4.6	26,590	0.99340
2.6	15,030	0.96909	9	52,020	0.99902
2.8	16,180	0.97455	19	109,800	0.99989

The fraction absorbed is then:

$$\alpha = \sum_{\substack{\text{absorbing} \\ \text{bands}}} \left(F_{0 \to (\lambda T_s)_{upper}} - F_{0 \to (\lambda T_s)_{lower}} \right)_{band}$$

$$= (0.95251 - 0.92195) + (0.97455 - 0.96909) + (0.99340 - 0.99028)$$

$$+ (0.99989 - 0.99902) = 0.040$$

Even though the gas layer is thick, only 4.0% of the incident energy is absorbed since the gas transmits well in the λ regions between the absorption bands, and much of the solar energy is at shorter wavelengths than the absorbing bands.

Hottel's original graphs of the total emittance $\varepsilon(pL_e, T)$ for CO_2 and H_2O were based on experimental data with extrapolations to high temperatures and large L_e-partial pressure regions based on theory (Hottel 1954; Hottel and Sarofim 1967). Alberti et al. (2018) provided new versions of the charts for the emittance of H_2O, CO_2, and CO and their mixtures with N_2, along with pressure correction and binary overlap correction charts. The correlations and charts are within +/– 1% of line-by-line evaluations based on HITEMP2010 data. They also provide a downloadable EXCEL spreadsheet for mixture calculation based on these correlations accessible at https://doi.org/10.1016/j.jqsrt.2018.08.008. Figures 8.7 and 8.8 show the emittance of water vapor and CO_2 from the Alberti et al. correlations. Observe that the emittance increases with the pressure-path length product as expected.

These correlations are for properties of the absorbing gases at essentially zero pressure mixed with air at a total pressure of one bar. If the total pressure differs considerably from one bar, then a pressure correction must be applied to the predicted one-bar emittance of the individual gases because of increased pressure broadening of the individual lines that make up the bands that are summed to obtain the total emittance.

The individual emittances calculated for H_2O and CO_2 in air must be modified when both gases are present in a mixture, which is commonly the case. This is because the individual spectral lines and absorption bands for these two gases overlap in some spectral regions, and simple addition of the individual emittances would overpredict the emittance of the mixture. In some cases, a simple addition predicts a gas absorptance and emittance that are greater than unity at certain wavelengths.

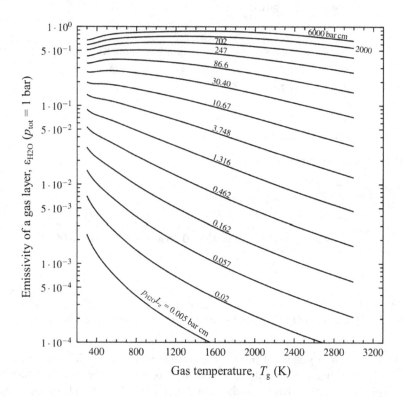

FIGURE 8.7 Computed emittance of water vapor from line-by-line data. Courtesy of Dr. Michael Alberti (2019).

If two absorbing gases have spectral absorption coefficients $\kappa_{\lambda,1}$ and $\kappa_{\lambda,2}$, then from Equation (8.12),

$$\varepsilon = \frac{1}{\sigma T_g^4} \int\limits_{\lambda=0}^{\infty} \left[1 - e^{-(\kappa_{\lambda,1}+\kappa_{\lambda,2})} \right] E_{\lambda b}(T_g)d\lambda$$

$$= \frac{1}{\sigma T_g^4} \int\limits_{\lambda=0}^{\infty} \left[1 - e^{-\kappa_{\lambda,1}} + 1 - e^{-\kappa_{\lambda,2}} - \left(1 - e^{-\kappa_{\lambda,1}}\right)\left(1 - e^{-\kappa_{\lambda,2}}\right) \right] E_{\lambda b}(T_g)d\lambda$$

(8.23)

The first four terms integrate to give the total emittances of the two individual gases, resulting in:

$$\varepsilon = \varepsilon_1 + \varepsilon_2 - \frac{1}{\sigma T_g^4} \int\limits_{\lambda=0}^{\infty} \left(1 - e^{-\kappa_{\lambda,1}}\right)\left(1 - e^{-\kappa_{\lambda,2}}\right) E_{\lambda b}(T_g)d\lambda = \varepsilon_1 + \varepsilon_2 - \Delta\varepsilon \quad (8.24)$$

The final term on the right is the band overlap correction. Alberti et al. (2018) present graphs of the band overlap correction (Appendix A at www.ThermalRadiation.net/appendix.html).

The final emittance equation, including both pressure and overlap correction $\Delta\varepsilon$ is:

$$\varepsilon(pL_e) = C_{H_2O}\varepsilon_{H_2O}(p_{H_2O}L_e) + C_{CO_2}\varepsilon_{CO_2}(p_{CO_2}L_e) - \Delta\varepsilon \quad (8.25)$$

The complete charts and spread sheet provided by Alberti et al. (2018) use the pressure correction and overlap correction relations proposed by Hottel, along with correlations to the emittance from line-by-line integration. Correlations of the emittances shown for H_2O (Figure 8.7), CO_2 (Figure 8.8), CO, and their mixtures modified by the pressure and overlap corrections are within +/–1% of the line-by-line calculations. Calculation of H_2O–CO_2–CO–N_2 mixture emittances are available in an EXCEL spreadsheet at https://doi.org/10.1016/j.jqsrt.2018.08.008.

Alberti et al. (2020) have also examined earlier methods for predicting the absorptance of uniform CO, CO_2, and H_2O mixtures when exposed to a source at another temperature and present a more accurate method based on correlations of line-by-line data from the HITRAN 2010 database.

Example 8.3

A container with an effective radiation thickness of $L_e = 2.4$ m contains a mixture of 15 volume percent of CO_2, 20% H_2O vapor, and the remainder air. The total pressure of the gas mixture is 1 atm, and the gas temperature is 1200 K. What is the emittance of the gas?

Using either the Alberti graphs in Appendix A at www.ThermalRadiation.net/appendix.html or the spreadsheet at https://doi.org/10.1016/j.jqsrt.2018.08.008 gives $\varepsilon=0.401$.

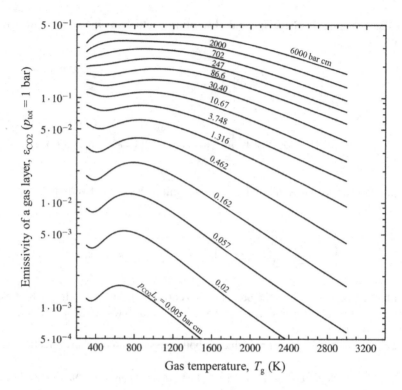

FIGURE 8.8 Computed emittance of CO_2 from line-by-line data. Courtesy of Dr. Michael Alberti (2019).

8.5   MEAN ABSORPTION COEFFICIENTS

The properties of many translucent materials vary considerably with the spectral variable. To try to overcome the need to include detailed property variations in an analysis, there have been attempts to use mean property values over a spectral range. A mean absorption coefficient that is spectrally averaged over all wavelengths provides a simplification that sidesteps carrying out a spectral analysis and then integrating the spectral energy over all wavelengths or using a k-distribution or SLWSGG method to obtain the total energy. It is difficult to decide in advance if using a mean absorption coefficient would give a sufficiently accurate solution for a particular situation.

Local total emission by a volume element dV in a material is given by the integral:

$$4dV\int_0^\infty \kappa_\lambda(T,P)\pi I_{\lambda b}(T)d\lambda = 4dV\int_0^\infty \kappa_\lambda(T,P)E_{\lambda b}(T)d\lambda \qquad (8.26)$$

For the integral term corresponding to emission, it is convenient to define the *Planck mean absorption coefficient* $\kappa_p(T, P)$ as:

$$\kappa_P(T,P) \equiv \frac{\int_0^\infty \kappa_\lambda(T,P)E_{\lambda b}(T)d\lambda}{\int_0^\infty E_{\lambda b}(T)d\lambda} = \frac{\int_0^\infty \kappa_\lambda(T,P)E_{\lambda b}(T)d\lambda}{\sigma T^4} \tag{8.27}$$

The κ_P is the mean of the spectral coefficient weighted by the blackbody emission spectrum. This Planck distribution is useful for considering emission from a volume and for certain special cases of radiative transfer. The Planck mean absorption coefficient κ_P is convenient since it depends only on the local properties at dV. It can be tabulated and is especially useful where the pressure is constant over the volume of a gaseous medium.

The *Rosseland mean attenuation coefficient* arises from treating radiation transfer in optically thick media, where transfer acts as a diffusion process (Section 10.5). For only absorption (no scattering),

$$\kappa_R(T,P) = \left[\int_0^\infty \frac{1}{\kappa_\lambda(T,P)} \frac{\partial E_{\lambda b}(T)}{\partial E_b(T)} d\lambda \right]^{-1} \tag{8.28}$$

At first glance, the Rosseland mean appears to be entirely different from κ_P, which is weighted by a spectral distribution of energy. However, for 1D diffusion, radiative spectral flux is found to depend only on the local blackbody emissive power gradient and, for a nonscattering medium (see Section 10.5, Equation (10.33)):

$$q_{\lambda,z}d\lambda = -\frac{4}{3\kappa_\lambda} \frac{dE_{\lambda b}(T)}{dz} d\lambda = -\frac{4}{3\kappa_\lambda} \frac{\partial E_b}{\partial z} \frac{\partial E_{\lambda b}}{\partial E_b} d\lambda \tag{8.29}$$

The $\partial E_{\lambda b}/\partial E_b$ is found by differentiating Planck's Law (Equation (2.19)) after letting $T = (E_b/\sigma)^{1/4}$. The result for evaluating κ_R is:

$$\frac{1}{\kappa_R} \equiv \int_{\lambda=0}^\infty \left(\frac{1}{\kappa_\lambda}\right)\left(\frac{\partial E_{\lambda b}}{\partial E_b}\right) d\lambda = \frac{\pi}{2} \frac{C_1 C_2}{\sigma T^5} \int_{\lambda=0}^\infty \left(\frac{1}{\kappa_\lambda}\right)\left(\frac{\exp(C_2/\lambda T)}{\lambda^6 \left[\exp(C_2/\lambda T)-1\right]^2}\right) d\lambda \tag{8.30}$$

Using Equation (8.29), $\int_0^\infty \kappa_\lambda q_{\lambda,z}d\lambda = -\frac{4}{3}\frac{\partial E_b}{\partial z}$ and $\int_0^\infty q_z d\lambda = -\frac{4}{3}\frac{\partial E_b}{\partial z}\int_0^\infty \frac{1}{\kappa_\lambda}\frac{\partial E_{\lambda b}}{\partial E_b} d\lambda$

Substituting into Equation (8.28) gives, for the *diffusion case*:

$$\kappa_R(T,P) = \frac{\int_0^\infty \kappa_\lambda q_{\lambda,z}d\lambda}{\int_0^\infty q_{\lambda,z}d\lambda} \tag{8.31}$$

The Rosseland absorption coefficient is thus a mean value of κ_λ weighted by the local spectral energy flux $q_{\lambda,z}d\lambda$ through the assumption that the local flux depends only on the local gradient of emissive power and the local κ_λ.

For a mythical gray gas, the absorption coefficient is independent of wavelength, $\kappa_\lambda(T,P) = \kappa(T,P)$, and the mean values reduce to $\kappa_P(T,P) = \kappa_R(T,P) = \kappa(T,P)$. Solution methods based on k-distributions (discussed above) have almost completely replaced the use of mean absorption coefficients for most analyses, as they are little more time-consuming and much more accurate.

8.6 TRANSLUCENT LIQUIDS AND SOLIDS

Many solids and liquids are opaque to thermal radiation, but there are important exceptions. Figure 8.9 shows the κ_λ spectrum for diamond (Garbuny 1967). Strong absorption peaks exist due to crystal lattice vibrational energy states at certain wavelengths. However, the spectral variations are more regular than for gases. This is true for most translucent solids and liquids.

Figure 8.10 shows the overall spectral transmittance of glass plates for normally incident radiation. Ordinary glasses typically have two strong cutoff wavelengths beyond which the glass becomes highly absorbing and T_λ decreases rapidly to near zero except for very thin plates.

Figure 8.10 shows the overall absorptance for various thicknesses of soda-lime glass. The effect of absorption is illustrated as the thickness increases. The

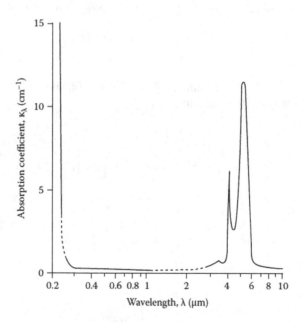

FIGURE 8.9 Spectral absorption coefficient of diamond. (From Garbuny, M., *Optical Physics*, Academic Press, New York, 1967.)

transmission behavior shows why glass windows have the important ability to trap solar energy. The sun radiates a spectral energy distribution very much like a blackbody at 5780 K (10,400°R). Considering the range $0.3 < \lambda < 2.7$ μm as being between the cutoff wavelengths in Figure 8.10, the blackbody characteristics show that 95% of the solar energy is in this range. Hence, glass has a low absorptance for solar radiation, and most of the solar radiation spectrum is transmitted through a glass window. The emission from objects at the ambient temperature inside an enclosure behind the glass is at long wavelengths and is trapped because of the high absorptance (poor transmission) of the glass in the long-wavelength spectral region. This trapping behavior is the well-known "greenhouse effect," which also occurs because of the strong infrared (IR) absorption of various gases (particularly H_2O and CO_2) in the Earth's atmosphere.

Clear ice has a low absorption coefficient in the visible range, and the absorption coefficient increases by a factor of the order of 10^3 as the radiation wavelength increases from about 0.55 to 1.2 μm (Warren and Brandt 2008). Radiation in the visible and near-visible range can therefore be transmitted through clear ice.

Silicon is a semitransparent material very important in modern technology. Pure solid silicon is used in manufacturing semiconductor chips, and the rapid thermal processing of silicon wafers may entail heating rates yielding temperature increase rates greater than 200 K/s. Intense radiant heaters are used to provide the energy for this process. In the required spectral and temperature range for processing, the wavelength-dependent complex refractive index of silicon is highly temperature-dependent. Hence, as expected from electromagnetic theory, the spectral reflectivity and transmissivity of the silicon wafers are also strong functions of wavelength. At low temperatures, the wafers are transparent in the infrared region, becoming opaque

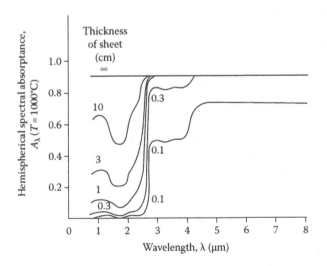

FIGURE 8.10 Absorptance (emittance) of sheets of window glass at 1000°C. (Data from Gardon, 1956.)

as the wafer temperature increases. The design of wafer thermal processing systems requires careful consideration of these factors for radiative transfer calculations.

8.7 ABSORPTION AND SCATTERING BY PARTICLES: DEFINITIONS

Absorption and scattering take place when light interacts with a particle, be it a raindrop, an agglomerate of soot particles, or a strand of cotton fiber. During this interaction, incident electromagnetic wave energy is either converted into internal energy of the particle via absorption or is scattered in all directions. Once absorbed, the radiative energy causes the temperature of a particle to increase. This means that after absorption, particle properties may change, and the particle may even go through a phase change. Scattering refers to the redirection of the incident energy into directions around the particle, and therefore does not affect the internal energy of particles.

The word "*scattering*" encompasses the optical phenomena known as reflection, refraction, transmission, and diffraction. A scattering event is *elastic* if the frequency of the incident radiation does not change during the process. Most of the problems in engineering radiative transfer fall into this category.

Absorption and scattering of a particle vary as a function of its size and shape and the wavelength of the incident EM wave. If there are many particles along the incident beam path, then they will have a cumulative effect on absorption and scattering. This increase is usually linear, particularly if particles are far from each other, at distances of at least several wavelengths. This regime is called *independent scattering*, where the effects can be added. However, when particles are in close proximity to each other or to a surface, then absorption and scattering depend on particle density in a nonlinear fashion. Then, the calculation of the interaction of incident radiation with particles must be performed by considering the effects of nearby particles and agglomerates simultaneously (*dependent scattering*). A rigorous treatment of this dependent effect requires a full solution by electromagnetic wave analysis, as shown by Mishchenko and Yurkin (2020).

8.7.1 ABSORPTION AND SCATTERING COEFFICIENTS, CROSS-SECTIONS, EFFICIENCIES

The absorption and scattering coefficients, κ_λ and σ_λ, are directly related to the decrease of intensity of a beam incident on a control volume along its path. They can be considered as proportionality constants indicating how much of the incident radiative intensity, I_λ, is reduced during the propagation along a small path interval dS (Equations (2.42) and (2.47)):

$$\frac{dI_\lambda}{dS} = -\kappa_\lambda I_\lambda \quad \text{or} \quad \frac{dI_\lambda}{dS} = -\sigma_{s\lambda} I_\lambda \qquad (8.32)$$

If a medium is both absorbing and scattering, then we use the *extinction coefficient* as the proportionality constant for attenuation of radiative energy by the medium: $\beta_\lambda = \kappa_\lambda + \sigma_{s\lambda}$. These proportionality constants reflect on what happens within an elemental control volume along the beam path. They have the units of inverse length, m^{-1}. The control volume may include different particles, agglomerates, and gases. The required properties are determined by accounting for their individual contributions in a linear fashion (assuming they are independent of each other, as discussed before):

$$
\kappa_\lambda = \sum_{i,\text{particles}} \kappa_{\lambda,i} + \sum_{j,\text{gases}} \kappa_{\lambda,j}
$$

$$
\sigma_{s\lambda} = \sum_{i,\text{particles}} \sigma_{s\lambda,i}
$$

$$
\beta_\lambda = \sigma_{s\lambda} + \kappa_\lambda = \sum_{i,\text{particles}} \sigma_{s\lambda,i} + \sum_{i,\text{particles}} \kappa_{\lambda,i} + \sum_{j,\text{gases}} \kappa_{\lambda,j}
$$

$$
\omega_\lambda = \frac{\sigma_{s\lambda}}{\beta_\lambda}
$$

(8.33)

where i and j refer to different particle types and gas species, respectively, and ω_λ is the spectral single scattering albedo.

For most engineering calculations, scattering by gas molecules is negligible. However, scattering by molecules and small particles explains the blue color of the sky and the red color of sunsets/moonsets. In these cases, the path length for radiation is so large that molecular and small particle scattering effects become significant.

Each particle absorbs or scatters radiation depending on its shape, size, and its optical properties. If these parameters are well defined, then one can determine how much of the incident energy will be absorbed and scattered by an *individual particle* by solving Maxwell's equations. The geometry of the problem considered for these solutions is in Figure 8.11 for a particle illuminated by a planar EM wave along with corresponding nomenclature for the incident and scattered intensity directions in terms of polar (θ) and azimuthal (ϕ) angles. We use the prime ($'$) to indicate the incident direction.

For simple shapes like spheres, the solution of Maxwell's equations can be obtained analytically. For complicated structures, more detailed numerical algorithms are needed. These solutions give us the absorption and scattering cross-sections, $C_{\lambda,a}$ or $C_{\lambda,s}$, which refer to an effective particle cross-sectional area that removes the EM energy from the path of the incident radiation; the cross-sections have the units of area, m^2. Following this, the *efficiency factors* $Q_{\lambda,a}$ and $Q_{\lambda,s}$ are obtained by dividing the absorption and scattering cross-sections with the actual geometric cross-section of the particle; therefore, efficiency factors have no units.

Particles in many practical applications have a wide range of sizes, shapes, and structures. They are called polydispersed particles if they have a size distribution; otherwise, they are called monodispersions. For a cloud of monodispersed spherical

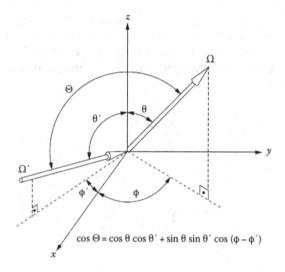

$$\cos \Theta = \cos \theta \cos \theta' + \sin \theta \sin \theta' \cos (\phi - \phi')$$

FIGURE 8.11 Nomenclature used for the incident and scattered intensities in Cartesian coordinates. Here, Θ is the scattering angle.

particles within a given control volume, the absorption and scattering coefficients are:

$$\kappa_\lambda = NC_{\lambda,a} = NQ_{\lambda,a}\pi \frac{D^2}{4}$$

$$\sigma_{s\lambda} = NC_{\lambda,s} = NQ_{\lambda,s}\pi \frac{D^2}{4}$$

(8.34)

where

D is the particle diameter

N is the number of particles per unit volume (with units of m^{-3})

N can be replaced by the volume fraction, f_v, of particles:

$$f_v = NV_{particle} = N\pi \frac{D^3}{6}$$

(8.35)

Here f_v has no units (m^3/m^3). These expressions are valid for a homogenous cloud of N *monodispersed* particles per volume with diameter D. Similar expressions can be written for a *polydispersed* particle cloud with different size particles, requiring the definition of size distributions. For simplicity, we only consider a monodispersed particle cloud.

Extinction cross-sections and extinction efficiency factors are obtained by following the addition rule we used in determining the extinction coefficient:

$$C_{\lambda,e} = C_{\lambda,a} + C_{\lambda,s}$$

$$Q_{\lambda,e} = Q_{\lambda,a} + Q_{\lambda,s}$$

(8.36a, 8.36b)

This rule is valid for most applications considered in radiative transfer, where absorption and scattering are *independent* in nature. We discuss *dependent* absorption and scattering behavior in Section 8.12.

8.7.2 SCATTERING PHASE FUNCTION

In addition to the values of the absorption and scattering coefficients, the solution of the radiative transfer equation depends on the directional distribution of scattering. This distribution is given by the *phase function*, the probability that radiation incident on a particle in the direction of (θ',ϕ') will be scattered into a direction of (θ,ϕ) within solid angle Ω. Then, as shown in Figure 8.11, the scattering angle Θ is defined as the angle between the incident and scattered directions:

$$\cos\Theta = \cos\theta\cos\theta' + \sin\theta\sin\theta'\cos(\phi - \phi')$$

(8.37)

If the incident beam is propagating in the z-direction, then the scattering is given in terms of polar or zenith angle: $\Theta = \theta$. For clarity, we assume the incident radiation is in the z-direction, and then give the expressions only in terms of the direction of the scattered beam in the zenith and azimuthal angles (θ,ϕ).

The intensity of the scattered light in the direction (θ,ϕ) at a far-field distance r is expressed using the differential scattering cross-section $dC_s/d\Omega$. The phase function is related to the scattering cross-section by:

$$\Phi(\theta,\phi) = \frac{1}{C_s}\frac{dC_s}{d\Omega}$$

(8.38)

Normalizing the phase function casts it as a probability function (Equation (2.69)) because its integral over all angles is 1:

$$\frac{1}{4\pi}\int_{\phi=0}^{2\pi}\int_{\theta=0}^{\pi}\Phi(\theta,\phi)\sin\theta d\theta d\phi = 1$$

(8.39)

Scattering profiles in nature are generally *anisotropic*. The angular profile of light scattered by matter varies with respect to both the polar and azimuthal angles. For many applications, azimuthal symmetry can be assumed and only the polar angle dependence is needed, and $\Phi(\theta,\phi) = \Phi(\theta)$.

Isotropic scattering means that scattered intensity is uniformly distributed in all directions. The integral term in the RTE is simplified by assuming isotropic scattering, which corresponds to $\Phi(\theta,\phi) = 1$. This approximation was extensively used during the early development of radiative transfer theory as it allowed analytical modeling of the radiative transfer problems when powerful computers were not available. In nature, however, no object scatters light isotropically and this simplification is simply a mathematical convenience. Some packed media, including sand and snow, tend to scatter light more uniformly in all directions, and isotropic scattering approximation may be appropriate in their analysis.

In general, the scattering phase function, along with values for the absorption and scattering cross-sections, is obtained from the solution of Maxwell's equations. Such a solution is obtained by specifying three parameters: (1) the wavelength of incident radiation, (2) the physical size/shape/structure of the scatterer, and (3) the complex index of refraction of the particle with respect to the surrounding medium. If particles are spherical, a dimensionless scaling parameter, called the *size parameter*, can be used to combine the first two parameters into one. The size parameter is defined as the ratio of the perimeter of a spherical particle with diameter D to the wavelength of incident radiation:

$$\xi = \frac{\pi D}{\lambda_m} \tag{8.40}$$

Here, the wavelength, λ_m, corresponds to that within the medium, such that $\lambda_m = \lambda_0/n$ with n being the real part of the complex index of refraction of the medium.

Lorenz–Mie theory (Lorenz 1898; Mie 1908a, 1908b) applies to all sizes of spherical particles and thus all values of the size parameter ξ. Its formulation leads to a series of analytical expressions in terms of complex functions of the size parameter ξ and therefore, it must be solved numerically. At the small and large ξ limits, relatively simple approximations may be used. If ξ is smaller than 0.1, the *Rayleigh theory* for small particles can be adapted. For particles much larger than the wavelength of incident radiation ($\xi \gg 1$), geometric optics approximation (GOA) can be used. They are discussed in the next two sections.

If the scattering behavior is determined from any of these models for a single particle, then scattering for a cloud of particles can be calculated using a simple summation rule. This assumes *independent scattering* applies, valid when the clearance between particles is sufficiently large relative to both the radiation wavelength and the particle diameter D. If particles are close together, such as in a packed bed, foams, granulated sugar, flour, sand, or snow, the scattering can differ from predictions using the addition of independent particle scattering. These *dependent absorption and scattering* regimes are discussed in Howell et al. (2021) and a rigorous formulation was provided by Mishchenko (2018).

8.8   SCATTERING BY SMALL PARTICLES

8.8.1   Rayleigh Scattering by Small Spheres

We now introduce expressions for *Rayleigh scattering*, attributed to Lord Rayleigh (John William Strutt), who was the first to explore molecular scattering to explain the color of the sky (1871). The scattering cross-section for Rayleigh scatterers is proportional to the fourth power of the size parameter, or to the inverse fourth power of the radiation wavelength. This suggests that when the incident radiation covers a wavelength spectrum, the shorter-wavelength radiation will be scattered much more strongly. For example, for the visible spectrum of solar radiation, where the wavelength changes from 400 nm (blue) to 700 nm (red), blue light will be scattered by almost ten times more, corresponding to the ratio of $(700/400)^4$. Since molecules in the atmosphere are much smaller than any particles, and because they preferentially scatter the blue wavelengths in all directions, we see the sky as blue. Without molecular scattering, the sky would appear black except for the direct view of the sun. As the sun is setting, the path length for direct radiation through the atmosphere becomes much longer than during the middle of the day. In traversing this longer path, more of the short-wavelength part of the visible spectrum is scattered away from the direct path of the sun's rays. As a result, at sunset, red color becomes dominant as the longer-wavelength red rays penetrate the atmosphere along the path to the observer with less attenuation than for the rest of the visible spectrum. If there are many dust particles or water droplets present, the sunset may be an even deeper red due to the forward scattering of longer-wavelength radiation.

If particles with a very limited range of sizes are present in the atmosphere, unusual scattering effects may be observed. The extensive scattering by an ash cloud may affect the radiation balance of the Earth. For example, the 1815 eruption of Mount Tambora on the island of Sumbawa in Indonesia was one of the most powerful in recorded history. The event lowered global temperatures, and some experts claim it led to global cooling and harvest failures across the world. Following the eruption of Krakatoa in Indonesia in 1883, the occurrence of blue and green suns and moons was noted for many years. This was attributed to particles in the atmosphere of such a size range as to scatter only the red portion of the visible spectrum. On September 26, 1950, a blue sun and moon were observed in Europe, believed due to finely dispersed uniform size smoke particles carried from forest fires in Canada. A green moon was seen after the 1982 eruption of El Chichon in Mexico. Zerefos et al. (2014) found a correlation between artists' rendering of sky color in their paintings and such events in the period between 1500 to 2000.

8.8.2 Cross-Section for Rayleigh Scattering

As mentioned above, the Rayleigh scattering approximation is used to model scattering by small particles. An approximate size limit for the applicability of Rayleigh scattering in terms of size parameter is $\xi \, (= \pi D/\lambda) < {\sim}0.3$. (This corresponds to the ratio of particle radius to the wavelength λ of radiation of less than 0.05.) However, light scattering studies conducted to show the phase function profiles suggest that this limit is about 0.1 for spherical particles, rather than 0.3. The limit depends on both ξ and the refractive indices of the particle and surrounding medium.

We can obtain an analytical expression for the scattering cross-section of small particles by considering the scattering of light from small nonabsorbing spherical particles within a nonabsorbing medium. In the formulation given below, the particle properties are designated by subscript 2 and the medium by subscript 1 so that $k_2 =$

0 and $k_1 = 0$. If n is the relative refractive index n_2/n_1, then the Rayleigh scattering cross-section for unpolarized incident radiation is:

$$C_{s,\lambda} = \frac{24\pi^3 V^2}{\lambda^4}\left(\frac{n^2-1}{n^2+2}\right)^2 = \frac{8}{3}\frac{\pi D^2}{4}\xi^4\left(\frac{n^2-1}{n^2+2}\right)^2 \tag{8.41}$$

where
  λ is the wavelength in the medium surrounding the particle
  D is the diameter of the spherical particle
  V is the volume of the spherical particle

The scattering efficiency for Rayleigh scattering is expressed by dividing the scattering cross-section with the physical cross-section of the small particle:

$$Q_{s\lambda} = \frac{C_{s,\lambda}}{\pi D^2/4} = \frac{8}{3}\xi^4\left(\frac{n^2-1}{n^2+2}\right)^2 \tag{8.42}$$

The scattering cross-section for particles in a medium may vary with λ, causing Rayleigh scattering to vary from the $1/\lambda^4$ dependence. This is shown in Figure 8.12, where the actual scattering dependence on wavelength is compared with the $1/\lambda^4$ approximation.

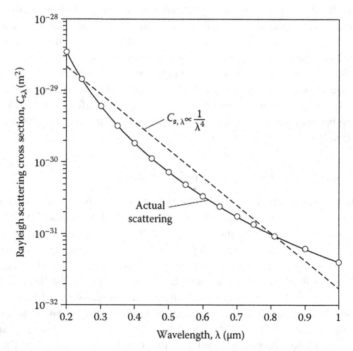

FIGURE 8.12 Comparison of actual Rayleigh scattering cross-section for air at standard temperature and pressure with $1/_\lambda^4$ variation. (From Goody and Yung, 1989.)

8.8.3 Phase Function for Rayleigh Scattering

For incident unpolarized radiation, the phase function can be obtained from the EM theory. The Rayleigh scattering approximation for particles much smaller than the wavelength is:

$$\Phi(\theta,\phi) = \frac{3}{4}(1+\cos^2\theta) \tag{8.43}$$

This expression is independent of the azimuthal angle ϕ. The phase functions for Rayleigh scattering and for *isotropic* scattering (a circle of unit radius) are shown in Figure 8.13. Note that for Rayleigh scattering, the scattered energy is symmetric with respect to the direction of the incident radiation and has both the forward and backward scattering lobes.

In Figure 8.13, a Rayleigh phase function is compared with the isotropic scattering phase function. A phase function can have a much more complicated angular pattern, as shown in four examples in Figure 8.14.

We now start with an intuitive approach for large spherical particles. After that, we provide simplifications for Rayleigh spheres. This is followed by a discussion of the general case of arbitrary-size spherical particles, where the Lorenz–Mie theory is outlined.

If the scattering behavior is determined from any of these models for a single particle, then scattering for a cloud of particles can be calculated using a simple summation rule (Equation 8.33). Again, this assumes *independent scattering* applies when the clearance between particles is sufficiently large relative to both the radiation wavelength and the particle diameter D. If particles are close together, such as in a packed bed, foams, granulated sugar, flour, sand, or snow, the scattering can

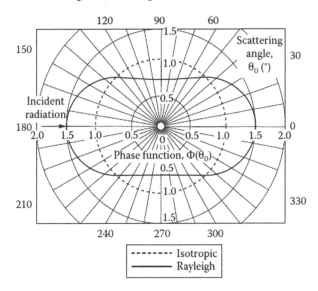

FIGURE 8.13 Comparison of isotropic scattering phase function with a Rayleigh phase function. The Rayleigh phase function is for small particles where the size parameter $\xi = \pi D/\lambda$ is much smaller than unity; the isotropic phase function is a mathematical idealization.

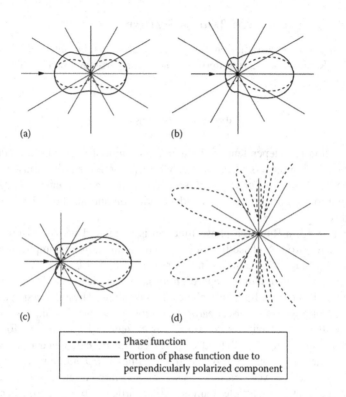

FIGURE 8.14 Phase function profiles for four different particles (on arbitrary scales). Lorenz–Mie profiles, for the corresponding size parameter and complex index of refraction values. (a) $\xi = _\pi D/_\lambda \rightarrow \infty$, metallic sphere, $\bar{n} = 0.57 - 4.29i$; (b) $_\xi = 9.15$, metallic sphere, $\bar{n} = 0.57 - 4.29i$; (c) $_\xi = 10.3$, metallic sphere, $\bar{n} = 0.57 - 4.29i$; and (d) $_\xi = 8$, dielectric sphere, $n = 1.25$.

differ from predictions using the addition of independent particle scattering. These *dependent absorption and scattering* regimes are discussed briefly in Section 8.12.

8.9   THE LORENZ–MIE THEORY FOR SPHERICAL PARTICLES

The Lorenz–Mie theory describes the interaction of a plane-parallel electromagnetic wave with a homogenous spherical particle. The theory is quite general in the sense that it can be used for any size of spherical particle.

For some applications, simpler analytical solutions for absorption and scattering cross-sections may be useful. Consider a simplified model for 'small particles', i.e., 'small' but somewhat larger than those covered by the Rayleigh theory. The general Lorenz–Mie solutions can be expanded into a power series in size parameter $\xi = \pi D/\lambda$, giving the scattering cross-section for particles as:

$$C_{s,\lambda} = Q_{s,\lambda} \frac{\pi D^2}{4} = \frac{8}{3} \frac{\pi D^2}{4} \xi^4 \left| \left(\frac{\bar{n}^2 - 1}{\bar{n}^2 + 2} \right) \left(1 + \frac{3}{5} \frac{\bar{n}^2 - 2}{\bar{n}^2 + 2} \xi^2 + \cdots \right) \right|^2 \qquad (8.44)$$

where D is the particle diameter. The first term in the series is identical to the formulation provided earlier for Rayleigh theory (see Equation 8.42). The second term in the series is then the first correction to the Rayleigh scattering relation.

In a similar fashion, with the addition of the second-order term, the *absorption cross-section* becomes,

$$
C_{a,\lambda} = Q_{a,\lambda} \frac{\pi D^2}{4}
$$

$$
= -4\xi \frac{\pi D^2}{4} \operatorname{Im}\left\{ \left(\frac{\bar{n}^2 - 1}{\bar{n}^2 + 2} \right) \left[1 + \frac{\xi^2}{15} \left(\frac{\bar{n}^2 - 1}{\bar{n}^2 + 2} \right) \left(\frac{\bar{n}^4 + 27\bar{n}^2 + 38}{2\bar{n}^2 + 3} \right) + \cdots \right] \right\}
$$

(8.45)

8.10 SCATTERING BY SPECIAL PARTICLES AND APPROXIMATIONS

Consider the simplified case of a large spherical particle illuminated by a collimated light beam, as shown in Figure 8.15. When incident radiation encounters the particle, some of the radiation may be reflected by its surface. The remainder penetrates the particle medium, where it can be partially absorbed. If the particle does not completely absorb the penetrated energy, some of this radiation leaves the particle in all directions after multiple internal *reflections* and *refractions*. When interacting with the particle boundary, radiation is refracted, and its direction is also changed by many internal reflection events. Scattering also occurs by *diffraction* that results from a slight bending of the paths for radiation passing near the edges of an obstruction. Scattering is the cumulative effect of all these physical phenomena that alter the path of the original beam, as described by Maxwell's equations for the propagation of EM-waves. If the particle is not absorbing, there will be no energy exchange between the incident radiation and the particle; then, the thermodynamic properties of the

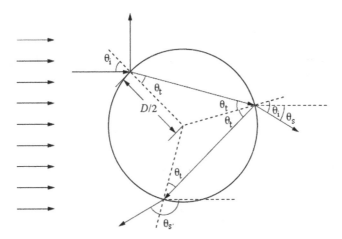

FIGURE 8.15 Geometric optics (ray tracing) approach through a transparent sphere.

medium are not affected. This means that for a completely scattering medium, there is no coupling between the RTE and the energy equation.

8.10.1 LARGE SPECULARLY REFLECTING SPHERES

Large, strongly reflecting metallic particles are extensively used in many engineering systems. For particles larger than 5 μm in diameter and for visible light, the corresponding size parameter is $\xi > 25$. For such metallic particles, reflection constitutes scattering, and their scattering efficiency factors can be calculated in a relatively simple manner by using reflection relations.

The particle scattering cross-section is:

$$C_{s,\lambda}(D) = \frac{\pi D^2}{4} Q_{s,\lambda} = \pi R^2 \rho_\lambda \tag{8.46}$$

and the scattering efficiency becomes:

$$Q_s = \rho_\lambda \tag{8.47}$$

The scattering cross-section is therefore equal to the particle projected area times the hemispherical reflectivity. For a cloud of independently scattering specular spheres with the same diameter D, the scattering coefficient $\sigma_{s,\lambda}$ is:

$$\sigma_{s,\lambda} = \rho_\lambda \frac{\pi D^2}{4} N \tag{8.48}$$

Similarly, the absorption coefficient for a cloud of specularly reflecting large spheres is

$$\kappa_\lambda = (1 - \rho_\lambda) \frac{\pi D^2}{4} N \tag{8.49}$$

Figure 8.16 shows that the energy reflected from the band of the specular sphere at angle θ is reflected into direction 2θ and into a solid angle $d\Omega_s = 2\pi\sin2\theta d(2\theta) = 8\pi\sin\theta\cos\theta d\theta$. The resulting phase function is:

$$\Phi(2\theta) = \frac{\rho_{\lambda,s}(\theta)}{\rho_\lambda} \tag{8.50}$$

and relative to the forward-scattering direction is:

$$\Phi(\theta_0) = \frac{\rho_{\lambda,s}\left(\dfrac{\pi - \theta_0}{2}\right)}{\rho_\lambda} \tag{8.51}$$

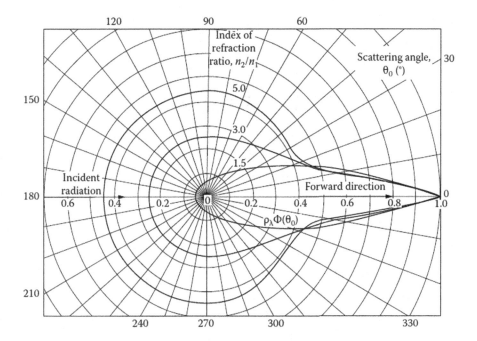

FIGURE 8.16 Scattering diagram for a specularly reflecting dielectric sphere that is large compared with radiation wavelength within the sphere, $_\xi \gg 1$, (n_2 for particle, n_1 for surrounding medium).

For unpolarized incident radiation, the specular reflectivity $\rho_{\lambda,s}(\theta_i)$ for a dielectric sphere is obtained from Equation (4.37). Also, the directional-hemispherical reflectivity is equal to one minus the emissivity. The quantity $\rho_\lambda \Phi(\theta_0)$ is plotted in Figure 8.16 for various refractive index ratios n_2/n_1 of the particle to surroundings. For a dielectric, the $\rho_{\lambda,s}(\theta_i)$ for normal incidence is usually small compared to that at grazing angles $\rho_\lambda(\theta_i)I_{\lambda,i}d\Omega_i d\lambda dA\cos\theta_i$. Consequently, in Figure 8.16 the forward scatter from the sphere (at $\theta_0 = 0$) is one, and the backward scatter (at $\theta_0 = \pi$) is small.

8.10.2 LARGE DIFFUSE SPHERES

If a large sphere has a *diffusely reflecting* surface rather than specularly reflecting, then each surface element that receives incident radiation will reflect it into the entire 2π solid angle above that element. Figure 8.17 shows the geometry considered for this case. The shaded portion of the sphere does not contribute any radiative energy in the direction of the observer because it either does not receive radiation or is hidden from the direction of the observer.

A simple analysis can be made by assuming $\rho_\lambda(\theta)$ is independent of the incidence angle and hence it is equal to the hemispherical reflectivity ρ_λ. This approximation results in the phase function for a *diffuse* sphere:

$$\Phi(\theta_0) = \frac{8}{3\pi}(\sin\theta_0 - \theta_0\cos\theta_0) \tag{8.52}$$

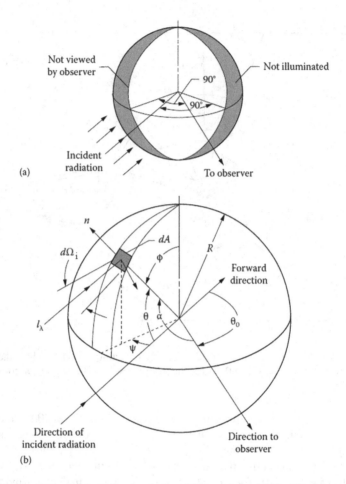

FIGURE 8.17 Scattering by reflection from a diffuse sphere. (a) Illuminated region visible to an observer; (b) geometry on sphere.

The scattering profiles based on this formulation are shown in Figure 8.18. Note that a large portion of energy is backscattered (reflected in this case) at $\theta_0 = 180°$, that is, toward an observer (back in the same direction as the source of the incident radiation). For this θ_0, the entire illuminated surface of the sphere is observed. (Consider the appearence of full moon on Earth, as discussed with Figures 4.13 and 4.14.)

8.10.3 LARGE IDEAL DIELECTRIC SPHERE WITH $n \approx 1$

For a large nonattenuating dielectric ($k = 0$) sphere with a refractive index $n \approx 1$, the reflectivity of the particle surface is almost zero if the index of refraction of the surrounding medium is $n = 1$ and $k = 0$. In this case, the incident radiation passes through the sphere without any change in its electric field amplitude. Therefore, there

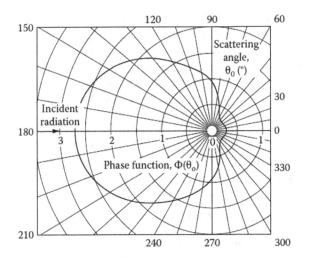

FIGURE 8.18 Scattering phase function for diffusely reflecting sphere, large compared with the wavelength of incident radiation and with uniform reflectivity.

is no scattering. However, the speed of electromagnetic waves within the sphere, $c = c_0/n$, is smaller than that in the surrounding medium. This means that radiation traveling through different portions of the sphere with different thicknesses would have different phase lags, as determined from the rigorous solution of the Maxwell equations. The resulting interference of the waves passing out of the sphere yields the scattering cross-section:

$$C_{s,\lambda} = \frac{\pi D^2}{4}\left[2 - \frac{4}{W}\sin W + \frac{4}{W^2}(1 - \cos W)\right] \qquad (8.53)$$

where $W = 2(\pi D/\lambda)(n-1)$. The scattering efficiency is:

$$Q_{s,\lambda} = \frac{C_{s,\lambda}}{\pi D^2/4} = 2 - \frac{4}{W}\sin W + \frac{4}{W^2}(1 - \cos W) \qquad (8.54)$$

8.10.4 Diffraction by a Large Sphere

In general, the scattering profile for any type and size particle needs to be obtained by solving Maxwell's equations. Such a solution is given in terms of the electric field vectors and accounts for interference, reflection, refraction, and diffraction. When the analysis is carried out for large spheres, as we have shown earlier, the diffraction is neglected. Diffraction always takes place in the forward direction and is a part of scattering. This means that it can be included in the radiative transfer calculations as if radiation energy were transmitted past the particle without any interaction. Hence, diffraction can often be neglected for energy exchange within a scattering medium.

However, if we want to have the total scattering pattern by a large particle, diffraction and reflection must be added. The entire projected area of a sphere takes part in the diffraction, and the scattering cross-section for diffraction is equal to the projected area $\pi D^2/4$. Because diffraction and reflection occur simultaneously, the total scattering cross-section can approach $2\pi D^2/4$ when a sphere is highly reflecting. This is the case for large particles in general, as the scattering efficiency Q_e asymptotically approaches to 2 as the size parameter ξ goes to infinity. This is known as the *extinction paradox*. This limit suggests that a large particle removes two times the incident energy from the incident beam compared to that cast by its own shadow. The reason for this additional extinction is due to the diffraction of the plane-parallel beam around the particle beyond what is incident on its physical structure.

8.10.5 GEOMETRIC OPTICS APPROXIMATION

In many applications, only the scattering by large spherical particles needs to be determined. For example, summer snow cover on Antarctic Sea ice is characterized as grains on the order of 1 cm diameter, which corresponds to size parameter ξ of about 10,000. The scattering behavior of spherical and even arbitrarily shaped particles can be determined using the geometric optics approximation, following a ray tracing approach (Section 10.7).

8.11 APPROXIMATE ANISOTROPIC SCATTERING PHASE FUNCTIONS

Because the exact phase functions obtained for the actual particles with arbitrary shapes, sizes, and size distributions are complicated and quite difficult to use in radiative transfer analyses, various approximate phase functions have been proposed. Usually, a phase function obtained from any of the LMT models discussed above is expressed in terms of a series expansion in Legendre polynomials to be used for the solution of the RTE. For particles much larger than the wavelength of radiation ($\xi > 5$ or so), the number of terms required in the series expansion becomes large, making the solution of the RTE quite difficult. For that reason, a phase function can be further simplified by replacing it with a delta function to represent the highly forward-scattering component and an isotropic component. Simpler phase functions attempt to retain the physics associated with the anisotropic characteristics of exact phase functions. They are expressed in terms of the cosine of the scattering angle $\mu_0 = \cos\theta_0$:

$$\mu_0 = \mu\mu' + (1-\mu^2)^{1/2}(1-\mu'^2)^{1/2}\cos(\phi-\phi') \tag{8.55}$$

where $\mu = \cos\theta$, (θ,ϕ) is the angle between the direction of the incident intensity and a coordinate, and $\mu' = \cos\theta'$. Note that the prime is used to denote the direction of incident radiation.

For a highly forward-scattering medium, the phase function can be approximated as:

$$\Phi_\delta(\theta,\phi,\theta',\phi') = 4\pi\delta(\cos\theta - \cos\theta')\delta(\phi - \phi') = 4\pi\delta(\mu - \mu')\delta(\phi - \phi') \quad (8.56)$$

where δ is the Dirac delta function, so that $\Phi_\delta(\theta,\phi,\theta',\phi')$ equals infinity when the argument is zero (when $\mu = \mu'$ and $\phi = \phi'$ for forward scattering) and Φ_δ equals zero otherwise.

A general phase function can be simplified using an isotropic scattering first term and a modification term. This is called linearly anisotropic scattering:

$$\Phi_{la}(\theta,\theta') = 1 + g\mu_0 \quad (8.57)$$

where g is a dimensionless asymmetry factor that can vary between ± 1. If $g = 0$, it corresponds to the isotropic phase function. This phase function is rotationally symmetric around the direction of travel of the incident intensity. Integration over all solid angles shows that this phase function is normalized for any value of g.

The delta-Eddington approximation uses a two-term Legendre polynomial expansion of the actual phase function plus a Dirac delta term to account for forward scattering, and Henyey and Greenstein (1940) proposed the phase function:

$$\Phi_{HG}(\mu_0) = \frac{(1 - g^2)}{(1 + g^2 - 2g\mu_0)^{3/2}} \quad (8.58)$$

where g is a dimensionless asymmetry factor that can vary from 0 (isotropic scattering) to 1 (for forward scattering). If g is negative, a backscattering phase function is produced. More details on these are in Howell et al. (2021).

8.12   DEPENDENT ABSORPTION AND SCATTERING

If particles, fibers, and other bodies are in close proximity, then their scattering behavior differs from predictions based on those of isolated individual particles. This so-called *dependent scattering* arises when the total scattering cross-section of a particle cloud is not simply obtained by multiplying the number of particles by the scattering cross-section of individual particles. As we have discussed above, the total and directional scattering of cross-sections can be obtained from the solution of the Lorenz–Mie theory, in the case of well-defined particle shapes, such as spheres and the cylinders. The underlying assumption in solving the LM theory is that a planar EM wave is incident on a particle. When particles are close to each other, then the scattered field from one particle can be considered 'planar' before incident on another particle if the distance between them is more than several wavelengths. Otherwise, either a full solution of the EM wave for the particle cloud needs to be

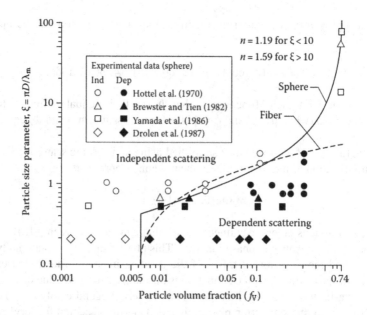

FIGURE 8.19 Regimes of dependent and independent scattering for nonabsorbing spheres and parallel cylinders (fibers); cylinders have normally incident radiation (λ_m is in the medium surrounding the spherical particles or cylindrical fibers). (From Lee, 1994.)

carried out (as done by Mishchenko (2018) or simple correlations are used, as outlined in Howell et al. (2021)).

For spherical particles, scattering is claimed to be independent if the following criteria are met (Kaviany and Singh 1993):

$$C + 0.1D > \frac{\lambda_m}{2} \quad \text{or} \quad \xi_c + 0.1\xi > \frac{\pi}{2} \tag{8.59}$$

where C is the clearance distance between particles, D is the particle diameter, the clearance parameter is $\xi_c = \pi C/\lambda_m$, and the size parameter is defined as usual, $\xi = \pi D/\lambda_m$ These correlations were derived for porosities (≈ 0.26) typical of rhombohedral packing.

A regime map of the approximate regions where dependent scattering occurs is in Figure 8.19. The map is based on experimental observations for nonabsorbing spheres and is only an approximate guide.

HOMEWORK (additional problems available in online Appendix I at www.ThermalRadiation.net/appendix.html)

8.1 Compute the half-width for Doppler broadening, γ_D, of neon at a wavelength of 0.75 μm and for $T = 400$ K.

    Answer: $\gamma_D = 0.0042$ cm^{-1}.

8.2   Two absorption lines have the same transition (centerline) wave number, $\eta_{ij} = 550$ cm^{-1}. Both have the same half-width, 0.30 cm^{-1}. One line has the Doppler profile; the other has the Lorentz profile. Draw the two line shapes, $\kappa_{\eta,ij}(\eta)/S_{ij}$, as a function of η, on the same plot.

8.3   A gas is composed of pure atomic hydrogen at a temperature of 1000 K. Calculate the half-width of the hydrogen Lyman alpha line (transition centerline frequency $= 2.4675 \times 10^{15}$ Hz) for the case of Doppler broadening. Then plot the line shape $\kappa_{\eta,ij}/S_{ij}$ for this line as a function of wave number. The mass of the hydrogen atom is 1.66×10^{-24} g.

Answer: $\gamma_D = 1.864$ cm^{-1}.

8.4   For the same gas and temperature as Homework Problem 8.3, compute the half-width of the line for collision broadening at a pressure of 1 atm. Assume the diameter of the hydrogen atom is about 1.06×10^{-8} cm. Plot $\kappa_{\eta,ij}(\eta)/S_{ij}$ for collision broadening on the same wave number plot as for Homework Problem 8.3.

Answer: $\gamma_c = 0.0178$ cm^{-1}.

8.5   From the spectral absorptance α_λ in Figure 8.2, estimate the total emittance of CO_2 for the temperature, pressure, and path length given in the figure caption. Do not assume that the spectral bands are black. Compare the result with the value obtained from the Alberti spreadsheet at https://DOI.org/10.1016/j.jqsrt.2018.08.008

Answers: $\varepsilon = 0.259$; $\varepsilon_{Alberti} = 0.279$.

8.6   Carbon dioxide is in a mixture with air. The CO_2 has a mole fraction of 0.4, and the gas mixture is at a temperature of 1250 K and a total pressure of 1 atm. Using the Alberti spreadsheet at https://DOI.org/10.1016/j.jqsrt.2018.08.008, determine the total emittance for a path length of 2 m.

Answer: $\varepsilon(X = pS = 0.8$ atm·m, $T = 1250$ K$) = 0.181$.

8.7   For water vapor at 1200 K with a mean beam length of 3 m and a partial pressure of 0.2 atm when mixed with air at a total pressure of 1 atm, compare the water vapor total emittance $\varepsilon(T, pL_e)$ using a) the predictions from the Alberti spreadsheet at https://DOI.org/10.1016/j.jqsrt.2018.08.008 and b) Figure 8.7.

Answer: $\varepsilon = 0.342$.

8.8   The spectral intensity of a laser beam ($\lambda \approx 0.484$ μm) along a path is to be attenuated by scattering. A proposed scheme is to use very small spherical particles of gold having a characteristic diameter of 200 Å (optical data for gold are in Table 4.2). (Absorption by the particles is being neglected.) The particles are to be suspended in a nonscattering, nonabsorbing medium. Assuming Rayleigh scattering is applicable, what is the particle scattering cross-section, C_λ? For the intensity to be 10% attenuated by scattering in a path length of 2.0 m, approximately what number density of particles would be required? What are the volume fraction of the particles and the mass of the particles per cubic centimeter of the scattering medium.

Answers: 1.02×10^{-6} μm^2; 5.13×10^{10} cm^{-3}; 2.15×10^{-7}; 4.15×10^{-6} g/cm^3.

8.9   The radiation intensity in Homework Problem 8.8 is changed to infrared ($\lambda \approx 8.06$ μm) while the scattering particle size and number density are kept the same. For the optical constants from Table 4.2 at this wavelength, what is the percent attenuation for this beam for a 2 m path length?

Answer: Insignificant

NOTE

1. These data can be accessed at the website hitran.org/data-index/. Advances in the field are periodically reported in *JQSRT*.

9 Radiative Transfer in Participating Media

Pierre Bouguer (1698–1758) *first discovered the law of exponential decay of light intensity through an absorbing medium (in his case, the atmosphere) in 1729. It is also called the Beer–Lambert Law after its later independent discoverers. Bouguer was an accomplished naval architect (known as "the Father of Naval Architecture"), beating out Euler for a prize given by the French Academy of Sciences for a paper on the masting of ships. Craters on the Moon and Mars are named after him.*

John William Strutt, Lord Rayleigh (1842–1919) *entered Cambridge in 1861, where he read mathematics and devoted full time to science. In 1859, John Tyndall had discovered that bright light scattering by nanoscopic particulates was faintly blue-tinted. He conjectured that a similar scattering of sunlight gave the sky its blue hue, but he could not explain the preference for blue, nor could atmospheric dust explain the intensity of the sky's color.*

In 1871, Rayleigh published two papers on the color and polarization of skylight to quantify Tyndall's effect in water droplets. In 1881 with the benefit of James Clerk Maxwell's 1865 proof of the electromagnetic nature of light, he showed that his scattering equations followed from electromagnetism.

9.1 INTRODUCTION

Consider a translucent medium that is hot and/or is subjected to external radiation. The energy transfer in such a case can be by radiation, conduction, and possibly convection. Then, the governing energy equation is derived by considering the local balance of energy arriving by all modes of energy transfer, internal energy stored, energy generated by local sources, and energy leaving by all modes of transfer. Radiative transfer plays an important role in such a translucent medium due to absorption and emission. The net energy deposited by radiation can be viewed as a local energy source for convection and conduction transfer in the same manner as an energy source provided by electrical dissipation or by nuclear or chemical reactions. The energy equation, including the radiative energy source is derived in this chapter, and it is to be solved to provide the temperature distribution in the translucent material and other energy transfer characteristics, such as local energy fluxes. Chapter 10 presents methods for determining the radiative flux divergence (the radiative source term in the energy equation). This is the quantity of interest in most energy transfer problems involving radiation transfer. The second portion of Chapter 10 then

DOI: 10.1201/9781003178996-9

presents solution methods for the energy equation, including the thermal radiation effects.

To obtain the radiative contribution to the conservation of energy equation, the radiative transfer equation (RTE) must be solved, which allows for finding the temperature distribution in the medium. Basic concepts in Sections 2.8 and 2.9 showed that radiation traveling along a path is attenuated by absorption and scattering and is enhanced by emission and by incoming scattering from other directions. These concepts are expanded here to develop an equation for the radiation intensity along a typical path through a translucent medium. This equation contains a first derivative of intensity with respect to the path coordinate, so a solution requires one boundary condition. This is usually the intensity at the origin of the radiation path being considered. Because the path often begins at a boundary of the radiating medium, the radiation at the boundary is thereby incorporated into the radiation distribution within the medium. The radiative boundary conditions for solving the RTE are in addition to the thermal boundary conditions that must be specified to solve the energy equation.

If the medium is scattering, the local radiative intensity is affected by radiation emitted and scattered throughout the medium. The scattering into a direction along a given path is often combined with the local emission into that direction to form the *source function* in the RTE, which is different from the radiative source term in the conservation of energy equation. For a solution of the RTE, including scattering, the equation for the source function is solved simultaneously with the energy equation to determine both the source function and temperature distributions. The formulations provided here do not consider interface reflection and refraction effects for translucent materials with a refractive index of $n > 1$. Additional analytical relations are provided in Chapter 11 for materials such as glass, water, translucent plastics, or translucent ceramics with an index of refraction $n > 1$.

9.2 CONSERVATION OF ENERGY EQUATION AND BOUNDARY CONDITIONS

An energy balance on a volume element within a material includes contributions by conduction, convection, and radiation energy transfer, internal energy sources such as by electrical dissipation and combustion, compression work, viscous dissipation, and energy storage during transients.

The net contribution of conduction heat transfer to a volume element can be written as the negative divergence of a conduction flux vector $-\nabla \cdot \mathbf{q}_c = \nabla \cdot (k\nabla T)$. Similarly, the net contribution by radiant energy per unit volume within a translucent medium can be written as the negative of the divergence of the radiant flux vector $\mathbf{q}_r$, and expressions for this term are obtained from the solution of the RTE. Thus, the conventional energy equation for a single-component translucent fluid can be modified for the effect of radiative transfer by adding $-\mathbf{q}_r$ to the $k\nabla T$ to obtain:

$$\rho c_p \frac{DT}{Dt} = \beta T \frac{DP}{Dt} + \nabla \cdot (k\nabla T - \mathbf{q}_r) + \dot{q} + \Phi_d \tag{9.1}$$

where $D\Psi / Dt$ is the substantial derivative:

$$\frac{D\Psi}{Dt} = \frac{\partial(\Psi)}{\partial t} + u\frac{\partial(\Psi)}{\partial x} + v\frac{\partial(\Psi)}{\partial y} + w\frac{\partial(\Psi)}{\partial z} \tag{9.2}$$

where

β is the thermal coefficient of volume expansion of the fluid.

$\dot{q}$ is the local energy source (electrical, chemical, nuclear) per unit volume and time.

Φ_d is the energy production by viscous dissipation.

Although traditionally referred to as the conservation of energy equation, each term in Equation (9.1) is really an energy rate (or power) per unit volume. For an incompressible fluid with constant properties in rectangular coordinates, this equation becomes:

$$\frac{\partial T}{\partial t} + u\frac{\partial T}{\partial x} + v\frac{\partial T}{\partial y} + w\frac{\partial T}{\partial z} = \frac{k}{\rho c}\left(\frac{\partial^2 T}{\partial x^2} + \frac{\partial^2 T}{\partial y^2} + \frac{\partial^2 T}{\partial z^2}\right) - \frac{1}{\rho c}\left(\frac{\partial q_{r,x}}{\partial x} + \frac{\partial q_{r,y}}{\partial y} + \frac{\partial q_{r,z}}{\partial z}\right)$$

$$+ \frac{1}{\rho c}\dot{q}(x,y,z,t) + \frac{1}{\rho c}\Phi_d$$

$$\tag{9.3}$$

To obtain the temperature distribution in the medium, Equations (9.1) or (9.3) need to be solved. For this, the divergence of radiative flux, $\nabla \cdot \mathbf{q}_r$ is required, which can be obtained by solving the radiative transfer equation separately (as discussed below). This means that absorbing and scattering particles or any type of heterogeneity or discontinuity in the medium also affect the solution of the RTE, and in turn the energy equation. If the radiative flux distribution in the medium is found, its divergence, $\nabla \cdot \mathbf{q}_r$ can readily be obtained; this is demonstrated in Chapter 10 for the specific geometry of a plane layer.

Another approach is to obtain $\nabla \cdot \mathbf{q}_r$ by considering the local radiative interaction within a differential volume in the medium. The $\nabla \cdot \mathbf{q}_r$ is found from the difference between the radiation emitted from a volume element and the absorption in that same element. This depends on the absorption coefficient and the intensity incident from all directions (note that for a medium with refractive index $n > 1$, an n^2 factor is in the emission term, as given in Chapter 11). The radiative energy required for the energy equation is the total radiation (includes all wavelengths, wave numbers, or frequencies). Hence, as will be developed in detail, relations are required that give the difference between the total radiation incident from all solid angles Ω_i that is locally absorbed and the locally emitted radiation.

A special case of the energy equation is obtained when radiation dominates over both conduction and convection. Then, for a steady state, the equilibrium temperature distribution is achieved only by radiative effects with or without energy sources,

$\dot{q}$, present. This condition is *radiative equilibrium*. This limit provides useful results for some high-temperature applications, and solutions for radiative equilibrium provide limiting results for comparison with solutions including conduction and/or convection combined with radiation. If a medium is considered with internal energy sources $\dot{q}$, such as chemical, nuclear, or electrical energy release, the energy equation for radiative equilibrium is:

$$\nabla \cdot \mathbf{q}_r(\mathbf{r}) = \dot{q}(\mathbf{r}) \tag{9.4}$$

where $\mathbf{r}$ is the position vector in the participating medium.

Solution of the energy equation requires boundary conditions. For the general energy relations given by Equations (9.1) and (9.3), the dependent variable is temperature. Because the equation is second order in the space coordinates and first order in time, the equation requires two boundary conditions for each independent space variable (x, y, and z in Cartesian coordinates) plus an initial condition. When radiation is present, there are additional boundary conditions for the radiative intensities. The radiative boundary conditions enter the results through the solution of the RTE that is used to determine the $\nabla \cdot \mathbf{q}_r$ term in the energy equation. Thus, the radiative boundary conditions are not explicitly stated for the energy equation, but are incorporated in the solution for the radiative flux divergence, $\nabla \cdot \mathbf{q}_r$.

In this chapter and Chapter 10, the discussion centers on cases with either radiative equilibrium or a prescribed internal source $\dot{q}(\mathbf{r})$. These cases are generally described by linear equations. Cases with conduction and/or convection become highly nonlinear in temperature. For these cases, refer to Howell et al. 2021.

9.3 RADIATIVE TRANSFER IN ABSORBING, EMITTING, AND SCATTERING MEDIA

9.3.1 RADIATIVE TRANSFER EQUATION

To obtain the radiative intensity throughout the translucent material, the *radiative transfer equation* is to be solved. This differential equation describes the radiation intensity along a path in a fixed direction through an absorbing, emitting, and scattering medium. Bouguer's law, Equation (2.45), accounts for attenuation of intensity by absorption and scattering. The RTE also includes augmentation of the radiation intensity by emission and scattering into the path direction. As derived in Chapter 2 (Equation (2.72)), the RTE is:

$$\frac{\partial I_\lambda(S,\Omega,t)}{c\partial t} + \frac{\partial I_\lambda(S,\Omega,t)}{\partial S} = \kappa_\lambda I_{\lambda b}(S,t) - \kappa_\lambda I_\lambda(S,\Omega,t) - \sigma_{s,\lambda} I_\lambda(S,\Omega,t)$$

$$+ \frac{\sigma_{s,\lambda}}{4\pi} \int_{\Omega_i=4\pi} I_\lambda(S,\Omega_i,t) \Phi_\lambda(\Omega_i,\Omega) d\Omega_i \tag{9.5}$$

We can combine the attenuation terms and define the *extinction coefficient* $\beta_\lambda = \kappa_\lambda + \sigma_{s,\lambda}$; these have the units of inverse meter. Equation (9.5) is written for a given propagation direction S and for steady conditions (e.g., over a time interval in which the radiation intensity varies insignificantly due to photon time-of-flight effects, $\partial I / c \partial t \approx 0$):

$$\frac{dI_\lambda}{dS} = -\beta_\lambda(S)I_\lambda(S) + \kappa_\lambda(S)I_{\lambda b}(S) + \frac{\sigma_{s,\lambda}}{4\pi} \int_{\Omega_i=0}^{4\pi} I_\lambda(S,\Omega_i)\Phi_\lambda(\Omega,\Omega_i)d\Omega_i \qquad (9.6)$$

The directional dependence of scattered radiation is expressed by the scattering phase function. The scattering coefficient $\sigma_{s,\lambda}$ itself is independent of the direction of incidence, Ω_i, except for some special cases such as filament wound structures, and the scattering coefficient must then be placed within the integral in Equation 9.6.

A detailed scattering profile from a particle is obtained by solving the EM-wave equation in detail to obtain the directional scattering cross section (Bohren and Huffman 1998).

We define the *single scattering albedo*, ω_λ, as the ratio of scattering coefficient to extinction coefficient,

$$\omega_\lambda = \frac{\sigma_{s,\lambda}}{\kappa_\lambda + \sigma_{s,\lambda}} = \frac{\sigma_{s,\lambda}}{\beta_\lambda}; \quad 1 - \omega_\lambda = \frac{\kappa_\lambda}{\kappa_\lambda + \sigma_{s,\lambda}} = \frac{\kappa_\lambda}{\beta_\lambda} \qquad (9.7)$$

If the medium is not absorbing, but scattering alone, $\omega_\lambda \to 1$. For a purely absorbing medium, $\omega_\lambda \to 0$. The *optical thickness* or *opacity* is defined in Equation (2.53) as:

$$\tau_\lambda(S) = \int_{S^*=0}^{S} \beta_\lambda(S^*)dS^* = \int_{S^*=0}^{S} \left[\kappa_\lambda(S^*) + \sigma_{s,\lambda}(S^*)\right] dS^* \qquad (9.8)$$

The RTE (Equation (9.6)) in terms of albedo and optical thickness becomes:

$$\frac{dI_\lambda}{d\tau_\lambda} = -I_\lambda(\tau_\lambda) + (1-\omega_\lambda)I_{\lambda b}(\tau_\lambda) + \frac{\omega_\lambda}{4\pi} \int_{\Omega_i=0}^{4\pi} I_\lambda(\tau_\lambda,\Omega_i)\Phi_\lambda(\Omega,\Omega_i)d\Omega_i \qquad (9.9)$$

where $d\tau_\lambda = \beta_\lambda(S)dS$ is the *differential optical thickness*; $\tau_\lambda = \tau_\lambda(S)$; and $\omega_\lambda = \omega_\lambda(S)$.

9.3.2 SOURCE FUNCTION

The final two terms in Equation (9.9) are sources of intensity along the direction S by emission and in-scattering. They are combined into the *source function*, $\hat{I}_\lambda(\tau_\lambda,\Omega)$,

$$\hat{I}_\lambda\left(\tau_\lambda,\Omega\right) = \left(1-\omega_\lambda\right)I_{\lambda b}\left(\tau_\lambda\right) + \frac{\omega_\lambda}{4\pi} \int_{\Omega_i=0}^{4\pi} I_\lambda\left(\tau_\lambda,\Omega_i\right)\Phi_\lambda\left(\Omega,\Omega_i\right)d\Omega_i \qquad (9.10)$$

For anisotropic scattering, $\hat{I}_\lambda(\tau_\lambda,\Omega)$ is a function of Ω (i.e., of the direction S). The RTE, Equation (9.9), then becomes:

$$\frac{dI_\lambda}{d\tau_\lambda} + I_\lambda\left(\tau_\lambda\right) = \hat{I}_\lambda(\tau_\lambda,\Omega) \tag{9.11}$$

where $\tau_\lambda = \tau_\lambda(S)$. This appears to be a differential equation; however, because I_λ is within the source function in Equation (9.10), Equation (9.11) is an integro-differential equation. It can be integrated after multiplying it first by the integrating factor e^{τ_λ}:

$$\frac{dI_\lambda}{d\tau_\lambda}e^{\tau_\lambda} + I_\lambda\left(\tau_\lambda\right)e^{\tau_\lambda} = \frac{d}{d\tau_\lambda}\left[I_\lambda\left(\tau_\lambda\right)e^{\tau_\lambda}\right] = \hat{I}_\lambda(\tau_\lambda,\Omega)e^{\tau_\lambda} \tag{9.12}$$

Carrying out the integral along the optical path from $\tau_\lambda = 0$ to $\tau_\lambda = \tau_\lambda$ (S) and rearranging gives:

$$I_\lambda(\tau_\lambda,\Omega) = I_\lambda(0,\Omega)e^{-\tau_\lambda} + \int_{\tau_\lambda=0}^{\tau_\lambda} I_\lambda\left(\tau_\lambda^*\right)e^{-(\tau_\lambda-\tau_\lambda^*)}d\tau_\lambda^*; \quad \tau_\lambda(S) = \int_{S^*=0}^{S} \beta_\lambda(S^*)dS^* \tag{9.13}$$

Here, τ_λ^* is a dummy optical variable of integration along S and $I_\lambda(0,\Omega)$ is the intensity in the direction of S at the boundary or location where $S = 0$.

Equation (9.13) is the *integrated form of the RTE* and is interpreted as the intensity at optical depth τ_λ, being composed of two terms. The first is the attenuated boundary intensity that arrives at τ_λ. The second is the intensity at τ_λ resulting from emission and incoming scattering in the S direction by all thickness elements along the path from 0 to S, reduced by exponential attenuation between each location of emission and incoming scattering τ_λ^* and the location τ_λ.

Without scattering (i.e., for absorption only), the source function defined in Equation (9.10) becomes $\hat{I}_\lambda\left(\tau_\lambda\right) = I_{\lambda b}\left(\tau_\lambda\right)$, and the RTE (Equation (9.13)) reduces to:

$$I_\lambda(\tau_\lambda,\Omega) = I_\lambda(0,\Omega)e^{-\tau_\lambda} + \int_{\tau_\lambda^*=0}^{\tau_\lambda} I_{\lambda b}\left(\tau_\lambda^*\right)e^{-(\tau_\lambda-\tau_\lambda^*)}d\tau_\lambda^*; \quad \tau_\lambda(S) = \int_{S^*=0}^{S} \kappa_\lambda(S^*)dS^* \tag{9.14}$$

For a purely scattering medium, the single scattering albedo $\omega_\lambda = 1$, and the source-function Equation (9.10) becomes:

$$\hat{I}_\lambda\left(\tau_\lambda,\Omega\right) = \frac{1}{4\pi}\int_{\Omega_i=0}^{4\pi} I_\lambda\left(\tau_\lambda,\Omega_i\right)\Phi_\lambda\left(\Omega,\Omega_i\right)d\Omega_i; \quad \tau_\lambda(S) = \int_{S^*=0}^{S} \sigma_{s,\lambda}(S^*)dS^* \tag{9.15}$$

The RTE becomes Equation (9.13) with this substitution for the source function. The local intensity for pure scattering is then no longer a function of medium temperature except through the possible temperature dependence of the scattering properties.

For isotropic scattering, $\Phi_\lambda = 1$, so the source-function Equation (9.10), including emission, becomes:

$$\hat{I}_\lambda(\tau_\lambda) = (1-\omega_\lambda)I_{\lambda b}(\tau_\lambda) + \frac{\omega_\lambda}{4\pi} \int_{\Omega_i=0}^{4\pi} I_\lambda(\tau_\lambda,\Omega_i)d\Omega_i \qquad (9.16)$$

The source function in this case is independent of direction (isotropic). In the limit of no absorption ($\omega_\lambda = 1$) and isotropic scattering, Equation (9.16) shows that the source function reduces to the local mean incident intensity,

$$\hat{I}_\lambda(\tau_\lambda) = \frac{1}{4\pi} \int_{\Omega_i=0}^{4\pi} I_\lambda(\tau_\lambda,\Omega_i)d\Omega_i \equiv \bar{I}_\lambda(\tau_\lambda) \qquad (9.17)$$

The boundary condition $I_\lambda(0,\Omega)$ is needed for the solution of the radiative transfer equation in Equation (9.13). For this, consider an opaque solid boundary; the radiative intensity at the boundary consists of the sum of emitted and reflected intensities. At a translucent boundary, the $I_\lambda(0,\Omega)$ depends on radiation entering from the exterior surroundings of the boundary.

For a medium with or without scattering, the radiative source term for the energy equation is the absorbed minus the emitted energy,

$$-\nabla \cdot \mathbf{q}_r(S) = 4\pi \int_{\lambda=0}^{\infty} \kappa_\lambda(\tau_\lambda) \left[\frac{1}{4\pi} \int_{\Omega_i=0}^{4\pi} I_\lambda(\tau_\lambda,\Omega_i)d\Omega_i - I_{\lambda b}(\tau_\lambda) \right] d\lambda;$$
$$\qquad (9.18)$$

$$\tau_\lambda(S) = \int_{S^*=0}^{S} \beta_\lambda(S^*)dS^*$$

For pure scattering of any type (no absorption), the radiative energy source is zero,

$$-\nabla \cdot \mathbf{q}_r(S) = 0 \qquad (9.19)$$

For isotropic scattering with absorption, the radiative energy source can be obtained from the temperature and source-function distributions by evaluating:

$$-\nabla \cdot \mathbf{q}_r(S) = 4\pi \int_{\lambda=0}^{\infty} \frac{\kappa_\lambda(\tau_\lambda)}{\omega_\lambda(\tau_\lambda)} \left[\hat{I}_\lambda(\tau_\lambda) - I_{\lambda b}(\tau_\lambda) \right] d\lambda; \quad \tau_\lambda(S) = \int_{S^*=0}^{S} \beta_\lambda(S^*)dS^* \qquad (9.20)$$

For absorption only without scattering, $\omega_\lambda = 0$, and Equation (9.20) becomes singular, and Equation (9.18) is used.

The source function for isotropic scattering with $\sigma_{s,\lambda}$ independent of Ω_i is obtained by solving an integral equation that requires the temperature distribution and the radiative boundary condition $I_\lambda(\tau_{\lambda,i} = 0, \Omega_i)$ for all paths $\tau_{\lambda,i}$ in the radiation field,

$$\hat{I}_\lambda(\tau_\lambda) = (1 - \omega_\lambda) I_{\lambda b}(\tau_\lambda)$$

$$+ \omega_\lambda \left[\frac{1}{4\pi} \int_{\Omega_i=0}^{4\pi} I_\lambda(\tau_{\lambda,i} = 0, \Omega_i) e^{-\tau_{\lambda,i}} d\Omega_i + \int_{\tau_{\lambda,i}*=0}^{\tau_{\lambda,i}} \hat{I}_\lambda(\tau_{\lambda,i}^*) e^{-\left(\tau_{\lambda,i} - \tau_{\lambda,i}^*\right)} d\tau_{\lambda,i}^* \right] \quad (9.21)$$

where $\omega_\lambda = \omega_\lambda(\tau_\lambda)$ and τ_λ and $\tau_{\lambda,i}$ are optical paths to the same location.

The $\tau_{\lambda,i}$ accounts for all paths to τ_λ through the medium. For absorption only (no scattering), $\omega_\lambda = 0$, and the source function is the local blackbody intensity:

$$\hat{I}_\lambda(\tau_\lambda) = I_{\lambda b}(\tau_\lambda) \quad (9.22)$$

For scattering only (no absorption), $\omega_\lambda = 1$, and Equation (9.21) becomes:

$$\hat{I}_\lambda(\tau_\lambda) = \frac{1}{4\pi} \int_{\Omega_i=0}^{4\pi} I_\lambda(\tau_{\lambda,i} = 0, \Omega_i) e^{-\tau_{\lambda,i}} d\Omega_i + \int_{\tau_{\lambda,i}*=0}^{\tau_{\lambda,i}} \hat{I}_\lambda(\tau_{\lambda,i}^*) e^{-\left(\tau_{\lambda,i} - \tau_{\lambda,i}^*\right)} d\tau_{\lambda,i}^* \quad (9.23)$$

If scattering is independent of incidence direction (as for most atmospheric, combustion and other engineering systems; however, see Section 12.4 for some situations where this is not the case). Equation (9.6) gives the change of intensity along a path. Equation (9.13) is then the integrated form of the RTE.

9.3.3 GRAY MEDIUM

In addition to isotropic scattering, if the medium is gray, the radiative energy source Equation (9.20) and source-function Equation (9.21) reduces to:

$$-\nabla \cdot \mathbf{q}_r(S) = 4\pi \frac{\kappa(\tau)}{\omega(\tau)} \left[I(\tau) - I_b(\tau) \right] = 4 \frac{\kappa(S)}{\omega(S)} \left[\pi I(S) - \sigma T^4(S) \right] \quad (9.24)$$

where $\tau(S) = \int_{S*=0}^{S} \beta(S^*) \, dS^*$ and

$$\hat{I}(\tau) = (1 - \omega) I_b(\tau) + \omega \left[\frac{1}{4\pi} \int_{\Omega_i=0}^{4\pi} I(\tau_i = 0, \Omega_i) e^{-\tau_i} d\Omega_i + \int_{\tau*=0}^{\tau_i} \hat{I}(\tau_i^*) e^{-\left(\tau_i - \tau_i^*\right)} d\tau_i^* \right] \quad (9.25)$$

where $\omega = \omega(\tau)$. For $\omega \rightarrow 0$ (no scattering), Equation (9.24) is singular, so Equation (9.18) (which is valid for $\omega \geq 0$) can be used:

$$-\nabla \cdot \mathbf{q}_r(S) = 4\pi\kappa(\tau) \left[\frac{1}{4\pi} \int_{\Omega_i=0}^{4\pi} I(\tau, \Omega_i) d\Omega_i - I_b(\tau) \right] = 4\kappa(\tau) \left[\pi\bar{I}(S) - \sigma T^4(S) \right]$$

(9.26)

where $\tau(S) = \int_{S^*=0}^{S} \kappa(S^*) dS^*$. For a uniform extinction coefficient β, Equation (9.25) in terms of the physical position S along a path is:

$$\hat{I}(S) = \frac{(1-\omega)}{\pi} \sigma T^4(S) + \omega \left[\frac{1}{4\pi} \int_{\Omega_i=0}^{4\pi} I(S_i = 0, \Omega_i) e^{-\beta S_i} d\Omega_i + \beta \int_{S^*=0}^{S_i} \hat{I}(S_i^*) e^{-\beta(S_i - S_i^*)} dS_i^* \right]$$

(9.27)

where the S_i are all incident paths to location S.

9.4 MEAN BEAM LENGTH APPROXIMATION

Some practical situations require evaluating the energy radiated from a volume of isothermal medium with uniform composition to all or part of its boundaries, without considering emission and reflection from the boundaries. This allows for assessing the relative importance of radiation in a complex problem before embarking on a more comprehensive analysis.

Consider radiation from hot furnace gases to walls that are cool enough for their emission to be small and that are rough and soot-covered, so they are essentially nonreflecting. In this section, the energy is considered in $d\lambda$; in the next section, an integration will include all λ to provide the total radiant energy. For the specified conditions, the spectral outgoing energy flux from any surface A_j, is negligible. The incoming spectral energy at A_k is $A_k G_{\lambda,k} d\lambda$, and only comes from each of the isothermal gas volume elements:

$$A_k G_{\lambda,k} d\lambda = \sum_{j=1}^{N} E_{\lambda b,g} d\lambda A_j F_{j-k} \bar{\alpha}_{\lambda,j-k}$$

(9.28)

If the geometry is a hemisphere of medium with radius R radiating to an area element dA_1 at the center of its base (Figure 9.1), then Equation (9.28) has an especially simple form. Since the hemispherical boundary A_2 is the only surface in view of dA_1, Equation (9.28) reduces to:

$$dA_1 G_{\lambda,1} d\lambda = E_{\lambda b,g} A_2 dF_{2-d1} \bar{\alpha}_{\lambda,2-d1} d\lambda = E_{\lambda b,g} dA_1 F_{d1-2} \bar{\alpha}_{\lambda,2-d1} d\lambda$$

(9.29a)

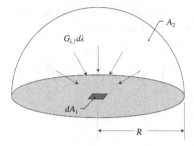

FIGURE 9.1 Geometry for irradiation of element dA_1 on the base of the hemisphere by a uniform medium.

or

$$G_{\lambda,1}d\lambda = E_{\lambda b,g}F_{d1-2}\bar{\alpha}_{\lambda,2-d1}d\lambda \qquad (9.29b)$$

where $\bar{\alpha}_{\lambda,2-d1} = 1 - \bar{t}_{\lambda,2-d1} = 1 - e^{-\kappa_\lambda R}$.

For radiation between dA_1 and the surface of a hemisphere, $F_{d1-2} = 1$, so Equation (9.29b) reduces to the following expression giving the *incident* energy flux from a hemisphere of isothermal medium emitting to the center of the hemisphere base:

$$G_{\lambda,1}d\lambda = \left(1 - e^{-\kappa_\lambda R}\right)E_{\lambda b,g}d\lambda \qquad (9.30)$$

The term $\left(1 - e^{-\kappa_\lambda R}\right)$ is the *spectral emittance* of the gas $\varepsilon_\lambda(T,P,R)$ for path length R (see Chapter 8). Then, Equation (9.30) becomes

$$G_{\lambda,1}d\lambda = \varepsilon_\lambda(\kappa_\lambda R)E_{\lambda b,g}d\lambda \qquad \varepsilon_\lambda(\kappa_\lambda R) = 1 - e^{-\kappa_\lambda R} \qquad (9.31)$$

The gas emittance and thus the incident spectral energy depend on the optical radius of the hemisphere, $\kappa_\lambda R$.

The simple form of Equation (9.31) would be convenient if it could be used to determine the value of $G_{\lambda,k}d\lambda$ on A_k for *any* geometry of radiating medium volume and for A_k being all or part of its boundary. Because the geometry of the medium enters Equation (9.31) only through $\varepsilon_\lambda(\kappa_\lambda R)$, we can define a fictitious equivalent value of R, say L_e, that would yield a value of $\varepsilon_\lambda(\kappa_\lambda L_e)$ such that Equation (9.31) would give the correct $G_\lambda d\lambda$ for a given geometry. This fictitious length L_e is called the *mean beam length*. Then, for an arbitrary geometry of gas, let

$$G_{\lambda,k}d\lambda = \varepsilon_\lambda(\kappa_\lambda L_e)E_{\lambda b,g}d\lambda = \left(1 - e^{-\kappa_\lambda L_e}\right)E_{\lambda b,g}d\lambda \qquad (9.32)$$

The mean beam length is the required radius of an equivalent hemisphere of a medium that radiates a flux to the center of its base equal to the average flux radiated to the area of interest by the actual medium volume. Note that this approach does not provide a detailed analysis of radiative transfer; it is just to evaluate the importance of radiative transfer before getting into a more detailed formulation and solution scheme.

9.4.1 Mean Beam Length for a Medium between Parallel Plates

Consider two black infinite parallel plates at $T_1 = T_2 = 0$ separated by a distance D. The plates enclose a uniform medium at T_g with absorption coefficient κ_λ. The rate at which spectral energy is incident upon A_k is, from Equation (9.29b),

$$G_{\lambda,k} A_k d\lambda = E_{\lambda b,g} d\lambda A_j F_{j-k} \bar{\alpha}_{\lambda,j-k} = E_{\lambda b,g} d\lambda A_j F_{j-k} \left[1 - 2E_3 \left(\kappa_\lambda D \right) \right] \quad (9.33)$$

where the exact expression $\bar{\alpha}_{\lambda,j-k} = [1 - 2E_3(\kappa_\lambda D)]$ for infinite parallel plates is from Table 11.1 of Howell et al. 2021. Here, E_3 is the exponential integral function, tabulated in online Appendix M at www.ThermalRadiation.net/text/html. For infinite parallel plates, $F_{j-k} = 1$, and by reciprocity, $F_{j-k} = A_k/A_j$, so Equation (9.33) reduces to:

$$G_{\lambda,k} d\lambda = \left[1 - 2E_3 \left(\kappa_\lambda D \right) \right] E_{\lambda b,g} d\lambda \quad (9.34)$$

Comparing Equations (9.33) and (9.34) defines the mean beam length for infinite parallel plates as:

$$L_e = -\frac{1}{\kappa_\lambda} \ln[2E_3(\kappa_\lambda D)] \quad (9.35)$$

or, in terms of optical thickness $\tau_\lambda = \kappa_\lambda D$,

$$\frac{L_e}{D} = -\frac{1}{\tau_\lambda} \ln[2E_3(\tau_\lambda)] \quad (9.36)$$

9.4.2 Radiation from the Entire Medium Volume to Its Entire Boundary for Optically Thin Media

Because of the integrations involved, the mean beam length for an entire medium volume radiating to all or part of its boundary is usually difficult to evaluate except for the simplest shapes. Practical approximations can be found by looking first at the optically thin limit. For a small optical path length $\kappa_\lambda S$, the transmittance becomes:

$$\lim_{\kappa_\lambda S \to 0} t_\lambda = \lim_{\kappa_\lambda S \to 0} e^{-\kappa_\lambda S} = \lim_{\kappa_\lambda S \to 0} \left[1 - \kappa_\lambda S + \frac{(\kappa_\lambda S)^2}{2!} - \cdots \right] \to 1$$

Each differential volume of the uniform temperature medium emits spectral energy $4\kappa_\lambda E_{\lambda b,g} d\lambda dV$. Since the transmittance in this case $t_\lambda \approx 1$, there is no attenuation of the emitted radiation, and it all reaches the enclosure boundary. For the entire radiating volume, the energy reaching the boundary is $4\kappa_\lambda E_{\lambda b,g}d\lambda V$, so the average spectral flux received at the entire boundary of area A is, in the optically thin limit,

$$G_\lambda d\lambda = 4\kappa_\lambda E_{\lambda b,g}\frac{V}{A}d\lambda \qquad (9.37)$$

By use of the mean beam length, the average spectral flux reaching the boundary is given by Equation (9.30). For the optically thin case, let L_e be designated by $L_{e,0}$. Then, from Equation (9.30), for small $\kappa_\lambda L_{e,0}$,

$$G_\lambda d\lambda = \left\{1-\left[1-\kappa_\lambda L_{e,0}+\frac{(\kappa_\lambda L_{e,0})^2}{2!}-\cdots\right]\right\}E_{\lambda b,g}d\lambda \approx \kappa_\lambda L_{e,0}E_{\lambda b,g}d\lambda \qquad (9.38)$$

Equating this to G_λ in Equation (9.37) gives the desired result for the mean beam length of an *optically thin* medium radiating to its entire boundary:

$$L_{e,0}=\frac{4V}{A} \qquad (9.39)$$

To give two examples, for a sphere of diameter D,

$$L_{e,0}=\frac{4V_s}{A_s}=\frac{4\pi D^3/6}{\pi D^2}=\frac{2}{3}D \qquad (9.40)$$

For an infinitely long circular cylinder of diameter D,

$$L_{e,0}=\frac{4\pi D^2/4}{\pi D}=D \qquad (9.41)$$

9.4.3 CORRECTION TO MEAN BEAM LENGTH WHEN A MEDIUM IS *NOT* OPTICALLY THIN

For a medium that is not optically thin, it would be convenient if L_e could be obtained by applying a simple correction factor to the $L_{e,0}$ from Equation (9.39). To do so, introduce a correction coefficient C so that L_e is given by:

$$L_e=CL_{e,0} \qquad (9.42)$$

Then, the incoming spectral flux in Equation (9.32) is:

$$G_\lambda d\lambda = \left(1-e^{-\kappa_\lambda CL_{e,0}}\right)E_{\lambda b,g}d\lambda \qquad (9.43)$$

Figure 9.2 shows the agreement between the prediction of Equation (9.43) and the exact relation of Equation (9.34).

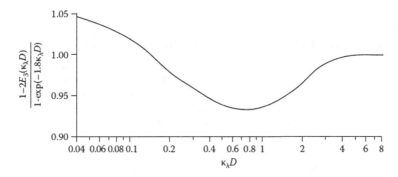

FIGURE 9.2 Ratio of emission by an infinite layer of medium to that calculated using a mean beam length, $L_e = 1.8D$. Deviation from 1.00 corresponds to an error in the approximation.

Table 9.1 gives the mean beam length $L_{e,0}$ for various geometries, along with L_e values that provide reasonably accurate radiative fluxes for nonzero optical thicknesses. The values of C are found for each case to be in a range near 0.9 (Hottel 1954, Eckert and Drake 1959, Hottel and Sarofim 1967). Hence, the approximation that is recommended for a geometry for which exact L_e values have not been calculated,

$$L_e \approx 0.9 L_{e,0} \approx 0.9 \frac{4V}{A} \tag{9.44}$$

for an entire uniform isothermal medium volume radiating to its entire boundary.

9.5 EXCHANGE OF TOTAL RADIATION IN AN ENCLOSURE BY USING A MEAN BEAM LENGTH

The mean beam length was obtained in the previous section at a single wavelength. The concept can also be applied to obtain the exchange of *total* energy within an enclosure. The use of the mean beam length simplifies the geometric considerations, but it must be spectrally integrated to obtain total energy transfer.

9.5.1 TOTAL RADIATION FROM THE ENTIRE MEDIUM VOLUME TO ALL OR PART OF ITS BOUNDARY

The mean beam length is *approximately independent* of κ_λ, as evidenced by Equation (9.44). This means that L_e can be used as a characteristic dimension of the gas volume and approximated as constant during integration over wavelength. The total energy flux from the medium that is incident on a surface is found by integrating Equation (9.32) over all λ:

$$G = \int_{\lambda=0}^{\infty} (1 - e^{-\kappa_\lambda L_e}) E_{\lambda b,g} d\lambda \tag{9.45}$$

TABLE 9.1
Mean Beam Lengths for Radiation from Entire Medium Volume

Geometry of Radiating System	Characterizing Dimension	Mean Beam Length for Optical Thickness, $\kappa_\lambda L_e \to 0, L_{e,0}$	Mean Beam Length Corrected for Finite Optical Thickness,[a] L_e	$C = L_e/L_{e,0}$
Hemisphere radiating to element at center of base	Radius R	R	R	1
Sphere radiating to its surface	Diameter D	(2/3)D	0.65D	0.97
Circular cylinder of infinite height radiating to concave bounding surface	Diameter D	D	0.95D	0.95
Circular cylinder of semi-infinite height radiating to:				
Element at center of base	Diameter D	D	0.90D	0.90
Entire base	Diameter D	0.81D	0.65D	0.80
Circular cylinder of height equal to diameter radiating to:				
Element at center of base	Diameter D	0.77D	0.71D	0.92
Entire surface	Diameter D	(2/3)D	0.60D	0.90
Circular cylinder of height equal to two diameters radiating to:				
Plane end	Diameter D	0.73D	0.60D	0.82
Concave surface	Diameter D	0.82D	0.76D	0.93
Entire surface	Diameter D	0.80D	0.73D	0.91
Circular cylinder of height equal to one-half the diameter radiating to:				
Plane end	Diameter D	0.48D	0.43D	0.90
Concave surface	Diameter D	0.52D	0.46D	0.88
Entire surface	Diameter D	0.50D	0.45D	0.90
Cylinder of infinite height and semicircular cross section radiating to element at center of plane rectangular face	Radius R		1.26 R	
Infinite slab of medium radiating to:				
Element on one face	Slab thickness D	2D	1.8D	0.90
Both bounding planes	Slab thickness D	2D	1.8D	0.90
Cube radiating to a face	Edge X	(2/3)X	0.6X	0.90
Rectangular parallelepipeds 1x1x4 radiating to:				
1 x 4 face	Shortest edge X	0.90X	0.82X	0.91
1 x 1 face	Shortest edge X	0.86X	0.71X	0.83

(Continued)

TABLE 9.1 (CONTINUED)
Mean Beam Lengths for Radiation from Entire Medium Volume

Geometry of Radiating System	Characterizing Dimension	Mean Beam Length for Optical Thickness, $\kappa_\lambda L_e \to 0, L_{e,0}$	Mean Beam Length Corrected for Finite Optical Thickness,[a] L_e	$C = L_e/L_{e,0}$
All faces	Shortest edge X	$0.89X$	$0.81X$	0.91
1 x 2 x 6 radiating to:				
2 x 6 face	Shortest edge X	$1.18X$		
1 x 6 face	Shortest edge X	$1.24X$		
1 x 2 face	Shortest edge X	$1.18X$		
All faces	Shortest edge X	$1.20X$		
Medium between infinitely long parallel concentric cylinders	Radius of outer cylinder R and of inner cylinder r	$2(R-r)$	See Anderson and Handvig (1989)	
Medium volume in the space between the outside of the tubes in an infinite tube bundle and radiating to a single tube:				
Equilateral triangular array:	Tube diameter D			
$S = 2D$	and spacing	$3.4(S{-}D)$	$3.0(S{-}D)$	0.88
$S = 3D$	between tube	$4.45(S{-}D)$	$3.8(S{-}D)$	0.85
Square array:	centers, S			
$S = 2D$		$4.1(S{-}D)$	$3.5(S{-}D)$	0.85

[a] Corrections are those suggested by Hottel (1954), Hottel and Sarofim (1967) or Eckert and Drake (1959). Corrections were chosen to provide maximum L_e where these references disagree.

where L_e is independent of λ. Now, define a *total emittance* ε_g for the medium such that:

$$G = \varepsilon_g \sigma T_g^4 \tag{9.46}$$

Equating the last two relations gives:

$$\varepsilon_g = \frac{\int_{\lambda=0}^{\infty} \left(1 - e^{-\kappa_\lambda L_e}\right) E_{\lambda b,g} d\lambda}{\sigma T_g^4} \tag{9.47}$$

The total gas emittance ε_g in Equation (9.47) is a convenient quantity that can be provided for each medium in terms of L_e and T_g. Values of ε_g are available for the important radiating gases, and analytical forms are also available that are convenient for computer use; these results for ε_g are in Section 8.4. Then, for a particular geometry and medium temperature and pressure, the ε_g is applied by use of Equation (9.46). An example illustrates how ε_g is obtained and used.

Example 9.1

A cooled right cylindrical tank 4 m in diameter and 4 m long has a black interior surface and is filled with hot gas at a total pressure of 1 atm. The gas is composed of CO_2 mixed with a transparent gas that has a partial pressure of 0.75 atm. The gas is uniformly mixed at $T_g = 1100$ K. Compute how much energy must be removed from the entire tank surface to keep it cool if the tank walls are all at a low temperature so that only radiation from the gas is significant.

The geometry is a finite circular cylinder of gas, and the radiation to its walls will be computed. Using Table 9.1, the corrected mean beam length is $L_e = 0.60D = 2.4$ m. The partial pressure of the CO_2 is 0.25 atm. Using the Alberti et al. (2018) spreadsheet from https://doi.org/10.1016/j.jqsrt.2018.08.008, $\varepsilon_{CO_2}(p_{CO_2}L_e, T_g) = 0.174$. From Equation (9.46), the energy to be removed is:

$$Q_i = GA = \varepsilon_{CO_2}\sigma T_g^4 A = 0.174 \times 5.6704 \times 10^{-8}(1100)^4 24\pi = 1089 \text{ kW}$$

9.5.2 RADIATIVE EXCHANGE BETWEEN THE ENTIRE VOLUME AND AN EMITTING BOUNDARY

In the previous section, the temperature of the black enclosure wall was low enough that emission from the wall was neglected. If wall emission is significant, the average energy flux removed at the wall is the emission of the medium to the wall, which is all absorbed because the wall is black, minus the average flux emitted from the wall that is absorbed by the medium. For a steady state, the total energy removed from the entire boundary equals the energy supplied to the medium by some other means, such as by combustion in a gas. For an enclosure with its entire boundary black and at T_w, the energy balance is:

$$-\frac{Q_w}{A} = \frac{Q_g}{A} = \sigma\left[\varepsilon_g(T_g)T_g^4 - \alpha_g(T_w)T_w^4\right] \quad (9.48)$$

Here, $\alpha_g(T_w)$ is the total absorptance by the medium for radiation emitted from the black wall at temperature T_w. It depends on the spectral properties of the medium and on T_w, as this determines the spectral distribution of the radiation received by the medium.

For radiative exchange calculations in furnaces, an approximate procedure for determining α_g is in Hottel and Sarofim (1967). The α_g is obtained from the gas total emittance values by using:

$$\alpha_g = \alpha_{CO_2} + \alpha_{H_2O} - \Delta\alpha \quad (9.49)$$

where

$$\alpha_{CO_2} = C_{CO_2}\varepsilon_{CO_2}^+ \left(\frac{T_g}{T_w}\right)^{0.5} \tag{9.50}$$

$$\alpha_{H_2O} = C_{H_2O}\varepsilon_{H_2O}^+ \left(\frac{T_g}{T_w}\right)^{0.5} \tag{9.51}$$

$$\Delta\alpha = (\Delta\varepsilon)_{at\,T_w} \tag{9.52}$$

The $\varepsilon_{CO_2}^+$ and $\varepsilon_{H_2O}^+$ are, respectively, ε_{CO_2} and ε_{H_2O} evaluated at T_w and at the respective parameters $p_{CO_2}L_e'$ and $p_{H_2O}L_e'$ where $L_e' = L_e T_w/T_g$. The curve fits and spreadsheet provided by Alberti et al. (2018) allow inclusion of CO in the analysis (see Appendix A in the online appendix at www.ThermalRadiation.net/appendix.html). Alberti et al. (2020) provide an alternative method for computing α_g based on their curve fits.

If the bounding walls are not black and hence are reflecting, radiation can pass through the gas by means of multiple reflections from the boundary.

9.6 RADIATIVE TRANSFER IN PLANE LAYERS WITH PARTICIPATING MEDIA

The energy rate conservation relation was introduced in Equations (9.1) and (9.3), along with relations for the associated radiative flux divergence (Equation (9.4)) and the radiative transfer equation (Equations (9.5) through (9.9)). In the rest of this section, these relations are applied to the cases of radiative transfer in plane layers, with and without internal generation. More general solution techniques are covered in Chapter 10.

To evaluate the influence of some of the many variables for energy transfer in a radiating gas or other translucent medium, we use a simple geometry such as a plane layer of a large extent relative to its thickness and with uniform conditions along each boundary. There is considerable literature for this one-dimensional (1D) geometry for engineering applications, atmospheric physics, and astrophysics since plane layers approximate the Earth's atmosphere and the outer radiating layers of the Sun. We outline a simple case below for illustration.

9.6.1 RADIATIVE TRANSFER EQUATION AND RADIATIVE INTENSITY

A plane layer between two boundaries is shown in Figure 9.3. The temperature and properties of the medium vary only in the x-direction for this 1D system. The boundaries of the plane layer form a two-surface enclosure, and all paths S and S_i for the radiation intensity originate at the lower or upper surface. The path direction is given by the angle θ measured from the positive x-direction. The superscript notation, +

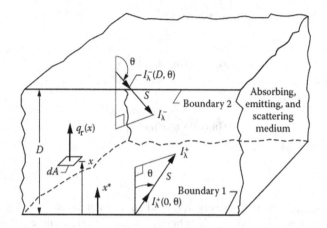

FIGURE 9.3   Plane layer between infinite parallel boundaries.

or −, indicates, respectively, the directions with positive or negative $\cos\theta$, so that I_λ^+ corresponds to $0 \leq \theta \leq \pi/2$ and I_λ^- to $\pi/2 \leq \theta \leq \pi$.

The optical depth $\tau_\lambda(x)$ within the layer is defined *along the x coordinate across the thickness* with the extinction coefficient β as:

$$\tau_\lambda(x) = \int_{x^*=0}^{x} \left[\kappa_\lambda(x^*) + \sigma_{s,\lambda}(x^*) \right] dx^* = \int_{x^*=0}^{x} \beta_\lambda(x^*) dx^* \tag{9.53}$$

which accounts for variable properties, $\beta_\lambda(x)$. For constant properties, $\tau_\lambda(x) = \beta_\lambda x$. For the positive x-direction, the relation between optical positions along the S- and x-directions is:

$$\tau_\lambda(S) = \int_{S^*=0}^{S} \beta_\lambda(S^*) dS^* = \frac{1}{\cos\theta} \int_{x^*=0}^{x} \beta_\lambda(x^*) dx^* = \frac{\tau_\lambda(x)}{\cos\theta} \tag{9.54}$$

For the negative direction, Equation (9.54) still applies because $dS = -dx/\cos(\pi - \theta) = dx/\cos\theta$. With $d\tau_\lambda(S) = d\tau_\lambda(x)/\cos\theta$, the radiative transfer Equation (9.11) becomes, for I_λ^+ and I_λ^-,

$$\cos\theta \frac{\partial I_\lambda^-}{\partial \tau_\lambda(x)} + I_\lambda^- \left[\tau_\lambda(x), \theta \right] = I_\lambda \left[\tau_\lambda(x), \theta \right] \quad \pi/2 \leq \theta \leq \pi \tag{9.55}$$

$$\cos\theta \frac{\partial I_\lambda^+}{\partial \tau_\lambda(x)} + I_\lambda^+ \left[\tau_\lambda(x), \theta \right] = I_\lambda \left[\tau_\lambda(x), \theta \right] \quad 0 \leq \theta \leq \pi/2 \tag{9.56}$$

The source function $\hat{I}_\lambda$, which consists of emission and in-scattering terms, depends on θ because of the angular dependence of the phase function Φ for anisotropic scattering, as in Equation (9.10) The partial derivatives in Equations (9.55) and (9.56) emphasize that I_λ^+ and I_λ^- depend on both $\tau_\lambda(x)$ and θ. Define $\mu \equiv \cos\theta$ and Equations (9.55) and (9.56) become:

$$\mu \frac{\partial I_\lambda^+}{\partial \tau_\lambda(x)} + I_\lambda^+ \left[\tau_\lambda(x), \mu\right] = I_\lambda \left[\tau_\lambda(x), \mu\right] \quad 1 \ge \mu \ge 0 \tag{9.57}$$

$$\mu \frac{\partial I_\lambda^-}{\partial \tau_\lambda(x)} + I_\lambda^- \left[\tau_\lambda(x), \mu\right] = I_\lambda \left[\tau_\lambda(x), \mu\right] \quad 0 \ge \mu \ge -1 \tag{9.58}$$

Using an integrating factor as in Equation (9.12), Equations (9.57) and (9.58) are integrated subject to the following boundary conditions:

$$I_\lambda^+ \left[\tau_\lambda, \mu\right] = I_\lambda^+ \left[0, \mu\right] \quad \text{at } \tau_\lambda = 0 \tag{9.59}$$

$$I_\lambda^- \left[\tau_\lambda, \mu\right] = I_\lambda^- \left[\tau_{\lambda,D}, \mu\right] \quad \text{at } \tau_\lambda = \tau_{\lambda,D} \tag{9.60}$$

where $\tau_{\lambda,D} = \int_{x^*=0}^{x} \beta_\lambda(x)dx = \int_{x=0}^{D} \left[\kappa_\lambda(x) + \sigma_{s,\lambda}(x)\right]dx$ is the layer optical thickness. Integration of this equation gives the intensities as a function of location and angle in the plane layer as:

$$I_\lambda^+ (\tau_\lambda, \mu) = I_\lambda^+ (0, \mu)e^{-\tau_\lambda/\mu} + \frac{1}{\mu} \int_{\tau_\lambda^*=0}^{\tau_\lambda} I_\lambda \left(\tau_\lambda^*, \mu\right)e^{-\left(\tau_\lambda - \tau_\lambda^*\right)/\mu} d\tau_\lambda^*; \quad 1 \ge \mu \ge 0 \tag{9.61}$$

$$I_\lambda^- (\tau_\lambda, \mu) = I_\lambda^- (\tau_{\lambda,D}, \mu)e^{-\tau_\lambda/-\mu} - \frac{1}{\mu} \int_{\tau_\lambda^*=0}^{\tau_{\lambda,D}} I_\lambda \left(\tau_\lambda^*, \mu\right)e^{-\left(\tau_\lambda^* - \tau_\lambda\right)/-\mu} d\tau_\lambda^*; \quad 0 \ge \mu \ge -1 \tag{9.62}$$

In Equation (9.62), the θ is between π/2 and π so that $\mu = \cos\theta$ is negative.

The use of the source function $\hat{I}_\lambda$ in Equations (9.61) and (9.62) applies for both scattering and nonscattering media; however, without scattering, Equation (9.22) gives $\hat{I}_\lambda (\tau_\lambda, \mu) = I_{\lambda b} (\tau_\lambda)$, which is isotropic, so Equations (9.61) and (9.62) simplify for a medium with absorption only to give:

$$I_\lambda^+ (\tau_\lambda, \mu) = I_\lambda^+ (0, \mu)e^{-\tau_\lambda/\mu} + \frac{1}{\mu} \int_{\tau_\lambda^*=0}^{\tau_\lambda} I_{\lambda b} \left(\tau_\lambda^*, \mu\right)e^{-\left(\tau_\lambda - \tau_\lambda^*\right)/\mu} d\tau_\lambda^*; \quad 1 \ge \mu \ge 0 \tag{9.63}$$

$$I_\lambda^- (\tau_\lambda,\mu) = I_\lambda^- (\tau_{\lambda,D},\mu) e^{-\tau_\lambda/-\mu} - \frac{1}{\mu} \int_{\tau_\lambda^*=0}^{\tau_{\lambda,D}} I_{\lambda b} (\tau_\lambda^*,\mu) e^{-(\tau_\lambda^*-\tau_\lambda)/-\mu} d\tau_\lambda^*; \quad 0 \geq \mu \geq -1 \quad (9.64)$$

9.6.2 LOCAL RADIATIVE FLUX IN A PLANE LAYER

At each x location, the total radiative energy flux in the positive x-direction is the integral of the spectral flux over all λ. With the notation that $q_{r\lambda} d\lambda$ is the spectral flux in $d\lambda$,

$$q_r (x) = \int_{\lambda=0}^{\infty} q_{r\lambda}(x) d\lambda = \int_{\lambda=0}^{\infty} q_{r\lambda}(\tau_\lambda) d\lambda \quad (9.65)$$

where $\tau_\lambda(x)$ is the optical depth at x. For a fixed x, the optical coordinate τ_λ depends on λ since β_λ is a function of λ.

The net spectral flux in the positive x-direction crossing dA in the plane at x in Figure 9.3 is obtained in two parts, one from I_λ^+ and one from I_λ^-. Since intensity is energy per unit solid angle crossing an area normal to the direction of I, the projection of dA must be considered normal to either I_λ^+ or I_λ^-. The spectral energy flux in the positive x-direction from I_λ^+ is then (using $d\Omega = 2\pi \sin\theta \, d\theta$ and $\mu = \cos\theta$):

$$q_{r\lambda}^+ (\tau_\lambda) d\lambda = 2\pi d\lambda \int_{\theta=0}^{\pi/2} I_\lambda^+ (\tau_\lambda,\theta) \cos\theta \sin\theta d\theta = 2\pi d\lambda \int_{\mu=0}^{1} I_\lambda^+ (\tau_\lambda,\mu) \mu d\mu \quad (9.66)$$

The total flux from I_λ^+ in the positive x-direction is:

$$q_r^+ (x) = J^+ (x) = \int_{\lambda=0}^{\infty} q_{r\lambda}^+ (\tau_\lambda) d\lambda \quad (9.67)$$

Similarly, in the negative x-direction,

$$q_{r\lambda}^- (\tau_\lambda) d\lambda = J_\lambda^- (\tau_\lambda) d\lambda = 2\pi d\lambda \int_{\pi-\theta=0}^{\pi/2} I_\lambda^- (\tau_\lambda,\theta) \cos(\pi-\theta) \sin(\pi-\theta) d(\pi-\theta)$$

$$\quad (9.68)$$

$$= -2\pi d\lambda \int_{\theta=\pi/2}^{\pi} I_\lambda^- (\tau_\lambda,\theta) \cos\theta \sin\theta d\theta = 2\pi d\lambda \int_{\mu=0}^{1} I_\lambda^- (\tau_\lambda,-\mu) \mu d\mu$$

The net spectral flux in the positive x-direction is then:

$$q_{r\lambda} (\tau_\lambda) d\lambda = \left[q_{r\lambda}^+ (\tau_\lambda) - q_{r\lambda}^- (\tau_\lambda) \right] d\lambda = 2\pi d\lambda \int_{\mu=0}^{1} \left[I_\lambda^+ (\tau_\lambda,\mu) - I_\lambda^- (\tau_\lambda,\mu) \right] \mu d\mu \quad (9.69)$$

The intensities can be substituted from Equations (9.61) and (9.62) to obtain:

$$q_{r\lambda}\left(\tau_\lambda\right)d\lambda = 2\pi\left[\int_{\mu=0}^{1} I_\lambda^+\left(0,\mu\right)e^{-\tau_\lambda/\mu}\mu d\mu - \int_{\mu=0}^{1} I_\lambda^-\left(\tau_{\lambda,D},-\mu\right)e^{-\left(\tau_{\lambda,D}-\tau_\lambda\right)/\mu}\mu d\mu \right.$$

$$\left. + \int_{\mu=0}^{1}\int_{\tau_\lambda^*=0}^{\tau_\lambda} \hat{I}_\lambda\left(\tau_\lambda^*,\mu\right)e^{-\left(\tau_\lambda-\tau_\lambda^*\right)/\mu}d\tau_\lambda^* d\mu + \int_{\mu=0}^{1}\int_{\tau_\lambda^*=\tau_\lambda}^{\tau_{\lambda,D}} \hat{I}_\lambda\left(\tau_\lambda^*,-\mu\right)e^{-\left(\tau_\lambda^*-\tau_\lambda\right)/\mu}d\tau_\lambda^* d\mu \right]d\lambda$$

(9.70)

Using the *exponential integral function* gives a convenient form. It is:

$$E_n\left(\xi\right) \equiv \int_0^1 \mu^{n-2}e^{-\xi/\mu}d\mu$$

(9.71)

Numerical values for $E_n\left(\xi\right)$ are available in handbooks and in computational software packages and are further explored in Appendix M of the online appendix at www. ThermalRadiation.net/text/html. With this substitution, Equation (9.70) becomes:

$$q_{r\lambda}\left(\tau_\lambda\right)d\lambda = 2\pi\left[\int_{\mu=0}^{1} I_\lambda^+\left(0,\mu\right)e^{-\tau_\lambda/\mu}\mu d\mu - \int_{\mu=0}^{1} I_\lambda^-\left(\tau_{\lambda,D},-\mu\right)e^{-\left(\tau_{\lambda,D}-\tau_\lambda\right)/\mu}\mu d\mu \right.$$

(9.72)

$$\left. + \int_{\tau_\lambda^*=0}^{\tau_\lambda} \hat{I}_\lambda\left(\tau_\lambda^*\right)E_2\left(\tau_\lambda-\tau_\lambda^*\right)d\tau_\lambda^* + \int_{\tau_\lambda^*=\tau_\lambda}^{\tau_{\lambda,D}} \hat{I}_\lambda\left(\tau_\lambda^*\right)E_2\left(\tau_\lambda^*-\tau_\lambda\right)d\tau_\lambda^* d\lambda \right]$$

The total net radiative flux at x is found by integrating Equation (9.72) from $\lambda = 0$ to $\lambda = \infty$ using the τ_λ at each λ corresponding to the fixed x.

9.6.3 DIVERGENCE OF THE RADIATIVE FLUX: RADIATIVE ENERGY SOURCE

As discussed at the beginning of the chapter, the radiative flux divergence $\nabla\cdot\mathbf{q}_r(x)$ is needed for the conservation of energy Equation (9.1). For a plane layer with uniform conditions over each boundary, the $\mathbf{q}_r(x)$ depends only on x so that:

$$\nabla\cdot\mathbf{q}_r\left(x\right) = \frac{dq_r\left(x\right)}{dx} = \frac{d}{dx}\int_{\lambda=0}^{\infty} q_{r\lambda}(x)d\lambda = \int_{\lambda=0}^{\infty}\frac{dq_{r\lambda}(x)}{dx}d\lambda = \int_{\lambda=0}^{\infty}\beta_\lambda\left(\tau_\lambda\right)\frac{dq_{r\lambda}(\tau_\lambda)}{d\tau_\lambda}d\lambda$$

(9.73)

where $d\tau_\lambda = \beta_\lambda(x)\,dx$ and, throughout the integration over λ, the τ_λ corresponds to the specified x. Differentiating Equation (9.72) with respect to τ_λ yields a different version of Equation (9.73) as:

$$\frac{dq_{r\lambda}(x)}{dx}d\lambda = \beta_\lambda(\tau_\lambda)\frac{dq_{r\lambda}(\tau_\lambda)}{d\tau_\lambda}$$

$$= -2\pi\beta_\lambda(\tau_\lambda)\left[\int_{\mu=0}^{1} I_\lambda^+(0,\mu)e^{-\tau_\lambda/\mu}d\mu + \int_{\mu=0}^{1} I_\lambda^-(\tau_{\lambda,D},-\mu)e^{-(\tau_{\lambda,D}-\tau_\lambda)/\mu}d\mu \right. \tag{9.74}$$

$$+ \int_{\mu=0}^{1}\frac{1}{\mu}\int_{\tau_\lambda^*=0}^{\tau_\lambda} I_\lambda\left(\tau_\lambda^*,\mu\right)e^{-\left(\tau_\lambda-\tau_\lambda^*\right)/\mu}d\tau_\lambda^* d\mu - \int_{\mu=0}^{1} I_\lambda(\tau_\lambda,\mu)d\mu$$

$$\left. + \int_{\mu=0}^{1}\frac{1}{\mu}\int_{\tau_\lambda^*=\tau_\lambda}^{\tau_{\lambda,D}} I_\lambda\left(\tau_\lambda^*,-\mu\right)e^{-\left(\tau_\lambda^*-\tau_\lambda\right)/\mu}d\tau_\lambda^* d\mu - \int_{\mu=0}^{1} I_\lambda(\tau_\lambda,-\mu)d\mu\right]$$

Another form of the divergence term starting from Equation (9.26) is:

$$\frac{dq_r(x)}{dx} = 4\pi\int_{\lambda=0}^{\infty} \kappa_\lambda(x)\left[I_{\lambda b}(x) - \bar{I}_\lambda(x)\right]d\lambda \tag{9.75}$$

which applies to cases with or without scattering. The forms in Equations (9.73) and (9.74) or in Equation (9.75) are equivalent because the $\bar{I}_\lambda(\lambda,x)$ in Equation (9.73) is found from: $\bar{I}_\lambda(x) = \frac{1}{2}\int_0^1\left[I_\lambda^+(x,\mu) + I_\lambda^-(x,-\mu)\right]d\mu$.

9.6.4 SOURCE FUNCTION IN A PLANE LAYER

The equation for the radiative source (that is, the divergence of radiative flux) is obtained from Equation (9.74). The substitutions are made that $\tau_\lambda(S) = \beta_\lambda(x)/\cos\theta$ and $d\Omega_i = 2\pi\sin\theta_i d\theta_i$. Then, following the procedure in the derivation of the previous plane layer relations, the source-function equation becomes (where $\omega_\lambda = \omega_\lambda(\tau_\lambda)$):

$$\hat{I}_\lambda(\tau_\lambda,\Omega) = (1-\omega_\lambda)I_{\lambda b}(\tau_\lambda) + \frac{\omega_\lambda}{2}\left[\int_{\mu_i=0}^{1} I_\lambda^+(0,\mu_i)e^{-\tau_\lambda/\mu_i}\Phi_\lambda(\mu,\mu_i)d\mu_i \right.$$

$$+ \int_{\mu_i=0}^{1} I_\lambda^+(\tau_{\lambda,D},\mu_i)e^{-(\tau_{\lambda,D}-\tau_\lambda)/\mu_i}\Phi_\lambda(\mu,-\mu_i)d\mu_i$$

$$\tag{9.76}$$

$$+ \int_{\mu_i=0}^{1}\frac{1}{\mu_i}\int_{\tau_\lambda^*=0}^{\tau_\lambda} \hat{I}_\lambda\left(\tau_\lambda^*,\mu_i\right)e^{-\left(\tau_\lambda-\tau_\lambda^*\right)/\mu_i}d\tau_\lambda^*\Phi(\mu,\mu_i)d\mu_i$$

$$\left. + \int_{\mu_i=0}^{1}\frac{1}{\mu_i}\int_{\tau_\lambda^*=\tau_\lambda}^{\tau_{\lambda,D}} \hat{I}_\lambda\left(\tau_\lambda^*,-\mu_i\right)e^{-\left(\tau_\lambda^*-\tau_\lambda\right)/\mu_i}d\tau_\lambda^*\Phi(\mu,-\mu_i)d\mu_i\right]$$

9.7 GRAY PLANE LAYER IN RADIATIVE EQUILIBRIUM

9.7.1 ENERGY EQUATION

In some situations, radiation transfer dominates over other energy transfer mechanisms. This provides a limiting case to compare with energy transfer by combined modes, where conduction and/or convection energy transfer is also present. In this section, a special case is considered for *steady state* without significant conduction, convection, viscous dissipation, or internal energy sources. For simplicity, the development is for a gray medium. When all energy sources and transfer mechanisms are negligible compared with radiation, the total energy emitted from each volume element must equal its total absorbed energy. This *radiative equilibrium* reflects steady-state energy conservation in the absence of any transfer but radiation. Then, Equation (9.1) becomes,

$$\nabla \cdot q_r = \frac{dq_r(x)}{dx} = 0 \quad \text{or} \quad \frac{dq_r(\tau)}{d\tau} = 0 \qquad (9.77)$$

Here q_r is the total energy flux being transferred since radiation is the only means of transfer.

9.7.2 RESULTS FOR GRAY MEDIUM WITH DIFFUSE-GRAY BOUNDARIES AT SPECIFIED TEMPERATURES

Equations (9.63), (9.64), and (9.77) for the case of a gray medium with no scattering with appropriate boundary conditions can be used to determine the temperature distribution and net radiation energy flux in a gray gas in radiative equilibrium contained between infinite parallel gray-diffuse plates (Figure 9.4).

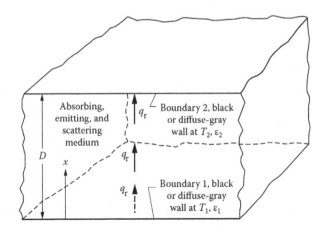

FIGURE 9.4 Plane layer of a medium in radiative equilibrium between infinite parallel diffuse surfaces.

Results are shown in Figures 9.5 and 9.6. The temperature distributions shown in Figure 9.5 indicate a peculiarity of radiation transfer in the absence of other energy transfer modes. The temperature in the medium does *not* go to the boundary temperature as the boundary is approached; rather, there is a discontinuity or "slip" in temperature at the boundary. The magnitude of the slip is shown in Figure 9.6.

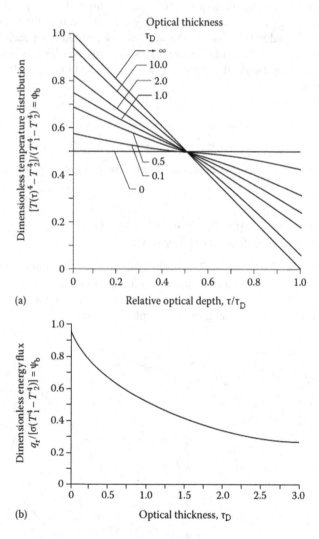

FIGURE 9.5 (a) Temperature distributions and (b) energy flux in a gray medium contained between infinite black parallel boundaries. (From Heaslet, M.A. and Warming, R.F., *Int. J. Heat Mass Trans.*, 8(7), 979, 1965.)

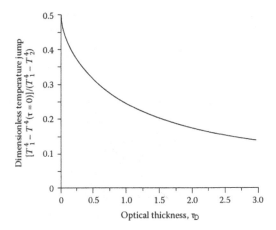

FIGURE 9.6 Discontinuity at the boundary between gray gas and black boundary temperatures for radiative equilibrium. (From Heaslet, M.A. and Warming, R.F., *Int. J. Heat Mass Trans.*, 8(7), 979, 1965.)

HOMEWORK (additional problems available in online Appendix I at www.ThermalRadiation.net/appendix.html)

9.1 A layer of isothermal gray gas with $T_g = 1000$ K and with uniform properties is 1 m thick. The layer boundaries are perfectly transparent. Intensity $I(x = 0)$ is normally incident on the left boundary; there is no incident intensity from the environment on the boundary at $x = 1$. For each set of conditions shown in the following figures, what is the value of the intensity in the x-direction normal to the layer at $x = 1$ m?

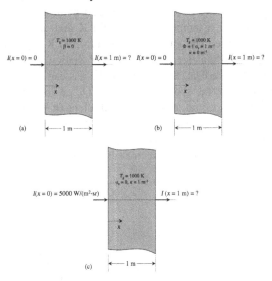

Answer: 0, 0, 13,250 W/(m²·sr)

9.2   A slab of nonscattering solid material has a gray absorption coefficient of $\kappa = 0.20$ cm^{-1} and refractive index $n \approx 1$. It is 5 cm thick and has an approximately linear temperature distribution within it, as established by thermal conduction.

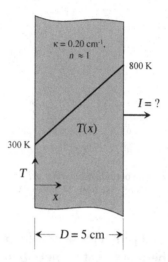

1.   For the condition shown, what is the emitted intensity normal to the slab at $x = D$? What average slab temperature would give the same emitted normal intensity?
2.   If the temperature profile is reversed (i.e., $T(x = 0) = 800$ K, $T(x = D) =$ 300 K), what is the normal intensity emitted at $x = D$?

Answers: (a) 1849 W/(m$^2 \cdot$ sr); 635 K (b) 1149 W/(m$^2 \cdot$ sr)

9.3   The radiation property of a gas is measured with the use of a blackbody radiation source at a temperature of 2000 K and an optical detector, as shown in the figure. The gas at temperature 300 K fills the 2 m spacing between the 1 cm diameter aperture of the radiation source and the 1.5 cm diameter detector surface. Scattering by the gas can be ignored. Derive an expression to relate the spectral radiation ($Q_{\lambda,abs}$) absorbed by the opaque detector with the spectral absorption coefficient (κ_λ) of the gas, the directional-hemispherical spectral reflectivity [$\rho_\lambda(\theta_i)$] of the detector surface, the angle θ_1 measured from the normal to the aperture, and the angle θ_2 measured from the normal to the detector surface.

Answer: $dQ_\lambda = \left[1 - \rho_\lambda(\theta_2)\right] I_\lambda(S) \cos(\theta_1) dA_1 \dfrac{dA_2 \cos(\theta_2)}{S^2}$

9.4 An isothermal enclosure is filled with a nonscattering absorbing and emitting medium. The medium and enclosure are at the same temperature. Show that $\nabla \cdot \mathbf{q}_r$ must be zero for this condition.

9.5 Pure carbon dioxide at 1 atm and 2000 K is contained between parallel plates 0.3 m apart. What is the radiative flux received at the plates from radiation by the gas? Use the Alberti et al. (2018) spreadsheet from https://doi.org/10.1016/j.jqsrt.2018.08.008 for CO_2 total emittance.

Answer: 99.8 kW/m²

9.6 A rectangular furnace of dimensions 0.6 × 0.5 × 2.2 m has soot-covered interior walls that are black and cold. The furnace is filled with well-mixed combustion products at a temperature of 2200 K, composed of 40% by volume CO_2, 30% by volume water vapor, and the remainder N_2. The total pressure is 1 atm. Compute the radiative flux and the total energy rate to the walls using the Alberti et al. (2018) spreadsheet from https://doi.org/10.1016/j.jqsrt.2018.08.008 for the total emittance of the CO_2 and H_2O mixture.

Answer: 914 kW

9.7 A 15 cm diameter pipe is carrying superheated steam at 1.5 atm pressure and at a uniform temperature of 1300 K. What is the radiative flux from the steam received at the pipe wall?

Answer: 64.5 kW/m²

9.8 A furnace at atmospheric pressure with an interior in the shape of a cube having an edge dimension of 1.0 m is filled with a 50:50 mixture by volume of CO_2 and N_2. The gas temperature is uniform at 1800 K, and the walls are cooled to 1100 K. The interior surfaces are black. At what rate is energy being supplied to the gas (and removed from the walls) to maintain these conditions?

Answer: 355 kW

9.9 Consider the same conditions and furnace volume as in Homework Problem 9.8. The furnace is now a cylinder with a height equal to two times its diameter. What is the energy rate supplied to the gas?

Answer: 317 kW

9.10 A black-walled cubical enclosure with 2 m edges contains a mixture of gases at $T = 1600$ K and a pressure of 2 atm. The gas has volume fractions of 0.4/0.4/0.2 for CO_2–H_2O–N_2. Cooling water is passed over one face of the cube. If the water is initially at 25°C and has a maximum allowable temperature rise of 10°C, what mass flow rate of water (kg/s) is required? Neglect reradiation from the wall, and assume all other walls are cool.

Answer: 13.1 kg/s

9.11 A large furnace contains parallel tubes arranged in an equilateral triangular array. The tubes are 2.5 cm in outer diameter, and the tube centers are spaced 7.5 cm apart. The furnace gas is composed of 75% CO_2 and 25% H_2O by volume, and the combustion process maintains the gas temperature at 1350 K and a total pressure of 0.9 atm. What is the radiative flux incident on a tube in the interior of the bundle (i.e., surrounded by other tubes) per meter of tube length?

Answer: 2.50 kW/m

10 Numerical Methods for Radiative Transfer Problems

Subrahmanyan Chandrasekhar (1910–1995) *did his graduate work at Cambridge under William A. Fowler. As part of his graduate work, Chandrasekhar hypothesized that stars may evolve into black holes. Despite a notorious public humiliation at the hands of Arthur Eddington, Chandrasekhar was eventually proven correct, and he and Fowler won the Nobel Prize in 1983 for their work on the structure and evolution of stars. His book* Radiative Transfer *(1960) outlines the discrete ordinates method (DOM) now used extensively in thermal radiation transfer and introduces methods for treating scattering using*

Stokes' parameters.

Photo courtesy of University of Chicago Photographic Archive, (apf1-09456), Special Collections Research Center, University of Chicago Library.

Stanislaw Ulam (1909–1984) *was a Polish-born scientist who worked at Los Alamos as part of the Manhattan Project during WWII. Following his recovery from a serious brain surgery in 1947, he conceived the idea of using statistical techniques to model the individual histories of neutrons as they travel through fissile material in order to calculate critical mass. He developed this approach, called the* Monte Carlo method, *with John von Neumann, Enrico Fermi, Nicholas Metropolis and others. Metropolis and Ulam (1949) published the first paper in the open literature on the method.*

Photo courtesy of Creative Commons-Oslo Museum

10.1 NUMERICAL METHODS FOR SOLVING PROBLEMS IN RADIATIVE TRANSFER

As discussed in Chapter 9, the integro-differential equations that govern radiative transfer are not analytically tractable except for very simple cases. This has motivated the development of many numerical techniques for solving these problems. Choosing the "right" technique depends on the nature of the underlying problem, the required level of accuracy, and the computational budget available.

DOI: 10.1201/9781003178996-10

Solution techniques can be categorized broadly as either deterministic or statistical. Deterministic schemes start with the governing equations (usually the RTE, Equation (9.6)) and approximate them with a set of linear equations that simplify the angular dependence. Well-known examples include the discrete ordinates method (DOM, or "S_N"), finite volume methods (which are related to the DOM method), finite element methods, the spherical harmonics (or "P_N") method, and, for problems involving optically thick media, the diffusion approximation.

Statistical schemes employ probability theory to solve the governing equations directly. These schemes are collectively labeled as "Monte Carlo" (MC) techniques. In a MC approach, the problem is formulated to generate a set of (typically millions) of random solutions by sampling from the probability distributions that define the emission location, direction, and path of energy bundles, and the expected value (or average) of these trials approximates the exact solution to the governing equation. These statistical schemes are often more intuitive, simpler to implement, and involve fewer approximations compared to deterministic methods, and therefore provide more accurate results. However, they are also very computationally demanding, and, until recently, have largely been confined to deriving high-accuracy solutions with well-defined uncertainty for benchmarking the performance of more efficient deterministic schemes. This picture is changing rapidly with the emergence of high-performance computers, multithreaded processors, quantum computers, and graphical processing units. Consequently, MC simulations that used to take days or weeks to run on dedicated supercomputers a decade ago can now be carried out in minutes or hours on laptops.

The chapter begins with a discussion that motivates the analysis of radiative transfer through 1D materials. We then present the Schuster–Schwarzschild method for solving this problem, and show how this technique can be generalized into the DOM. Next, we introduce the spherical harmonics (P_N) method, followed by the diffusion approximation. Finally, we consider MC techniques for calculating view factors between surfaces, radiative transfer between diffuse-gray surfaces and transparent media, and, finally, radiative transfer within absorbing, emitting, and scattering media.

For simplicity, this chapter focuses on gray media bounded by diffuse-gray surfaces, but these techniques can be extended to consider spectrally varying properties, e.g., by solving the RTE over spectral bands over which the properties may be approximated as gray and summing the results, or, in the case of molecular gases, using the methods discussed in Section 8.3.

10.2 MOTIVATION FOR 1D RADIATIVE TRANSFER MODELS

The earliest numerical solution schemes for the RTE focused on 1D planar media. Early work was motivated by modeling the transport of sunlight through cloud layers (Schuster 1905) and radiative transfer through stars (Schwarzschild 1906, Milne 1930, Eddington 1926). Radiative transfer plays a critical role in the structure and evolution of stars since inwards-acting gravitational forces are counterbalanced by outwards-acting photon pressure. While both cases involve curved media, the radius

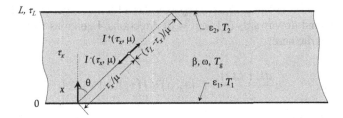

FIGURE 10.1 Problem geometry for radiative heat transfer through a 1D slab containing absorbing, emitting, and isotropically scattering media.

of curvature is often large relative to the optical depth or thickness of the surface layer, so the radiative transfer can often be modeled as occurring through 1D planar media, as shown in Figure 10.1.

In the engineering world, 1D models are often sufficient to model radiative transfer through coatings, paints, and oxide layers. For example, the performance of paints used to enhance the absorption or reflection of solar energy (see Chapter 4) may be predicted using a 1D model, as can the propagation of radiation through layers of human tissue. Some researchers have used radiation models to assess the performance of cosmetics and sunscreens.

Another important application is thermal barrier coatings (TBCs), thin ceramic layers applied to protect metallic surfaces in high-temperature environments, such as turbine blades in gas turbine engines. The turbine blades are actively cooled by flowing air through passages within the blades. The blade temperature is further suppressed by adding a thin ceramic layer on the outer surface of the blade that scatters according to the "snow effect" (Section 4.5.1). Given the temperatures involved, heat is transferred by both conduction and radiation through the TBC to the blade, so modeling radiation through the coating is crucial to predict the effectiveness of the TBC.

In these applications, scattering is often assumed to be isotropic, in which case the relevant form of the RTE is:

$$\mu \frac{dI}{d\tau_x} = -I\left(\tau_x,\mu\right) + \underbrace{\left(1-\omega\right)I_b\left(\tau_x\right) + \omega\bar{I}\left(\tau_x\right)}_{\hat{\imath}(\tau_x)} \tag{10.1}$$

where $\mu = \cos(\theta)$ is the cosine of the angle defined from the x-direction, τ_x is the optical depth at x, Equation (9.53), and

$$\bar{I}\left(\tau_x\right) = \frac{1}{4\pi}\int_{4\pi} I\left(\tau_x,\Omega_i\right)d\Omega_i = \frac{1}{2}\int_{\mu_i=-1}^{1} I\left(\tau_x,\mu_i\right)d\mu_i \tag{10.2}$$

is the average incident flux at τ_x. The in-scattering and emission terms are often combined into a single radiant source term, $\hat{\imath}\left(\tau_x\right) = \left(1-\omega\right)I_b\left(\tau_x\right) + \omega\bar{I}\left(\tau_x\right)$.

It is also often convenient to write separate RTEs for intensity fields in the upwards $(0 < \mu < 1)$ and downwards $(-1 < \mu < 0)$ directions, Equations (9.57) and (9.58), which we rewrite here:

$$\mu \frac{dI^+(x,\Omega)}{d\tau_x} + I^+(\tau_x,\mu) = \hat{I}(\tau_x), \quad 1 \geq \mu \geq 0 \tag{10.3}$$

and

$$-\mu \frac{dI^-(x,\Omega)}{d\tau_x} + I^-[\tau_x,\mu] = \hat{I}(\tau_x), \quad 0 \geq \mu \geq -1 \tag{10.4}$$

Equations (10.3) and (10.4) are partial differential equations that depend on both x and μ. The intensity in one direction is coupled to the intensities in other directions by the in-scattering integral in Equation (10.2), as well as the reflected irradiation at the boundaries, which we assume are diffuse:

$$I^-(\mu,\tau_L) = \frac{\varepsilon_2 \sigma T_2^4}{\pi} + 2\rho_2 \int_{\mu=0}^{1} I^+(\mu,\tau_L)\mu d\mu \tag{10.5}$$

and

$$I^+(\mu,0) = \frac{\varepsilon_1 \sigma T_1^4}{\pi} + 2\rho_1 \int_{\mu=-1}^{0} I^-(\mu,0)\mu d\mu \tag{10.6}$$

Because of this coupling, analytical solutions are only possible for nonscattering media bounded by black, isothermal surfaces. Therefore, numerical schemes are needed to solve this problem.

10.3   DISCRETE ORDINATES METHOD

10.3.1   Schuster–Schwarzschild (Two-Flux) Method

In order to solve Equations (10.3) and (10.4), it is necessary to make an assumption about the angular distribution of the intensity at each location, x, or, equivalently, each optical depth, τ_x. In his 1905 study on the propagation of sunlight through a cloud layer, Arthur Schuster (1905) proposed that the intensity at any point within the medium be modeled as piecewise constant over each hemisphere aligned with the x-axis, $I^+(\tau_x)$ and $I^-(\tau_x)$. This approach was subsequently adopted by Karl Schwarzschild (1906) to model radiative equilibrium in a stellar atmosphere.

In this scenario, the in-scattering term is simply $\omega[I^+(\tau_x) + I^-(\tau_x)]/2$. Each intensity equation is then multiplied by $d\Omega = 2\pi d\mu$ and integrated over the corresponding hemisphere:

$$2\pi \frac{dI^+}{d\tau_x} \int\limits_{\mu=0}^{1} \mu d\mu = -2\pi I^+\left(\tau_x\right) \int\limits_{\mu=0}^{1} d\mu + 2\pi\left(1-\omega\right) I_b\left(\tau_x\right) \int\limits_{\mu=0}^{1} d\mu$$

$$\qquad\qquad\qquad\qquad + \omega\pi\left[I^+\left(\tau_x\right)+I^-\left(\tau_x\right)\right] \int\limits_{\mu=0}^{1} d\mu \qquad (10.7)$$

and

$$-2\pi \frac{dI^-}{d\tau_x} \int\limits_{\mu=-1}^{0} \mu d\mu = -2\pi I^-\left(\tau_x\right) \int\limits_{\mu=-1}^{0} d\mu + 2\pi\left(1-\omega\right) I_b\left(\tau_x\right) \int\limits_{\mu=-1}^{0} d\mu$$

$$\qquad\qquad\qquad\qquad + \pi\omega\left[I^+\left(\tau_x\right)+I^-\left(\tau_x\right)\right] \int\limits_{\mu=-1}^{0} d\mu \qquad (10.8)$$

Carrying out the integrations, and noting that the radiant fluxes in the upwards and downwards directions are related to the corresponding intensities by $q_r^+(\tau_x) = \pi I^+(\tau_x)$ and $q_r^-(\tau_x) = \pi I^-(\tau_x)$ because the intensities are uniform over each hemisphere, results in:

$$\frac{dq_r^+}{d\tau_x} = -\left(2-\omega\right)q_r^+ + \omega q_r^- + 2\left(1-\omega\right)E_b\left(\tau_x\right) \qquad (10.9)$$

and

$$-\frac{dq_r^-}{d\tau_x} = \omega q_r^+ - \left(2-\omega\right)q_r^- + 2\left(1-\omega\right)E_b\left(\tau_x\right) \qquad (10.10)$$

Equations (10.9) and (10.10) are solved simultaneously with radiosity boundary conditions derived from Equations (10.5) and (10.6), $q_r^-(\tau_{\lambda L}) = \varepsilon_2\sigma T_2^4+\rho_2 q_r^+(\tau_{\lambda L})$ and $q_r^+(0) = \varepsilon_1\sigma T_1^4+\rho_1 q_r^-(0)$. The fact that two radiant flux components are being solved for gives this technique its alternate name, the "two-flux" or "S_2" method. In practice, these ordinary differential equations are usually solved using a finite difference method. For this, the spatial is divided into discrete nodes, including the boundaries, and an initial solution is assumed for q_r^-. The solution then proceeds through k iterations in successive upwards and downwards sweeps, as shown in Figure 10.2, until the solution converges. The radiant flux leaving each boundary is a combination of the emitted radiation and the reflected irradiation calculated from the most recent sweep.

The net radiant heat flux is found by subtracting the downwards component from the upwards component, $q_r(\tau_x) = q_r^+(\tau_x) - q_r^-(\tau_x)$. For gray media, this quantity may be differentiated numerically with respect to τ_x to obtain the radiant source term,

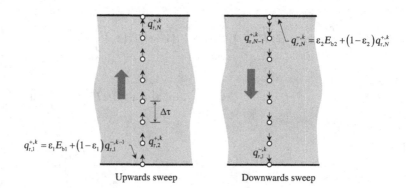

FIGURE 10.2 Schuster–Schwarzschild algorithm for a 1D slab. The upper and downward components of the radiant flux vector (or intensities) are solved through a succession of upwards and downwards sweeps.

$$\dot{q}_r(x) = -\frac{dq_r}{dx} = -\frac{dq_r}{d\tau_x}\frac{d\tau_x}{dx} = -\beta(x)\frac{dq_r}{d\tau_x} \tag{10.11}$$

Example 10.1

Consider a homogeneous nonscattering gray medium contained between two black plates. The bottom plate is at $T_1 = 500$ K, while the top plate is at $T_2 = 750$ K, and the uniform temperature medium is at $T_g = 600$ K. Calculate $\dot{q}_r/\beta$ for $\tau_L = 0.1$, 1, and 10 using the Schuster–Schwarzschild method. Plot the corresponding intensity distributions at $\tau_x = 0.5\tau_L$. Compare these answers with the exact solutions.

The problem domain is discretized using 41 nodes. The gradients in Equations (10.9) and (10.10) are approximated using a second-order finite difference scheme. At the bottom-most node, the radiant flux is known, $q_{r,1}^+ = E_{b1} = \sigma T_1^4$. Starting from $j = 2$, a backward difference approximation is used for the other nodes:

$$\left.\frac{dq_r^+}{d\tau_x}\right|_{\tau_{x,j}} \approx \frac{q_{r,j}^+ - q_{r,j-1}^+}{\Delta\tau_x} = -2q_{r,j}^+ + 2E_{b,g}$$

with similar equations for q_r^- starting from the topmost node and working downwards. These equations are solved for the radiant flux in one upwards sweep for the $\{q_r^+\}$ terms, followed by one downwards sweep for the $\{q_r^-\}$ terms, starting from an assumed initial solution. Per Figure 10.2, successive upwards and downwards sweeps are carried out until the radiant fluxes converge. In this example, the upwards and downwards sweeps are independent since there is no scattering and the boundaries are black. It may be necessary to employ under-relaxation to obtain a converged solution. The radiant source term may then be found from the net radiant heat flux, again using finite differences.

The exact solution for the intensity distribution is given by:

$$I^+(\tau_x,\mu) = \frac{E_{b1}}{\pi}\exp(-\tau_x/\mu) + \frac{E_{b,g}}{\pi}\left[1-\exp(-\tau_x/\mu)\right], \quad 0 \le \mu \le 1$$

and

$$I^-(\tau_x,\mu) = \frac{E_{b2}}{\pi}\exp\left(\left[\tau_L - \tau_x\right]/\mu\right) + \frac{E_{b,g}}{\pi}\left\{1-\exp\left[\left(\tau_L - \tau_x\right)/\mu\right]\right\}, \quad -1 \le \mu \le 0$$

The corresponding radiant flux terms are found by integrating each component over the corresponding hemisphere,

$$q_r^+(\tau_x) = 2\pi\int_0^1 I^+(\tau_x,\mu)\mu d\mu = 2(E_{b1} - E_{b,g})E_3(\tau_x) + 2E_{b,g}$$

and

$$q_r^-(\tau_x) = 2\pi\int_{-1}^0 I^-(\tau_x,\mu)\mu d\mu = 2(E_{b2} - E_{b,g})E_3(\tau_L - \tau_x) + 2E_{b,g}$$

where E_3 is the exponential integral function, Equation (9.71). The net radiant flux is found by $q_r(\tau_x) = q_r^+(\tau_x) - q_r^-(\tau_x)$ and the radiant source term is given by:

$$\frac{\dot{q}_r(\tau_x)}{\beta} = -\frac{dq_r(\tau_x)}{d\tau_x} = 2\left[(E_{b1} - E_{b,g})E_2(\tau_x) + (E_{b2} - E_{b,g})E_2(\tau_L - \tau_x)\right]$$

Solutions from the Schuster–Schwarzschild method and the exact solution are plotted in Figure 10.3. The radiant source term predicted by the Schuster–Schwarzschild method, Figure 10.3 (a), becomes more accurate as the medium

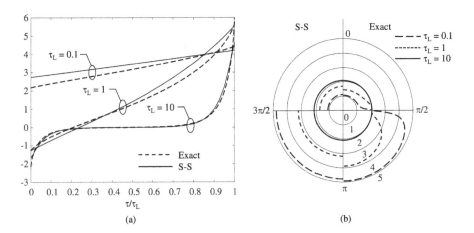

FIGURE 10.3 (a) Radiant source term through the medium and (b) intensities calculated at $\tau_x = 0.5\tau_L$ for Example 10.1. (Radiant intensities are normalized by I_{b1}.)

becomes more opaque, since the intensity distribution approaches that of a hemi-sphere, as shown in Figure 10.3(b).

Example 10.2

A 1 m thick conducting and nonscattering medium is bound between two plates at $T_1 = 600$ K and $T_2 = 800$ K. The medium has an absorption coefficient of 5 m^{-1}, and a thermal conductivity of 1 W/(m·K). Each plate has an emissivity of 0.5. Calculate the temperature distribution through the medium, and the heat transfer across the medium, using the Schuster–Schwarzschild method.

The temperature distribution between the two plates is governed by the heat equation,

$$\frac{d^2T}{dx^2} = -\frac{\dot{q}_r(x)}{k}$$

The solution proceeds by solving the heat conduction problem for the temperature using a numerical scheme (e.g., finite difference method), and then using the temperature field to obtain the radiant flux vector using the Schuster–Schwarzschild algorithm. The radiant flux vector is then post-processed to obtain the radiant source term, which is substituted back into the heat equation. The procedure continues until the temperature field has converged. The net heat flux at the boundaries may be calculated by summing the conductive and radiative components; the net radiative flux into the medium is found by subtracting irradiation from the radiosity.

Figure 10.4 compares the temperature profiles obtained with and without radiative transfer, along with the radiant source term. In general, the temperature field

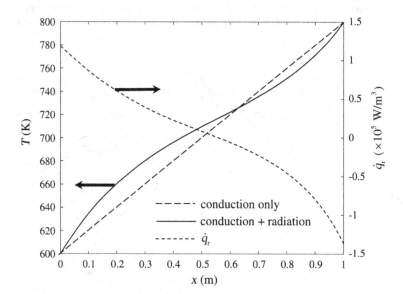

FIGURE 10.4 Temperature distribution through the 1D slab problem of Example 10.3 with and without radiative transfer, and volumetric radiative source term.

TABLE 10.1

Conductive, Radiative, and Net Fluxes at Upper and Lower Surfaces for Example 10.2

	Surface 1 (T = 600 K)	Surface 2 (T = 800 K)
Conductive	−4.00 kW/m²	3.42 kW/m²
Radiative	−4.00 kW/m²	4.58 kW/m²
Total	−8.00 kW/m²	8.00 kW/m²

(Positive denotes into the medium)

becomes more uniform within the medium and produces sharper gradients at the surface, since hotter regions of the medium are cooled by radiation, while colder regions are heated by radiation.

The conductive, radiative, and net fluxes at the boundaries are summarized in Table 10.1. Note that the combined fluxes at the upper and lower boundaries sum to zero, indicating that the net energy entering the medium equals zero.

10.3.2 DISCRETE ORDINATES METHOD FOR A 1D SLAB

Example 10.1 highlights that the main inaccuracy in the Schuster–Schwarzschild method arises from modeling the intensities as isotropic over each hemisphere. In the case of a cold medium bounded by hot surfaces, the intensity will have a strong forward or backward peak (a "lobular" distribution) at $\mu = 1$ or $\mu = -1$, since this direction corresponds to the shortest pathlength, and therefore the least attenuation, between a location in the medium and the hot boundary. The extent of this non-uniformity depends on the optical distance between the medium and the boundary and the temperature difference between the medium and the boundary. The assumption of an isotropic intensity only holds when the medium and boundary are at thermal equilibrium with each other, or when the medium is extremely optically thick, in which case any point in the medium only "sees" the immediate vicinity and is therefore in a state of local thermal equilibrium.

The accuracy of this method can be improved by subdividing each hemisphere into $N/2$ finite solid angles over each of which the intensity is modeled as uniform. In this approach, the RTE is written for each of the N intensity components, which are coupled by the in-scattering term, Equation (10.2), and the irradiation term in the wall boundary conditions, Equations (10.5) and (10.6). In the case of isotropic scattering, the in-scattering integral is approximated by:

$$\int_{4\pi} I\left(\tau_x, \mu_i\right) d\Omega = 2\pi \int_{\mu_i=-1}^{1} I\left(\tau_x, \mu_i\right) d\mu_i \approx \sum_{m=1}^{N} w_m I_m\left(\tau_x\right) \qquad (10.12)$$

where w_m are the weights of the integration. The choice of weights depends on the quadrature used to approximate the integration. Were the domain of μ divided so that each intensity component corresponded to the same finite solid angle $\Delta\Omega$, the weights would be equal to $4\pi/N$. It is common practice to use Gaussian quadrature to obtain a more accurate solution, in which case the ordinates and weights are dictated by the quadrature points.

Likewise, the integral for the upper wall boundary condition, Equation (10.5) is approximated by:

$$I_{m=3,4}\left(\tau_L\right) = \frac{\varepsilon_2\sigma T_2^4}{\pi} + \rho_2 \int\limits_{\mu=0}^{1} I\left(\mu,\tau_L\right)\mu d\mu \approx \frac{\varepsilon_2\sigma T_2^4}{\pi} + \frac{\rho_2}{\pi} \sum\limits_{\substack{m'=1 \\ \mu_{m'}>0}}^{N} w_{m'}I_{m'}\left(\tau_L\right)\mu_{m'} \quad (10.13)$$

The notation below the summation term denotes that only the intensities in the upward direction ($\mu > 0$) should be included in the summation.

Table 10.2 shows the Gaussian quadrature points and weights for the upper hemisphere of a 1D slab using four and six quadrature points; the corresponding directions are shown in Figure 10.5.

TABLE 10.2
Gaussian Quadrature Points for Upper Hemisphere in 1D Slab Geometry

DOM order, S_N	Weights, w_m	Ordinate directions, μ_m
S_2	2π	0.5
S_4	4.1887902	0.2958759
	2.0943951	0.9082483
	2.7382012	0.183867
S_6	2.9011752	0.6950514
	0.6438068	0.9656013

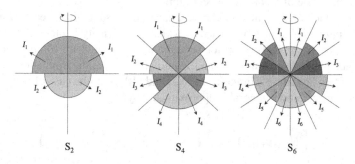

FIGURE 10.5 Solid angles and ordinate directions used in the DOM method for 1D planar media.

When the intensity is written in this way, the method is called the "S_N" method, where the "N" denotes the number of quadrature points over each hemisphere. It is always an even number since $N/2$ directions are taken over each hemisphere.

10.3.3 GENERALIZED DISCRETE ORDINATES METHOD

The approach developed for a 1D slab can be generalized to solve radiative transfer in two- or three-dimensions using any coordinate system. In the case of a 3D Cartesian system, the S_N scheme amounts to solving the intensities over directions distributed over 2π sr, at a set of spatial grid points distributed over the 3D problem domain and along the boundary. The "N" denotes the number of different direction cosines to be used for each principal direction, and all the possible combinations of these cosines generate the set of ordinates, $\{\Omega_m, m = 1, 2,...N_m\}$. In a 3D problem, each octant of a sphere around a grid point contains $N/2+[N/2-1]+[N/2-2]+...+ 1 = N\cdot(N+2)/8$ ordinates, resulting in $N_m = N\cdot(N+2)$ discrete ordinates over a sphere.

As is the case for 1D media, the set of ordinates is chosen according to the integration scheme used to compute the in-scattering integral and irradiation integral for the wall boundary condition. Gaussian quadrature is the most common choice, but other options are available. Let $\hat{s}_m$ be a unit vector pointing in the Ω_m direction. The RTE is written for the intensity in each direction at a given location in the medium, $\mathbf{r}$, and the chain rule is applied to relate the rate at which the intensity changes with distance in the Ω_m direction to corresponding changes in the x-, y-, and z-directions. Noting that $dx/dS_m = \hat{i} \cdot \hat{s}_m$ and with similar expressions for the y- and z-components,

$$\frac{dI_m}{dS_m} = \frac{\partial I_m}{\partial x}\frac{dx}{dS_m} + \frac{\partial I_m}{\partial y}\frac{dy}{dS_m} + \frac{\partial I_m}{\partial z}\frac{dz}{dS_m} = (\nabla \cdot I_m)\cdot\hat{s}_m$$

$$= \kappa(\mathbf{r})I_b(\mathbf{r}) - \beta(\mathbf{r})I_m(\mathbf{r},\Omega_m) + \frac{\sigma_s(\mathbf{r})}{4\pi}\int_{4\pi} I(\mathbf{r},\Omega_i)\Phi(\mathbf{r},\Omega_m,\Omega_i)d\Omega_i$$

(10.14)

The in-scattering integral is then approximated according to:

$$\int_{4\pi} I(\mathbf{r},\Omega_i)\Phi(\mathbf{r},\Omega_m,\Omega_i)d\Omega_i \approx \sum_{m'=1}^{N_m} w_{m'}I_{m'}(\mathbf{r})\Phi_{m,m'}(\mathbf{r})$$

(10.15)

while the intensity at a location $\mathbf{r}_w$ on the wall is given by:

$$I(\mathbf{r}_w) = \frac{\varepsilon(\mathbf{r}_w)\sigma T^4(\mathbf{r}_w)}{\pi} + \frac{\rho(\mathbf{r}_w)}{\pi}\sum_{\substack{m'=1 \\ \hat{n}\cdot\hat{s}_{m'}<0}}^{N_m} w_{m'}I_{m'}(\mathbf{r}_w)(-\hat{n}\cdot\hat{s}_{m'})$$

(10.16)

where $\hat{n}$ is the local surface normal vector at $\mathbf{r}_w$.

The spatial derivatives in Equation (10.14) are then approximated using a finite difference or, more often, a finite volume scheme based on the conservation of radiant energy over a control volume centered on each grid point. The solution proceeds in "sweeps" over each ordinate direction in a manner similar to the "up/down" sweeps used in the 1D example. Oftentimes the radiant intensity field must be solved as part of a larger CFD simulation that also considers fluid flow, conductive and convective heat transfer, and, potentially, chemical reactions. In this context the DOM algorithm may exploit the underlying structure of a finite volume computer code, and for this reason, it is often the preferred method for CFD simulations involving radiative transfer.

10.4   SPHERICAL HARMONICS (P_N) METHODS

As noted in Section 10.3.3, to solve the RTE it is necessary to assume a general shape for the angular distribution of intensity at every grid point. In other words, the angular distribution of intensity must be *parameterized* in some manner. In DOM, the intensity is parameterized by modeling it as uniform over finite solid angles centered on a set of discrete ordinates.

An alternative approach is to approximate the intensity field as the sum of basis functions multiplied by unknown coefficients to be solved for. In the spherical harmonics method, the intensity at each position **r** is expressed as an expansion of orthogonal harmonic functions:

$$I(\mathbf{r},\Omega) = \sum_{n=0}^{\infty} \sum_{m=-n}^{n} A_n^m(\mathbf{r}) Y_n^m(\Omega) \qquad (10.17)$$

where $A_n^m(\mathbf{r})$ are position-dependent (unknown) coefficients and $Y_n^m(\Omega)$ are the basis functions, which depend on direction. The basis functions (spherical harmonics) are:

$$Y_n^m(\Omega) = \begin{cases} (-1)^m \sqrt{2} \sqrt{\dfrac{2n+1}{4\pi} \dfrac{(n-|m|)!}{(n+|m|)!}} \cos(m\phi) P_n^m(\mu), & m \geq 0 \\[4mm] (-1)^m \sqrt{2} \sqrt{\dfrac{2n+1}{4\pi} \dfrac{(n-|m|)!}{(n+|m|)!}} \sin(|m|\phi) P_n^{|m|}(\mu), & m < 0 \end{cases} \qquad (10.18)$$

where $P_n^m(\mu)$ are associated Legendre polynomials of the first kind, of degree n and order m. These basis functions have several important attributes. First, Figure 10.6 shows that the basis functions become "wavier" as the degree and order increase. This means that the basis functions of Equation (10.18) can capture more complicated "higher frequency" angular distributions for larger values of m (and hence, larger values of n). Indeed, these "spherical harmonics" basis functions provide an exact representation for any distribution as $n \rightarrow \infty$, but in practice the summation in Equation (10.17) is capped at a finite number, N. The improvement realized from even numbers of N is small relative to the next-smallest odd number, so N is always chosen as an odd number; the choice of N is the "order" of the method, and the

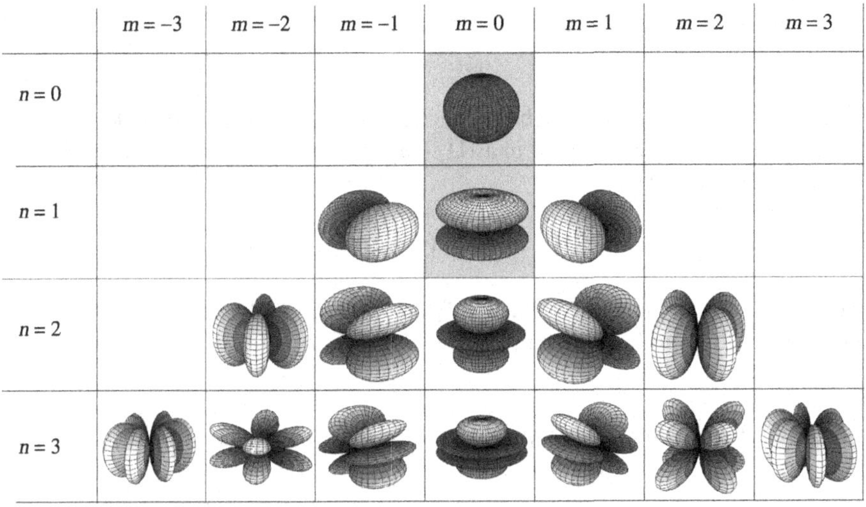

	$m=-3$	$m=-2$	$m=-1$	$m=0$	$m=1$	$m=2$	$m=3$
$n=0$							
$n=1$							
$n=2$							
$n=3$							

FIGURE 10.6 Spherical harmonics basis functions. Note that the complexity of the basis function increase with n and m. The basis functions used for the P_1 solution of a 1D plane slab have shaded backgrounds.

technique is called the "P_N" method. In practice, $N = 1$ (P_1) and $N = 3$ (P_3) are the most common choices for radiation problems. Even higher orders have been used, up to P_9 for highly forward-scattering problems (Mengüç and Viskanta 1983).

Determining the $A_n^m(\mathbf{r})$ at discrete grid points requires a system of $(N+1)^2$ differential equations in space. These equations are formed by multiplying the RTE by successive powers of direction cosines and then integrating over the spherical domain $\mu = [-1,1]$, in what is called "the method of moments." Each integral of the intensity is considered as a moment; if the integral is obtained after the intensity is multiplied by 1 (unity) and integrated, it is called the zeroth moment. If the intensity is multiplied by a direction cosine in one of the orthogonal directions and then integrated, then it is called the first moment in that direction. Physically, the first moment is the radiative flux in the corresponding direction. As expected, the accuracy of the model increases if a large number of moments (the N in P_N) are considered.

Originally, Arthur Eddington (1926) and E.A. Milne (1930)[1] developed models for radiative transfer through 1D layers of stellar atmospheres, i.e., the same problem as Schwarzschild tackled with the two-flux method. Eddington and Milne parameterized the intensity field in the same manner as did Schwarzschild, i.e., as uniform over each hemisphere, and they solved the resulting differential equations using the method of moments. The "moment" methods are directly related to the elegant spherical harmonics approximation discussed by Davison (1957), who developed it for the solution of the neutron transport equation.

Applying the P_1 method to solve the 1D RTE for plane layers (Equation (10.1)), Equation (10.17) becomes:

$$I(\tau_x,\mu) = A_0^0(\tau_x)\underbrace{Y_0^0(\Omega)}_{1} + A_1^{-1}(\tau_x)\underbrace{Y_1^{-1}(\Omega)}_{\propto\sin\phi\cos\theta} + A_1^0(\tau_x)\underbrace{Y_1^0(\Omega)}_{\propto\cos\theta} + A_1^1(\tau_x)\underbrace{Y_1^1(\Omega)}_{\propto\cos\phi\sin\theta} \quad (10.19)$$

where the coefficients from the basis functions are absorbed into the unknown coefficients. Since the intensity field in a 1D slab is independent of ϕ, it follows A_1^{-1} and A_1^1 must be zero, so:

$$I(\tau_x,\mu) = A_0^0(\tau_x) + A_1^0(\tau_x)\mu \quad (10.20)$$

The unknown coefficients, A_0^0 and A_1^0, are determined by substituting Equation (10.20) into the zeroth and first moment of the RTE,

$$\int_{-1}^{1}\mu\frac{dI}{d\tau_x}d\mu = -\int_{-1}^{1}I(\tau_x,\mu)d\mu + (1-\omega)I_b(\tau_x)\int_{-1}^{1}d\mu + \omega\bar{I}(\tau_x)\int_{-1}^{1}d\mu \quad (10.21)$$

and

$$\int_{-1}^{1}\mu^2\frac{dI}{d\tau_x}d\mu = -\int_{-1}^{1}I(\tau_x,\mu)\mu d\mu + (1-\omega)I_b(\tau_x)\int_{-1}^{1}\mu d\mu + \omega\bar{I}(\tau_x)\int_{-1}^{1}\mu d\mu \quad (10.22)$$

where the average incident intensity at any location τ_x is:

$$\bar{I}(\tau_x) = \int_{4\pi}I(\Omega_i)d\Omega_i = 2\pi\int_{-1}^{1}\left[A_0^0(\tau_x) + A_1^0(\tau_x)\mu_i\right]d\mu_i = 4\pi A_0^0(\tau_x) \quad (10.23)$$

Carrying out the integrations results in:

$$\frac{2}{3}\frac{dA_1^0}{d\tau_x} = -2A_0^0 + 2(1-\omega)I_b(\tau_x) + 8\pi\omega A_0^0(\tau_x) \quad (10.24)$$

and

$$\frac{dA_0^0}{d\tau_x} = -A_1^0(\tau_x) \quad (10.25)$$

Herein we see a second key attribute of the spherical harmonics basis functions is the fact that integrations of odd powers of μ (μ, μ^3, etc.) over the domain $\mu \in [-1, 1]$ go to zero, which simplifies the higher-order moment equations.

Solving these differential equations requires boundary conditions at $\tau_x = 0$ and $\tau_x = \tau_L$, which, unlike the S_2 method, do not have an obvious physical interpretation due to the spherical harmonics parameterization. The most common of these are Marshak's boundary conditions (Marshak 1947). For further details, consult Howell et al. (2021) or Modest and Mazumder (2021).

The P_N approximation is generally regarded as one of the most computationally-efficient ways in which to solve the RTE, due to the properties of the spherical harmonics basis functions. However, a major drawback is the complexity of the scheme, particularly for complex geometries and anisotropic scattering, where higher-order P_N approximations become necessary.

10.5   DIFFUSION APPROXIMATION

In Example 10.1 and Figure 10.3(b), we showed that the intensity may be modeled as nearly isotropic when the medium is optically thick (i.e., $\tau_L \gg 1$). This is because any point in the medium can only "see" its immediate surroundings, which are nearly at the same temperature. In this scenario, the local intensity field only depends on the local temperature and temperature gradient and is not influenced by radiant exchange from more distant parts of the medium and boundaries that may be at different temperatures. Radiative transport thus assumes a diffusive nature, much like conduction heat transfer. The radiative transfer equation may then be simplified considerably.

First, since the intensity field only depends on local conditions, it makes sense to discard the global coordinate system and instead define a local coordinate system, as shown in Figure 10.7(a). Starting from the integrated form of the RTE, Equation (9.13), transforming coordinates, and ignoring the influence of the boundary intensities, the intensities in the upward and downward directions are given by:

$$I^+\left(\tau_x,\mu\right)=\frac{1}{\mu}\int_0^{\tau_x} I\left(\tau_x-\tau_x^*\right)e^{-\tau_x^*/\mu}\,d\tau_x^*,\quad 0\le\mu\le 1 \qquad (10.26)$$

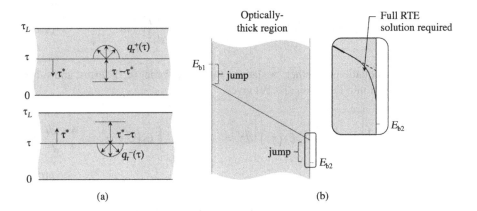

FIGURE 10.7   (a) Coordinate system transformation used to derive the diffusion approximation; (b) "jump" boundary conditions.

where the integral is carried out "downwards" from τ_x, and

$$I^-(\tau_x, \mu) = -\frac{1}{\mu} \int_0^{\tau_L - \tau_x} I(\tau_x + \tau_x^*) e^{\tau_x^*/\mu} d\tau_x^*, \quad -1 \le \mu \le 0 \qquad (10.27)$$

where the integral is carried out "upwards" from τ_x.

Multiplying these integrals by $\mu d\Omega$ and integrating over their respective hemispheres gives the radiant heat flux in the upwards and downwards directions, which are then combined into the net radiant flux:

$$q_{r,x}(\tau_x) = 2\pi \left[\int_0^{\tau_x} I(\tau_x - \tau_x^*) E_2(\tau_x^*) d\tau_x^* - \int_0^{\tau_L - \tau_x} I(\tau_x + \tau_x^*) E_2(\tau_x^*) d\tau_x^* \right] \qquad (10.28)$$

where E_2 is the exponential integral function, Equation (9.71). For optically thick media, $E_2(\tau_x^*)$ is expected to decay rapidly to a negligible value before τ_x^* reaches a boundary, so the upper bounds of both integrals may be replaced with infinity. Thus, Equation (10.28) may be rewritten as:

$$q_{r,x}(\tau_x) = 2\pi \int_0^\infty \left[I(\tau_x - \tau_x^*) - I(\tau_x + \tau_x^*) \right] E_2(\tau_x^*) d\tau_x^* \qquad (10.29)$$

For sufficiently optically thick media, the local variation of the intensity source term with respect to τ_x can be approximated by a Taylor series:

$$\hat{I}(\tau_x - \tau_x^*) = \hat{I}(\tau_x) - \tau_x^* \left. \frac{d\hat{I}}{d\tau_x} \right|_{\tau_x} + (\tau_x^*)^2 \left. \frac{d^2\hat{I}}{d\tau_x^2} \right|_{\tau_x} + \ldots \qquad (10.30)$$

$$\hat{I}(\tau_x + \tau_x^*) = \hat{I}(\tau_x) + \tau_x^* \left. \frac{d\hat{I}}{d\tau_x} \right|_{\tau_x} + (\tau_x^*)^2 \left. \frac{d^2\hat{I}}{d\tau_x^2} \right|_{\tau_x} + \ldots \qquad (10.31)$$

Adding these equations together, neglecting the higher-order $(\tau_x^*)^2$ terms, and substituting the result into Equation (10.29) gives:

$$q_{r,x}(\tau_x) = 2\pi \int_0^\infty \left[-2\tau_x^* \left. \frac{dI}{d\tau_x} \right|_{\tau_x} E_2(\tau_x^*) \right] d\tau_x^* = -4\pi \left. \frac{dI}{d\tau_x} \right|_{\tau_x} \int_0^\infty E_2(\tau_x^*) d\tau_x^* = \frac{-4\pi}{3} \left. \frac{dI}{d\tau_x} \right|_{\tau_x}$$

$$(10.32)$$

In the special case of a gray, purely absorbing and emitting media, the source term is equal to the blackbody intensity, so:

$$q_{r,x}\left(\tau_x\right)=-\frac{4\pi}{3}\frac{dI_b}{d\tau_x}=-\frac{4}{3}\frac{dE_b}{dx}\frac{dx}{d\tau_x}=-\frac{16}{3\beta}\sigma T^3\frac{dT}{dx}=-k_R\left(T\right)\frac{dT}{dx} \qquad (10.33)$$

where k_R is a temperature-dependent radiation diffusion coefficient. It can also be shown that Equation (10.33) applies to an isotropically scattering medium.

Under conditions of local radiative equilibrium, the divergence (or derivative) of $q_{r,x}$ is zero, in which case E_b should vary linearly with respect to τ_x and x.

In the case of nongray media, Equation (10.33) is written using the Rosseland mean diffusion coefficient:

$$\frac{1}{\beta_R}=\int_{\gamma=0}^{\infty}\left(\frac{1}{\beta_\lambda}\right)\frac{\partial E_{\lambda b}}{\partial E_b}d\lambda \qquad (10.34)$$

which was derived by Rosseland (1923, 1936). This treatment is appropriate if the medium is optically thick over all wavelengths that involve an appreciable radiative transfer. For example, this is the case for radiative transfer through plasmas, such as those of stellar atmospheres. It should not be used for gases and some semiconductors that have transparent "windows."

The diffusion approximation can only be applied at large optical distances from the boundaries, but, as shown in Figure 10.7(b), it fails at locations closer to the boundaries, under which conditions the assumptions used to derive Equation (10.32) no longer apply. Conceptually, this is much like walking down a sidewalk on a foggy day. If one is sufficiently far away from other objects, all one sees is fog; however, objects gradually appear out of the fog as we move closer to them, and the optically thick approximation breaks down.

Accurately modeling the intensity field close to the boundaries requires a more detailed solution of the RTE. In the absence of conduction and convection, the local thermodynamic temperature within the medium adjacent to the wall is generally not equal to the wall temperature because the temperature in the medium is influenced by everything it "sees." This discontinuity may be approximated by "jump" boundary conditions (Deissler 1964). For the upper surface of a 1D slab:

$$E_{b2}-E_{bw2}=\left(\frac{1}{\varepsilon_2}-\frac{1}{2}\right)q_{r,x}\left(\tau_L\right)-\frac{1}{2}\frac{\partial^2 E_b}{\partial\tau_x^2}\bigg|_{\tau_L} \qquad (10.35)$$

and at the lower surface:

$$E_{bw1}-E_{b1}=\left(\frac{1}{\varepsilon_1}-\frac{1}{2}\right)q_{r,x}\left(0\right)+\frac{1}{2}\frac{\partial^2 E_b}{\partial\tau_x^2}\bigg|_{0} \qquad (10.36)$$

where $E_{bw1}=\sigma T_1^4$, $E_{bw2}=\sigma T_2^4$ (the true blackbody emissive powers of the walls), and E_{b1} and E_{b2} are the "effective" emissive powers of the medium projected next to the wall. When solving diffusive radiative transfer through the medium, Equations (10.35) and (10.36) act as boundary conditions for Equation (10.32).

When conduction and convection are present along with radiation, the temperature distribution will be influenced by all three modes, and Equations (10.35) and (10.36) should not be used. It is also important to appreciate that the diffusion approximation only holds far from the boundaries, where the local intensity field depends only on the local temperature gradient.

Example 10.3

A homogeneous absorbing and scattering medium is bounded by two black surfaces at E_{bw1} and E_{bw2}, respectively. Conduction is negligible. Using the diffusion approximation, calculate the radiative flux between the plates and the temperature distribution within the medium for optical thicknesses of $\tau_L = 0.1, 0.5, 1, 2,$ and 5, and compare the result with the analytical solution. Express the answer in terms of $\psi_b = q_r/(E_{bw1}-E_{bw2})$ and $\phi_b(\tau) = [E_b(\tau)-E_{bw2}]/[E_{bw1}-E_{bw2}]$.

Since conduction is negligible, the radiation absorbed at any location within the medium must balance the emitted radiation, so the medium is in radiative equilibrium. It follows that q_r is constant, so $dq_r/d\tau = 0$ and $E_b(\tau)$ follows a linear profile. The radiant flux and temperature are found by solving Equation (10.33) with boundary conditions given by Equations (10.35) and (10.36). The solution is:

$$q_r = \frac{E_{bw1} - E_{bw2}}{1+(3/4\tau_L)}, \quad \psi_b = \frac{1}{1+(3/4\tau_L)}$$

The corresponding distribution for $E_b(\tau)$ in dimensionless form is:

$$\phi_b = \frac{(3/4)(\tau_L - \tau) + (1/2)}{1+(3/4\tau_L)}$$

The exact solution for the emissive power is found by solving the Fredholm integral equation of the second-kind:[2]

$$\phi_b(\tau) = \frac{1}{2}\left[E_2(\tau) + \int_{\tau^*=0}^{\tau_L} \phi_b(\tau^*)E_1(|\tau-\tau^*|)d\tau^*\right]$$

and the dimensionless heat flux is obtained by substituting this solution into:

$$\psi_b(\tau) = 1 - 2\int_{\tau=0}^{\tau_L} \phi_b(\tau)E_2(\tau)d\tau$$

The distributions for $\phi_b(\tau)$ obtained using the exact solution and the diffusion approximation are plotted in Figure 10.8, while the corresponding ψ_b estimates are in Table 10.3. The results show that the diffusion approximation becomes progressively more accurate as the optical thickness increases.

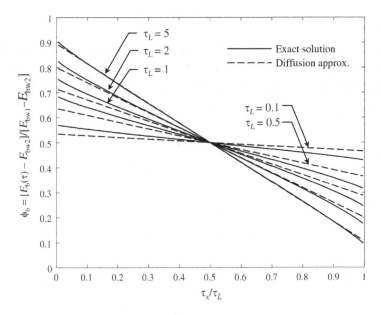

FIGURE 10.8 Dimensionless emissive power for a 1D slab with black boundaries in radiative equilibrium, comparison of the exact solution and diffusion approximation.

TABLE 10.3
Dimensionless Radiant Flux between Two Black Plates, ψ_b, Example 10.2

τ_L	Diffusion approx.	Exact solution[1]
0.1	0.9302	0.9157
0.5	0.7273	0.704
1	0.5714	0.5532
2	0.4	0.39
5	0.2105	0.2077

Example 10.4

The temperature at a location within a porous ceramic burner is 1200 K, and the temperature gradient is $dT/dx = 100$ K/m. The Rosseland mean extinction coefficient, β_R, is 5 m^{-1}, while the thermal conductivity, k, is 15 W/(m·K). Calculate the local conductive and radiative heat fluxes.

The local conduction heat flux is found from Fourier's Law, $q_{cond} = -k\partial T/\partial x = 500$ W/m^2. The radiative conductivity is given by $k_r = 16/(3\beta_R)\sigma T^3 = 105$ W/(m·K), with a corresponding radiative flux of $q_r = -k_r\partial T/\partial x = 1.05\times10^4$ W/m^2. This result shows that, at such high temperatures, radiative heat transfer far exceeds conduction heat transfer.

TABLE 10.4
Diffusion Predictions of Energy Transfer and Temperature Distribution for an Absorbing-Emitting Gray Gas with Isotropic Scattering in Radiative Equilibrium between Gray Walls and without Internal Heat Sources

Geometry	Relations[a]
Infinite parallel plates	$$\psi = \frac{1}{(3\beta D/4) + \overline{E}_1 + \overline{E}_2 + 1}$$ $$\phi(z) = \psi\left[\frac{3\beta}{4}(D-z) + \overline{E}_2 + \frac{1}{2}\right]$$
Infinitely long concentric cylinders	$$\psi = \frac{1}{\frac{3}{8}\left[\beta D_1 \ln\left(\frac{D_2}{D_1}\right) + \frac{1-(D_1/D_2)^2}{\beta D_1}\right] + \left(\overline{E}_1 + \frac{1}{2}\right) + \frac{D_1}{D_2}\left(\overline{E}_2 + \frac{1}{2}\right)}$$ $$\phi(r) = \psi\left\{-\frac{3}{8}\left[\beta D_1 \ln\left(\frac{D}{D_2}\right) + \frac{D_1}{\beta D_2^2}\right] + \left(\overline{E}_2 + \frac{1}{2}\right)\frac{D_1}{D_2}\right\}$$
Concentric spheres	$$\psi = \frac{1}{\frac{3}{8}\left[\beta D_1\left(1 - \frac{D_1}{D_2}\right) + 2\frac{1-(D_1/D_2)^3}{\beta D_1}\right] + \left(\overline{E}_1 + \frac{1}{2}\right) + \frac{D_1^2}{D_2^2}\left(\overline{E}_2 + \frac{1}{2}\right)}$$ $$\phi(r) = \psi\left\{-\frac{3}{8}\left[\beta D_1\left(\frac{D_1}{D_2} - \frac{D_1}{D}\right) + \frac{2D_1^2}{\beta D_2^3}\right] + \left(\overline{E}_2 + \frac{1}{2}\right)\frac{D_1^2}{D_2^2}\right\}$$

[a] Definitions $\overline{E}_N = (1-\varepsilon_{wN})/\varepsilon_{wN}$, $\psi = Q_1/\sigma(T_{w1}^4 - T_{w2}^4)$, $\phi(\xi) = [T^4(\xi) - T_{w2}^4]/(T_{w1}^4 - T_{w2}^4)$, $D = 2r, \beta = \kappa + \sigma_s$.
Note: ϕ is defined only if $\kappa > 0$.

For other 1D geometries, the results for radiative energy transfer and temperature distributions at radiative equilibrium are given in Table 10.4.

10.6 ZONAL METHOD

The zonal method has been used extensively for engineering problems with radiating gases. In this method, a nonisothermal enclosure filled with nonisothermal gas or other translucent medium (with $n \approx 1$ for this development) is subdivided into boundary surface areas and volumes (called "zones") that are each approximated as isothermal. Geometric exchange factors are precomputed between each pair of zones. An energy balance is written for each zone. This provides a set of simultaneous equations for the unknown energy fluxes or temperatures. The method is somewhat an extension of enclosure theory, and Hottel and Sarofim (1967) and Hottel et al. (1970) discuss it in detail. The zonal technique has largely been superseded by the other methods in this chapter. It is further developed in Howell et al. (2021).

10.7 MONTE CARLO METHODS

Transforming the integro-differential radiative transfer equations into a set of linear equations that can be solved numerically is often far from straightforward, particularly for problems involving non-diffuse surfaces and anisotropic scattering. Moreover, the assumptions and simplifications needed for this transformation introduce "model errors" into the solution, which can often be large and hard to quantify. In contrast to deterministic techniques, MC approaches (for the most part) avoid the need to simplify the governing equations. Instead, many trial solutions are generated from random processes, which, combined, are equivalent to solving the governing analytical equations. These results are then analyzed en masse to develop a solution along with an estimate of the statistical uncertainty introduced by the finite number of samples. Accordingly, the main advantages of MC over deterministic techniques are: (1) MC provides a much more accurate estimate of the quantity-of-interest, particularly for problems involving complicated phenomena, provided that enough trials are used; and (2) the uncertainty of the estimate can be calculated readily from a statistical analysis.

The MC technique was conceived at Los Alamos National Laboratory by Stanislav Ulam to model neutron scattering in fissile material. To make this calculation deterministically, one would need to integrate the effects of millions of possible paths and outcomes as neutrons propagate through the material and interact with atoms. While convalescing from a head injury, Ulam was playing the card game Solitaire, and began to ponder whether it would be possible to predict the expected outcome of the games if one were to play many hands. Ulam shared his idea with John von Neumann and Nicholas Metropolis, and collectively they realized that they could exploit recently developed, general-purpose electronic computers to carry out these calculations, specifically the Electronic Numerical Integrator and Computer (ENIAC) at the University of Pennsylvania. These scientists named this approach "the Monte Carlo method" in homage to Metropolis's uncle, an enthusiastic gambler who frequented the famous casino in Monte Carlo.

Since then, MC methods have been applied to virtually every discipline imaginable. In the context of thermal radiation, researchers at the NASA Lewis (now Glenn) Research Centre were early users of MC methods to model advanced rocket propulsion systems in the 1960s (Howell and Perlmutter 1964a, 1964b; Perlmutter and Howell 1964). MC algorithms are particularly suited to problems involving highly complex phenomena, and, consequently, the MC implementation is often derived through phenomenological arguments, as opposed to a rigorous mathematical process. In the context of thermal radiation, MC simulations are often interpreted as tracing individual "photon bundles" from their point of emission, through intermediate interactions with surfaces and the medium, until they are ultimately absorbed, although this is not strictly correct from the perspective of physics. It is important to appreciate that MC methods are really numerical tools for solving a set of governing equations, instead of a high-fidelity model for solving the problem of physics. Such a misapprehension can lead to a faulty interpretation of the physics underlying radiative transport (Howell and Daun 2021).

10.7.1 MONTE CARLO AS AN INTEGRATION TOOL

In the context of thermal radiation, Monte Carlo is mainly used to carry out integrations, e.g., over the wavelength spectrum, over a solid angle, over a surface, or over path lengths. Consider, for example, calculation of the total hemispherical absorptivity from the spectral hemispherical absorptivity. As we show in Equation (3.25), we can envision the spectral absorptivity as the probability with which an incident energy bundle having a wavelength λ will become absorbed by the surface, and the total absorptivity as the marginalization of this probability.

Suppose that a discrete number of photons, all of which share a common wavelength, are grouped into N bundles so that the bundles contain an equal amount of energy, e. (As discussed in Chapter 1, real photons behave in a far more complex way, so it is important to appreciate that this is purely a "thought model.") Bundle wavelengths are sampled randomly in a way that reflects the spectral distribution of incident irradiation, p_λ. In this case, we can show:

$$\alpha = \frac{\int_{\lambda=0}^{\infty} \alpha_\lambda G_\lambda d\lambda}{\int_{\lambda=0}^{\infty} G_\lambda d\lambda} = \int_{\lambda=0}^{\infty} \alpha_\lambda p_\lambda d\lambda \approx \frac{\sum_{k=1}^{N} \alpha_{\lambda k} e}{\sum_{k=1}^{N} e} = \frac{1}{N} \sum_{k=1}^{N} \alpha_{\lambda k} \qquad (10.37)$$

The probability of drawing a random wavelength λ_j between λ and $\lambda+\Delta\lambda$ should equal the irradiation rate contained over this interval (W/m²) divided by the total irradiation rate. Taking the limit as $\Delta\lambda \to 0$ gives:

$$P_{\lambda \to \lambda + d\lambda} = p_\lambda d\lambda = G_\lambda d\lambda \bigg/ \int_0^\infty G_\lambda d\lambda \qquad (10.38)$$

How do we generate a set of bundle wavelengths $\{\lambda_j\}$? The most efficient approach is by inverting the cumulative distribution function (CDF), defined by:

$$\text{CDF}(\lambda) = \int_0^\lambda p_{\lambda^*} d\lambda^*, \quad 0 \leq \text{CDF}(\lambda) \leq 1 \qquad (10.39)$$

A set of $\{\lambda_j\}$ may be generated by sampling a random number distributed uniformly between zero and one, $R_\lambda^k \sim U(0,1)$, and then calculating $\lambda_j = \text{CDF}^{-1}(R_\lambda^k)$. If G_λ is a blackbody distribution, it is easiest to work with the blackbody fraction $\text{CDF}(\lambda T)$. One would then generate a set of $\{(\lambda T)_j\}$ by sampling random numbers between 0 and 1 and then obtain the corresponding wavelengths for a given temperature.

To see why this is the case, imagine that the domain of λT is divided into N uniformly-spaced segments that are mapped onto N (generally non-uniformly-spaced)

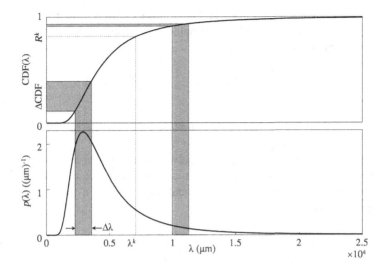

FIGURE 10.9 Sampling a PDF by inverting the CDF. The two gray bars show equally wide intervals for $\Delta\lambda$, but are mapped to different widths on the CDF axis. The probability of sampling within this interval is proportional to ΔCDF.

intervals on the CDF axis, as shown in Figure 10.9. The probability of a random draw $R_\lambda^k \sim U(0,1)$ falling onto a particular segment ΔCDF_j is given by:

$$\Delta\text{CDF}_j = \int_{(\lambda T)_j}^{(\lambda T)_j + \Delta(\lambda T)} p_{\lambda T} d(\lambda T) \tag{10.40}$$

Accordingly, the probability of a randomly-drawn value R_λ^k falling over an infinitesimal interval $[(\lambda T)_j, (\lambda T)_j + d(\lambda T)_j]$ comes from taking the derivative of Equation (10.40),

$$d\text{CDF}\left[(\lambda T)_j\right] = p_{(\lambda T)_j} d(\lambda T)_j \tag{10.41}$$

A key attribute of the MC method is that one can calculate the uncertainty of the estimate from a statistical analysis of the samples. According to the Central Limit Theorem, the uncertainty of our estimate for α due to the finite number of samples N is equal to the standard deviation of our estimates divided by the square root of the number of samples, $\sigma_\alpha/N^{1/2}$.

This procedure can be generalized to carry out integrations over multiple variables:

$$\int_{x_1} \int_{x_2} \cdots \int_{x_n} f(x_1, x_2, \ldots x_n) p(x_1, x_2, \ldots x_n) dx_1 dx_2 \ldots dx_n \approx \frac{1}{N_k} \sum_{k=1}^{N_k} f(x_1, x_2, \ldots x_n) \tag{10.42}$$

where $p(x_1, x_2,..., x_n)$ is a joint PDF for the variables of integration. Individual PDFs for each variable may be found by marginalizing, or "integrating out" the influence of the other variables from the joint PDF:

$$p(x_i) = \int_{x_1} \int_{x_2} \cdots \int_{x_{i-1}} \int_{x_{i+1}} \cdots \int_{x_n} p(x_1, x_2,... x_n) dx_1 dx_2 ... dx_{i-1} dx_{i+1} ... dx_n \quad (10.43)$$

Each marginalized PDF is then used to draw samples for the variables, ideally by inverting the corresponding CDF.

Monte Carlo is particularly useful for solving integrals that involve complicated integrands or integral domains. For example, consider the case of calculating the configuration factor between an infinitesimal area dA_1 and a finite area A_2. This amounts to solving:

$$F_{d1-2} = \frac{dQ_{d1-2}}{dQ_{d1}} = \frac{1}{\pi} \int_{\Omega_{2-d1}} \cos\theta d\Omega \quad (10.44)$$

where Ω_{2-d1} denotes the solid angle subtended by A_2 when viewed from dA_1. The difficulty in carrying out this integral lies in the complexity of the integration domain for θ and ϕ. Consequently, the integral is usually recast in terms of A_2 via Equation (5.11), but this can also be difficult to carry out for all but the simplest of geometries.

Instead, we modify the integrand by adding an "intercept factor."

$$b(\Omega) = \begin{cases} 1, & \Omega \in \Omega_{2-d1} \\ 0, & \text{otherwise} \end{cases} \quad (10.45)$$

which may be determined by ray tracing. Equation (10.44) can then be written:

$$F_{d1-2} = \frac{1}{\pi} \int_{2\pi} b(\Omega) \cos\theta d\Omega = \frac{1}{\pi} \int_0^{2\pi} \int_0^{\pi/2} b(\Omega) \cos\theta \sin\theta d\theta d\phi \quad (10.46)$$

where $p(\theta,\phi) = \cos\theta \cdot \sin\theta /\pi$. The PDFs for θ and ϕ are then found by marginalizing out the other variable:

$$p(\theta) = \int_0^{2\pi} \frac{\cos\theta \sin\theta}{\pi} d\phi = 2\cos\theta\sin\theta, \quad p(\phi) = \int_0^{\pi/2} \frac{\cos\theta\sin\theta}{\pi} d\theta = \frac{1}{2\pi} \quad (10.47)$$

Note that the PDF for ϕ is uniform, meaning that all azimuthal directions are equally probable, while the PDF for θ is zero at $\theta = 0$ and $\theta = \pi/2$, and a maximum value of $\theta = \pi/4$. This is a consequence of the fact that the emitting solid angle depends on $\sin\theta$, and the view factor also contains a $\cos\theta$ to account for the projected area of dA_1. The CDFs for these variables are found by integration, and may be used to generate random values of θ and ϕ:

TABLE 10.5
Monte Carlo Sampling Relations for Radiative Transfer between Surfaces and within Participating Media

Phenomena	Variables	Relations	Conditions
Emission from a diffuse surface.	Cone angle, θ_e Circumferential angle, ϕ_e	$\sin\theta_e = R_{\theta_e}^{1/2}$ $\phi_e = 2\pi R_{\phi_e}$	
Emission within a volume element with absorption coefficient κ.	Cone angle, $\mu_e = \cos\theta_e$ Circumferential angle, ϕ_e	$\mu_e = 2R_{\mu_e} - 1$ $\phi_e = 2\pi R_{\phi_e}$	
Attenuation by medium with extinction coefficient β.	Path length, S	$S = -(1/\beta)\ln R_s$	Gray, uniform medium
Scattering in medium with phase function Φ	Cone angle, $\mu_s = \cos\theta_s$ Circumferential angle, ϕ_s	$\mu_s = 2R_{\mu_s} - 1$ $\phi_s = 2\pi R_{\phi_s}$	Isotropic scattering

$$\text{CDF}(\theta) = \int_0^\theta 2\cos\theta^* \sin\theta^* d\theta^* = \sin^2\theta \quad \text{CDF}(\phi) = \int_0^\phi \frac{1}{2\pi} d\phi^* = \frac{\phi}{2\pi} \qquad (10.48)$$

Finally, the integral in Equation (10.46) is approximated by:

$$F_{d1-2} = \int_{2\pi} b(\Omega)\cos\theta d\Omega \approx \frac{1}{N_1} \sum_{k=1}^{N_1} b(\Omega_k) = \frac{1}{N_1} \sum_{k=1}^{N_1} b(\Omega_k) = \frac{N_{1-2}}{N_1} \qquad (10.49)$$

where the set of directions $\{\Omega^k\}$ are drawn from $(\sin\theta)^k = (R^k_{\sin\theta})^{1/2}$ and $\phi^k = 2\pi R_\phi^k$, and N_{1-2} are the number of bundles emitted by dA_1 that are intercepted by A_2. Ray tracing typically requires the sine or cosine of the angle and not the angle itself, so more often $(\sin\theta)^k$ is sampled and then $(\cos\theta)^k = [1-(\sin\theta^k)^2]^{1/2}$ is computed to avoid the extra computational expense of evaluating the inverse sine of $(R_{\sin\theta})^{1/2}$. The relations for sampling random circumferential and cone angles are summarized in Table 10.5.

The uncertainty of the integral may be obtained from a statistical analysis of the trials. Since evaluating the integral amounts to a counting process of discrete events, the number of samples usually obeys a Poisson distribution, so the variance of N_{1-2} equals N_{1-2} itself, and the standard deviation of the mean varies with $N_{1-2}^{1/2}$. Therefore, F_{d1-2} can be approximated as normally-distributed with a mean μ equal to N_{1-2}/N_1 and a standard deviation of $\sigma_\mu = N_{1-2}^{1/2}/N_1$. This result can be used to find an uncertainty interval, e.g., F_{d1-2} is contained within $\mu \pm 2\sigma_\mu$ with 95% probability.

Example 10.5

Consider two horizontally opposed, coaxial disks shown in Figure 10.10. The disks are $R = 0.5$ m in radius and separated by a distance $H = 1$ m. Estimate the configuration factor F_{1-2} by MC, along with the associated uncertainty.

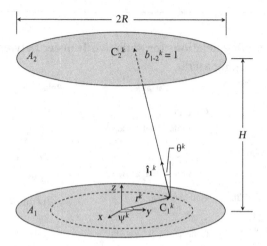

FIGURE 10.10 Geometry used to calculate the configuration factor between two horizontally opposed coaxial disks, Example 10.5.

Any location on the lower disk is specified by a radius $r_1 \in [0,R]$ and angle $\psi_1 \in [0,2\pi]$ so that $dA_1 = r_1 dr_1 d\psi_1$. The configuration factor integral is then written as:

$$F_{1-2} = \frac{1}{\pi R^2} \underbrace{\int_0^{2\pi} \int_0^R \int_0^{2\pi} \int_0^{\pi/2} b_{1-2}(\theta_1,\phi_1,r_1,\psi_1) p(\theta_1) d\theta_1 p(\phi_1) d\phi_1}_{F_{d1-2}(r_1,\psi_1)} r_1 dr_1 d\psi_1$$

where $p(\theta_1)$ and $p(\phi_1)$ are defined from Equation (10.48). Since the variables are independent, probability density functions for r_1 and ψ_1 can be found by noting that $p(r_1) \propto r_1$ and $p(\psi_1) \propto 1$. To satisfy the Law of Total Probability:

$$p(r_1) = \frac{2r_1}{R^2}, \quad \text{CDF}(r_1) = \frac{r_1^2}{R^2}, \quad r_1^k = \text{CDF}^{-1}(R_r^k) = R\sqrt{R_r^k}$$

and

$$p(\psi_1) = \frac{1}{2\pi}, \quad \text{CDF}(\psi_1) = \frac{\psi_1}{2\pi}, \quad \psi_1^k = \text{CDF}^{-1}(R_\psi^k) = 2\pi R_\psi^k$$

where R_r^k and ψ^k are uniformly-distributed random iterates. The integral becomes:

$$F_{1-2} = \int_0^{2\pi} \int_0^R \int_0^{2\pi} \int_0^{\pi/2} b_{1-2}(\theta_1,\phi_1,r_1,\psi_1) p(\theta_1) d\theta_1\, p(\phi_1) d\phi_1\, p(r_1) dr_1\, p(\psi_1) d\psi_1 \approx \frac{1}{N_1} \sum_k^{N_1} b_{1-2}(\theta_1^k,\phi_1^k,r_1^k,\psi_1^k) = \frac{N_{1-2}}{N_1}$$

where N_{1-2} is the total number of rays emitted from A_1 that intercept A_2. Once an initial emission location (r_1^k,ψ_1^k) and direction (θ_1^k,ϕ_1^k) are sampled, the intercept factor b_{1-2} can be calculated by projecting the ray into the plane of the second

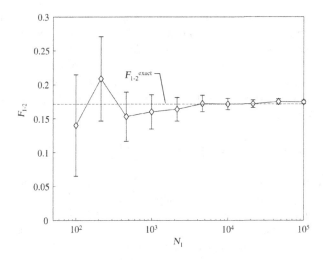

FIGURE 10.11 Monte Carlo estimates of F_{1-2} for Example 10.5, along with 90% uncertainty intervals.

disk and comparing the radial distance of the intercept point to the disk radius. If the emission location on A_1 is given by vector $\mathbf{C}_1 = [r_1{}^k\cos_1{}^k), r_1{}^k\sin\theta_1{}^k), 0]^T$ in the direction $\hat{\imath}_1 = [\sin(\theta_1{}^k)\cos(\phi_1{}^k), \sin(\theta_1{}^k)\sin(\phi_1{}^k), \cos(\theta_1{}^k)]^T$, the ray must travel a distance $H/\cos(\theta_1{}^k)$ to intercept the plane $z = H$. Accordingly, $\mathbf{C}_2 = [C_{1,x} + \cdot H\tan(\theta_1{}^k)\cos(\phi_1{}^k), C_{1,y} + \cdot H\tan(\theta_1{}^k)\sin(\phi_1{}^k), H/\cos(\theta_1{}^k]^T$, and

$$b_{1-2}\left(\theta_1^k,\phi_1^k,r_1^k,\psi_1^k\right)=\begin{cases}1, & \left[C_{1,x}+H\tan\left(\theta_1^k\right)\cos\left(\phi_1^k\right)\right]^2 +\left[C_{1,y}+H\tan\left(\theta_1^k\right)\sin\left(\phi_1^k\right)\right]^2 \le R^2 \\ 0, & otherwise\end{cases}$$

Figure 10.11 shows MC estimates for F_{1-2} compared to the analytical solution. Since N_{1-2} is a counting process, it obeys a Poisson distribution and has a variance equal to N_{1-2}. Accordingly, a $\pm 90\%$ confidence interval is found from $N_{1-2}/N_1\pm2\ (N_{1-2})^{1/2}/N_1$ which contains the analytical solution to F_{1-2} with 90% probability.

10.7.2 Radiation between Nonblack Surfaces

Monte Carlo can also be used to calculate radiative transfer between nonblack surfaces. As was the case in Example 10.5, this calculation involves an integration over four dimensions: two dimensions specify a location over the emitting surface (e.g., x_e and y_e), while two other dimensions specify the direction of the emitted energy packet, (θ_e,ϕ_e). In this scenario, however,

$$Q_{i-j} = Q_i\iint_{A_i\ 2\pi} b_{i-j}\left(x_e,y_e,\theta_e,\phi_e\right)p\left(\theta_e\right)p\left(\phi_e\right)p\left(x_e\right)p\left(y_e\right)d\theta_e d\phi_e dx_e dy_e \quad (10.50)$$

where b_{i-j} expresses the probability that an energy bundle emitted from the ith surface at (x_e, y_e) in the direction (θ_e, ϕ_e) is ultimately absorbed by the jth surface. This probability depends on the energy bundle undergoing an indeterminate number of interactions with other intervening surfaces. An important property of this process is that each intermediate interaction, or "event," depends only on the preceding event—the energy bundle carries no "memory" beyond its current state. These attributes define a *Markov Chain*:

$$P\left(\mathbf{x}^k \middle| \mathbf{x}^0\right) = P\left(\mathbf{x}^k \middle| \mathbf{x}^{k-1}\right) P\left(\mathbf{x}^{k-1} \middle| \mathbf{x}^{k-2}\right) P\left(\mathbf{x}^{k-2} \middle| \mathbf{x}^{k-3}\right) \cdots P\left(\mathbf{x}^1 \middle| \mathbf{x}^0\right) \quad (10.51)$$

where $\mathbf{x}$ specifies the location and direction of the energy bundle and each conditional probability describes what may happen to the energy bundle after it intercepts an intervening surface at a given location and incident angle (e.g., is it absorbed or reflected? If it is reflected, in which direction is it reflected?). These probabilities may be derived from the radiative properties of the intervening surface, as described in Chapter 3. Calculating the individual probabilities directly in Equation (10.51) would require marginalizing over every conceivable outcome, which would quickly become computationally intractable. Instead, we use MC to approximate these probabilities by launching many particles and tracking their trajectories through multiple random events until they are ultimately absorbed. This is the *Markov Chain Monte Carlo* (MCMC) procedure.

Consider a system of N_s diffuse-gray surfaces that completely envelop a transparent medium. In this case, the emission locations and directions are sampled in the same way as they are for calculating view factors. The bundles are then traced to an intervening surface, surface m, where they may be either absorbed or reflected, with a probability equal to the absorptivity of the intercepting surface. To determine if a bundle is absorbed, we sample a uniform random iterate, α. If $\alpha < \alpha_m$, the bundle is absorbed by surface m, and the ray tracing is terminated. Otherwise, a new reflected direction is sampled using the relations in Equation (10.48), defined relative to the local surface normal vector at the intercept location. Each one of these events amounts to sampling a probability from a step in the Markov Chain.

If N_i bundles are emitted by the ith surface, and a counter N_{i-j} is incremented each time one of these bundles is absorbed by the jth surface, the expected value of b_{i-j} is given by $E(b_{i-j}) \approx N_{i-j}/N_i$. By carrying out a similar procedure for the jth surface, the net radiative transfer between the ith and jth surface is found by:

$$Q_{i-j} \approx \frac{N_{i-j}}{N_i} \varepsilon_i A_i E_{b,i} - \frac{N_{j-i}}{N_j} \varepsilon_j A_j E_{b,j} \quad (10.52)$$
$$\text{net}$$

The above approach focuses on calculating radiative transfer between two surfaces. When analyzing an enclosure of N_{surf} surfaces, the objective is usually to calculate the net radiative transfer between all surface combinations. In the "bundle energy"[3] approach, the total number of bundles used in the simulation, N_b, is defined based on the desired accuracy or computational budget. The "energy" per bundle is then calculated by:

$$e = \frac{1}{N_b} \sum_{i=1}^{N_s} \varepsilon_i A_i E_{b,i} \tag{10.53}$$

where the summation term represents the total rate of emission by all surfaces (W). The number of bundles emitted per surface is then calculated according to $N_{i,\text{emit}} = \varepsilon_i A_i E_{b,i}/e$. The bundles are emitted from each surface and ray-traced until they are ultimately absorbed by another surface, at which point the corresponding counter, N_{i-j}, is incremented. Once all the bundles have been emitted and absorbed, the net radiation between two pairs of surfaces is calculated by:

$$\underset{\text{net}}{Q_{i-j}} \approx e\left(N_{i-j} - N_{j-i}\right) \tag{10.54}$$

which is equivalent to Equation (10.52) for two surfaces. The net rate at which a surface emits radiation is simply:

$$Q_i \approx e\left(N_{i,\text{emit}} - N_{i,\text{abs}}\right) \tag{10.55}$$

where $N_{i,\text{abs}}$ is the number of bundles absorbed by the ith surface element. An estimate for the surface heat flux may then be obtained by dividing Q_i by the area of the element. Uncertainty estimates for these quantities may be derived from a statistical analysis of the results as described in Section 10.7.1. An additional discretization error may be introduced by treating the temperature and radiative properties over each element as uniform.

This procedure can also be used to estimate the *exchange factor* between pairs of surfaces. The exchange factor $\mathfrak{F}_{i-j}$ defines the fraction of radiation leaving the ith surface that is absorbed by the jth surface relative to the net emitted radiation by the ith surface were it a blackbody. The exchange factor is analogous to the view factor, except: (1) it accounts for the finite probabilities of emission and absorption by the ith and jth surfaces, and (2) it considers radiative transfer through both direct and indirect paths (i.e., reflection from intervening surfaces).

$$\mathfrak{F}_{i-j} = \varepsilon_i \lim_{N_i \to \infty} \frac{N_{i-j}}{N_i} = P\left(\text{emit}_i\right)P\left(\text{abs}_j \mid \text{emit}_i\right) \approx \varepsilon_i \frac{N_{i-j}}{N_i} \tag{10.56}$$

Exchange factors must satisfy a version of the summation rule for view factors, since:

$$Q_{i,\text{emit}} = \varepsilon_i A_i E_{b,i} = \sum_{j=1}^{N_s} A_i \mathfrak{F}_{i-j} E_{b,i} \to \sum_{j=1}^{N_s} \mathfrak{F}_{i-j} = \varepsilon_i \tag{10.57}$$

It follows that this equation is satisfied exactly, since:

$$\sum_{j=1}^{N_s} N_{i-j} = N_i \qquad (10.58)$$

The net rate of radiative transfer between the *ith* and *jth* surfaces is given by:

$$\underset{\text{net}}{Q_{i-j}} = A_i \mathfrak{F}_{i-j} E_{b,i} - A_j \mathfrak{F}_{j-i} E_{b,j} \qquad (10.59)$$

If we consider the special case of two surfaces at the same temperature, the Clausius statement of the Second Law of Thermodynamics imposes a reciprocity relationship between the exchange factors analogous to the one for view factors, $A_i\mathfrak{F}_{i-j} = A_j\mathfrak{F}_{j-i}$. However, this relationship is not satisfied exactly for MC estimates of exchange factors (or view factors) due to the finite number of bundles used in the estimate. Several techniques have been proposed to improve the accuracy of MC-estimated exchange factors by finding the smallest correction that satisfies both reciprocity and summation rules (Daun, Howell, and Morton 2005; Dupoirieux et al. 2006; Zhang et al. 2012).

10.7.3 RADIATION WITHIN ABSORBING, EMITTING, AND SCATTERING MEDIA

The above approach can be readily extended to analyze enclosures containing a participating medium. The participating medium is most often discretized into N_v finite volumes, each of which is assumed to be homogenous. The procedure for determining the number of emitting bundles, Equation (10.53), must be modified to account for the additional energy emitted by the medium. According to Equation (2.60), each infinitesimal volume element emits radiation at the rate of $4\kappa E_b dV$. It follows, then, that the energy per bundle is given by:

$$e = \frac{1}{N_b}\left[\sum_{i=1}^{N_s} \varepsilon_i A_i E_{b,i} + 4\sum_{j=1}^{N_v} \kappa_j V_j E_{b,j}\right] \qquad (10.60)$$

Bundles are then emitted from surface and volume elements: the number of bundles emitted per unit time by the *j*th volume element is given by $N_{j,\text{emit}} = 4\kappa_j V_j E_{b,j}/e$.

Bundle emission locations and directions are sampled from appropriate PDFs. In the case of a parallelepiped volume element centered at $[x_j,y_j,z_j]^T$ and with sides $\Delta x, \Delta y$, and Δz, energy is emitted at a rate of:

$$Q_{\text{emit},j} = \iint_{V\ 4\pi} \kappa_j I_{b,j} d\Omega dV = \kappa_j I_b(T_j) \int_{z_j-\Delta x/2}^{z_j+\Delta z/2} \int_{y_j-\Delta y/2}^{y_j+\Delta y/2} \int_{x_j-\Delta x/2}^{x_j+\Delta x/2} \int_0^{2\pi} \int_0^{\pi} \sin\theta d\theta d\phi dx dy dz \qquad (10.61)$$

$$= 4\pi\kappa_j I_{b,j} \Delta x \Delta y \Delta z$$

so the joint PDF for bundle emission direction and location is given by:

$$\underbrace{\int_{z_j-\Delta z/2}^{z_j+\Delta z/2} \int_{y_j-\Delta y/2}^{y_j+\Delta y/2} \int_{x_j-\Delta x/2}^{x_j+\Delta x/2} \int_0^{2\pi} \int_0^{\pi} \frac{\sin\theta}{4\pi\Delta x\Delta y\Delta z} d\theta d\phi dx dy dz}_{p(x_e,y_e,z_e,\theta_e,\phi_e)} = 1 \qquad (10.62)$$

(Note that the integral domain for θ ranges from 0 to π, as opposed to 0 to $\pi/2$ for surface elements.) Using the marginalization technique described in Section 10.7.1, the PDF for the x coordinate emission location, x_e, is $1/\Delta x$, and the CDF is:

$$CDF(x_e) = \frac{1}{\Delta x} \int_{x_j-\Delta x/2}^{x_e} dx^* = \frac{x_e - x_j}{\Delta x} + \frac{1}{2} \qquad (10.63)$$

This CDF may be inverted to obtain:

$$x_e^k = x_j - \Delta x\left(R_x^k - 1/2\right) \qquad (10.64)$$

with similar relations for y_e^k and z_e^k. Likewise, the marginalized PDF for θ_e is $1/2\sin(\theta_e)d\,\theta_e$, so:

$$CDF(\theta_e) = \frac{1}{2}\int_0^{\theta} \sin\theta_e^* d\theta_e^* = \frac{1-\cos\theta_e}{2} \qquad (10.65)$$

which may be inverted to obtain:

$$\cos\theta_e^k = 2R_{\cos\theta}^k - 1 \qquad (10.66)$$

Finally, the PDF for ϕ_e is $1/\pi$ (as it is for emission from a surface element) so $\phi_e^k = 2\pi R_\phi^k$. Note that all emission directions are equally probable, but this does not mean that all values of θ_e are equally probable; values of θ_e close to $\pi/2$ are sampled more frequently, since an infinitesimal arc length $d\theta_e$ carves out a larger $d\Omega_e$ near this value.

Once an emission location and direction have been sampled for each bundle emitted from a surface or volume element, the bundle is ray-traced until it intercepts a surface, or it is absorbed or scattered within the medium. The probability that a bundle is either absorbed or scattered after traveling a given distance S through the medium (not necessarily in a straight line) can be intuited from the Bouguer law,

$$P_{\text{ext} \atop 0\to S} = 1 - \frac{I(S)}{I_0} = 1 - \exp\left(-\int_0^S \beta(S^*)dS^*\right) = 1 - \exp\left[-\tau(S)\right] = CDF\left[\tau(S)\right] \quad (10.67)$$

since the ratio $I(S)/I_0$ defines the fraction of bundles that persist along the path over a distance S. Accordingly, each time a bundle is emitted, an optical distance τ^k may be sampled according to:

$$\tau^k = -\ln\left(R_\tau^k\right) \tag{10.68}$$

In the case of a homogeneous medium, the optical thickness may be transformed directly into a path length, $S^k = \tau^k/\beta$, while, in the case of a nonuniform medium discretized into homogenous elements, a piecewise path is summed as the bundle traverses different elements until the total optical pathlength traveled equals τ^k. The bundle may be absorbed or reflected by an intercepting surface as defined in Section 10.7.2.

Once a bundle has traveled an optical distance τ^k, assuming it has not been absorbed by an intervening surface, it is either absorbed or scattered by the medium, with the odds of scattering equal to the local albedo, $\omega = \sigma_s/\beta$. A uniform random iterate, R_ω, is drawn. If $R_\omega > \omega$, the bundle is absorbed, and the bundled energy is added to a counter attached to the absorbing volume element. Otherwise, the bundle is scattered, and a new scattering direction is sampled according to the scattering phase function; in the case of isotropic scattering, all scattering directions are equally probable, so the scattered direction is sampled in the same manner as the volumetric emission direction. The process continues until the bundle is ultimately absorbed by a surface element or a volume element, at which point a new bundle is emitted.

Once all the bundles have been emitted and absorbed, the net radiant energy emitted by each surface and volume element is found by multiplying the difference between the bundles emitted and absorbed by the element by the energy per bundle, e. Uncertainty estimates may also be derived by noting that the variance of each estimate equals the number of absorbed bundles, although, again, an additional error is introduced in cases where the medium temperature and radiative properties are approximated as piecewise constant. A more exact analysis may be employed that accounts for nonuniform properties and temperatures by accounting for these effects in the PDFs used to sample bundle emission locations, and it is even possible to employ a "meshless" approach that avoids the need to consider the mesh altogether (Galtier et al. 2013).

Example 10.6

A homogeneous absorbing and isotropically scattering medium is contained within two semi-infinite diffuse plates at 1000 K (lower) and 500 K (upper), both having an emissivity of 0.5. The medium is at radiative equilibrium and has a scattering albedo of 0.5. Calculate the energy transfer and temperature distribution within the medium for various optical thicknesses. Compare the MC simulation results with those found using the diffusion approximation.

The problem is solved in terms of the dimensionless variables introduced in Example 10.3, in which the emissive power is expressed as $\pi_b(\tau) = [E_b(\tau)-E_{bw2}]/[E_{bw1}-E_{bw2}]$, where E_{bw1} is taken to be the hotter of the two surfaces. Therefore, the dimensionless emissive power on the lower and upper surfaces are 1 and 0, respectively, and bundles are only emitted from the bottom plate.

The volume is discretized into N_v elements. If N_1 bundles are emitted from surface 1, each bundle has an "energy" of $e = \varepsilon_1 A/N_1$. (Note the units are m².) Because of symmetry, the circumferential direction does not matter, and only the cosine of the cone angle, μ, needs to be considered. Since $\sin^2\theta^k = 1-\cos^2\theta^k = R_\theta$ and $1 - R_\theta$ is also a uniformly-distributed random number between 0 and 1, it follows that $\pi^k = -(R_\mu^k)^{1/2}$ for the top (downwards facing) surface and $\mu^k = (R_\mu^k)^{1/2}$ for the bottom (upwards facing) direction. Bundles are ray-traced through absorption, emission, and scattering events using an MCMC procedure. Each time a bundle is emitted, a path length is sampled using Equation (10.68).

Once a bundle has traveled its path length without being absorbed by a boundary, it is either scattered or absorbed in the medium depending on whether $R_\omega < \omega$. If the bundle is absorbed by the jth volume element, a counter is incremented,

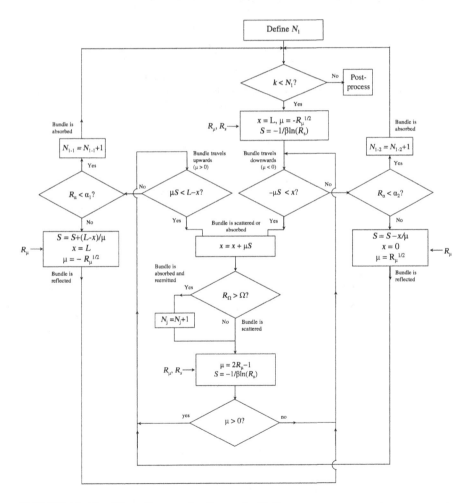

FIGURE 10.12 Flow diagram for Example 10.6.

and, because the medium is at radiative equilibrium, it is immediately reemitted. Each time a bundle is scattered or reemitted within the medium a new direction is sampled from $\mu^k = 1 - 2R_\mu^k$ and a new path length is drawn from Equation (10.68). The bundle is terminated once it is absorbed by the boundary. The flow chart for this procedure is shown in Figure 10.12.

The net rate at which energy is emitted by surfaces 1 and 2 are $Q_1 = e \cdot (N_1 - N_{1-2})$ and $Q_2 = e \cdot (-N_{1-2})$. The corresponding dimensionless radiant fluxes, $\psi_b = q_r / (E_{bw1} - E_{bw2})$, are:

$$\psi_1 \approx \varepsilon_1 \left(1 - N_{1-2}/N_1\right), \quad \psi_2 \approx \varepsilon_1 N_{1-2}/N_1$$

Since the medium is in radiative equilibrium, the dimensionless fluxes must be equal and opposite for energy to be conserved.

The radiation emitted by each volume element, in terms of the dimensionless emissive power, is $4\Delta x \phi_{b,j}$, which equals eN_j. Substituting $e = \varepsilon_1 A/N_1$ results in:

$$\phi_{b,j} \approx \frac{\varepsilon_1 N_j}{N_1 4 \kappa \Delta x} = \frac{\varepsilon_1 N_j}{N_1 4 (1 - \omega) \Delta \tau}$$

The corresponding diffusion approximation is found using Deissler's jump boundary condition. Since the medium is at radiative equilibrium, $dE_b/d\tau = 0$, and since $\varepsilon_1 = \varepsilon_2 = 0.5$, Equations (10.35), (10.36), and (10.33) simplify to:

$$\phi_{b1} = \frac{2\tau_L + 3}{2\tau_L + 6}, \quad \phi_{b2} = \frac{3}{2\tau_L + 6}, \quad \psi = \frac{1}{(\tau_L + 3)}$$

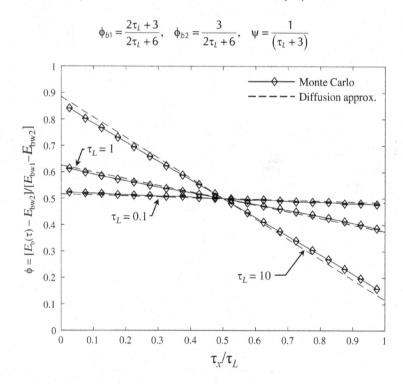

FIGURE 10.13 Dimensionless emissive power profiles for Example 10.6.

TABLE 10.6

Dimensionless Radiant Flux between Two Diffuse Grey Plates, ψ, Example 10.6

τ_L	Monte Carlo	Diffusion Approximation
0.1	$0.3234 \pm 1 \times 10^{-4}$	0.3226
0.5	$0.2925 \pm 1 \times 10^{-4}$	0.2857
1	$0.2626 \pm 1 \times 10^{-4}$	0.25
5	$0.1466 \pm 1 \times 10^{-4}$	0.125
10	$0.0946 \pm 1 \times 10^{-4}$	0.0769

Uncertainty estimates correspond to 95% probability.

The MC solution is found by discretizing the slab into 20 elements and emitting a total of 2×10^7 bundles. The dimensionless emissive power profile obtained using the MC and diffusion approximation are plotted in Figure 10.13, while the heat flux is shown in Table 10.6. Because the medium is in radiative equilibrium, the result is independent of scattering albedo.

10.8   CONCLUSIONS

The complexity of the equations that govern radiative transfer means that numerical techniques must be deployed for all but the simplest problems. Some of these techniques predate the advent of the general-purpose electronic computers. Early 1D applications focused on radiative transfer through stellar atmospheres and cloud layers. In the present day, radiative transfer through a 1D slab of absorbing and emitting media is important for applications that include atmospheric modeling, thermal barrier coatings, and exposure of human tissue to radiation and laser beams.

Numerical solution schemes may be categorized as deterministic techniques, which transform the radiative transfer equation into a system of linear equations, and Monte Carlo methods, which approximate the solution of the RTE through many statistical trials. Choosing the right numerical technique depends on the nature of the underlying problem, the available computational resources, and the required level of accuracy. Broadly speaking, deterministic schemes, like DOM (S_N) and the spherical harmonics (P_N) approximation, are more efficient and can be more easily accommodated into the architecture of existing linear solvers, while MC techniques can be highly computationally expensive due to the large number of draws required to obtain statistically-significant solutions. The advantage that deterministic schemes hold over MC diminishes as the problem becomes more complex, both in terms of geometry as well as spatially, directionally, and spectrally varying properties. For example, the application of S_N or P_N schemes to problems involving specular (mirror-like) surfaces can be very difficult, whereas in MC it is easy to incorporate such reflections into the ray-tracing algorithm.

Recent advances in high-performance computing (e.g., massively parallel architecture, GPU clusters), combined with the development of innovative algorithms for reducing variance and improving computational efficiency, have changed this picture and opened new horizons for the MC technique. Given the promise of quantum computing and further developments in MC techniques, this trend is sure to continue.

HOMEWORK (additional problems available in online Appendix I at www.ThermalRadiation.net/appendix.html)

10.1   A gray, absorbing and isotropic scattering gas is contained between large gray parallel plates, both of which have an emissivity of 0.3. Plate 1 is maintained at a temperature of $T_1 = 1150$ K, and plate 2 is at $T_2 = 525$ K. The gray medium between the plates has a uniform extinction coefficient of $\beta = 0.75$ m^{-1} and heat conduction is negligible. Using the Schuster–Schwarzschild method, predict the net radiative heat flux between the surfaces (W/m^2) and plot the dimensionless temperature distribution in the gas. Compare your results with the solution using the first-order diffusion method.

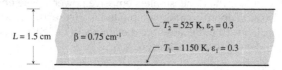

Answers: Diffusion: $q = 14,750$ W/m^2, $\Phi(\tau) = 0.5648–0.1152(\tau)$. Schuster–Schwarzschild: 13,970 W/m^2.

10.2   A fusion reactor core is contained within a spherical shell of radius R_1 blanketed by a concentric layer of gray absorbing and isotropically scattering gas of thickness L so that the outer radius is $R_2 = R_1 + L$. The spherical surfaces that contain the gray–gas blanket are black. The outer spherical surface is at T_{w2}, and the gas has extinction coefficient $\beta = \kappa + \sigma_s$. If the reactor generates power Q, find expressions for the inner surface temperature T_{w1} using the diffusion methods (conduction is neglected). Plot the ratio $A_1(E_{b1}–E_{b2})/Q = 1/\psi$ as a function of $\tau = \beta L$ and R_1/R_2.

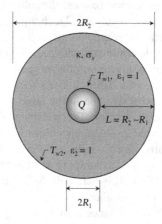

$$Answer:\ T_{w1}^4 = T_{w2}^4 + \frac{Q}{\pi D_1^2 \sigma}\left(\frac{3}{4}\left\{\beta L\frac{D_1}{D_2} + \frac{\left[1-\left(\frac{D_1}{D_2}\right)^3\right]}{\beta D_1}\right\} + \frac{1}{2}\left[1+\left(\frac{D_1}{D_2}\right)^2\right]\right)$$

10.3 The hemispherical spectral emissivity of a surface is approximated by $\varepsilon_\lambda = 0.15\lambda^3$ for $\lambda \le 5$ μm. Derive the inverse CDF used to sample wavelengths from the surface.

$$\left[\text{Hint: } \int \frac{dx}{e^x - 1} = -x + \ln\left(e^x - 1\right) = \ln\left(1 - e^{-x}\right)\right]$$

$$Answer:\ \lambda = \frac{-C_2}{T}\frac{1}{\ln\left\{1-\left[1-\exp\left(-\frac{C_2}{5T}\right)\right]^R\right\}}$$

10.4 An area element dA_1 has directional emissivity given by $\varepsilon_1(\theta) = 0.659\cos\theta$. For the geometry pictured, find the fraction of energy leaving dA_1 that is absorbed by the black disk A_2. Use the Monte Carlo method. Compare your result with the analytical solution.

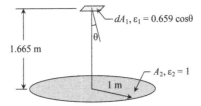

Answer: $F_{d1-2} = 0.37$

10.5 Implement the solution for Example 10.6 for $\tau_L = 1$ but vary the scattering albedo in the medium between zero and unity, using 10^5 bundles and 20 volume elements. Provide an uncertainty estimate for your results. How does changing the scattering albedo impact the uncertainty?

10.6 Derive the CDFs for the scattering angle for: (a) isotropic scattering; (b) the Rayleigh phase function, $\Phi(\theta) = (3/4)(1 + \cos 2\theta)$; and (c) the Heyney–Greenstein phase function, Equation (8.58).
Answers: (a) $\mu = (1-2R)$; (b) Not invertible;

$$(c)\ \mu = \frac{1}{2g}\left[1 + g^2 - \left(\frac{1-g^2}{1+g-2gR}\right)^2\right]$$

10.7 A gray, nonscattering gas with constant absorption coefficient κ is within a rectangular enclosure of finite width L, height D, and infinite length normal to the cross-section shown. All the boundaries are black. The lower and upper boundary temperatures are T_1 and T_2, the gas is at a constant temperature T_g, and the side boundaries are at zero temperature as shown. Construct a Monte Carlo flowchart to obtain the radiative energy transfer to the upper boundary at T_2.

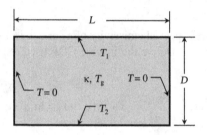

NOTES

1. See Chandrasekhar, S.: The 1979 Milne Lecture—Edward Arthur Milne: His part in the development of modern astrophysics, *Q. Jl. R. Astr. Soc.*, 21, 93–107, 1980.
2. For details, see Equation (12.72) in Howell et al., *Thermal Radiation Heat Transfer*, 7th Ed., 2021.
3. The "energy" per bundle has units of *W*. It is important to remember that the bundles represent discrete statistical particles for a rate of energy transfer, so they are not a true "energy" in a thermodynamic sense.

11 Radiation in Solids, Windows, and Coatings

Sir Isaac Newton (1643–1727) *established the fundamentals of calculus, mechanics, gravitation, and the behavior of light. He built the first practical reflecting telescope, and developed a theory of color based on observations of the prismatic visible spectrum. Newton argued that the geometric nature of reflection and refraction of light could only be explained if light was made of particles, referred to as* corpuscles, *because waves do not tend to travel in straight lines. Refraction was caused by the acceleration of the corpuscles due to attraction by a denser medium. He explained why the particles were partially refracted and partially reflected at a glass (prism) surface by noting that the particles had "fits of attraction and fits of repulsion."*

Christiaan Huygens (1629–1695) *came from Holland to England in 1689 to make the acquaintance of Newton, whose* Principia *had been published in 1687.*

On his return in 1690, Huygens published his Treatise on Light, *in which the wave theory was expounded and explained. Newton's reputation led to the rejection of any theory that contradicted him, and Newton's corpuscular theory held sway over wave ideas for many years.*

11.1 INTRODUCTION

In the previous chapters, most of the discussion of radiative transfer was for materials with a simple refractive index $n \approx 1$. This assumption is valid when the absorbing, emitting, and scattering medium is a gas, because most gases have a refractive index very close to 1 (Table 11.1). However, many important effects result when the refractive index $n > 1$ as is the case for many common materials. These are our focus in this chapter. Attention here is to relatively geometrically thick systems where nanoscale effects, covered briefly in Chapter 12, are not important. This is the case for many engineering systems. A key application is for predicting reflection and refraction profiles at an interface, as formulated in Chapter 3. This is very important in predicting radiative transfer in translucent coatings, thin films, and multiple windows. Also, local blackbody emission within a medium is increased by a factor of n^2, as noted in Section 2.7.

The quantification of reflection, transmission, and absorption of solar and environmental radiation through single- and multilayered windows is important for

DOI: 10.1201/9781003178996-11

TABLE 11.1

Refractive Indices of Some Common Substances

Material	Refractive Index n
Air (at $\lambda = 0.589$ μm)	1.0002924
Liquids (at $\lambda = 0.589$ μm)	
Chlorine	1.3834
Ethyl alcohol	1.3614 (25°C)
Oxygen	1.2243
Water	1.33432–1.31861 (0°C–100°C)
Solids (at λ given)	
Glass:	
Crown	1.5454–1.50091 (0.3126–1.060 μm)
Light flint	1.61926–1.56594 (0.365–1.060 μm)
Flint	1.64606–1. 58636(0.365–1.060 μm)
Ice	1.3082 (0.6328 μm, -3°C)
α-Quartz	1.65–1.5369(0.200–0.750 μm)
Rock salt	1.7898-1.382 (0.150-0.750 μm)

Source: Rumble, 2019.

---

determining heating and cooling loads of buildings, designing flat-plate solar collectors, and many optical devices. Multiple reflections from window surfaces can be appreciable in reducing transmission, especially if more than one window layer is present, as in a two-cover-plate solar collector. Transient temperature distributions are important during manufacturing operations for glass and other translucent materials where there is internal absorption and emission of radiation in the heated materials combined with energy conduction.

If reflecting translucent layers are directly attached to an opaque surface or to other reflecting translucent layers, a coating system can be formed that has desirable radiative properties. A coating can be either thick or thin compared with the wavelength of the incident radiation. Thin films produce wave interference effects between the incident and reflected waves in the film, thus changing their reflection and transmission characteristics. Thin-film coating systems with high or low reflectivity can be fabricated. A composite window can be formed that is selectively transmitting, or a coated surface that is selectively absorbing can be produced. Examples are transparent radiation mirrors and solar selective surfaces.

The formulations in this chapter provide the necessary analytical tools for treating radiation transfer in enclosures with windows and for determining the radiative behavior of coatings. Since windows and coatings are often thin (that is, have relatively small differences in temperature between their surfaces), any temperature variations within them are not considered here. There are applications, however, in which the temperature distribution must be obtained within a translucent medium, including the effects of interface reflections. This requires solving the RTE and energy equations within the medium, as discussed in Chapters 9 and 10, with the addition of reflection boundary conditions at the interfaces. Some important applications are heat treating of glass plates, temperature distributions in glass-melting tanks, high-temperature solar components, laser heating of windows and lenses, heating of spacecraft and aircraft windows, radiation-scattering heat shields, and some thermal barrier coatings.

11.2 TRANSMISSION, ABSORPTION, AND REFLECTION FOR WINDOWS

Enclosures can have windows that are partially transparent to radiation. A window can be of a single material, or it can have one or more transmitting coatings on it. The transparency of a window is a function of the glass, plastic, and coating thicknesses and can be strongly wavelength-dependent, as illustrated by the absorptance of glass in Figure 8.10. Some applications are glass or plastic cover plates for flat-plate solar collectors and thick- and thin-film coatings to provide modified reflection and transmission properties for camera lenses and solar cells. This section is concerned with the transmission of incident radiation as modulated by surface reflections and absorption within the window. Scattering is assumed to be negligible in these applications.

As illustrated in Figure 11.1, radiation incident on a surface is reflected and refracted. For a layer of thickness D, the refracted portion travels a distance $D/\cos\chi$ and is then partially reflected from the inside of the second surface. To analyze the radiative behavior of a layer, such as in Figure 11.1, the reflectivities are needed for radiation striking the outside and the inside of an interface. For a smooth interface, reflectivity relations are given by Equations (4.33) through (4.37), since windows are often dielectrics with a small extinction (absorption) index, k. Since these relations depend only on the square of the terms containing the angles $(\theta-\chi)$, the θ and χ can be interchanged, and hence along the incident and refracted paths, the reflectivity is

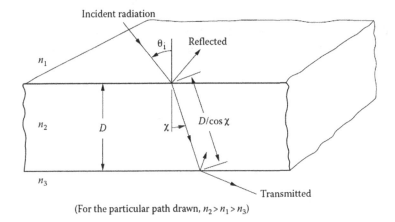

(For the particular path drawn, $n_2 > n_1 > n_3$)

FIGURE 11.1 Reflection, refraction, and transmission of incident radiation by a partially transmitting layer.

the same for radiation incident on an interface either from the outside or from within the material. For a constant absorption coefficient κ within the material, the transmittance t along the path length within the window is $t = e^{-\kappa D/\cos\chi}$, where κ is a function of wavelength as shown by the transmission spectrum of glass in Figures 8.10 (to simplify notation, the functional dependence on λ is omitted here as well as in most of the formulations that follow).

Fresnel reflection relations in Equations (4.33) through (4.36) show that the reflectivity at a smooth interface is not only a function of incidence angle but is also different for each of the two components of polarization. In what follows, the path of radiation will be traced through partially transparent windows where the interface reflections are specular (i.e., mirror-like). For precise results, the resulting formulas are applied at each incidence angle and for each component of polarization. If the incident radiation is nonpolarized, half the energy is in each component of polarization. For diffuse incident radiation, the fraction of energy incident in each θ direction within the increment $d\theta$ is $2\sin\theta\cos\theta d\theta$. Then, when integrating to find the total reflectivity, the results for each direction are weighted in the integration according to the amount of incident energy in each $d\theta$ at θ and in each component of polarization.

11.2.1 SINGLE PARTIALLY TRANSMITTING LAYER WITH THICKNESS $D \gg \lambda$

When a window or transmitting layer is very thin, so that its thickness is comparable to the radiation wavelength, there can be interference between incident and reflected waves. This is discussed later. Here, consider a windowpane where D is at least several wavelengths thick so that the interference effects need not be considered (Figure 11.2). Then,

$$J_3 = \rho G_3 + (1-\rho)G_4 = \rho G_3 \qquad (11.1)$$

$$J_4 = (1-\rho)G_3 + \rho G_4 = (1-\rho)G_3 \qquad (11.2)$$

The transmittance of the layer is used to relate the internal incident radiative flux G and outgoing radiative flux (radiosity) J to give $G_2 = J_3\tau$ and $G_3 = J_2\tau$. These are used

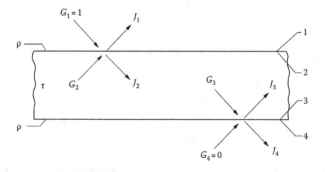

FIGURE 11.2 Net-radiation method applied to partially transmitting layer.

to eliminate the G from Equations (11.1) and (11.2), and the resulting equations are solved for the J to yield:

$$J_1 = \rho\left[1 + \frac{(1-\rho)^2\tau^2}{1-\rho^2\tau^2}\right]; \quad J_2 = \frac{1-\rho}{1-\rho^2\tau^2}$$

$$J_3 = \frac{\rho\tau(1-\rho)}{1-\rho^2\tau^2}; \quad J_4 = \frac{\tau(1-\rho)^2}{1-\rho^2\tau^2}$$

Normalizing to $G_1 = 1$, the fractions reflected and transmitted by the plate are J_1 and J_4 so that:

$$R = J_1 = \rho\left[1 + \frac{(1-\rho)^2\tau^2}{1-\rho^2\tau^2}\right] = \rho(1+\tau T) \tag{11.3}$$

$$T = J_4 = \frac{\tau(1-\rho)^2}{1-\rho^2\tau^2} = \tau\left(\frac{1-\rho}{1+\rho}\right)\left(\frac{1-\rho^2}{1-\rho^2\tau^2}\right) \tag{11.4}$$

The fraction absorbed is:

$$A = (J_2 + J_3)(1-\tau) = \frac{(1-\rho)(1-\tau)}{1-\rho\tau} \tag{11.5}$$

If the ρ at the upper and lower surfaces are not equal, the results for R and T are left as an exercise in Homework Problem I.11.1 in online Appendix I at www.ThermalRadiation.net/appendix.html.

Example 11.1

What is the fraction of externally incident unpolarized radiation that is transmitted through a glass window in air? The window is $D = 0.75$ cm thick, and radiation is incident at $\theta = 50°$, $n_{glass} = 1.53$, and $\kappa_{glass} = 0.1$ cm^{-1}.

To find the path length through the glass, evaluate $\chi = \sin^{-1}(\sin\theta/n) = \sin^{-1}(\sin 50°/1.53) = 30°$. The path length is $S = 0.75/\cos\chi = 0.866$ cm. The transmittance is $\tau = \exp(-\kappa S) = \exp(-0.1 \times 0.866) = 0.917$. The surface reflectivities for the two components of polarization are:

$$\rho_{\parallel} = \frac{\tan^2(\theta-\chi)}{\tan^2(\theta+\chi)} = 0.00412; \quad \rho_{\perp} = \frac{\sin^2(\theta-\chi)}{\sin^2(\theta+\chi)} = 0.1206$$

Then, the overall transmittance for each component is:

$$T_{\parallel} = \tau\frac{1-\rho_{\parallel}}{1+\rho_{\parallel}}\frac{1-\rho_{\parallel}^2}{1-\rho_{\parallel}^2\tau^2} = 0.917\frac{0.9959}{1.0041}\frac{1-0.00002}{1-0.00001} = 0.9095$$

$$T_\perp = \tau \frac{1-\rho_\perp}{1+\rho_\perp} \frac{1-\rho_\perp^2}{1-\rho_\perp^2 \tau^2} = 0.917 \frac{0.8794}{1.1206} \frac{1-0.0145}{1-0.0122} = 0.7175$$

For unpolarized incident radiation, one-half the energy is in each component. Hence, $T = (T_\parallel + T_\perp)/2 = 0.814$.

11.2.2 MULTIPLE PARALLEL WINDOWS

Consider now more complicated multilayer formulations. We analyze a system of m and n plates by the net-radiation method. Different layers with different properties separated by air or a vacuum are analyzed, such as for double- or triple-glazed windows. Using the notation in Figure 11.3, the outgoing radiation terms J are written in terms of the incoming radiation fluxes G as:

$$J_{m1} = R_m + G_{m2}T_m \tag{11.6}$$

$$J_{m2} = G_{m2}R_m + T_m \tag{11.7}$$

$$J_{n1} = G_{n1}R_n \tag{11.8}$$

$$J_{n2} = G_{n1}T_n \tag{11.9}$$

The G are further related to the J by using the relations $G_{m2} = J_{n1}$ and $G_{n1} = J_{m2}$. The G are then eliminated, and solving for the J yields:

$$T_{m+n} = J_{n2} = \frac{T_m T_n}{1 - R_m R_n} \tag{11.10}$$

$$R_{m+n} = J_{m1} = R_m + \frac{R_n T_m^2}{1 - R_m R_n} \tag{11.11}$$

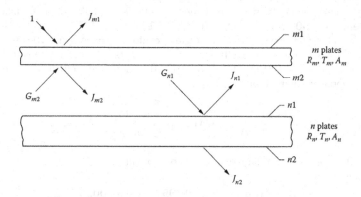

FIGURE 11.3 Net-radiation method for the system of multiple parallel transmitting plates.

The fractions of energy absorbed in the group of m plates and in the group of n plates within the system of $m + n$ plates are:

$$A_m^{(m+n)} = A_m + G_{m2}A_m = A_m\left(1 + \frac{T_m R_n}{1 - R_m R_n}\right) \tag{11.12}$$

$$A_n^{(m+n)} = G_{n1}A_n = \frac{T_m A_n}{1 - R_m R_n} \tag{11.13}$$

The $A_m^{(m+n)}$ is the absorption in the system of m plates when it is a part of a total system of $m + n$ plates. The A_m is the absorption in a system of m plates that is by itself and is not a part of a larger system.

Note that T_{m+n} shows symmetry; that is, the m and n subscripts can be exchanged, and the expression remains the same; hence, $T_{m+n} = T_{n+m}$. The T_{m+n} is the transmission for the system of $m + n$ plates for radiation incident first on the m plates, while T_{n+m} is for incidence first on the n plates. From Equation (11.11), however, $R_{m+n} \neq R_{n+m}$; the system reflectance depends on whether the radiation is incident first on the group of m plates or on the group of n plates.

If there is a stack of N plates that can all be different, as in Figure 11.4, then the fraction of incident energy absorbed by the top plate (a system of 1 plate) is, from Equation (11.12),

$$A_1^{(N)} = A_1\left(1 + \frac{T_1 R_n}{1 - R_1 R_n}\right) \tag{11.14}$$

where n is the system composed of plates: 2, 3,…N. The fraction absorbed in plate 2 is then:

$$A_{\text{in plate 2}}^{(N)} = A_m^{(N)} - A_1^{(N)} = A_m\left(1 + \frac{T_m R_n}{1 - R_m R_n}\right) - A_1^{(N)} \tag{11.15}$$

where
 m is the system composed of plates: 1 and 2
 n is the system composed of plates: 3, 4,…N

This process can be continued to determine the amount of radiant energy absorbed in each plate.

11.2.3 Transmission through Multiple Parallel Glass Plates

Transmission through multiple glass plates is needed in the design of flat-plate solar collectors and multi-pane windows. The glass surface reflectivity, as given by Equations (4.35) and (4.36), depends on the angle of incidence and the component of polarization. Since reflections from within the glass are assumed to be specular, the same angles of reflection and refraction are maintained throughout the multiple

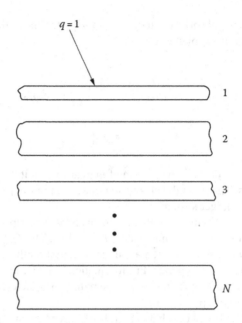

FIGURE 11.4   A stack of N partially transmitting parallel plates.

reflection process. Equation (11.10) can be used to calculate the overall transmittance through one and three parallel glass plates with an index of refraction $n = 1.5$. Neglecting absorption within the glass, results are in Figure 11.5 for the two components of polarization and as a function of the radiation incidence angle θ. As $\theta \rightarrow 90°$, the transmission goes to zero; this is because dielectrics have perfect reflectivity at grazing incidence (Figure 4.7). For incidence at Brewster's angle, the overall transmittance for the parallel component becomes 1, as there is zero reflection at this angle and losses by absorption are being neglected.

Incident solar radiation is unpolarized and hence has equal energy in the parallel and perpendicular components. The transmittance is then the average of the two transmittance values computed by individually using $\rho_{\parallel}$ and $\rho_{\perp}$. This is shown in Figure 11.6 for the limiting case of nonabsorbing plates and for absorbing plates having a product of absorption coefficient and thickness of 0.0524 per plate. For angles at near-normal incidence, the effect of absorption reduces transmission by about 5% for each plate.

Reflection and transmission as a function of wavelength are important for the proper choice of window glasses for buildings. Both architectural constraints (effects of window spectral properties as they affect both glass color and the changes in observed color when looking through the windows) and thermal engineering for energy efficiency dictate the need for such information.

11.2.4   FLAT-PLATE SOLAR COLLECTORS

A flat-plate solar collector usually consists of one or more parallel transmitting windows (for reducing convective losses) covering an opaque absorber plate,

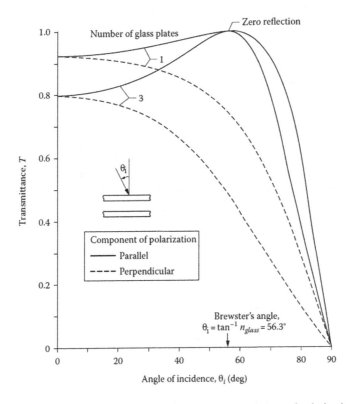

FIGURE 11.5 Overall transmittance of radiation in two components of polarization for non-absorbing parallel glass plates; $n_{glass} = 1.5$. (From Shurcliff, 1974.)

as in Figure 11.7. It is desired to obtain the fraction A_c of incident energy that is absorbed by the opaque collector plate. At the collector plate, the absorbed flux is:

$$A_c = G_c - J_c \tag{11.16}$$

$$J_c = (1 - \alpha_c)G_c \tag{11.17}$$

Across the space between the transmitting plates and the opaque collector plate:

$$J_c = G_n \tag{11.18}$$

$$J_n = G_c \tag{11.19}$$

For the system of n transmitting plates,

$$J_n = T_n + G_n R_n \tag{11.20}$$

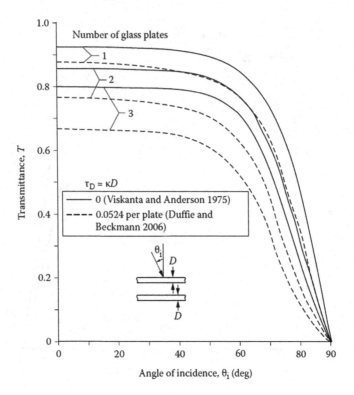

FIGURE 11.6 Effect of incidence angle and absorption on overall transmittance of multiple parallel glass plates; $n_{glass} = 1.5$.

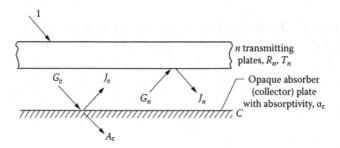

FIGURE 11.7 Interaction of transmitting windows and an absorbing collector plate.

The system of Equations (11.16) through (11.20) is solved to yield the fraction of incident energy absorbed by the opaque collector plate:

$$A_c = \frac{\alpha_c T_n}{1-(1-\alpha_c)R_n} \tag{11.21}$$

This is an important relation for determining the efficiency of solar absorbers with glass cover plates.

Regarding radiative transfer between parallel surfaces, there are additional effects when reflecting layers have a spacing on the order of a wavelength. These near-field effects are briefly discussed in Chapter 12.

11.2.5 Thin Film with Wave Interference Effects

11.2.5.1 Dielectric Thin Film on Dielectric Substrate

When a film coating is very thin, in the order of the wavelength λ of the incident radiation, interference effects may occur between waves reflected from the first and second surfaces of the film.

The discussion here is limited to *normal incidence* on the thin film. Figure 11.8 shows the radiation reflected from the first and second interfaces (note: for clarity in showing each path, the paths are drawn at an angle).

Consider a thin nonabsorbing film of refractive index n_1 on a substrate with index n_2 and neglect the possibility of interference among the reflected beams. For a normally incident wave of unit amplitude, the reflected radiation is shown in Figure 11.8. Accounting for the phase relationships caused by the beam reflected from surface 2 traveling a distance $2D$ more than a wave reflected from the first interface, and defining $\gamma_1 \equiv 4\pi n_1 D/\lambda_0$, the reflected amplitude is:

$$R_M = r_1 + t_1 t_1' r_2 e^{-i\gamma_1} - t_1 t_1' r_1 r_2^2 e^{-2i\gamma_1} + t_1 t_1' r_1^2 r_2^3 e^{-3i\gamma_1} - \cdots = r_1 + \frac{t_1 t_1' r_2 e^{-i\gamma_1}}{1 + r_1 r_2 e^{-i\gamma_1}} \qquad (11.22)$$

Note that $t_1 t_1' = 1 - r_1^2$, so this can be reduced to:

$$R_M = \frac{r_1 + r_2 e^{-i\gamma_1}}{1 + r_1 r_2 e^{-i\gamma_1}} \qquad (11.23)$$

An important application for a thin coating is to obtain low reflection from a surface to reduce reflection losses during transmission through a series of lenses in optical

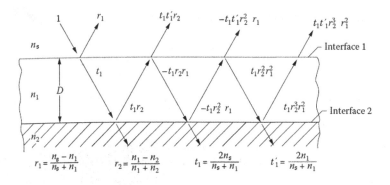

FIGURE 11.8 Multiple reflections within a thin nonattenuating film for normal incidence; paths are drawn at an angle for clarity.

TABLE 11.2

Reflectivity of Thin Films of Various Materials on Glass to Provide Dielectric Mirrors

Film	n_1	n_2	R
None	1	1.5	0.04
ZnS	2.3	1.5	0.31
Ge	4.0	1.5	0.69
Te	5.0	1.5	0.79

---

equipment. To have zero reflected amplitude, $R_M = 0$, requires $r_1 = -r_2 e^{-i\gamma_1}$. This can be obtained if $r_1 = r_2$ and $e^{-i\gamma_1} = -1$. Since $e^{-i\pi} = -1$, this yields $D = \lambda_0/4n_1$. The quantity λ_0/n_1 is the wavelength of the radiation within the film. Hence the film thickness for zero reflection at normal incidence is one-quarter of the wavelength within the film. The required condition is $(n_s-n_1)/(n_s + n_1) = (n_1-n_2)/(n_1 + n_2)$, which reduces to $n_1 = \sqrt{n_s n_2}$. Thus, for normal incidence onto a quarter-wave film from a dielectric medium with index of refraction n_s, the index of refraction of the film for zero reflection should be $n_1 = \sqrt{n_s n_2}$. The thin film provides better performance than the thick film, as it is possible to achieve zero reflectivity. However, this result is only for normal incidence at a single wavelength. To obtain more than one condition of zero reflectivity, it is necessary to use multilayer films. For a system of two nonabsorbing quarter-wave films, zero reflection is obtained for $n_1^2 n_3 = n_2^2 n_s$, where n_1 and n_2 are for the coatings (the coating with n_2 is next to the substrate) and n_3 is for the substrate.

For a quarter-wave film, the reflectivity at normal incidence becomes zero when $n_1 = \sqrt{n_s n_2}$. If n_1 is high, reflectivity is increased, and this behavior can be used to obtain *dielectric mirrors*. For various film materials of refractive index n_1 on glass ($n_2 = 1.5$), the reflectivity for incidence in air ($n_s \approx 1$) is given in Table 11.2.

Multilayer films can be used to obtain reflectivities very close to 1. One application is for the reflection of high-intensity laser beams where the fractional absorption of energy must be very small to avoid damage from mirror heating.

11.3 LIGHT PIPES AND FIBER OPTICS

Perfect internal reflection at an interface within medium 2 where $n_2 > n_1$ is predicted to occur (Equation (4.35) et seq.) whenever the incident angle $\theta > \tan^{-1}\left(\dfrac{n_2}{n_1}\right)$. The radiation within the higher refractive index material will undergo 100% reflection at the interface. This means that once radiation enters a perfectly transmitting material (a perfect dielectric), such as a pure optical fiber in which no radiation absorption occurs, the radiation will propagate without loss. The perfect wall reflectivity allows no losses by refraction through the fiber surface. This phenomenon is observed in a planar geometry by swimmers (and fish), who can only see through the water surface above them within a cone of angles described by:

$$\theta < \tan^{-1}\left(\frac{n_{\text{air}}}{n_{\text{water}}}\right) = \tan^{-1}\left(\frac{1}{1.33}\right) = \tan^{-1}(0.75) = 36.9°$$

At greater angles, the water surface acts like a mirror; at lesser angles, although all directions in the hemisphere are viewed from within medium 2, the refraction through the surface produces a considerable distortion of the view. You, the angler, probably appears quite weird to a fish.

For a light pipe or fiber optic, radiation enters the flat end of a long circular cylinder. If the refractive index ratio n_2/n_1 exceeds $\sqrt{2}$, all radiation entering the light-pipe end will encounter the internal cylindrical surface at greater than the critical angle. An example of the pattern of radiation transmitted through a light pipe is shown in Figure 11.9. Because of the increasing reflectivity (decreasing transmissivity) of dielectrics with incident angle (Figures 4.6 and 4.7), the radiation entering the light-pipe end is chiefly from near-normal angles.

The signal entering the detector at the end of a light-pipe radiation thermometer (LPRT) consists of emitted plus reflected energy from the object being viewed. If the environment is cold relative to the viewed object so that reflected energy can be neglected and the detector is only sensitive in a small wavelength range around a particular wavelength λ, then the detected emission is proportional to, from Equation (3.7),

$$E_\lambda(T_{\text{act}}) = \varepsilon_\lambda E_\lambda(T_{\text{act}}) = \varepsilon_\lambda \frac{2\pi C_1}{\lambda^5 \left[\exp\left(\dfrac{C_2}{\lambda T_{\text{act}}}\right) - 1\right]} \tag{11.24}$$

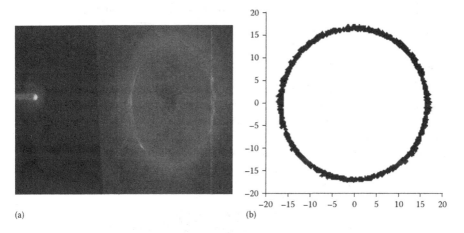

(a) (b)

FIGURE 11.9 (a) Experimental and (b) predicted patterns of laser energy at the exit of a fused quartz light pipe ($n = 1.45843$) for laser energy incident on the inlet end at 20° to the light-pipe axis. Light-pipe diameter 4 mm and length 16 cm. (From Qu, 2007.)

However, rather than the actual temperature T_{act}, the detector reads an apparent temperature T_{app} that would appear to originate from a blackbody:

$$E_\lambda(T_{app}) = \frac{2\pi C_1}{\lambda^5 \left[\exp\left(\dfrac{C_2}{\lambda T_{app}} \right) - 1 \right]} \tag{11.25}$$

Equating the two emitted energy rates relates the actual and apparent temperatures as:

$$\varepsilon_\lambda \frac{2\pi C_1}{\lambda^5 \left[\exp\left(\dfrac{C_2}{\lambda T_{act}} \right) - 1 \right]} = \frac{2\pi C_1}{\lambda^5 \left[\exp\left(\dfrac{C_2}{\lambda T_{app}} \right) - 1 \right]} \tag{11.26}$$

For most engineering conditions, the factor of (-1) in the denominator can be neglected relative to the exponential term (resulting in Wien's spectral distribution Equation (2.25)), and the actual temperature in terms of the apparent absolute temperature then becomes:

$$T_{act} = \frac{T_{app}}{\left[1 + \dfrac{\lambda T_{app}}{C_2} \ln \varepsilon_\lambda \right]} \tag{11.27}$$

As $\varepsilon_\lambda \to 1$, the apparent and actual temperatures approach one another. Equation (11.27) can be used for spectrally based temperature measurements, remembering the proviso that the environmental temperature must be low to get an accurate temperature measurement using an optical pyrometer or light-pipe radiation thermometer.

11.4 FINAL REMARKS

In this chapter, we have discussed several models and formulations for radiative transfer through translucent materials, windows, and coatings. Because of space constraints, many other radiative transfer applications and the corresponding details are omitted. A natural extension would be to paints, for example, which are important for energy efficiency applications in buildings. Similarly, there are many other dispersed media, from foams to food preparation coatings (cheese making or baking), where radiation transfer plays a crucial role. The concepts discussed in this book are a starting point for the analysis of such systems.

HOMEWORK (additional problems available in online Appendix I at www.ThermalRadiation.net/appendix.html)

11.1 A horizontal glass plate 0.225 cm thick is covered by a plane layer of water 0.625 cm thick. Radiation is incident from within air onto the upper surface of the water at an incidence angle of 60°. What is the path length of this radiation through the glass? (Let $n_{H2O} = 1.33$, and $n_{glass} = 1.57$.) *Answer:* 0.270 cm.

11.2 A radiation flux q_s is incident from the normal direction on a series of two glass sheets, each 0.35 cm thick, in air. What is the fraction T that is transmitted? ($n_{glass} = 1.53$, and $\kappa_{glass} = 9.5$ m^{-1}.) What is the fraction transmitted for a single plate 0.7 cm thick in air?

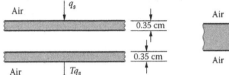 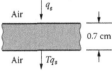

Answers: 2 plates, $T = 0.790$; 1 plate, $T = 0.857$.

11.3 Prove that for an absorbing–emitting layer with $L \gg \lambda$, $T + A + R = 1$.

11.4 Derive an analytical expression that shows whether the total absorption in a system of n plates and m plates, $A(m + n)$, is independent of whether radiation is the first incident on the n or the m plates.

11.5 A double-panel glass window consists of two 5 mm parallel glass plates separated by a 10 mm air gap. The outside glass plate is made of crown glass with a refractive index of 1.52. The inside glass plate is made of flint glass with a refractive index of 1.62. Absorption in the glass plates can be ignored. Calculate the transmittance for normal incidence on the outer plate. *Answer:* 0.828

11.6 For the two-plate system in Homework Problem 11.5, determine the maximum transmittance for parallel-polarized radiation and the incidence angle that gives rise to that maximum transmittance. Also, find the transmittance through the plate assembly for unpolarized incident radiation at this angle. Compare with the transmittance found for normal incidence in Homework Problem 11.5. *Answer:* $T_{top + bottom} = 0.725$.

11.7 The two glass plates in Homework Problem 11.2 are followed by a collector surface with absorptivity $\alpha_c = 0.95$. What is the fraction of A_c absorbed by the collector surface for normally incident radiation?

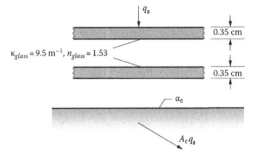

Answer: $A_c = 0.756$.

11.8   Do Homework Problem I.10.1 in the online Appendix at www. ThermalRadiation.net/appendix.html with the index of refraction for glass equal to 1.53, and for the liquid equal to 1.35.
Answer: 503 W/m².

11.9   A semitransparent layer of fused silica has a complex index of refraction of $n - ik = 1.42 - i \times 1.45 \times 10^{-4}$ at the wavelength $\lambda = 5.20$ μm. Unpolarized diffuse radiation in air at this wavelength is incident on the layer. What is the value of the overall transmittance of the layer for this diffuse incident radiation?
Answer: $T = 0.148$.

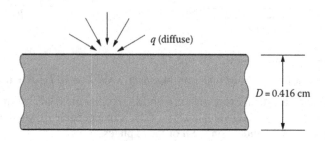

11.10 A layer of transmitting glass is over a parallel opaque cold substrate with gray absorptivity α. The glass has a surface reflectivity of ρ, and the glass layer has a transmittance of τ. Derive an expression for the reflectance R of the glass/substrate assembly.

Answer: $R = \rho + \dfrac{(1-\rho)^2 \tau \left[1 - \rho - \alpha + 2\rho\alpha\right]}{1 - \rho + \rho\alpha - \rho\tau^2 \left[1 - \rho - \alpha + 2\rho\alpha\right]}$

12 Emerging Areas

Sergei Mikhailovich Rytov (1908–1996) *advanced near-field radiation transfer concepts with his pioneering studies on fluctuational electrodynamics. Earlier in his career in Soviet Russia, he developed an approximate solution of Maxwell's equations to describe the propagation of an electromagnetic wave through a turbulent atmosphere, and was a pioneer in the field of radiophysics. His work in inverse techniques is widely used for determining the properties (phase function and scattering coefficient) of scattering media from remote signals.*

Andrey Nikolayevoch Tikhonov (1906–1993) *was a Russian mathematician. He received his PhD from Moscow State University in 1927. Upon graduation, he immediately made important contributions to topology and mathematical physics, including a uniqueness proof for solutions to the heat equation. He developed one of the most widely used regularization techniques for ill-posed problems, which bears his name.*

12.1 INTRODUCTION

In this chapter, some important emerging research and application areas of radiation energy transfer are explored. A complete exposition of the developing state-of-the-art is outside the scope of this text, but the flavor is given. The four chosen emerging areas are radiation effects at the nanoscale, inverse design applications for radiation systems, radiation transfer in emerging manufacturing processes, and the use of radiation for diagnostics and remote measurement.

12.2 NANOSCALE RADIATION ENERGY TRANSFER

Throughout the following discussion it will be assumed that the linear dimensions of all parts of space considered, as well as the radii of curvature of all surfaces under consideration, are large compared with the wavelengths of the rays considered.

In his seminal book, *The Theory of Heat Radiation* (1913), Max Planck presents the above condition in his analysis that the dimensions of the surfaces and geometries are always large compared with the primary wavelength of radiation considered. This assumption shows that his analysis of thermal radiation was developed for the *far field*, i.e., at distances several times longer than Wien's peak wavelength of emission, λ_{max}. As we discussed in Chapter 2, the peak blackbody emission wavelength is obtained from Wien's expression, $\lambda_{max}T = (hc/k_B)T = 2897.8/T$, where T is in Kelvin (K) and λ_{max} is in micrometers (Equation (2.29)).

DOI: 10.1201/9781003178996-12

For objects around 700 K (or 800°F), Planck's analysis applies at distances greater than about 4 micrometers from the source of emission. With the advances in nanotechnology and nanoscale engineering, engineers can now work with objects as well as spacings much smaller than λ_{max}. If the gap size between objects is below a few hundred nanometers, *near-field* radiative energy transfer can be enhanced by several orders of magnitude over that predicted by standard radiation transfer analysis. These additional effects are quantified by using the theory of fluctuational electrodynamics. Originally suggested by Rytov in 1953, this approach shows that the magnitude of radiative energy exchange between bodies that are very closely spaced can be significantly different from that predicted using Planck's formulation.

The second exception to the classical Planck relations used in radiative exchange occurs when the objects exchanging radiative energy are themselves smaller than the governing wavelength. Experiments show that significant enhancements over blackbody exchange can exist between very small objects, even in the far field.

This section provides a brief overview of near-field radiation transfer either between nano-size and larger objects, or large objects separated by nano-size gaps. We briefly discuss tailored metamaterials and their importance in near-field radiative transfer. These are sometimes called "designer" materials, and their electromagnetic properties and corresponding resonances, which are critical for near-field radiative transfer exchange, can be tuned to some extent.

Near-field radiation transfer analysis must account for the sensible effects between two objects caused by electron oscillations and/or lattice vibrations. These oscillations are enhanced in a medium either due to its energy, measured by its temperature, or with the help of an external source (e.g., an incident laser beam). These fluctuations in an electron field (plasmons) or in vibrating lattices (phonons) can be tunneled into a second nearby medium, as predicted by Maxwell's equations. The role of the fluctuating near field was not considered in classical radiation theory until it received the attention of researchers such as Rytov (1959).

Consider two nonmagnetic, isotropic, homogenous, parallel plates separated by a narrow gap of a few nanometers to a few tens of micrometers (Figure 12.1). One plate is at a higher temperature than the other. We know that radiation transfer takes place in both directions, but net radiative flux will be from the hotter to the colder plate. Radiative energy emanating from deep inside the hot medium travels to the surface, and if the angle at which it is incident on the surface is smaller than the critical angle, it will be transferred through the interface. Here, the critical angle refers to the one obtained from Brewster's angle at interfaces (Equation (4.35)). If the angle of incidence is greater than the critical angle, then our conventional analysis shows there would be no transferred energy; however, *evanescent* waves are present that will traverse the interface even for total internal reflection (TIR).

Evanescent waves are naturally oscillating standing waves. However, they can tunnel energy from the first into a second medium if the latter is sufficiently close. They decay rapidly with distance and are unable to carry energy to the other medium if it is in the far field (Figure 12.1b).

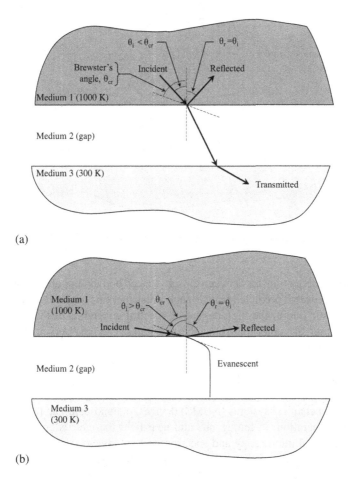

FIGURE 12.1 a) Far-field representation of EM-wave propagation; b) Representation of evanescent wave.

Surface waves are the standing waves resulting from the total internal reflectance of electromagnetic waves or due to the random oscillations of phonons and/or electrons. In classical radiative energy transfer, far-field modes are the primary carrier in energy transport, and it is in this regime where Planck's blackbody limit is formulated. However, when two bodies are closely spaced, it is possible for surface wave modes to tunnel across the surface to the second medium, as depicted in Figure 12.2. The resulting enhancement in the radiative transfer can exceed Planck's blackbody limit.

In near-field analysis, both the phase and the polarization variations must be accounted for.

In the far-field regime, thermal radiation emission can be treated as a broadband phenomenon with quasi-isotropic angular distribution. In the near field, thermal sources can exhibit high spatial and temporal coherences due to the presence of surface waves.

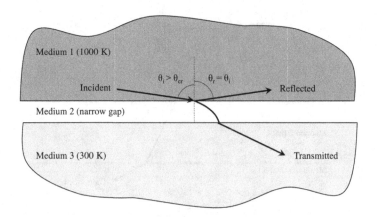

FIGURE 12.2 Configuration considered for near-field thermal radiation transfer via evanescent waves.

Radiative energy transfer between closely spaced bodies supporting surface waves not only exceeds that predicted from Planck's distribution but can also become quasi-monochromatic due to the high degree of spectral coherence of these waves. This means that laser-like emission from a thermal source can be observed at distances far from the emitting body.

Recent advances in nanotechnology and nanopatterning procedures allow exploitation of near-field radiation effects. Near-field thermal radiation problems are important in the thermal management of microelectromechanical systems (MEMS) and nanoelectromechanical systems (NEMS) devices, in nanoscale-gap thermophotovoltaic power generation, in tuning far- and near-field thermal radiation emission, in near-field thermal microscopy and spectroscopy, in advanced nanofabrication techniques, and in production of near-blackbody absorbers to name only a few. The rich physics behind near-field thermal radiation is at the interfaces of quantum mechanics, electrodynamics, and statistical thermodynamics.

12.2.1 SPATIAL AND TEMPORAL COHERENCE OF THERMAL RADIATION

The coherence properties of an emitted thermal field are directly related to the wave nature of radiation. The coherence length of blackbody radiation is about $\lambda/2$, so that coherence properties of thermal radiation are not observed in the far field. However, by structuring nanoscale surfaces, near-field coherence can be transmitted to the far field, and it is possible to achieve quasi-monochromatic and directional thermal radiative sources.

The radiative energy flux in the near field can exceed the values predicted for two blackbodies by several orders of magnitude. Since there is a significant increase in emission in a very narrow frequency range, the spectral nature of emission shows quasi-monochromatic behavior. The emitted near-field radiative energy flux versus the gap thickness d smaller than the dominant wavelength increases proportionally to $1/d^2$.

12.2.2 Computational Studies of Near-Field Thermal Radiation

A computational technique that can consider all the near-field modes of any given problem is of crucial importance with increasing interest in nanoscale energy-harvesting devices and nanoscale manufacturing and sensing. The original near-field radiative transfer calculations were carried out for two parallel surfaces.

Figure 12.3 summarizes recent results from Didari and Mengüç (2018) to demonstrate how near-field thermal radiation transfer calculations for complicated structures lead to the development of new concepts. The inset in the figure shows a two-plate geometry for SiC, and two additional nanostructures inspired by nature, specifically from butterfly wings. Spectral radiative energy flux profiles in the near field for standing thin films for two different geometries are compared against the benchmark: A standing tree structure with a thin film and a 90° rotated "simple" structure with a thin film. The thin film is assumed to have the same thickness of 65 nm as the thickness of the tree branches, and it is placed at 65 nm above the first tree-like structure. The results show that the near-field radiative energy flux is strongly dependent on the geometric features and given dimensions. The development of new structures using new or modified materials may allow us to develop new processes for advanced radiative cooling applications.

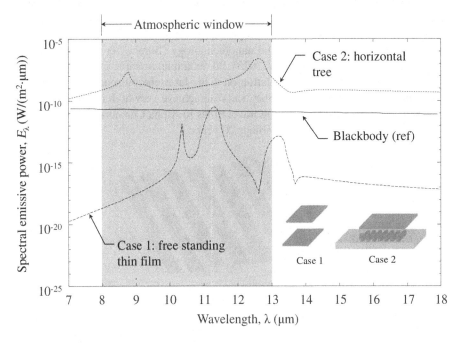

FIGURE 12.3 Spectral radiative energy flux profiles in the near field for a benchmark of standing SiC thin films, standing tree structure in the presence of an upper thin film, and 90° rotated structure in the presence of the upper free-standing thin film (Adapted from Didari, A. and Mengüç, M. P.: "A biomimicry design for nanoscale radiative cooling applications inspired by Morpho didius butterfly," *Scientific Reports*, 8, 16891, 2018.).

12.2.3 Experimental Studies of Near-Field Thermal Radiation

12.2.3.1 Near-Field Effects on Radiative Properties and Metamaterials

Near-field effects are important not only in the way they affect the energy transfer rates between closely spaced bodies but also in the way they can be manipulated to change the medium properties. By nanoscale structuring of a surface supporting surface waves, it is possible to transmit this coherence into the far field. This allows directional emittance by the surfaces, which has potential application to thermophotovoltaic studies and spectroscopic sensors.

Greffet et al. (2002) show how the angular profile of emissivity can be altered using a diffraction grating. They fabricated gratings on a SiC substrate (Figure 12.4). Diffraction by the grating leads to high emissivity in narrow solid angles, as shown in Figure 12.5. This emissivity is about 20 times larger than from a flat source. This extraordinary behavior is due to the coherence properties only observed and manipulable in the near field.

The observations made by Greffet et al. (2002) show that radiative properties can be significantly modified by nano-structuring the surface of a metamaterial. For the specific case of SiC, they showed that a reflectivity of 94% can be reduced to almost zero in the infrared region due to the presence of the grating that couples surface waves into propagating waves. In addition, nanostructures on surfaces have an impact on far-infrared properties in the far field, which allow application to radiation cooling.

Near-field radiation studies have burgeoned since the early 2000s because of the emphasis on nanoscale sciences and nanotechnology (Baranov et al. 2019, Pascale et al. 2022, Mengüç and Francoeur, 2023). For example, research is being sponsored by the US National Science Foundation on using laser energy sources to accelerate very light weight nanopatterned solar "sails" to achieve very high interstellar spacecraft velocities, potentially reducing travel times to Alpha Centauri from the 80,000 years by conventional rockets to perhaps 20 years.

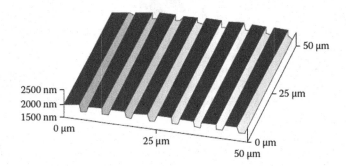

FIGURE 12.4 Schematic image of grating obtained by atomic force microscopy. Period $d = 0.55\lambda$ ($\lambda = 11.36$ μm) chosen so that a surface wave propagating along the interface is coupled to a propagating wave in the range of frequencies of interest. Depth: $h = \lambda/40$ was optimized so that the peak emissivity is $\varepsilon = 1$ at $\lambda = 11.36$ μm. Fabricated on SiC by standard optical lithography and reactive ion etching. (Courtesy of J.-J. Greffet.)

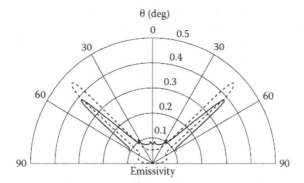

FIGURE 12.5 Emissivity of a SiC grating in *p*-polarization. Curve at 45° for $\lambda = 11.86$ μm. Emissivity is deduced from specular reflectivity measurements using Kirchhoff's Law. Data were taken at ambient temperature using FTIR spectrometer as a source and a detector mounted on a rotating arm. The dashed line with a peak at 45° is based on theoretical calculations; the solid line is from the experimental data. (Courtesy of J.-J. Greffet.)

12.3 RADIATION TRANSFER IN INVERSE APPLICATIONS

Most problems considered so far in the book can be categorized as *forward problems*, in which a well-defined system is analyzed to obtain an unambiguous solution. There is a growing appreciation, however, that many (if not most) practical radiative transfer problems are *inverse problems*, which, broadly speaking, involve inferring some parameter or quantity from information that is only indirectly related to the unknown. In the context of radiative transfer, most inverse problems are either *parameter estimation* or *inverse design* problems. The goal of parameter estimation problems is to infer some property or property distribution from indirect measurements, e.g., to reconstruct the absorption coefficient from spectral intensity measurements made along the domain periphery. Inverse design focuses on inferring a design configuration that corresponds to some ideal outcome, such as a furnace geometry that must uniformly irradiate a material being processed.

Inverse problems are *ill-posed* to the extent that they depend on the "indirectness" of the relationship between the known information and the unknown variables; this makes inverse problems much more difficult to solve compared with forward problems. Traditional solutions were by trial and error. Newer approaches provide much better insight and solutions.

Ill-posedness arises from an information deficit; the available information is either insufficient to specify a unique solution, or there may be multiple solutions that "almost" provide the parameter values or design characteristics being sought. In parameter estimation problems, this deficit sensitizes the inference process to small amounts of measurement noise and model errors, while inverse design problems are ill-posed if multiple candidate designs exist that provide similar outcomes. All inverse solution schemes, without exception, address this information deficit by incorporating additional information into the inference process.

Inverse problems involving parameter estimation have been around for a long time, although they are sometimes not recognized as inverse problems. Examples include absorption and emission tomography and light scattering measurements used to infer particle morphologies.

This section starts with an overview of inverse analysis and the mathematical properties of ill-posed inverse problems, followed by a discussion of solution schemes, and concludes with a summary of parameter estimation and inverse design problems involving radiative transfer. A monograph explores this field in much more detail (Ertürk et al. 2023).

12.3.1 INVERSE ANALYSIS AND ILL-POSED PROBLEMS

Inverse problems are challenging to solve because they are ill-posed. This term originates from Hadamard (1902), who identified well-posed problems as those that have solutions that: (i) exist; (ii) are unique; and (iii) are stable to small perturbations in the problem definition. Problems that do not satisfy these criteria are ill-posed. Hadamard focused on the properties of differential equations and viewed those derived from "real-world" problems, such as the heat equation, as being inherently well-posed. Ill-posed problems, in contrast, were interpreted as "artificial" problems having only theoretical interest. There has been a growing appreciation that this is far from true: inverse problems are ubiquitous in science, engineering, and daily life. While parameter estimation problems satisfy the existence criterion, Bohren and Huffman (1998) provide a helpful example of why they may violate the uniqueness or stability criterion. Suppose a knight who is hunting dragons comes across a set of footprints, which could have been left by one of several species of dragons (Figure 12.6). The knight could easily predict which sort of footprints would be left by a particular species, e.g., from a *Field Guide to Dragons*; this is the *forward problem*. On the other hand, inferring the dragon species from an observed footprint is potentially more difficult if each species leaves a similar footprint; this is the *inverse problem*. This process is even more difficult if the fine features of the footprint that distinguish the species become smeared in the mud, leading the knight to misidentify the species. In this analogy, the "smearing" represents measurement noise and inconsistencies between the physical system and the measurement model, which is already difficult because different species leave similar footprints. To resolve this problem, the knight must provide additional information, e.g., a particular dragon species is known to inhabit the forest.

12.3.2 MATHEMATICAL PROPERTIES OF INVERSE PROBLEMS

The first step when solving an inverse problem is to classify the type of inverse problem; this is critical, since this determines both the ill-posed characteristics, as well as the techniques that should be used to solve the problem. Most broadly, inverse problems can be categorized as either *linear* or *nonlinear*.

Forward problem
What kind of footprints does each dragon produce?

Inverse problem
What dragon left the footprints?

FIGURE 12.6 A knight hunting dragons; an example of an inverse parameter estimation problem. (Courtesy of T. Sipkens.)

12.3.2.1 Linear Inverse Problems

In many problems, the unknown variables are related to the known parameters in a linear way through an integral equation of the first kind (IFK):

$$f(u) = \int_a^b g(v) K(u, v) dv \tag{12.1}$$

where $f(u)$ is the known function, $g(v)$ is the unknown function, and $K(u,v)$ is the kernel. This frequently arises in radiative transfer. In a radiant enclosure design problem, for example, $f(u)$ could represent a desired irradiation of a surface being heated (a "design surface"), $g(v)$ would be the radiosity distribution over a heater surface to be designed (the "heater surface"), and $K(u,v)dv$ would be the infinitesimal view factor, dF_{u-v} between these locations (e.g., Equation (6.30)). Alternatively, in parameter estimation problems, $g(v)$ is an unknown parameter to be inferred, $f(u)$ represents measured data, and the kernel function is derived from the measurement equations that relate these two quantities.

In the forward problem, $g(v)$ is known, and the objective is to calculate the system response, $f(u)$, by carrying out the integration in Equation (12.1). Because of the smoothing action of the kernel, the features of $g(v)$ are blended by the integration into smaller variations in $f(u)$. When solving the inverse problem, in principle one could reconstruct $g(v)$ if $f(u)$ were known perfectly through the reverse procedure. However, if $f(u)$ is contaminated by small amounts of noise, the inverse approach amplifies this noise into large variations in the inferred $g(v)$. In the case of inverse design, the blending action of the kernel means that the irradiation on each point of the design surface is influenced by the radiosity over the entire heater surface, and many different radiosity distributions $g(v)$ could be substituted into the integral to provide very similar distributions for $f(u)$.

Example 12.1

A rapid thermal processing (RTP) chamber heats a silicon wafer, as shown in Figure 12.7(a). All other surfaces in the chamber are black, and all surfaces are cold except for the heater. Calculate the emissive power distribution over the heater surface that is required to produce a uniform irradiation of 10 kW/m² over the wafer.

This is an inverse design problem in which the wafer is the design surface. Because the surrounding surfaces are cold and black, the irradiation of the design surface is related to the heater emissive power by a Fredholm IFK:

$$G_{DS}(r_1) = G_{target} = \int_0^{R_s} E_{b,HS}(r_2) dF_{d1-d2}(r_1,r_2) = \int_0^{R_s} E_{b,HS}(r_2) K(r_1,r_2) dr_2 \qquad (12.1.1)$$

where dF_{d1-d2} is the infinitesimal view factor from a ring element dA_1 on the wafer to ring element dA_2 on the heater surface, and $K(r_1,r_2) = dF_{d1-d2}/dr_2$ is the kernel.

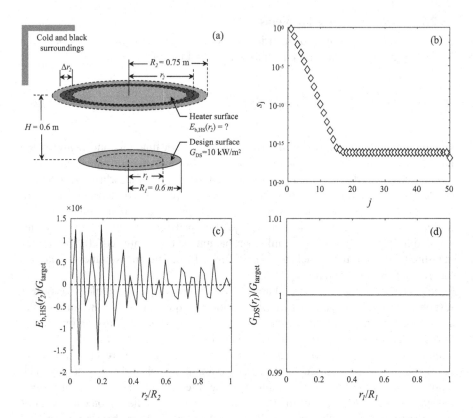

FIGURE 12.7 (a) Inverse design problem for the emissive power distribution over a heater used to irradiate a silicon wafer; (b) singular values from the **A** matrix shows that it is ill-conditioned; (c) the emissive power distribution over the heater surface recovered from $\mathbf{x} = \mathbf{A}^{-1}\mathbf{b}$ is nonphysical.

This problem is ill-posed because each point on the wafer can "see" the entire heater. Consequently, multiple emissive power distributions over the heater surface may exist that can produce the desired irradiation.

In practice, the IFK is solved by discretizing the heater surface into ring elements, and

$$G_{DS}(r_i) = G_{target} = b_i = \sum_{j=1}^{n} E_{b,HS,j} F_{di-j} = \sum_{j=1}^{n} x_j a_{ij} \qquad (12.1.2)$$

where F_{di-j} is the view factor between an infinitesimal ring element on the design surface at $r_1 = r_i$ and the jth ring element on the heater. Writing this equation for n discrete locations on the design surface produces an $n \times n$ matrix equation. Figure 12.7(b) shows the singular values for $n = 50$, which decay continuously over several orders of magnitude. This is typical of matrixes generated from IFKs. In principle, the emissive power distribution over the heater can be found from $x = A^{-1}b$, which is plotted in Figure 12.7(c). Although substituting this result into $Ax = b$ produces the desired irradiation, Figure 12.7(d), and constitutes a mathematical solution to the problem, the emissive power distribution plotted in Figure 12.7(c) is nonphysical with some negative values.

12.3.2.2 Nonlinear Inverse Problems

Both parameter estimation and inverse design problems may involve nonlinear measurement models, i.e., the knowns and unknowns are not related in a linear way. For example, if the objective of the design problem in Example 12.1 were to optimize the RTP chamber geometry, the problem would be governed by a *nonlinear* IFK having the form:

$$f(u) = \int_a^b g(v) K\left[u, v, h(v)\right] dv \qquad (12.2)$$

where u and v indicate locations on the design and heater surfaces, $g(v)$ would be the emissive power over the heater surface (which may or may not be known), and the kernel would depend on some intermediate geometric variable $h(v)$, e.g., the local curvature of the heater surface, which must also be solved for.

Nonlinear problems are usually discretized into $a(x) = b$, where a is a vector of nonlinear model equations that relate the unknowns in x to the knowns in b (a discrete set of measured data or design outcomes). Since this nonlinear equation cannot be inverted directly, it is usually solved as a minimization problem:

$$x^* = \arg\min_x \left[F(x)\right] = \arg\min_x r(x)^T r(x) \qquad (12.3)$$

where $r(x) = a(x) - b$ is the residual between the measured and modeled quantity, and "arg min $F(x)$" means "the value of x that minimizes $F(x)$."

12.3.3 Solution Methods for Inverse Problems

The algorithm chosen to solve the inverse problems depends on the type of inverse problem (parameter estimation or design optimization), the mathematical properties of the problem (e.g., linear versus nonlinear, types, and the number of unknowns), the prior information available to mitigate the ill-posedness of the inverse problem, the desired outcomes of the problem, and the computational budget available to solve the problem. A few of the most common techniques for solving inverse problems involving radiative transfer are summarized below.

12.3.3.1 Linear Regularization

Linear regularization techniques form approximate solutions to $\mathbf{Ax} = \mathbf{b}$ by numerically stabilizing the inversion of $\mathbf{A}$; these can be thought of as solutions to problems that approximate $\mathbf{Ax} = \mathbf{b}$ but are less ill-conditioned. Most often, regularization works like a bandpass filter, where high-frequency solution components that most likely correspond to amplified measurement noise or model error are filtered out, leaving an approximation to the true solution. In all cases, the frequency cut-off is controlled by a *regularization parameter*. Regularization amounts to incorporating prior information, since the solution is not expected to have high-frequency components, and the regularization parameter determines the relative influence of the original measurement equations, $\mathbf{Ax} = \mathbf{b}$, and the prior information that the solution is smooth.

In the case of singular value decomposition (SVD), $\mathbf{x}$ is reconstructed like a Fourier series; the first few summation terms "carve out" the basic shape of $\mathbf{x}$, while the latter terms are responsible for the high-frequency components. The smallest singular values amplify the measurement noise, so an obvious strategy is to truncate the summation at $n-p$ terms. This is called Truncated Singular Value Decomposition (TSVD), and p is the regularization parameter.

Other methods (Ertürk et al. 2023) are Tikhonov regularization (Tikhonov 1963, Santos et al. 2020) and conjugate gradient (CG) regularization (Beckman 1960, Ertürk, Ezekoye, and Howell 2002).

12.3.3.2 Nonlinear Programming

Nonlinear inverse problems may be written as least-squares minimization problems, $\mathbf{x}^* = \arg\min[F(\mathbf{x})] = \arg\min[\mathbf{r}(\mathbf{x})^{\mathrm{T}}\mathbf{r}(\mathbf{x})]$, where $\mathbf{r}(\mathbf{x}) = \mathbf{a}(\mathbf{x}) - \mathbf{b}$ is the residual between a modeled outcome, $\mathbf{a}(\mathbf{x})$ and some specified vector, $\mathbf{b}$ (e.g., measurement data, or a desired design outcome). If $F(\mathbf{x})$ is convex, the most efficient way to solve this type of problem is through nonlinear programming (NLP). Conjugate gradient regularization is an example.

12.3.3.3 Metaheuristics

If $F(\mathbf{x})$ is nonconvex, an NLP algorithm can only find one of the multiple local minima, typically the one closest to $\mathbf{x}^0$. Metaheuristics avoid this problem by incorporating a random element that prevents the algorithm from becoming "stuck" in the vicinity of a shallow local minimum. In contrast to NLP, metaheuristics do not require the calculation of a derivative or sensitivity, and they can more easily accommodate discrete variables (e.g., the number of heaters in a furnace).

Many metaheuristics are inspired by optimization processes in the natural world. All metaheuristics contain heuristics, or "knobs", that must be adjusted to balance a fast convergence rate with the odds of finding a strong minimum; this is usually done by trial and error. Many metaheuristic algorithms have been deployed to solve inverse radiative transfer problems, including simulated annealing, genetic algorithms, and particle swarm optimization.

Additional techniques include Bayesian methods and artificial neural networks (Ertürk et al. 2023).

12.4 RADIATION TRANSFER IN EMERGING MANUFACTURING PROCESSES

Many manufacturing processes require precise temperature control to obtain the desired properties at the end of the process. In the case of rapid thermal processing chambers, for example, nonuniform radiative heating of silicon wafers during lithography may lead to thermally induced stresses within the wafer, and defective integrated circuits. Advanced high-strength steel alloys require precise temperature control during annealing to obtain the microstructures responsible for their superior mechanical properties. In these and many other processes, it is paramount to accurately model the radiative transfer to the product being processed. Other manufacturing processes, such as photolithography, also involve irradiating a surface, although with the objective of inducing photochemical, as opposed to thermal, transformations.

Design and control of radiation-based manufacturing processes require accurate radiation models. These, in turn, require detailed knowledge of the radiative properties of the surfaces, which may be elusive, particularly since the properties often evolve as the surface is being processed. Accordingly, inverse design and control strategies, including techniques discussed in Section 12.3, should account for uncertainty in these properties.

As an example, fiber-wound composite tanks and shells usually require lengthy curing of the resin in an autoclave after winding is completed. Chern et al. (2002a, 2002b) studied the use of infrared heating during the winding process to provide curing "on the fly." This requires careful integration of the incident radiation intensity and the resulting IR-induced resin curing rate with the winding rate so that successive winding layers bond with previous layers as the winding process proceeds. This is a case where both dependent scattering and angle-of-incidence-dependent scattering coefficient and the scattering phase function of the fiber matrix (Section 8.12) must be considered.

While many manufacturing processes exploit the uniform heating that can be achieved through radiative transfer, in other cases it is desirable to irradiate the product in a highly nonuniform way, as is the case in laser welding and powder-based laser additive manufacturing. Particularly in the case of additive manufacturing, the final product quality (e.g., minimized porosity) requires precise temperature control using a detailed laser heating model. A major challenge is knowing precisely the radiative properties of the product surface, which change dramatically as the surface evolves from powder to liquid to solid, as well as the potential masking of the product surface by any laser-induced surface plasma.

Many manufacturing processes incorporate some sort of radiometric feedback control, such as through pyrometers or infrared cameras. These sensors often have much higher spatial and temporal resolution than can be achieved through contact measurements, and, in many cases contact measurements are impractical or impossible, e.g., a moving steel strip or molten glass. Again, however, a major challenge is the uncertain spectral properties of the product surface, which often evolve with processing.

12.5 RADIATION TRANSFER FOR DIAGNOSTICS

The use of remote sensing of/or by radiation is widespread in industry and academia (Zhang et al. 2010). Monitoring of greenhouse gas concentrations is an obvious need for understanding global warming and climate change. Measurements can be by remote sensing of gas emissions from point sources such as power plant stacks, methane leaks from oil wells and processing plants, and global sensing of temperature distributions, ice coverage, and flora patterns from satellite measurements. Additional predictions of climate effects include measuring and understanding enhanced absorption of solar radiation because of deposited soot and pollution particles in snow masses and the resulting increased melting of ice caps. Improved radiation spectroscopy techniques have helped greatly in enhancing models for climate change.

Biomedical thermal radiation-based diagnostic and treatment methods (cancer diagnosis and treatment, laser eye surgery, rapid UV curing of dental filling materials, and detection of circulatory deficits) are developing areas, described in Section 1.4.7.

12.6 CONCLUDING REMARKS

The theory of thermal radiation energy transfer that began with the corpuscular theory of light (Descartes 1637, expanded by Newton in 1672 and published 1704) arguably predates serious studies of the other energy transfer modes—convection (Newton's Law of Cooling, 1701) and conduction (Fourier's Law, 1807–1822). In addition, the material in Chapter 1 shows that radiation and its many contemporary engineering applications and environmental effects greatly affect our present everyday life. In this chapter, we have briefly outlined contemporary basic research and applications that promise to continue to make radiative energy transfer an important and exciting field of research and application.

HOMEWORK (additional problems available in online Appendix I at www.ThermalRadiation.net/appendix.html)

12.1 Consider a SiC grating like that reported by Greffet et al. (2002) (see Figure 12.5). Comment on the structure of the grating and the emission wavelengths. List your suggestions for the use of such structures.

12.2 A 2D enclosure consists of two opposing black surfaces with cold surroundings. The overall objective of the inverse boundary condition design problem is to find the required temperature distribution on the upper heater surface that produce the desired temperature and energy flux over the lower design surface, T_1 and $q_1(x_1)$.

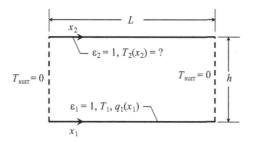

(a) Show that the boundary conditions on the heater surface and design surface are related by a Fredholm IFK,

$$\int_{x_2=0}^{W} E_{b2}(x_2)k(x_1,x_2)\,dx_2 = E_{b1} - q_1(x_1)$$

where $E_{b2}(x_2) = \sigma T_2^4(x_2)$, or, in dimensionless form,

$$\int_{X_2=0}^{1} \Theta_2(X_2)k(X_1,X_2)dX_2 = 1 - \Psi_1(X_1)$$

with $\Theta = E_b/\sigma T_1^4$, $\Psi = q/\sigma T_2^4$, and $X = x/L$. The kernel of the integral is given by:

$$k(X_1,X_2) = \frac{1}{2} \frac{H^2}{\left[H^2 + (X_2 - X_1)^2\right]^{3/2}}$$

where $H = h/L$.

(b) Discretize the integral equation so that the upper and lower surfaces are split into 30 uniform elements. Following Equation 12.2, transform the integral equation into a matrix equation $\mathbf{Ax} = \mathbf{b}$, assuming that the

desired non-dimensional energy flux distribution over the design surface is $\Psi_1 = AX_1^2 - BX_1 - C$ with $A = B = 16$ and $C = 6$, and the plates are
separated by a dimensionless gap height of $H = 0.4$.

(c) Carry out a singular value decomposition on the **A** matrix and plot the
singular values.

(d) Use TSVD to obtain Θ_2 for different values of p and plot the corresponding values of Θ_1 if Ψ_1 is known.

(e) Repeat Part (c) for different values of H and provide a physical expla-
nation for the trends in the singular values.

(f) Repeat Part (d) for $H = 0.6$, but this time using a constant energy flux of
$\Psi = -1$. How do these results compare to the temperature distribution
predicted for a parabolic energy flux profile?

12.3 Four experiments are performed to find the radiative flux on the three walls
of a triangular enclosure at measured absolute temperatures T_1, T_2, and T_3.
The energy fluxes on the three surfaces are q_1, q_2, and q_3. From the four
experiments, the matrix of equations that relate the measurements to the
energy fluxes is found by the net radiation method to be

$$\mathbf{A \cdot q} = \begin{pmatrix} 93.48 & 10.20 & -28.83 \\ 1.96 & 32.82 & 62.41 \\ 26.82 & 36.82 & 57.23 \\ 23.21 & -86.40 & 44.69 \end{pmatrix} \begin{pmatrix} q_1 \\ q_2 \\ q_3 \end{pmatrix} = \begin{pmatrix} 34.72 \\ 70.92 \\ 82.93 \\ -26.22 \end{pmatrix} = \mathbf{b}$$

(a) Believing that only three equations are needed for three unknowns, a
grad student deletes the last row of the **A** matrix and the **b** vector and
solves for the unknowns. What does the student predict? (Make sure
the determinant of the reduced 3×3 **A** matrix is nonzero.)

(b) Knowing that measurements of T at high temperatures are generally
accurate to three significant figures at best, the student rounds the values in the **b** matrix to three significant figures and again solves for the
unknowns using the three-row matrixes. Compare with the answer to
part (a).

(c) Find the singular values and condition number of the 3×3 **A** matrix
and see whether they might indicate problems in finding the answer to
parts (a) and (b).

(d) Now believing that you should use all the data you can get, determine
the singular values and condition number of the complete **A** matrix.

(e) Determine the values of **q** using the complete **A** matrix and the four-
row **b** vector for both the complete and rounded **b** values.

12.4 A plate is coated with a material combining directional and spectral selec-
tivity so that the plate has a normal total solar absorptivity of 0.92 and an
infrared hemispherical emissivity of 0.040 at long wavelengths. When
placed in sunlight normal to the Sun's rays, what temperature will the plate
reach (neglecting conduction and convection and with no heat losses from

the unexposed side of the plate)? What assumptions did you make in reaching your answer? The incident solar flux ("insolation") at the Earth's surface is 1050 W/m².
Answer: 652 K

12.5 An uncovered Styrofoam pan filled with water is placed outdoors on a cloudy night. The air temperature is 5°C. There is almost no wind, so the energy transfer coefficient between the air and water is $h = 10$ W/(m²·K). The high cirro-stratus clouds in the night sky act as blackbody surroundings at $T_e = 210$ K. Water is opaque for long-wavelength radiation. The index of refraction of water is 1.33.
(a) Will the water start to freeze? Show calculations to prove your answer.
(b) What minimum air temperature will prevent freezing?
Answers: (a) freezing does begin; (b) 20.5°C.

12.6 Stephen Hawking (1974) predicted that black holes should emit a perfect blackbody radiation spectrum in proportion to the surface gravity of the black hole. Further predictions are that the surface gravity of a black hole is inversely proportional to its mass. Thus, very small black holes should emit large blackbody radiation. If a black hole with 1 Earth mass has a predicted blackbody temperature of 10^{-7} K, what fraction of an Earth-mass black hole will emit at $T = 2.7$K (equal to the background radiation temperature of space and thus possibly be detectable by its blackbody emission)?
Answer: $3.7 \times 10^{-8} \, m_{Earth}$

12.7 Observers at NASA noticed an unexpected small deceleration of the early deep space probes Pioneer 10 and 11 as they moved away from Earth. This was known as "the Pioneer Anomaly." After much speculation and many false starts, researchers determined that thermal radiation pressure exerted by emitted radiation waste energy from the electronics was the culprit. Assuming a projected radiating area of 0.2 m² facing forward along the spacecraft trajectory, a spacecraft mass of 258 kg, and a mean radiating temperature on the outward-facing surface (effective emissivity = 0.3) of 375 K, what deceleration was exerted on the spacecraft? (A much more detailed analysis can be found in Turyshev et al. (2012)). Their detailed thermal model predicts a deceleration of 7 to 10 x10^{-10} m/s².)
Answer: 8.7×10^{-10} m/s

Appendix
Radiation Constants and Blackbody Functions

Table A.1 lists some fundamental constants of radiation physics. In Table A.2, values of various radiation constants are given in both SI and US conventional system (USCS) engineering units. All values in Tables A.1 and A.2 are based on values of the physical constants given by Tiesinga et al. (2020).

The emission properties of a blackbody are presented in Chapter 2. Table A.3 is included for quick reference and for help in checking calculations for blackbody properties. It lists blackbody emission properties as functions of λT in both SI and USCS units. The $F_{0 \rightarrow \lambda T}$ were calculated using the series obtained by integrating the blackbody radiation function using integration by parts as in Chapter 2. This series was used to evaluate the blackbody integrals as given by Chang and Rhee (1984). The relation is repeated here for convenience:

$$F_{0 \rightarrow \lambda T} = \frac{15}{\pi^4} \sum_{m=1}^{\infty} \left[\frac{e^{-m\zeta}}{m} \left(\zeta^3 + \frac{3\zeta^2}{m} + \frac{6\zeta}{m^2} + \frac{6}{m^3} \right) \right]$$

where $\zeta = C_2 / \lambda T$. This series is quite useful for computer solutions to provide calculation of the $F_{0 \rightarrow \lambda T}$ function. A calculator for the blackbody functions is available at www.ThermalRadiation.net/blackbody.html.

TABLE A.1

Values of the Fundamental Physical Constants

Definition	Symbol and Values

Bohr electron radius $\quad a_0 = (4\pi / \mu_0 c_0^2)(\hbar^2 / m_e e^2)$

$\qquad\qquad\qquad\qquad = 0.529\,177\,210\,90 \times 10^{-10}\,\text{m}$

Speed of light in vacuum $\quad c_0 = 2.999\,792\,458 \times 10^8$ m/s

Electronic charge $\qquad e = 1.602\,176\,634 \times 10^{-19}$ C

Planck's Constant $\qquad h = 6.626\,070\,15 \times 10^{-34}$ J·s

$\qquad\qquad\qquad\qquad \hbar = h/2\pi = 1.054\,571\,800 \times 10^{-34}$ J·s

Boltzmann's constant $\qquad k = 1.380\,649 \times 10^{-34}$ J/K

Electron rest mass $\qquad m_e = 9.109\,383\,70 \times 10^{-31}$ kg

Avogadro's number $\qquad N_a = 6.022\,140\,76 \times 10^{26}$ particles/kg·mol

Classical electron radius $\quad r_0 = (1/4\pi\gamma_0)(e^2/m_e c_0^2) = 2.817\,940\,3227 \times 10^{-15}$ m

Thomson cross-section $\qquad \sigma_T = 8\pi r_0^2/3 = 0.665\,245\,873\,21 \times 10^{-28}$ m^2

Permittivity of vacuum $\qquad \gamma_0 = 8.854\,187\,8128 \times 10^{-12}$ F/m

Permeability of vacuum $\qquad \mu_0 = 4\pi \times 10^{-7}\,\text{N}\cdot\text{s}^2/\text{C}^2$

Electron volt $\qquad\qquad 1\text{ eV} = 1.602\,176\,6208 \times 10^{-19}$ J

Ionization potential of hydrogen $\quad R = \left(c_0^2\mu_0 / 4\pi\right)\left(e^2 / 2a_0\right) = \left[c_0^4\mu_0^2 / \left(4\pi\right)^2\right]\left(e^4 m_e / 2\hbar^2\right)$
atom (Rydberg's constant)

$\qquad\qquad\qquad = 13.605\,693\,123\,\text{eV}$

Source: Tiesinga et al.,2021

TABLE A.2
Radiation Constants

Symbol	Definition	Value
C_1	Constant in Planck's spectral energy (or intensity) distribution	0.18878×10^8 Btu·µm⁴/(h·ft²·sr) $0.595\ 522\ 00 \times 10^8$ W·µm⁴/(m²·sr) $0.595\ 522\ 00 \times 10^{-16}$ W· m²/sr
C_2	Constant in Planck's spectral energy (or intensity) distribution	$25{,}897.9924$ µm·°R $14{,}387.7688$ µm·K $0.014\ 387\ 7688$ m·K
C_3	Constant in Wien's displacement law	$5{,}215.9912$ µm·°R $2{,}897.7720$ µm·K $0.002\ 897\ 7720$ m·K
C_4	Constant in equation for maximum blackbody intensity	6.8761×10^{-14} Btu/(h·ft²·µm·°R⁵·sr) 4.09570×10^{-12} W/(m²·µm·K⁵·sr)
σ	Stefan-Boltzmann constant	0.17123×10^{-8} Btu/(h·ft·°R⁴) $5.670\ 367 \times 10^{-8}$ W/(m²·K⁴)
q_{solar}	Solar constant	433 Btu/(h·ft²) $1{,}366$ W/m²
T_{solar}	Effective surface radiating temperature of the Sun	$5{,}780$ K $10{,}400$°R

Source: Tiesinga et al.,2021

TABLE A.3A
Blackbody Functions

Wavelength–Temperature Product, λT		Blackbody Hemispherical–Spectral Emissive Power Divided by the Fifth Power of Temperature, $E_{\lambda b}/T^5$		Blackbody Fraction
$\mu m \cdot K$	$\mu m \cdot °R$	$W/(m^2 \cdot \mu m \cdot K^5)$	$Btu/(h \cdot ft^2 \cdot \mu m \cdot °R^5)$	$F_{0 \rightarrow \lambda T}$
600	1,080	185.40E–18	311.04E–20	9.29E–8
800	1,440	176.59E–16	296.26E–18	1.64E–5
1,000	1,800	211.13E–15	354.20E–17	3.21E–4
1,200	2,160	933.41E–15	156.59E–16	0.00213
1,400	2,520	239.45E–14	401.72E–16	0.00779
1,600	2,880	443.82E–14	744.56E–16	0.01972
1,800	3,240	669.05E–14	112.24E–15	0.03934
2,000	3,600	879.01E–14	147.47E–15	0.06673
2,200	3,960	105.04E–13	176.22E–15	0.10089
2,400	4,320	117.37E–13	196.90E–15	0.14026
2,500	4,500	121.72E–13	204.19E–15	0.16136
2,600	4,680	124.93E–13	209.58E–15	0.18312
2,700	4,860	127.09E–13	213.21E–15	0.20536
2,800	5,040	128.30E–13	215.24E–15	0.22789
2,897.7720[a]	5,216	128.67E–13	215.86E–15	0.25006
2,900	5,220	128.67E–13	215.86E–15	0.25056
3,000	5,400	128.30E–13	215.24E–15	0.27323
3,100	5,580	127.30E–13	213.56E–15	0.29578
3,200	5,760	125.76E–13	210.98E–15	0.31810
3,300	5,940	123.77E–13	207.63E–15	0.34011
3,400	6,120	121.41E–13	203.67E–15	0.36173
3,500	6,300	118.75E–13	199.21E–15	0.38291
3,600	6,480	115.86E–13	194.36E–15	0.40360
3,800	6,840	109.59E–13	183.85E–15	0.44337
4,000	7,200	102.97E–13	172.75E–15	0.48087
4,200	7,560	962.6 IE–14	161.49E–15	0.51600
4,400	7,920	896.45E–14	150.39E–15	0.54878
4,600	8,280	832.48E–14	139.66E–15	0.57926
4,800	8,640	771.49E–14	129.43E–15	0.60754
5,000	9,000	713.96E–14	119.78E–15	0.63373
5,200	9,360	660.11E–14	110.74E–15	0.65795
5,400	9,720	609.99E–14	102.33E–15	0.68034
5,600	10,080	563.56E–14	945.44E–16	0.70102
5,800	10,440	520.67E–14	873.49E–16	0.72013
6,000	10,800	481.16E–14	807.21E–16	0.73779
6,400	11,520	411.45E–14	690.26E–16	0.76920
6,800	12,240	352.71E–14	591.71E–16	0.79610

(Continued)

TABLE A.3A (CONTINUED)
Blackbody Functions

Wavelength–Temperature Product, λT		Blackbody Hemispherical–Spectral Emissive Power Divided by the Fifth Power of Temperature, $E_{\lambda b}/T^5$		Blackbody Fraction
$\mu m \cdot K$	$\mu m \cdot °R$	$W/(m^2 \cdot \mu m \cdot K^5)$	$Btu/(h \cdot ft^2 \cdot \mu m \cdot °R^5)$	$F_{0 \to \lambda T}$
7,200	12,960	303.27E–14	508.78E–16	0.81918
7,600	13,680	261.65E–14	438.96E–16	0.83907
8,000	14,400	226.55E–14	380.07E–16	0.85625
8,400	15,120	196.87E–14	330.28E–16	0.87116
8,800	15,840	171.71E–14	288.06E–16	0.88413
9,200	16,560	150.30E–14	252.14E–16	0.89547
9,600	17,280	132.02E–14	221.48E–16	0.90541
10,000	18,000	116.37E–14	195.22E–16	0.91416
11,000	19,800	860.92E–15	144.43E–16	0.93185
12,000	21,600	649.08E–15	108.89E–16	0.94505
13,000	23,400	497.78E–15	835.10E–17	0.95509
14,000	25,200	387.67E–15	650.37E–17	0.96285
15,000	27,000	306.13E–15	513.58E–17	0.96893
16,000	28,800	244.80E–15	410.68E–17	0.97377
18,000	32,400	161.78E–15	271.41E–17	0.98081
20,000	36,000	111.03E–15	186.26E–17	0.98555
25,000	45,000	492.47E–16	826.18E–18	0.99217
30,000	54,000	250.2 IE–16	419.76E–18	0.99529
35,000	63,000	140.12E–16	235.07E–18	0.99695
40,000	72,000	844.11E–17	141.61E–18	0.99792
45,000	81,000	538.22E–17	902.93E–19	0.99852
50,000	90,000	359.11E–17	602.45E–19	0.99890
60,000	108,000	117.57E–17	297.90E–19	0.99935
70,000	126,000	975.65E–18	163.68E–19	0.99959

[a] $\lambda_{max}T$.

References

Note: The following abbreviations are used in the references:

HTE: Heat Transfer Engineering
ICHMT: International Communications in Heat and Mass Transfer
IJHMT: International Journal of Heat and Mass Transfer
IJTS: International Journal of Thermal Sciences
IPSE: Inverse Problems in Science and Engineering (formerly *Inverse Problems in Engineering*)
JHT: ASME Journal of Heat Transfer
JTHT: AIAA Journal of Thermophysics and Heat Transfer
JOSA: Journal of the Optical Society of America
JQSRT: Journal of Quantitative Spectroscopy and Radiative Transfer

Alberti, M., Weber, R., and Mancini, M.: Gray gas emissivities for H_2O-CO_2-CO-N_2 mixtures, *JQSRT*, 219, 274–291, 2018.
Alberti, M., Weber, R., and Mancini, M.: New formulae for gray gas absorptivities of H_2O, CO_2, and CO, *JQSRT*, 255, 107277, 2020.
André, F., Solovjov, V. P., and Webb, B. W.: The ω-absorption line distribution function for rank correlated SLW model prediction of radiative transfer in non-uniform gases, *JQSRT*, 280, 108081, 2022.
Baranov, D. G., Xiao, Y., Nechepurenko, I. A., Krasnok, A., Alù, A., and Kats, M. A.: Nanophotonic engineering of far-field thermal emitters, *Nat. Mater.*, 18, 920–930, 2019.
Beckman, F. S.: The solution of linear equations by the conjugate gradient method, in *Mathematical Methods for Digital Computers*, A. Ralston and H. S. Wilf eds., 62–72, John Wiley and Sons, New York, 1960.
Beer, A.: Bestimmung der Absorption des rothen Lichts in farbigen Flüssigkeiten [Determination of the absorption of red light in colored liquids], *Annalen der Physik und Chemie (in German)*, 86(5), 78–88, 1852.
Bernath, P. and Rothman, L.: HITRAN2016 special issue, *JQSRT*, 203, 1–582, 2017.
Bohren, C. and Huffman, D.: *Absorption and Scattering of Light by Small Particles*, New edition, Wiley-Interscience, New York, 1998
Boltzmann, L.: Ableitung des Stefan'schen Gesetzes, betreffend die Abhängigkeit der Wärmestrahlung von der Temperatur aus der electromagnetischen Lichttheorie, *Ann. Phys., Ser. 2*, 22, 291–294, 1884.
Bouguer, Pierre: *Essai d'Optique, sur la gradation de la lumiere*, Claude Jombert, Paris, 16–22, 1729.
Branstetter, J. R.: *Radiant Heat Transfer between Nongray Parallel Plates of Tungsten*, NASA TN D-1088, Washington, DC, 1961.
Buckley, H.: On the radiation from the inside of a circular cylinder, part I, *Philos. Mag.*, 4, 753–762, 1927; Part II, *Philos. Mag.*, 6, 447–457, 1928.
Case, K. M.: Transfer problems and the reciprocity principle, *Rev. Mod. Phys.*, 29, 651–663, 1957.
Chambers, R. L. and Somers, E. V.: Radiation fin efficiency for one-dimensional heat flow in a circular fin, *JHT*, 81(4), 327–329, 1959.
Chandrasekhar, S.: *Radiative Transfer*, Dover, New York, 1960.

Chandrasekhar, S.: The 1979 Milne Lecture – Edward Arthur Milne: His part in the development of modern astrophysics, *Q. J. R. Astr. Soc.*, 21, 93–107, 1980.

Chang, S. L. and Rhee, K. T.: Blackbody radiation functions, *ICHMT*, 11, 451–455, 1984.

Charle, M.: *Les Manuscripts de Léonard de Vinci, Manuscripts C, E, et K de la Bibliothèque de l'Institute Publiés en Facsimilés Phototypiques*, Ravisson-Mollien, Paris, 1888. (Referenced in Knowles Middleton, W. E.: Note on the invention of photometry, *Am. J. Phys.*, 31(3), 177–181, 1963.)

Chern, B.-C., Moon, T. J., and Howell, J. R.: On-line processing of unidirectional fiber composites using radiative heating: I. Model, *J. Compos. Mater.*, 36(16), 1905–1934, 2002a.

Chern, B.-C., Moon, T. J., and Howell, J. R.: On-line processing of unidirectional fiber composites using radiative heating: II. Radiative properties, experimental validation and process parameter selection, *J. Compos. Mater.*, 36(16), 1935–1665, 2002b.

Chin, J. H., Panczak, T. D., and Fried, L.: Spacecraft thermal modeling, *Int. J. Numer. Methods Eng.*, 35, 641–653, 1992.

Christiansen, C.: Absolute determination of the heat emission and absorption capacity, *Ann. Phys. Wied.*, 19, 267–283, 1883.

Cumber, P. S.: View factors-when is ray tracing a good idea?, *IJHMT*, 189, 122698, 2022.

d'Aguillon, F., S. J.: Opticorum Libri Sex, Antwerp, 1613. (Referenced in Knowles Middleton, W. E.: Note on the invention of photometry, *Am. J. Phys.*, 31(3), 177–181, 1963.)

Daun, K. J., Howell, J. R., and Morton, D. P.: Smoothing Monte Carlo exchange factors through constrained maximum likelihood estimation, *JHT*, 127(10), 1124–1128, 2005.

Davison, B.: Spehrical-harmonics method for neutron-transport problems in cylindrical geometry, *Can. J. Phys.*, 35(5), 576–593, 1957.

De Vos, J. C.: A new determination of the emissivity of tungsten ribbon, *Physica*, 20, 690–714, 1954.

Deissler, R. G.: Diffusion approximation for thermal radiation in gases with jump boundary condition, *JHT*, 86(2), 240–246, 1964.

Denison, M. K. and Webb, B. W.: The absorption-line blackbody distribution function at elevated pressure, in *Radiative Transfer—I, Proceedings of the First International Symposium on Radiation Transfer*, M. P. Mengüç, ed., 228–238, Begell House, New York, 1996a.

Denison, M. K. and Webb, B. W.: The spectral line weighted-sum-of-gray-gases model—A review, in *Radiative Transfer—I, Proceedings of the First International Symposium on Radiative Transfer*, M. P. Mengüç, ed., 193–208, Begell House, New York, 1996b.

Descartes, R.: *Meditations on First Philosophy*, Library of Liberal Arts, Prentice Hall, Englewood Cliffs, NJ, 1951 [originally published 1641].

Didari, A. and Mengüç, M. P.: A biomimicry design for nanoscale radiative cooling applications inspired by Morpho didius butterfly, *Sci. Rep.*, 8, 16891, 2018.

Draper, J. W.: On the production of light by heat, *Phil. Mag., Ser. 3*, 30, 345–360, 1847.

Drude, P.: Zur Elektronentheorie der Metalle, *Annalen der Physik*, 306(3), 566–613, 1900a.

Drude, P.: Zur Elektronentheorie der Metalle; II Teil: Galvanomagnetische und Thermomagnetische Effecte, *Annalen der Physik*, 308(11), 369–402, 1900b.

Duffie, J. A. and Beckman, W. A.: *Solar Energy Thermal Processes*, 3rd edn., Wiley, New York, 2006.

Dupoirieux, F., Tesse, L., Alvia, S., and Taine, J.: An optimized reciprocity Monte Carlo method for the calculation of radiative transfer in media of various thicknesses, *IJHMT*, 49, 1310–1319, 2006.

Eckert, E.: Das strahlungsverhaltnis von flachen mit einbuchtungen und von zylindrischen bohrungen, *Arch. Waermewirtsch.*, 16(5), 135–138, 1935.

Eckert, E. R. G. and Drake, R. M. Jr.: *Heat and Mass Transfer*, 2nd edn., McGraw–Hill, New York, 1959.

Eddington, A. S.: *The Internal Constitution of the Stars*, Cambridge University Press, 1926.

Edwards, D. K.: Molecular gas band radiation, in *Advances in Heat Transfer*, vol. 12, T. F. Irvine, Jr. and J. P. Hartnett, eds., 115–193, Academic Press, New York, 1976.

Einstein, A.: Über einen die Erzeugung und Verwandlung des Lichtes betreffenden heuristischen Gesichtspunkt, *Annalen der Physik*, 17(6), 132–148, 1905.

Ertürk, H., Daun, K., França, F. H. R., Hajimirza, S., and Howell, J. R.: Inverse methods in thermal radiation analysis and experiment, *JHT*, Special issue in honor of Raymond Viskanta, 145, April, 2023. Published on-line 3 Dec. 2022.

Ertürk, H., Ezekoye, O. A., and Howell, J. R.: Comparison of three regularized solution techniques in a three-dimensional inverse radiation problem, *JQSRT*, 73, 307–316, 2002.

Feynman, R.: *The Feynman Lectures on Physics*, vol. 1, Eq. 45.17, California Institute of Technology, Pasadena, CA, 1963.

Galtier, M., El Hafi, M., Caliot, C., et al.: Integral formulation of null collision Monte-Carlo algorithms, *JQSRT*, 125, 57–68, 2013.

Garbuny, M.: *Optical Physics*, 2nd printing, Academic Press, New York, 1967.

Gardon, R.: The emissivity of transparent materials, *J. Am. Ceram. Soc.*, 39(8), 278–287, 1956.

Gardon, R.: Calculation of temperature distributions in glass plates undergoing heat-treatment, *J. Am. Ceram. Soc.*, 41(6), 200–209, 1958.

Gardon, R.: A review of radiant heat transfer in glass, *J. Am. Ceram. Soc.*, 44(7), 305–312, 1961.

Gebhart, B.: Surface temperature calculations in radiant surroundings of arbitrary complexity—For gray, diffuse radiation, *IJHMT*, 3(4), 341–346, 1961.

Gebhart, B.: *Heat Transfer*, 2nd edn., McGraw-Hill, New York, 150–163, 1971.

Goody, R. M. and Yung, Y. L.: *Atmospheric Radiation*, 2nd edn., Oxford University Press, New York, 1989.

Gordon, I. E., Dothe, H., Rothman, L. S.: The resurrection of the HITEMP database and its application to the study of stellar and planetary atmospheres, *Geophys Res. Abst.*, 9, 01799, 2007.

Gordon, I. E., et al., eds.: The HITRAN 2016 molecular spectroscopic database, *JQSRT*, 203, 3–69, 2017.

Gordon, I. E., et al., eds.: The HITRAN 2020 molecular spectroscopic database, *JQSRT*, 277, 107949, 2022.

Greffet, J.-J., Carminati, R., Joulain, K., Mulet, J.-P., Mainguy, S., and Chen, Y.: Coherent emission of light by thermal sources, *Nature*, 416, 61–64, 2002.

Hadamard, Jacques: Sur les problèmes aux dérivées partielles et leur signification physique, *Princeton University Bulletin*, 49–52, 1902.

Hagen, E. and Rubens, H.: Metallic reflection, *Ann. Phys.*, 1(2), 352–375, 1900. (See also E. Hagen and H. Rubens: Emissivity and electrical conductivity of alloys, *Deutsch. Phys. Ges. Verhandl.*, 6(4), 128–136, 1904).

Hajimirza, S. and Howell, J. R.: Computational and experimental study of a multi-layer absorptivity enhanced thin film silicon solar cell, *JQSRT*, 143, 56–62, 2014.

Hargreaves, R. J., Gordon, I. E., Rey, M., Nikitin, A. V., Tyuterev, V. G., Kochanov, R. V., and Rothman, L. S.: An accurate, extensive, and practical line list of methane for the HITEMP database, *Astrophys. J. Supp. Ser.*, 247, 55, 2020.

Haslinger, P., Jaffe, M., Xu, V., Schwartz, O., Sonnleitner, M., Ritsch-Marte, M., Ritsch, H., and Müller, H.: Attractive force on atoms due to blackbody radiation, *Nat. Phys.*, 14, 257–260, 2018.

Hau, L. V., Taming light with cold atoms, *Physics World*, 14, 35–40, 2001.

Hau, L. V., Harris, S. E., Dutton, Z., and Behroozi, C. H.: Light speed reduction to 17 metres per second in an ultracold atomic gas, *Nature*, 397, 594–598, 1999.

Hawking, S. W.: Black hole explosions? *Nature*, 248 (5443), 30–31, 1974.

Heaslet, M. A. and Warming, R. F.: Radiative transport and wall temperature slip in an absorbing planar medium, *IJHMT*, 8(7), 979–994, 1965.

Heaviside, O.: On the electromagnetic effects due to the motion of electrification through a dielectric, *Phil. Mag. S.5*, 27, 324, 1889.

Helfenstein, P., Veverka, J., and Hillier, J: The Lunar opposition effect: A test of alternative models, *Icarus*, 128(1), 1997.

Henyey, L. G. and Greenstein, J. L.: Diffuse radiation in the galaxy, *Astrophys. J.*, 88, 70–83, 1940.

Hering, R. G. and Smith, T. F.: Surface radiation properties from electromagnetic theory, *IJHMT*, 11(10), 1567–1571, 1968.

Hibbard, R. R.: Equilibrium temperatures of ideal spectrally selective surfaces, *Sol. Energy*, 5(4), 129–132, 1961.

Hively, L., Loebl, A., and Stoll, J.: Comment appended to Chris Mattmann: A vision for data science, *Nature*, 493(7433), 473–5, 2013.

Hottel, H. C.: Radiant heat transmission, Chapter 4, in *Heat Transmission*, 3rd edn., W. H. McAdams, ed., McGraw-Hill, New York, 1954.

Hottel, H. C. and Sarofim, A. F.: *Radiative Transfer*, McGraw-Hill, New York, 1967.

Hottel, H. C., Sarofim, A. F., Vasalos, I. A., Dalzell, W. H.: Multiple scatter: Comparison of theory with experiment, *JHT*, 92(2), 285–291, 1970.

Howell, J. R., ed.: Editorial on nomenclature and symbols in heat transfer, *JHT*, 121(4), 770–773, 1999.

Howell, J. R.: The effect of bifurcation on numerical calculation of conjugate heat transfer with radiation, *JQSRT*, 196, 242–245, 2017.

Howell, J. R. and Daun, K.: The past and future of the Monte Carlo method in thermal radiation transfer, *JHT*, 143, 100801-1–100801-9, 2021.

Howell, J. R., Mengüç, M. P., Daun, K., and Siegel, R.: *Thermal Radiation Heat Transfer*, 7th edn., CRC Press-Taylor and Francis, Boca Raton, FL, 2021.

Howell, J. R. and Perlmutter, M.: Monte Carlo solution of thermal transfer through radiant media between gray walls, *JHT*, 86(1), 116–122, 1964a.

Howell, J. R. and Perlmutter, M.: Monte Carlo solution of radiant heat transfer in nongrey nonisothermal gas with temperature dependent properties, *AICHE J.*, 10(4), 562–567, 1964b.

IPCC: *Climate Change 2022: Impacts, Adaptation and Vulnerability*, The Working Group II contribution to the Sixth Assessment Report, The Intergovernmental Panel on Climate Change (IPCC), Geneva, Switzerland, 27 February 2022.

Jeans, Sir. J.: On the partition of energy between matter and the ether, *Phil. Mag.*, 10, 91–97, 1905.

Jonasson, J., Billur, E., and Ormaetxea, A.: A hot stamping line: From a technological and business perspective, in *Hot Stamping of Ultra High-Strength Steels*, E. Billur, ed., Springer, Cham, 2019.

Karman, T., et al.: Update of the HITRAN collision-induced absorption section, *Icarus*, 328, 160–175, 2019.

Kaviany, M. and Singh, B. P.: Radiative heat transfer in porous media, in *Advances in Heat Transfer*, J. P. Hartnett and T. F. Irvine Jr., eds., 23, 133–186, Elsevier, 1993.

Kirchhoff, G.: Appendix, Über das Verhältniß zwischen dem Emissionsvermögen und dem Absorptionsvermögen der Körper für Wärme und Licht, in *Untersuchungen über das Sonnenspectrum und die Spectren der chemischen Elemente*, 22–39, Ferd. Dümmler's Verlagsbuchhandlung, Berlin, 1862.

Koba, Y. and Matsumoto, S., Recently Monte Carlo calculation in radiation medicine, *Hoshasen*, 43(2), 69–72, 2017.

Kuo, L., Labrie, D., and Chylek, P.: Refractive indices of water and ice in the 0.65- to 2.5-μm spectral range, *Appl. Opt.*, 32(19), 3531–3540, 1993.

Kurtscheid, C., Dung, D., Busley, E., Vewinger, F., Rosch, A., and Weitz, M.: Thermally condensing photons into a coherently split state of light, *Science*, 366, 894–897, 2019.

Lacis, A. A. and Hansen, J. E.: A parameterization for the absorption of solar radiation in the Earth's atmosphere, *J. Atm. Sci.*, 31, 118–133, 1973.

Lambert, J. H.: *Photometrie: Photometria, sive de mensura et gradibus luminus, colorum et umbrae*, (1760) republished by Verlag von Wilhelm Engelmann, Leipzig, 1892.

Langley, S. P.: Experimental determination of wave-lengths in the invisible prismatic spectrum, *Mem. Natl. Acad. Sci.*, 2, 147–162, 1883.

Larsen, M. E. and Howell, J. R.: Least squares smoothing of direct exchange areas in zonal analysis, *JHT*, 18(1), 239–242, 1986.

Lebedev, P.: Untersuchungen über die Druckkräfte des Lichtes, *Annalen der Physik*, 4th Series, 6, 433–458, 1901.

Lee, S.-C.: Dependent vs independent scattering in fibrous composites containing parallel fibers, *JTHT*, 8, 641–646, 1994.

Lide, D. R., ed.: *Handbook of Chemistry and Physics*, 88th edn., CRC Press, Boca Raton, FL, 2008.

Lin, E. I., Stultz, J. W., and Reeve, R. T.: Effective emittance for Cassini multilayer insulation blankets and heat loss near seams, *JTHT*, 10(2), 357–363, 1996.

Lin, S. H. and Sparrow, E. M.: Radiant interchange among curved specularly reflecting surfaces—Application to cylindrical and conical cavities, *JHT*, 87(2), 299–307, 1965.

Long, R. L.: A review of recent air force research on selective solar absorbers, *J. Eng. Power*, 87(3), 277–280, 1965.

Lorentz, H. A.: The motion of electrons in metallic bodies. I. *Proc. Koninklijke Akademie van Wetenschappen*, 7, 438–453, 1905a.

Lorentz, H. A.:The motion of electrons in metallic bodies. II. *Proc. Koninklijke Akademie van Wetenschappen*, 7, 585–593, 1905b.

Lorentz, H. A.: The motion of electrons in metallic bodies. III. *Proc. Koninklijke Akademie van Wetenschappen*, 7, 684–691, 1905c.

Lorentz, H. A.: On the radiation of heat in a system of bodies having a uniform temperature, *Proc. Koninklijke Akademie van Wetenschappen*, 8, 401–421, 1905d.

Lorentz, H. A.: *The Theory of Electrons and its Applications to the Phenomena of Light and Radiant Heat*, B. G. Teubner, Liepzig, 1st edn., 1909; 2nd edn., 1916.

Lorenz, L: Sur la lumière réfléchie et réfractée par une sphère (surface) transparente, in *Oeuvres scientifiques de L. Lorenz." revues et annotées* par H. Valentiner. Tome Premier, Librairie Lehmann & Stage, Copenhague, Denmark, 403–529, 1898 (see Classical Papers at http://www.t-matrix.de/, accessed April 26, 2015.).

Marshak, R. E.: Note on the spherical harmonics methods as applied to the Milne problem for a sphere, *Phys. Rev.*, 71, 443–446, 1947.

Maruyama, S.: Concept design of linear-motor-accelerated projectile for nanoparticle dispersion in stratosphere, *Therm. Sci. Eng. Prog.*, 15, 100437, 2020.

Maslowski, W., Kinney, J. C., Higgens, M., and Roberts, A.: The future of Arctic sea ice, *Ann. Rev. Earth Planet. Sci.*, 40, 625–654, 2012.

Mason, J., Cone, M. T., and Fry, E.: Ultraviolet (250–550 nm) absorption spectrum of pure water, *Appl. Opt.*, 55(25), 7163–7172, 2016.

Maxwell, J. C.: A dynamical theory of the electromagnetic field, in W. D. Niven, ed., *The Scientific Papers of James Clerk Maxwell*, vol. 1, Cambridge University Press, Cambridge, UK, 1890.

Meng, S., Long, L., Wu, Z., Denisuk, N., Yang, Y., Wang, L., Cao, F., and Zhu, Y.: Scalable dual-layer film with broadband infrared emission for sub-ambient daytime radiative cooling, *Sol. Energy Mater. Sol. Cells*, 208, 110393, 2020.

Mengüç, M. P., Francoeur, M., eds, *Light, Plasmonics and Particles*, Nanophotonics Series, Elsevier, Amsterdam, 2023.

Mengüç, M. P. and Viskanta, R.: Comparison of radiative transfer approximations for highly forward scattering planar medium, *JQSRT*, 29(5) 381–394, 1983.

Mie, G.: Beiträge zur Optik trüber Medien, speziell kolloidaler Metallösungen, *Ann. Phys.*, 330, 377–445, 1908a.

Mie, G.: Optics of turbid media, *Ann. Phys.*, 25(3), 377–445, 1908b.

Milne, F. A.: Thermodynamics of the Stars, in *Handbuch der Astrophysik*, G. Eberhard, A. Kohlschütter, H. Ludendorff, E. A. Milne, A. Pannekoek, S. Rosseland, and W. Westphal eds. 3, 65–255, Springer-Verlag, OHG, Berlin, 1930.

Mishchenko, M. I.: "Independent" and "dependent" scattering by particles in a multi-particle group, *OSA Continuum*, 1(1), 243–260, 2018.

Mishchenko, M. I. and Yurkin, M. A.: Co- and counter-propagating wave effects in an absorbing medium, *JQSRT Special Issue, Electromagnetic and Light Scattering VIII*, 242, 106688, 2020.

Modest, M. F.: The weighted-sum-of-gray-gases model for arbitrary solution methods in radiative transfer, *JHT*, 113(3), 650–656, 1991.

Modest, M. F.: The treatment of nongray properties in radiative heat transfer: From past to present, *JHT*, 135(6), 061801–1–061801-12, 2013.

Modest, M. F. and Mazumder, S.: *Radiative Heat Transfer*, 4th edn., Academic Press, New York, 2021.

Moore, S. W.: Solar absorber selective paint research, *Sol. Energy Mater.*, 12, 435–447, 1985a.

Moore, S. W.: Progress on solar absorber selective paint research, *Sol. Energy Mater.*, 12, 449–460, 1985b.

Newton, I.: *Opticks: or, a Treatise of the Reflexions, Refractions, Inflexions and Colours of Light*, 1 edn., Sam. Smith, and Benj. Walford, London, 1704.

Nicodemus, F. E., Richmond, J. C., Hsia, J. J., Ginsberg, I. W., and Limperis, T.: *Geometrical Considerations and Nomenclature for Reflectance*, NBS Monograph 160, National Bureau of Standards, United States Department of Commerce, Washington, DC, 1977.

Nobili, L. and Melloni, M.: New experiments in caloric, performed by means of the thermomultiplier, *Am. J. Sci. and Arts*, 1, 185–180, 1833. (Trans. from *Annales de Chim. et de Phys.*, October 1831.)

Nusselt, W.: Graphische bestimmung des winkelverhaltnisses bei der wärmestrahlung, *VDI Z.*, 72, 673, 1928.

Ocean Studies Board. *Climate Intervention: Reflecting Sunlight to Cool the Earth*, National Research Council, National Academies Press, 260 pp., Washington, DC, 2015.

Orlova, N. S.: Photometric relief of the lunar surface, *Astron. Z.*, 33(1), 93–100, 1956.

Öztürk, F. E., Lappe, T., Hellmann, G., Schmitt, J., Klaers, J., Vewinger, F., Kroha, J., and Weitz, M.: Observation of a non-Hermitian phase transition in an optical quantum gas, *Science*, 372(6537), 88–91, 2021.

Palik, E. D., ed.: *Handbook of Optical Constants of Solids*, vols. I–IV, Elsevier, New York, 1998.

Panczak, T. D., Rickman, S., Fried, L., and Welch, M..: Thermal synthesizer system: An integrated approach to spacecraft thermal analysis, *SAE Trans. J. Aerospace*, 100(1), 1851–1867, 1991.

Parker, W. J. and Abbott, G. L.: Theoretical and experimental studies of the total emittance of metals, in *Symposium on Thermal Radiation of Solids*, 11–28, NASA SP-55, San Francisco, CA, 1964.

Pascale, M., Giteau, M., Papadaki, G. T.: Perspective on near-field radiative heat transfer, *arXiv*:2210.00929, Oct. 2022.

Pearson, J. T., Webb, B. W., Solovjov, V. P., and Ma, J.: Efficient representation of the absorption line blackbody distribution function for H_2O, CO_2, and CO at variable temperature, mole fraction, and total pressure, *JQSRT*, 138, 82–96, 2014.

Perlmutter, M. and Howell, J. R.: Radiant transfer through a gray gas between concentric cylinders using Monte Carlo, *JHT*, 86(2), 169–179, 1964.

Perlmutter, M. and Siegel, R.: Heat transfer by combined forced convection and thermal radiation in a heated tube, *JHT*, 84(4), 301–311, 1962.

Planck, M.: Distribution of energy in the spectrum, *Ann. Phys.*, 4(3), 553–563, 1901.

Planck, M.: *The Theory of Heat Radiation*, 2nd edn., Dover, New York, 1959 (first published in German, 1913).

Poljak, G.: Analysis of heat interchange by radiation between diffuse surfaces, *Tech. Phys. USSR*, 1(5–6), 555–590, 1935.

Pope, R. and Fry, E.: Absorption spectrum (380–700 nm) of pure water. II. Integrating cavity measurements, *Appl. Opt.*, 36(33), 8710–8723, 1997.

Qu, Y., Puttitwong, E., Howell, J. R., and Ezekoye, O. A.: Errors associated with light-pipe radiation thermometer temperature measurements, *IEEE Trans. Semicond. Manuf.*, 20(1), 26–38, 2007.

Rayleigh, Lord: On scattering of light by small particles, *Phil. Mag.*, 41, 447–454, 1871.

Rayleigh, Lord: The law of complete radiation, *Phil. Mag.*, 49, 539–540, 1900.

Rayleigh, Lord: The dynamical theory of gases and of radiation, *Nature*, 72, 54–55, 1905.

Robbins, W. H. and Finger, H. B.: An historical perspective of the NERVA nuclear rocket engine technology program, *NASA CR 187154 and AIAA Paper 91-3451*, July 1991.

Robock, A.: Benefits and risks of stratospheric solar radiation management for climate intervention (Geoengineering), *The Bridge*, U. S. Natl. Academy of Engineering, 59–67, Spring 2020.

Rosseland, S.: Zur Quantentheorie der radioaktiven Zerfallsvorgänge, *Z. Physik*, 14, 173–181, 1923.

Rosseland, S.: *Theoretical Astrophysics: Atomic Theory and the Analysis of Stellar Atmospheres and Envelopes*, Clarendon Press, Oxford, 1936.

Rothman, L. S. Editorial, *HITRAN Newslett.*, 5(4), 1–4, 1996.

Rothman, L. S., Gordon, I. E., Barber, R. J., Dothe, H., Gamache, R. R., Goldman, A., Perevalov, V., Tashkun, S. A., and Tennyson, J.: HITEMP, the high-temperature molecular spectroscopic database, *JQSRT*, 111(15), 2139–2150, 2010. Available on-line at https://hitran.org/hitemp/

Rothman, L. S., Gordon, I. E., Barbe, A., et al.: The HITRAN 2012 molecular spectroscopic database, *JQSRT*, 130, 4–50, 2013. (For most recent version (2012) see http://www.cfa .harvard.edu/hitran//. Accessed April 27, 2015).

Rumble, J. R., ed.: *Handbook of Chemistry and Physics*, 100th edn., CRC Press, Boca Raton, FL, 2019.

Rytov, S. M.: *A Theory of Electric Fluctuations and Thermal Radiation*, USSR Academy of Sciences, Moscow, 1953, republished Air Force Cambridge Research Center, Bedford, MA, 1959.

Santos, T. M. R., Tavares, C. A., Lemes, N. H. T., dos Santos, J. P. C., and Braga, J. P.: Improving a Tikhonov regularization method with a fractional-order differential operator for the inverse black body radiation problem, *IPSO*, 28(11), 1513–1527, 2020.

Sapritsky, V. and Prokhorov, A.: *Blackbody Radiometry*, Springer Series in Measurement Science and Technology, Springer Nature, Switzerland, 2020.

Saunders, O. A.: Notes on some radiation heat transfer formulae, *Proc. Phys. Soc. Lond.*, 41, 569–575, 1929.

Schuster, A.: Radiation through a foggy atmosphere, *Astrophys. J.*, 21, 1–22, 1905.

Schwarzschild, K.: Equilibrium of the Sun's atmosphere, *Ges. Wiss. Gottingen Nachr., Math-Phys. Klasse*, 1, 41–53, 1906.

Seban, R. A.: *Thermal Radiation Properties of Materials*, pt. III, WADD-TR-60-370, University of California, Berkeley, CA, 1963.

Shaffer, L. H.: Wavelength-dependent (selective) processes for the utilization of solar energy, *J. Solar Energy Sci. Eng.*, 2(3–4), 21–26, 1958.

Shouman, A. R.: An exact solution for the temperature distribution and radiant heat transfer along a constant cross-sectional area fin with finite equivalent surrounding sink temperature, in *Proceedings of the Ninth Midwestern Mechanics Conference*, 175–186, Madison, WI, August 16–18, 1965.

Shouman, A. R.: *Nonlinear Heat Transfer and Temperature Distribution through Fins and Electric Filaments of Arbitrary Geometry with Temperature-Dependent Properties and Heat Generation*, NASA TN D-4257, Washington, DC, 1968.

Shurcliff, W. A.: Transmittance and reflection loss of multi-plate planar window of a solar-radiation collector: Formulas and tabulations of results for the case of $n = 1.5$, *Sol. Energy*, 16, 149–154, 1974.

Sister, I., Leviatan, Y. and Schäcter, L.: Evaluation of blackbody radiation emitted by arbitrarily shaped bodies using the source model technique, *Opt. Express*, 25(12), A589, 2017.

Solovjov, V. P., Webb, B. W., and André, F.: Radiative properties of gases, in *Handbook of Thermal Science and Engineering*, Kulacki, F. ed., 1–74, Springer, New York, 2017.

Sowell, E. F. and O'Brien, P. F.: Efficient computation of radiant-interchange configuration factors within an enclosure, *JHT*, 94(3), 326–328, 1972.

Sparrow, E. M. and Albers, L. U.: Apparent emissivity and heat transfer in a long cylindrical hole, *JHT*, 82(3), 253–255, 1960.

Stark, J.: Der Dopplereffekt bei den kanalstrahlen und die spektra der positiven atomionen, *Physikalische Zeitschrift*, 6, 892–897, 1905.

Stefan, J.: Über die beziehung zwischen der wärmestrahlung und der temperatur, *Sitzber. Akad. Wiss. Wien*, 79(2), 391–428, 1879.

Tam, W. C. and Yuen, W. W.: *OpenSC – An Open-source Calculation Tool for Combustion Mixture Emissivity/Absorptivity*, NIST Technical Note 2064, National Institute of Standards and Technology, September 2019.

Taylor, R. P. and Luck, R.: Comparison of reciprocity and closure enforcement methods for radiation view factors, *JTHT*, 9(4), 660–666, 1995.

Taylor, R. P., Luck, R., Hodge, B. K., and Steele, W. G.: Uncertainty analysis of diffuse-gray radiation enclosure problems, *JTHT*, 9(1), 63–69, 1995.

Thompson, D., Zhu, L., Mittapally, R., Sadat, S., Xing, Z., McArdle, P., Mumtaz Qazilbash, M., Reddy, P., and Meyhofer, E.; Hundred-fold enhancement in far-field radiative heat transfer over the blackbody limit, *Nature*, 561, 216–221, 2018. (Also see Correction, *Nature*, 567, E12, 2019).

Tiesinga, E., Mohr, P. J., Newell, D. B., and Taylor, B. N.: CODATA recommended values of the fundamental physical constants: 2018, *Rev. Mod. Phys.*, 93, 025010, 2021.

Tikhonov, A. N.: Solution of incorrectly formulated problems and the regularization method, *Soviet Math. Dokl.*, 4, 1035–1038, 1963. [Engl. trans. *Dokl. Akad. Nauk. SSSR*, 151, 501–504, 1963.]

Touloukian, Y. S. and Ho, C. Y., eds.: Thermophysical properties of matter, TRPC data services, in *Thermal Radiative Properties: Metallic Elements and Alloys*, vol. 7, Y. S. Touloukian and D. P. DeWitt, eds., vol. 8, 1972a; vol. 9, 1972b, Plenum Press, New York, 1970.

Tsai, C.-C., Childers, R. A., Shi, N. N., Ren, C., Pelaez, J. N., Bernard, G. D., Pierce, N. E., Yu, N.: Physical and behavioral adaptations to prevent overheating of the living wings of butterflies, *Nat. Commun.*, 11, 551, 2020.

Turyshev, S. G., Toth, V. T., Kinsella, G., Lee, S.-C, Lok, S. M., and Ellis, J.: Support for the thermal origin of the pioneer anomaly, *Phys. Rev. Lett.*, 108, 241101, 2012.

Usiskin, C. M., Siegel, R.: Thermal radiation from a cylindrical enclosure with specified wall heat flux, *JHT*, 82(4), 369–374, 1960.

van Leersum, J.: A method for determining a consistent set of radiation view factors from a set generated by a nonexact method, *Int. J. Heat Fluid Flow*, 10(1), 83–85, 1989.

Varanasi, P.: A critical appraisal of the current spectroscopic databases used in atmospheric and other radiative transfer applications, in *Radiation III: Third International Symposium on Radiative Transfer*, M. P. Mengüç and N. Selçuk, eds., Begell House, New York, 2001.

Vercammen, H. A. J. and Froment, G. F.: An improved zone method using Monte Carlo techniques for the simulation of radiation in industrial furnaces, *IJHMT*, 23(3), 329–336, 1980.

Viskanta, R. and Anderson, E. E.: Heat transfer in semi-transparent solids, in *Advances in Heat Transfer*, vol. 11, J. P. Hartnett and T. F. Irvine, Jr., eds., 317–441, Academic Press, New York, 1975.

Warren, S. G. and Brandt, R. E.: Optical constants of ice from the ultraviolet to the microwave: A revised compilation, *J. Geophys. Res.*, 113, D14220, 2008.

Wien, W.: Über die Energievertheilung im Emissionsspectrum eines schwarzen Körpers, *Annalen der Physik, Ser. 3*, 58, 662–669, 1896.

World Meteorological Organization Statement on the State of the Global Climate in 2019, WMO-No. 1248, WMO, Geneva, Switzerland, 2020.

World Meteorological Organization, *United in Science 2022: A multi-organization high-level compilation of the most recent science related to climate change, impacts and responses*, WMO, Geneva, Switzerland, 2022.

Zerefos, C. S., Tetsis, P., Kazantzidis, A., Amiridis, V., Zerefos, S. C., Luterbacher, J., Eleftheratos, K., Gerasopoulos, E., Kazadzis, S., and Papayannis, A.: Further evidence of important environmental information content in red-to-green ratios as depicted in paintings by great masters, *Atmos. Chem. Phys.*, 14, 2987–3015, 2014.

Zhang, Y. F., Gicquel, O., and Taine, J.: Optimized emission-based reciprocity Monte Carlo method to speed up computation in complex systems, *IJHMT*, 55, 8172–8177, 2012.

Zhang, Z. M., Tsai, B. K., and Machin, G., eds.: Radiometric temperature measurements I. Fundamentals and II applications, *Experimental Methods in Thermal Sciences* (series), vol. 42, Elsevier, Amsterdam, 2010.

Zheng, S., Sui, R., Sun, Y., Yang, Y, and Lu, Q.: A review on the applications of non-gray gas radiation models in multi-dimensional systems, *ES Energy Environ.*, 12, 4–45, 2021.

Index

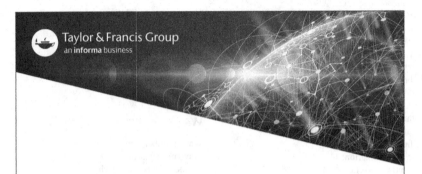

Printed in the United States
by Baker & Taylor Publisher Services